W9-BRC-119

1988

Science Year

The World Book Annual Science Supplement

A Review of Science and Technology During the 1987 School Year

World Book, Inc.
a Scott Fetzer company
Chicago London Sydney Toronto

Staff

Publisher
William H. Nault

Editor in Chief
Robert O. Zeleny

Editorial

Executive Editor
A. Richard Harmet

Managing Editor
Wayne Wille

Associate Editor
Darlene R. Stille

Senior Editors
David L. Dreier
Jinger Griswold
Mary A. Krier
Barbara A. Mayes
Jay Myers
Joan Stephenson
Rod Such

Contributing Editor
Sara Dreyfuss

Research Editor
Irene B. Keller

Index Editor
Claire Bolton

Editorial Assistant
Ethel Matthews

Art

Executive Art Director
William Hammond

Art Director
Roberta Dimmer

Assistant Art Director
Scot McIntosh

Senior Artists
Nikki Conner
Alexandra Kalantzis

Artists
Alice F. Dole
Melanie J. Lawson
Lucy Smith

Photography Director
John S. Marshall

Senior Photographs Editor
Sandra M. Ozanick

Photographs Editor
Barbara A. Bennett

Assistant Photographs Editor
Geralyn Swietek

Research and Services

Director of Research Services
Mary Norton

Library Services
Mary Kayaian, Head

Cartographic Services
H. George Stoll, Head
Wayne K. Pichler

Product Production

Executive Director
Peter Mollman

Director of Manufacturing
Joseph C. La Count

Research and Development Manager
Henry Koval

Pre-Press Services
Jerry Stack, Director
Lori Frankel
Madelyn Krzak
Alfred J. Mozdzen
Barbara Podczerwinski

Proofreaders
Ann Dillon
Marguerite Hoye
Esther Johns
Daniel Marotta

Copyright © 1987 World Book, Inc. Merchandise Mart Plaza, Chicago, Illinois 60654
Portions of the material contained in this volume are taken from
The World Book Encyclopedia, Copyright © 1987 by World Book, Inc.
All rights reserved.

ISBN 0-7166-0588-0
ISSN 0080-7621
Library of Congress Catalog Number: 65-21776
Printed in the United States of America.

Editorial Advisory Board

Charles L. Drake is Albert Bradley Professor of Earth Sciences at Dartmouth College. He received the B.S.E. degree from Princeton University in 1948 and the Ph.D. degree from Columbia University in 1958. He is past-President of the Geological Society of America and the American Geophysical Union and President of the 28th International Geological Congress, to be held in Washington, D.C., in 1989.

Richard G. Klein is Professor of Anthropology and Evolutionary Biology at the University of Chicago. He received the A.B. degree from the University of Michigan in 1962 and the A.M. and Ph.D. degrees from the University of Chicago in 1964 and 1966. He is a member of the Editorial Boards of *Quaternary Research* and *The Journal of Human Evolution* and Editor of the *Journal of Archaeological Science.* His principal research interests are the evolution of human behavior and the ecology of early people.

Leon M. Lederman is Director of Fermi National Accelerator Laboratory (Fermilab) and Eugene Higgins Professor of Physics at Columbia University. He received the B.S. degree in 1943 from City College of New York and the M.A. and Ph.D. degrees from Columbia University in 1948 and 1951. He was awarded the National Medal of Science in 1965 and the Elliot Cresson Medal from the Franklin Institute in 1976 and was a co-winner of the Wolf Prize in physics in 1982.

Gene E. Likens is Vice President of the New York Botanical Garden and Director of the Garden's Institute of Ecosystem Studies and Mary Flagler Cary Arboretum, Millbrook, N.Y. He received the B.S. degree from Manchester College in 1957. He received the M.S. degree in 1959 and the Ph.D. degree in 1962, both from the University of Wisconsin, Madison. He is a member of the National Academy of Sciences and holds professorships at Cornell University, Rutgers University, and Yale University.

Margaret L. A. MacVicar is Dean for Undergraduate Education at Massachusetts Institute of Technology (M.I.T.), where she is also Professor of Physical Science and Cecil and Ida Green Professor of Education. She received the S.B. degree in 1964 and the Sc.D. degree in 1967, both from M.I.T. She is co-chair of the American Association for the Advancement of Science's Project 2061 and vice-chairman of the Advisory Committee to the National Science Foundation's Directorate of Science and Engineering Education.

Douglas S. Reynolds is Chief of the Bureau of Science Education for the New York State Education Department. He received the B.S. and M.S. degrees in science education from the State University of New York, Plattsburgh, in 1960 and 1963. He serves on the National Advisory Board for the Second International Science Study. He has served on numerous state and national advisory committees for the improvement of science education.

Lucille Shapiro is chairman of the Department of Microbiology and Higgins Professor of Microbiology at the College of Physicians and Surgeons, Columbia University. She received the A.B. degree from Brooklyn College in 1962 and the Ph.D. degree from the Albert Einstein College of Medicine in 1966. She serves on the Science Board of the Helen Hay Whitney Foundation, the Advisory Board for the National Science Foundation's Biological and Behavior Sciences Directorate, and the Editorial Board of the journal *Genes and Development.*

Contents

See page 45.

See page 66.

See page 195.

See page 204.

See page 367.

Contributors

Adelman, George, M.S.
Editorial Consultant and Editor.
[*Neuroscience*]

Alderman, Michael H., M.D.
Chairman,
Department of Epidemiology
& Social Medicine,
Albert Einstein College of Medicine.
[*Public Health*]

Anderson, Ian, B.A.
Free-Lance Writer.
[*Anthropology* (Close-Up)]

Andrews, Peter J., M.S.
Free-Lance Writer and Chemist.
[*Chemistry*]

Asa, Richard Warner, B.S.
Media Manager,
American Dental Association.
[Science You Can Use: *Sorting Out Trends in Oral Health Care*]

Bell, William J., Ph.D.
Professor of Biology,
University of Kansas. [*Zoology*]

Belton, Michael J. S., Ph.D.
Astronomer,
National Optical Astronomy
Observatories.
[*Astronomy, Solar System*]

Bierman, Howard, B.E.E.
President,
Paje Consultants, Inc.
[*Computer Hardware; Computer Software; Electronics; Electronics* (Close-Up)]

Black, John H., Ph.D.
Associate Professor of Astronomy,
Steward Observatory,
University of Arizona.
[*Astronomy, Galactic*]

Brett, Carlton E., Ph.D.
Associate Professor,
Department of Geological Sciences,
University of Rochester.
[*Paleontology*]

Comis, Donald L., M.S.
Associate Editor,
USDA Agricultural Research Service.
[Science You Can Use: *Choosing the Right Insecticide*]

Conkey, Margaret W., Ph.D.
Associate Professor,
Department of Anthropology,
State University of New York.
[Special Report: *Images from the Ice Age*]

Covault, Craig P., B.S.
Space Technology Editor,
Aviation Week magazine.
[*Space Technology*]

Cunliffe, Barry, Ph.D.
Professor of European Archaeology,
University of Oxford.
[Special Report: *Excavating Roman Bath*]

Day, Lucille, Ph.D.
Free-Lance Writer.
[People in Science: *Bruce N. Ames*]

Diamond, Jared M., Ph.D.
Professor of Physiology,
UCLA Medical School.
[*Ecology*]

Duerig, Tom, Ph.D.
Free-Lance Writer.
[*Chemistry* (Close-Up)]

Fisher, Arthur, M.A.
Science and Engineering Editor,
Popular Science magazine.
[Special Report: *The Microchip—A Miniature Marvel;*
Science You Can Use: *Automatic 35-mm Cameras: More Help for the Amateur*]

Fritz, Sandy
Free-Lance Writer.
[Special Report: *High Schools of Science*]

Gates, W. Lawrence, Sc.D.
Professor and Chairman,
Department of Atmospheric
Sciences,
Oregon State University.
[*Meteorology*]

Goldhaber, Paul, D.D.S.
Dean and Professor of
Periodontology,
Harvard School of Dental Medicine.
[*Dentistry*]

Grey, Jerry, Ph.D.
Publisher,
Aerospace America magazine.
[Special Report: *Airplane Design Takes Off Toward Tomorrow*]

Hartl, Daniel L., Ph.D.
Professor of Genetics,
Washington University School of
Medicine.
[*Genetic Sciences*]

Hay, William W., Ph.D.
Professor of Natural History and
Director,
University of Colorado Museum,
University of Colorado, Boulder.
[*Geology*]

Hellemans, Alexander, B.S.
Editor,
Physics Today magazine.
[*Physics, Fluids and Solids*]

Hester, Thomas R., Ph.D.
Professor of Anthropology and
Director,
Center for Archaeological Research,
University of Texas, San Antonio.
[*Archaeology, New World*]

Jones, William G., A.M.L.S.
Assistant University Librarian,
University of Illinois at Chicago.
[*Books of Science*]

Kalson, David, M.A.
Manager, Public Information
Division,
American Institute of Physics.
[*Science Education* (Close-Up)]

Katz, Paul, M.D.
Associate Professor of Medicine,
Georgetown University Medical
Center.
[*Immunology*]

Kies, Constance Virginia, Ph.D
Professor of Human Nutrition,
University of Nebraska.
[*Nutrition*]

King, Lauriston R., Ph.D.
Deputy Director,
Sea Grant Program,
Texas A&M University.
[*Oceanography*]

Kling, George W., Ph.D.
Research Associate,
Duke University.
[*Geology* (Close-Up)]

Lamberg, Lynne, M.A.
Free-Lance Writer.
[Special Report: *Learning to Use Our Inner Clocks;*
Science You Can Use: *Facts About Acne Treatment*]

Latamore, G. Berton, B.A.
Free-Lance Writer.
[Science You Can Use: *How Much Home Security Do You Need?*]

Maran, Stephen P., Ph.D
Senior Staff Scientist,
NASA-Goddard Space Flight Center.
[*Books of Science* (Close-Up)]

March, Robert H., Ph.D.
Professor of Physics,
University of Wisconsin.
[Special Report: *Exploring the States of Matter;*
Physics, Subatomic]

Meier, Mark F., Ph.D.
Director,
Institute of Arctic and Alpine Research,
University of Colorado.
[Special Report: *Glaciers On the Go*]

Merbs, Charles F., Ph.D.
Professor of Anthropology,
Arizona State University.
[*Anthropology*]

Merz, Beverly, A.B.
Associate Editor,
Journal of the American Medical Association.
[*Medical Research; Medicine*]

Meyer, B. Robert, M.D.
Chief, Division of Clinical Pharmacology,
North Shore University Hospital.
[*Drugs*]

Murray, Stephen S., Ph.D.
Astrophysicist,
Harvard/Smithsonian Center for Astrophysics.
[*Astronomy, Extragalactic*]

Olson, Maynard V., Ph.D.
Professor,
Department of Genetics,
Washington University School of Medicine.
[*Molecular Biology*]

Patrusky, Ben, B.E.E.
Free-Lance Science Writer.
[Special Report: *New Hope for the Everglades*]

Pennisi, Elizabeth, M.S.
Free-Lance Science Writer.
[Special Report: *Addiction and the Brain; Zoology*]

Rabinowitz, Philip D., Ph.D.
Professor of Oceanography,
Texas A&M University.
[Special Report: *Drilling Under the Sea*]

Rodgers, Joann Ellison, M.S.
Free-Lance Science Writer.
[Special Report: *Cancer: A Genetic Time Bomb?*]

Salisbury, Frank B., Ph.D
Professor of Plant Physiology,
Utah State University.
[*Botany*]

Samz, Jane, M.A.
Editor,
Science World magazine.
[Special Report: *Your House Can Make You Sick*]

Schefter, Jim, M.B.A.
Editor,
Popular Science magazine.
[*Physics, Fluids and Solids* (Close-Up)]

Soderblom, Laurence A., Ph.D.
Geophysicist,
United States Geological Survey.
[Special Report: Voyager's *Close Look at Uranus*]

Trotter, Robert J., B.S.
Senior Editor,
Psychology Today magazine.
[*Psychology*]

Vietmeyer, Noel D., Ph.D.
Associate,
National Research Council.
[Special Report: *New Uses for "The Ship of the Desert"*]

Visich, Marian, Jr., Ph.D.
Associate Dean of Engineering,
State University of New York.
[*Energy*]

Watson, Robert,
Research Scientist,
NASA Office of Earth Science and Applications.
[*Environment* (Close-Up)]

Wenke, Robert J., Ph.D.
Associate Professor,
Department of Anthropology,
University of Washington.
[*Archaeology, Old World*]

Westman, Walter E., Ph.D.
Research Associate,
NASA Ames Research Center.
[*Environment*]

Wittwer, Sylvan H., Ph.D
Director Emeritus,
Agricultural Experiment Station,
Michigan State University.
[*Agriculture*]

Yager, Robert E., Ph.D
Professor of Science Education,
University of Iowa.
[*Science Education*]

Special Reports

Fourteen Special Reports give in-depth treatment to major advances in science and technology. The subjects were chosen for their current importance and lasting interest.

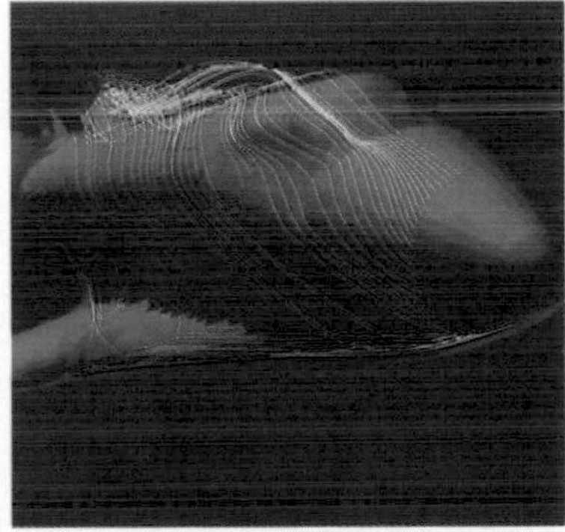

Scientists and public officials are solving water-control problems that had imperiled this unique marshland.

New Hope for the Everglades

BY BEN PATRUSKY

Suppose you picked up your newspaper one morning and read:

Bulletin: *Florida's historic Everglades have disappeared. Once the largest freshwater marsh on the United States mainland, the Everglades succumbed to the effects of drainage, dredging, and other means of water diversion. The saw grass and tree islands that lent a unique character to the Everglades have been replaced by farms, condominiums, and shopping malls. The marsh's hardwood trees have been replaced by dense tangles of melaleuca trees—trees native to Australia—and other "foreign" pest vegetation. The animals that made the Everglades their home—beautiful wading birds, the American alligator, the Florida panther, and many others—have become extinct or moved to new habitats.*

Fortunately, this death notice is fictitious, and with luck it will never actually be written. Just a few years ago, however, the Everglades were in serious trouble. Decades of tinkering with the waterflow of this vast marsh for the benefit of farm-

A flock of roseate spoonbills feeds in a shallow pond in Florida's Everglades.

ers and city dwellers had severely disrupted the Everglades' natural ecological cycles. By the 1970's, many of the Everglades' plants and animals were disappearing. Although ecologists had for many years been predicting the demise of the Everglades—predictions that had not been borne out—this time the crisis seemed to be real.

In response to the worsening situation, the National Park Service and the state of Florida took action in the early 1980's to restore the natural ebb and flow of water in the Everglades. According to Gary Hendrix, research director of Everglades National Park, the major water-control changes came barely in the nick of time. Without them, the Everglades would almost certainly have passed the point of no return in their march to destruction.

The Everglades will never again be as they were in the 1800's before Florida became developed; the need to protect the state's farms and urban communities from flooding prevents that. But the corrective measures that have been taken so far are a big step in the right direction. If the restoration program succeeds, the Everglades' waterflow patterns—though still largely controlled by artificial means—will be similar to what they were originally.

Aptly called the "grassy waters" by their early Indian inhabitants, the Everglades are a swath of shallow wetlands some 30 to 130 kilometers (20 to 80 miles) wide and 160 kilometers (100 miles) long. From their present upper limit, several kilometers south of Lake Okeechobee in south-central Florida, the Everglades stretch southward to Florida Bay and the Gulf of Mexico.

The Everglades were born more than 10,000 years ago during the most recent emergence of the Florida lowlands from beneath the sea, the last of several such appearances occurring over the past 1 million years. Each surfacing coincided with an advance of glaciers during the Ice Age. The glaciers tied up huge amounts of water in the form of ice and snow. That water came originally from the ocean, so the expansion of the glaciers lowered the sea level and exposed low-lying areas such as the Florida Peninsula. With the last retreat of the ice, the ocean did not rise quite as high as it did in earlier warm periods, so the lowlands remained slightly above sea level. The result was a wetland area of some 23,300 square kilometers (9,000 square miles). In their original state, the Everglades extended to central Florida, about 160 kilometers north of Lake Okeechobee. There, hundreds of small lakes met and formed the Kissimmee River, a twisting channel that meandered for 148 kilometers (92 miles) and ultimately emptied into the lake.

Lake Okeechobee marked the highest point of the Everglades, about 5.2 meters (17 feet) above sea level. From there, it was all downhill to the sea, like a slightly tilted table. A wide *sheet flow* of water, fed by overflow from the lake and by rainfall, coursed no more than 0.5 kilometer (1/3 mile) a day along a slope that dropped barely 4.8 centimeters per kilometer (3 inches per mile).

The author:
Ben Patrusky is a free-lance science writer and a media consultant to several scientific institutions.

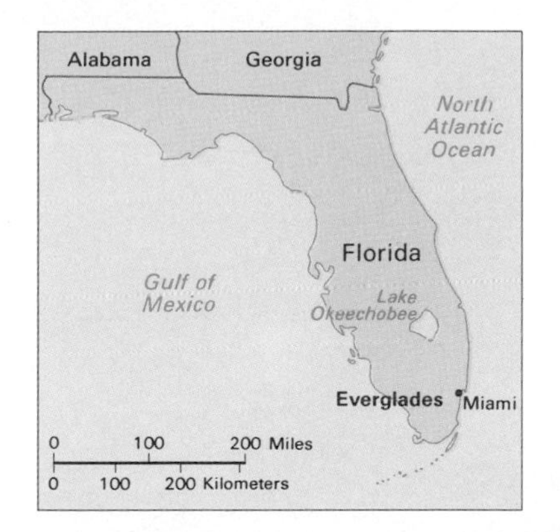

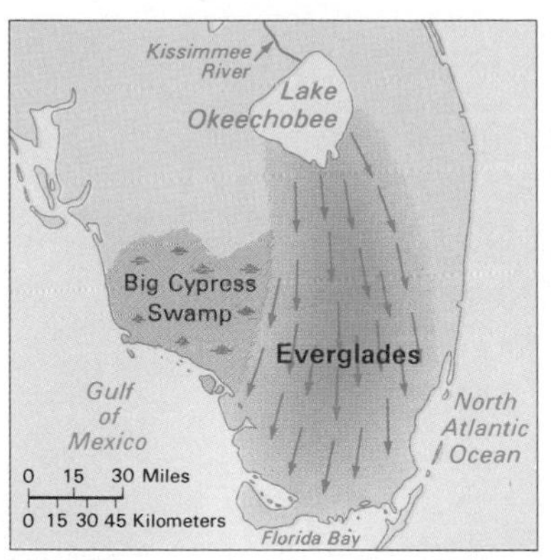

Until the early 1900's, the shallow waters of the Everglades flowed slowly south in a wide sheet from Lake Okeechobee to the Gulf of Mexico and Florida Bay, *above left and center*. The natural level of the water varied during the rainy and dry seasons. Then, the construction of an elaborate water-control system, including canals, *above,* and control gates, *left,* permanently cut off this sheet flow and altered the cycles of high and low water in the Everglades. The marshland's ecology suffered as a result.

The waters of the Everglades have never run very deep. Even in the wettest times, water levels, on average, have ranged from a few centimeters to no more than 2 meters (6 feet). Sprouting from the shallow water are vast areas of *saw grass*, a tall, grasslike plant with slender saw-toothed leaves. The saw grass grows out of a bed of *muck*, a dark, rich soil produced by the slow, centuries-long buildup of decaying plants.

After saw grass, the most prominent features of the Everglades are the *tree islands*. Shaped like teardrops, these small, tree-covered outcrops of bedrock dot the "river of grass" all along its length, much like an armada of ships sailing southward. The islands teem with a variety of hardwood trees, including live oak and mahogany.

The river of grass gives rise to still other kinds of plant communities, each linked intimately to the season-to-season ebb and flow of water. Biologists have observed that just a few centimeters' variation in water depth can spell the difference between one kind of plant community and another. Sloughs (pronounced *slews*), for example—areas with somewhat deeper water—are largely free of saw grass and are dominated instead by floating-plant species such as the white water lily. *Wet prairies*, stretches of marsh that often form tran-

Plant Life of the Everglades

Ecologists have learned that the tremendous variety of life in the Everglades is closely tied to the natural ebb and flow of water. Much of the Everglades is covered by vast expanses of saw grass growing in shallow water. Here and there are *tree islands,* outcrops of bedrock supporting mainly hardwood trees, *right*. Wet periods favor the growth of water lilies, *bottom*. Other plants, such as the wax myrtle, *below,* flourish in drier times when water levels are low.

sition zones between sloughs and the saw-grass areas and tree islands, support several varieties of grasslike plants.

The Everglades are always in flux. From May through October, in a typical year, rains drench lower Florida, and the wetlands flow with an abundance of water. The rainy season is followed by a six-month dry season. Before the rains return, much of the Everglades turns to mud and then to dry, cracked earth.

Some years are wetter than average; in other years, the Everglades are gripped by drought. And sometimes there are several unusually wet or dry years in a row. A period of drought, scientists have found, is likely to promote a more abundant growth of woody, dry-land plants. A succession of excessively wet years tends to reverse the process, resulting in more saw grass and white water lilies.

The Everglades' wide spectrum of habitats is able to support an extraordinary diversity of animal species, each finely adapted to the wetlands' watery rhythms. A good example is the white-tailed Everglades deer, one of some 40 species of mammals that are native to the region. The deer eat aquatic vegetation during the night and early morning and retire to tree islands during the day to rest.

The Everglades' other residents include at least 18 species of

Water Levels and Water Birds

The variety of birds in the Everglades depends on naturally rising and falling water levels. Wading birds, such as wood storks, *right,* and herons and egrets, *opposite page, top,* need low water levels during their breeding period. Fish trapped in shallow pools near the birds' nests provide ample, easily obtainable food for the birds and their offspring. On the other hand, the snail kite, *opposite page, bottom,* thrives during high-water periods. The kite dines exclusively on apple snails, found in deep water. Frequent flooding caused by human tampering with the Everglades' normal cycle of waterflow led to a decline in the number of wading birds and an increase in snail kites.

frogs and a variety of reptiles. Among the latter are 26 kinds of snakes, of which 4 are poisonous, and 16 species of turtles. But the most famous Everglades reptile is the American alligator. This formidable predator is often referred to as the "keeper of the Glades" because biologists and naturalists have found that the alligator's life style helps to sustain the ecology of the marsh. Alligators rout out mud and plants from naturally occurring holes in the Everglades' underlying bed of limestone, creating deep ponds called " 'gator holes." During the dry winter season, 'gator holes provide refuge to fish, turtles, snails, and other water-dwelling animals. Those animals, in turn, are preyed on by birds and mammals—and alligators. With the return of the rains in late May, the survivors among both predator and prey depart to repopulate the marshland.

Just before the wet season begins, female alligators lay their eggs in nests atop mounds built high enough, by the alligator's guess, to be above that year's high-water mark. In especially wet seasons, water may rise above the nests, killing the offspring before they hatch.

It is for the diversity of their bird population, however, that the Everglades are best known. In fact, Everglades National Park was established in 1947 mainly as a way of preserving the more than 100 species of birds that go to the Everglades to breed and nest. Among the most colorful and exotic are the 20 or so species of wading birds—such as herons, egrets, ibises, and the wood stork. These long-legged birds gather at shallow pools to dine on fish.

Breeding birds need an adequate supply of fish near their nesting

Here are your

1988 Science Year Cross-Reference Tabs

For insertion in your World Book

Each year Science Year, The World Book Annual Science Supplement, adds a valuable dimension to your World Book set. The Cross-Reference Tab System is designed especially to help students and parents alike link Science Year's Special Reports, its People in Science articles, and its new or revised World Book articles to the related World Book articles they update.

How to Use These Tabs

The top Tab on this page is AIRPLANE. Turn to the A volume of your World Book and find the page with the AIRPLANE article on it. Affix the AIRPLANE Tab to that page. The second tab is for the scientist BRUCE N. AMES. There is no article about him in World Book, so put the tab in your A volume on the same page as the article about AMES, ADELBERT, JR.

Now put the remainder of the Tabs in your World Book volumes, and your new Science Year will be linked to your World Book set.

Special Report
AIRPLANE
1988 Science Year, p. 170

People in Science
AMES, BRUCE N.
1988 Science Year, p. 338

Special Report
BATH (England)
1988 Science Year, p. 82

New Article
BEETLE
1988 Science Year, p. 381

Special Report
BIOLOGICAL CLOCK
1988 Science Year, p. 26

New Article
BIOLOGY
1988 Science Year, p. 366

Special Report
CAMEL
1988 Science Year, p. 56

Special Report
CANCER
1988 Science Year, p. 156

New Article
CHEMISTRY
1988 Science Year, p. 372

Special Report
DEEP SEA DRILLING PROJECT
1988 Science Year, p. 128

Special Report
DRUG ADDICTION
1988 Science Year, p. 116

Special Report
EVERGLADES
1988 Science Year, p. 10

Special Report
GLACIER
1988 Science Year, p. 68

People in Science
HIGH SCHOOL
1988 Science Year, p. 350

Special Report
MATTER
1988 Science Year, p. 184

Special Report
MICROPROCESSOR
1988 Science Year, p. 198

Science Year Close-Up
NEWTON, SIR ISAAC
1988 Science Year, p. 236

Special Report
PREHISTORIC PEOPLE
1988 Science Year, p. 40

Special Report
RADON
1988 Science Year, p. 142

Special Report
URANUS
1988 Science Year, p. 100

sites to enable them to feed their hungry offspring. Biologists have found, however, that water levels in the Everglades do not always cooperate with the requirements of parenthood. If water remains unseasonably high, fish have a larger area to swim in and are thus harder for wading birds to find and catch. With less food available nearby, the birds may abandon their efforts to establish nests that year. On the other hand, high water levels work to the advantage of another bird, the snail kite, a type of hawk. The snail kite feeds only on apple snails, which are found in deeper water.

The Everglades deer, *above,* relies on a normal cycle of water levels for its survival. It feeds on vegetation in the marshes, while tree islands provide dry land for resting and breeding. When flooding occurs, some deer become stranded on the islands and begin to starve. Those that brave the deep water in search of food may drown. Poor water management by human beings has sometimes caused prolonged periods of flooding, resulting in the death of many deer.

Throughout this century, the ecology of the Everglades has had another species to reckon with—human beings. Human intrusion upon the wetlands was spurred by dreams of draining large parts of the insect-infested marsh for farms and cities. By 1920, four major canals had been dug from Lake Okeechobee to the Atlantic Ocean. The canals drew off excess water from the lake, and rain water from the wetlands, and carried it to the sea.

A few years later, disaster struck. In 1926 and 1928, killer hurricanes pushed the water of the lake up into its northern end. When the winds died down, the lake water—swelled by heavy rainfall—rushed south and smashed through a protective dike, flooding the farm communities below. Together, the two hurricanes took the lives of more than 2,500 people.

To prevent a recurrence of those tragedies, the U.S. Army Corps of Engineers in 1930 all but walled off Lake Okeechobee with the Hoover Dike, an earthen embankment 8 meters (25 feet) high completely circling the lake. The dike, together with the developing farmlands, once and for all severed Lake Okeechobee and the network of lakes emptying into it from the Everglades.

The 1930's brought a new problem: drought. With rainfall at a minimum, less water seeped downward through the bed of the Everglades to replenish the Biscayne aquifer, a natural underground reservoir that was southeastern Florida's main source of fresh water. As a result, salt water from the ocean began to penetrate inland.

In the 1940's, the population of south Florida began to swell. Many newcomers bought land, cheap, on the parched swamp plain and settled there. But the marsh was not to stay drought-ridden forever. In 1947, a hurricane drenched the Everglades, causing severe flooding of low-lying farms south of Lake Okeechobee and communities along the coast. The same thing happened in 1948.

Enough was enough. People demanded that something be done. In response, the Florida legislature in 1949 established a state agency, the Central and Southern Florida Flood Control District—now the South Florida Water Management District—to manage water levels in the Everglades.

The U.S. government also got involved. In 1948, Congress gave the go-ahead to the U.S. Army Corps of Engineers to plan and construct a water-control system for the marsh and adjoining areas.

The alligators of the Everglades, *above left,* were also endangered by human-engineered flooding. Female alligators construct and lay their eggs in nests atop mounds, *above,* built high enough—by the alligator's estimate—to be above the year's high-water level. Biologists in the 1970's and early 1980's noted that many alligator nests in Everglades National Park were wiped out by flooding, and this provided an important measurement of how much damage the water management program was doing to the Everglades ecosystem.

That project was conceived with several major goals in mind, including the protection of the Everglades' fish and wildlife. In reality, the corps' design was mainly intended for flood control and farmland drainage, a bias that would eventually cause grief to Everglades National Park and come back to plague the project. But the original plans were drawn up at a time when there were still big gaps in knowledge about the Everglades' ecology and waterflow patterns.

By the time it was completed, in 1962, the project had produced an Everglades that were a far cry from what they had been historically. The original wetlands were now divided into three distinct parts. At the northern end of the system, the area just south of Lake Okeechobee, was the Everglades Agricultural Area, a region devoted to dairy farms and the growing of vegetable crops and sugar cane. The southernmost portion of the Everglades, encompassing some 81,000 hectares (200,000 acres), belonged to the national park. The 405,000 hectares (1 million acres) between the park and the agricultural area were subdivided into three Water Conservation Areas, each of which was surrounded by *levees*—embankments similar to dikes. Because of the natural slope of the Everglades and the dividing levees, the conservation areas were arranged in steplike fashion. When one conservation area was full, water could flow to the next area, but only through control gates.

During the wet season, the job of the conservation areas was to collect and store rain water, thus keeping it from moving across the marsh surface in a sheetlike flow, as it once did, and flooding the coastal communities. The conservation areas also stored water pumped into them from the agricultural area through a canal system. With flood control the cornerstone of the project, the Corps of Engineers set maximum water levels for each conservation area.

The southern end of Conservation Area 3 was closed off by construction of a levee and flood-control gates. The levee was built just north of the Tamiami Trail, a highway dating from the 1920's. It blocked water from flowing through hundreds of culverts under the highway and into the Shark River Slough, a swath of wetlands where the now disrupted sheet flow narrowed to a width of about 32 kilometers (20 miles). The slough was the main drainage route through the national park. The purpose of the levee was to give flood protection to the growing agricultural and residential communities of a region known as the East Everglades, between the park and the coastal cities. Four control gates were erected to allow water into the park along 16 kilometers (10 miles) of slough to the west. The result of all this building of canals and gates and levees was that the park's waterflow patterns were dramatically altered. Water to the park would now be funneled through a much narrower spigot, making uninhibited sheet flow a thing of the past.

In 1962, with the project finally completed, southern Florida was beset by a severe drought that lasted three years. Water levels in the conservation areas stayed low, and as a result, the park received no water deliveries. The public, increasingly conservation-minded and stirred by news photos of the parched, mud-caked Everglades, protested the situation. So did the National Park Service. Congress responded in 1970 with a law guaranteeing that the park would receive a minimum annual water delivery, no matter how little water was stored in the conservation areas. The minimum added up to 260,000 acre-feet annually, delivered on a monthly basis. (One acre-foot is the amount of water that would cover 1 acre [0.4 hectare] to a depth of 1 foot [30.5 centimeters].)

By that time, however, the park's water-supply system had undergone yet another radical change. In 1967, an extension to a canal was dug into Conservation Area 3 to serve as a new water route into the park. This channel, called L-67 extended, ran south for 15.2 kilometers (9½ miles) at the midpoint of the Shark River Slough. Designed to get water to the park quickly and in large quantities, L-67—in effect, a giant water cannon—all but eliminated any remaining vestige of the wetlands' historic sheet-flow patterns.

Research conducted by Everglades National Park biologists in the late 1970's and early 1980's clearly showed that the park was suffering the consequences of something wholly unexpected—not drought conditions, but flooding. Water management was playing havoc with the natural wet-and-dry rhythms of the Everglades, a situation that threatened the wildlife, particularly the birds, alligators, and deer. Bird-count data provided the most compelling evidence that the park was having problems. Biologist Bill Robertson, for example, found that the breeding population of all bird species combined had declined at least 50 per cent from 1962 to 1981.

The park scientists believed they had the problem pretty well diagnosed. The park was getting too much water too often and in too concentrated a stream. Since 1967, huge amounts of water had been cannonading into the park through a huge "hose" just a few meters wide—L-67. Moreover, much of the water was released into the park just at the start of the dry season, when birds were establishing their nests. It was also being released early in the wet season, when alligators were building theirs. For birds, the sudden surge of water meant a loss of prey because fish were no longer concentrated in shallow pools. For alligators, fooled in their appraisal of how high to build their egg-bearing mounds, the rising water caused the death of offspring. As for the deer, flooding sometimes stranded them on the tree islands, where they quickly began to starve.

Park scientists recognized another source of trouble as well: Water was coming in much too regularly. Each month, a fixed amount was delivered, regardless of variations in rainfall. That delivery schedule was nothing like the Everglades' historical seesawing between extended periods of dryness and bouts of excessive wetness. Such variations had been the basis of the Everglades ecosystem, providing habitats for a diverse spectrum of wildlife. In contrast, unvarying waterflow patterns tend, over a period of years, to favor some species at the expense of others—another factor that threatened to reduce the park's great variety of wildlife. A prime example of this phenomenon is the snail kite, which was near extinction because of drought in the early 1960's. By the mid-1980's, after years of flooding in the park, the kite was thriving, finding plenty of apple snails in the high water—at the expense of other species that depend on drier conditions.

Other regions of the Everglades ecosystem had their own share of woe. In Conservation Area 3, for instance, the construction of L-67 and improvements in other canals accelerated the flow of water. Water drained quickly away from the northern section and "stacked up" at the walled-off southern end. As a result, the northern section dried out, and in many places the flammable muck caught fire. Meanwhile, in the south, the rising water did serious damage to tree islands.

The East Everglades, too, were besieged by severe muck fires. But even more worrisome, that area was being overrun by exotic plants—trees such as the melaleuca, the Australian pine, and the Brazilian pepper. Introduced in the early 1900's as ornamental plants or as windbreaks around farms, these trees can take over disturbed sites, pushing out the native plant species that provide food and a home for the Everglades wildlife. The exotic trees have spread to other parts of the Everglades and could pose the greatest long-term menace to the wetlands' imperiled ecosystem.

In the winter of 1982-1983, two developments in the Everglades—one of human origin, the other an act of nature—brought matters to a head. The human contribution was the *eutrophication* of Lake Okeechobee. Eutrophication is the increase of nutrient-rich

The Problem and the Solution

The natural sheet flow of water in the Everglades was cut off by an extensive system of canals, locks, dikes, and levees, *below,* that severed Lake Okeechobee and the Kissimmee River from the Everglades ecosystem and split the Everglades into three main parts: the Everglades Agricultural Area, Everglades National Park, and a large expanse of wetlands subdivided into three Water Conservation Areas. Flooding in the park and other problems resulted from the changes. Plugging some canals and cutting gaps in levees is restoring a more normal waterflow.

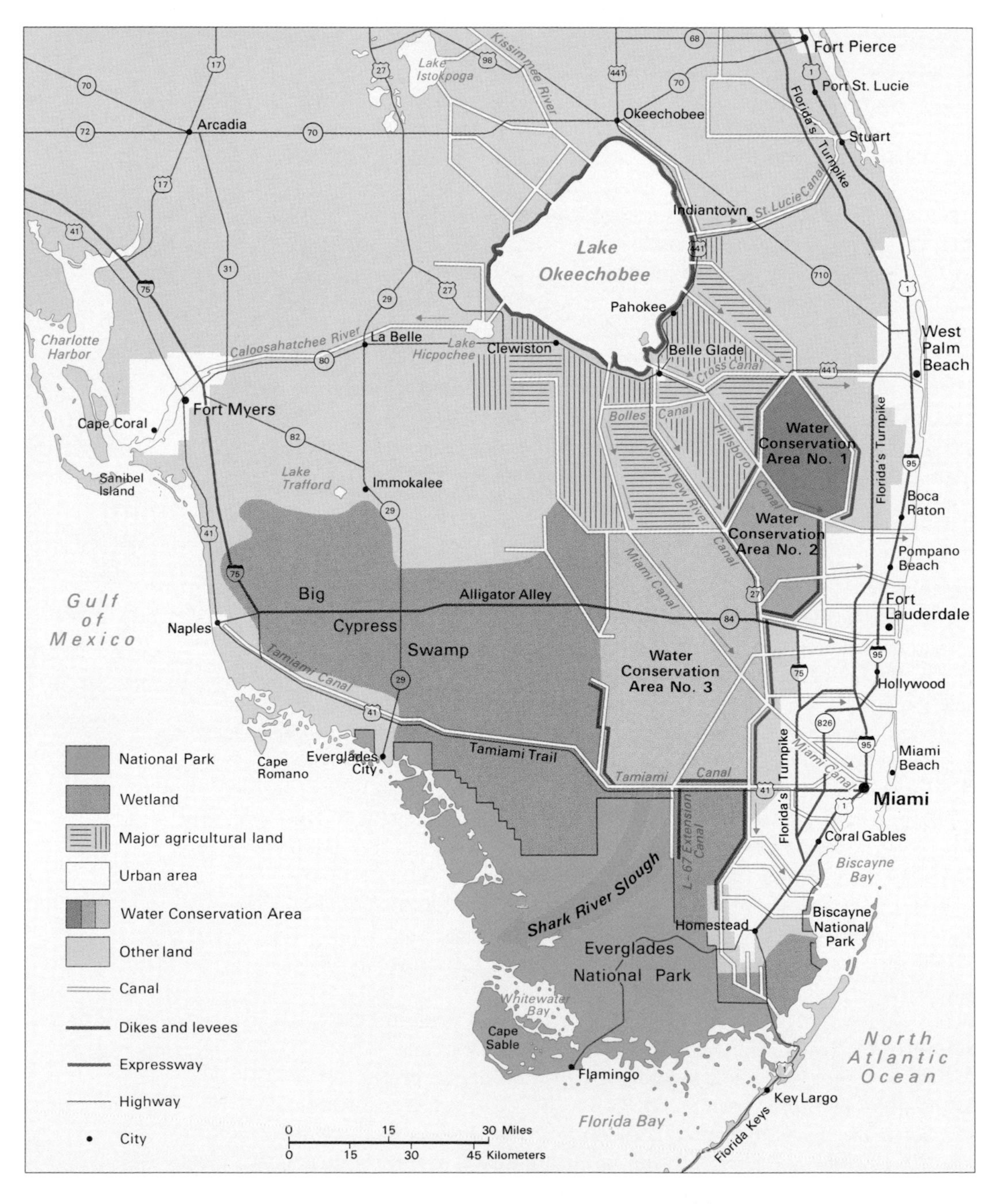

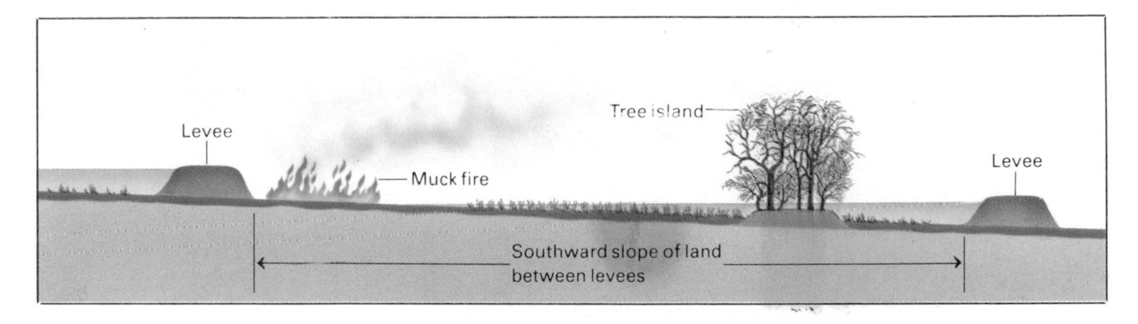

Artificial barriers to waterflow in the Everglades resulted in flooding in some areas and drought in others. These problems were especially serious in the Water Conservation Areas. Water in the Everglades flows southward along the natural slope of the land, but when stopped by a levee, *above,* it "stacks up," creating flood conditions that often swamp tree islands, damaging trees. During periods of sparse rainfall, the northern sections of these areas dry out, and the *muck* (soil composed of decayed plants) often catches fire, *right*.

Park rangers canoe through Everglades National Park, *left,* to monitor the health of the ecosystem. Rangers and park scientists check frequently on the numbers of birds in the park, the health of various kinds of plants, and the condition of alligator nests. Such observations provide valuable information on how the park's wildlife is faring.

The Everglades still face problems, but the future looks brighter than it did in the early 1980's. Biologists and public officials now have a better understanding of the marshland's delicate ecology—and a commitment to preserve it for future generations.

pollutants, primarily fertilizers and treated sewage, in a body of water, resulting in an explosive growth of algae. As the algae die, the bacteria that decompose them use up most of the oxygen in the water. The depletion of oxygen causes the death of fish.

The eutrophication of Lake Okeechobee was caused by fertilizer-laden water entering the lake from farmland north and south of the lake. Although the problem came to a head in the 1980's, it began in the early 1960's when the Army Corps of Engineers straightened the Kissimmee River to improve agricultural drainage. This project turned the meandering river into an arrow-straight ditch 84 kilometers (52 miles) long that carried pollutants rapidly to the lake. Adding to the problem, the South Florida Water Management District had, from time to time, pumped surplus water from the agricultural areas south of Lake Okeechobee back into the lake. For fear of making lake conditions worse, this backpumping was halted. But as a result, still more water was released into the park.

The deteriorating situation threatened to grow even worse. Nature's contribution was El Niño, an unusually warm Pacific Ocean current that appears every few years, setting off changes in world weather patterns. During the winter and spring of 1983, Florida received nearly as much rainfall as during a normal summer. The rains, coming just after the normal rainy season, led to extensive flooding in Everglades National Park. "That was the final straw," said Gary Hendrix. "We could see it was going to be downhill for the Everglades from then on unless something was done."

The prospect of a dying Everglades prompted park officials in March 1983 to recommend that the South Florida Water Management District plug the L-67 "water cannon" and restore the sheet flow of water in the eastern half of the Shark River Slough. The park authorities also urged the water management district to replace the minimum-amount water delivery schedule with a more ecologically compatible system pegged to the park's seasonal needs. Whenever flood conditions exist in the conservation areas, the officials said, every effort should be made to divert the water elsewhere—perhaps to Big Cypress Swamp, a huge preserve lying on the western edge of Conservation Area 3. At the time, Big Cypress Swamp was walled off from the conservation areas by a levee and an adjacent canal.

A few months later, Florida Governor Bob Graham, now a U.S. senator, kicked off a highly publicized "Save Our Everglades" campaign. The program endorsed the park's recommendations and added several other goals, including restoring the Kissimmee River to its former meandering self; buying back large tracts of privately owned wetlands and reflooding them; and making a special effort to save the Florida panther— its number reduced to about 30 by the encroachments of civilization.

By mid-1987, the joint crusade by Graham and park officials had led to some significant changes in the Everglades:

- The Big Cypress canal and L-67 had been plugged, and gaps had been punched into adjacent levees to allow water to flow through.
- Water was once again coming into the park across the entire 32-kilometer (20-mile) width of the Shark River Slough.
- An experimental water-release schedule for the park, called the Rainfall Plan, was put into operation in 1985. The scheme bases water delivery to the park on current rainfall and evaporation patterns in Conservation Area 3.
- Substantial private acreage lying just below the Everglades Agricultural Area had been purchased by the state, and work was underway to reestablish sheet flow from the newly acquired sections to the overdrained, fire-ridden northern end of Conservation Area 3.
- A 19-kilometer (12-mile) stretch of the Kissimmee River had been restored to its former banks. After a study of the project's effects, more of the river may be "dechannelized."
- Thirty-six underpasses were being built under Alligator Alley—a state highway being converted to a federal interstate—to help protect panthers and other animals from automobiles.
- Biologists were testing a technique to get rid of exotic trees by chopping into their bark and applying lethal chemicals. They hope that this method, together with the revitalization of the Everglades ecosystem, will eventually eliminate the unwanted plants.

In all, since 1983 there has been a startling turnabout in the long-term outlook for the Everglades. The people of Florida, farmers and city dwellers alike, have come to see the wisdom of cooperating with state and federal efforts to protect this magnificent resource, one of America's greatest natural treasures. So while the Everglades may not yet be off the critical list, there is now considerable hope for recovery—and certainly no reason to post any death notices.

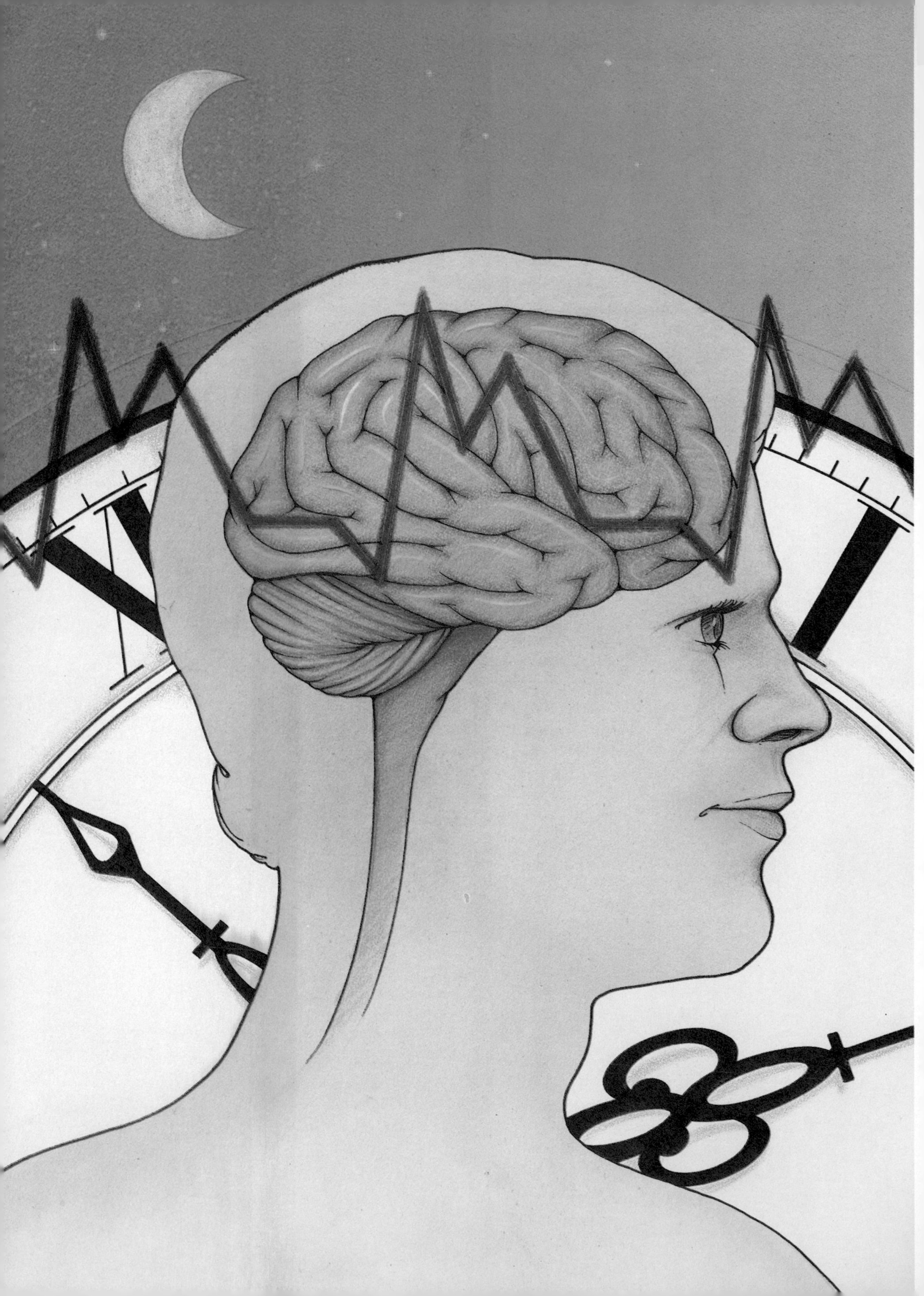

BY LYNNE LAMBERG

Learning to Use Our Inner Clocks

As scientists learn more about the body's rhythms, they are finding that these biological clocks may be used to ease jet lag, schedule better work shifts, and even treat physical and emotional illnesses.

Sixteen-year-old Ryan seldom goes to sleep before 11:30 P.M. on school nights. In order to wake up on school days, he needs the blare of rock music from his radio-alarm and, sometimes, a parent's firm hand on his shoulder. Often he has trouble staying awake during his early afternoon classes. On Fridays and Saturdays, Ryan may not go to sleep until 1 or 2 o'clock in the morning and may sleep until noon. Not surprisingly, he finds it difficult to get up at 7 o'clock on Monday mornings.

By keeping such late and irregular hours, Ryan has upset his body's natural biological rhythms of regular periods of sleeping and being awake. This daily cycle is the most easily recognized of more than 100 known biological rhythms involving the processes and activities of the human body. Most of the time, these rhythms work together in harmony to keep the body well adjusted to its surroundings and functioning smoothly. But when these rhythms are disrupted, a person may have trouble functioning as well as usual, as illustrated by Ryan's sleepiness in class.

A variety of rhythms

Human biological rhythms include not only the cycles of sleeping and waking but also rhythms of alertness and sleepiness during waking hours, and cycles of hunger, thirst, cell division, heart rate, blood pressure, body temperature, and all other functions of body and mind that have ever been studied.

Many of these rhythms follow a pattern that is about 24 hours long. Such biological rhythms are called *circadian* rhythms, from the Latin words *circa*, meaning *about*, and *dies*, meaning *day*. In someone keeping a regular schedule, each circadian rhythm rises and falls at roughly the same time each day. And the high point of one circadian cycle may be the low point or the midpoint of another. For example, body temperature usually reaches its lowest point late in sleep, while the concentration of growth hormone in the bloodstream peaks near the start of sleep.

Circadian rhythms vary from person to person. Individual differences in body-temperature rhythms make about 1 person in 10 a "morning person" and 1 in 10 a "night person"; the rest of the population falls somewhere in between. Body temperature rises earlier and peaks earlier in morning people—who wake early and function better in the morning. The opposite is true of night people.

Human beings also have biological rhythms with cycles that are much shorter or much longer than 24 hours. These cycles range from just a fraction of a second for electrical activity in the brain to about 1 second for heartbeats, about 6 seconds for each breath, and about 28 days for menstruation. Some people even experience a yearly cycle of depression that begins every autumn when the period of daylight shortens. Their depression lifts in the spring when daylight lengthens.

Scientists have made great progress in identifying the body's "bi-

The author:
Lynne Lamberg is a freelance medical writer and author of the *American Medical Association Guide to Better Sleep.*

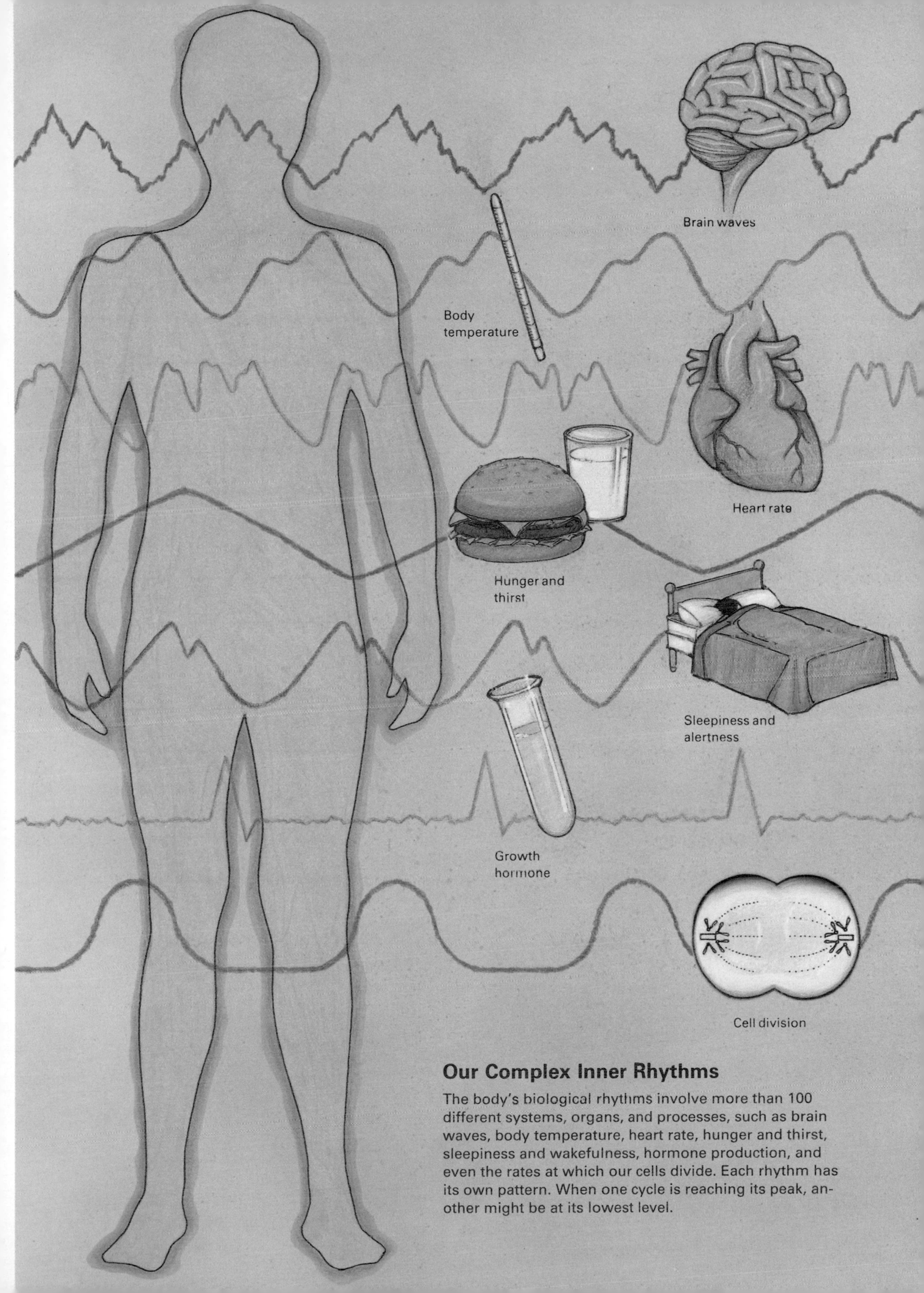

Our Complex Inner Rhythms

The body's biological rhythms involve more than 100 different systems, organs, and processes, such as brain waves, body temperature, heart rate, hunger and thirst, sleepiness and wakefulness, hormone production, and even the rates at which our cells divide. Each rhythm has its own pattern. When one cycle is reaching its peak, another might be at its lowest level.

ological clocks"—the structures and processes that keep track of time and control all of the body's biological rhythms. This field is now called *chronobiology*.

The scientific study of biological rhythms has nothing to do with the fad theory of biorhythms, which claims to predict by the date of a person's birth when that person is more prone to suffer accidents or to experience success in love and work. Scientists dismiss the theory of biorhythms as a pseudoscience, like astrology.

Recent advances in chronobiology are leading to practical applications in such fields as business and medicine. The young science of chronobiology is providing new ways to improve learning and productivity. It is helping people minimize the physical problems caused by jet travel across several time zones and is also helping workers adjust to abrupt changes in work shifts. In addition, knowledge of biological rhythms is helping physicians diagnose and treat a long list of maladies, including sleep disorders, depression, cancer, and heart disease.

The discovery of biological clocks

Since ancient times, people have observed rhythms in living things. Many kinds of plants bloom at the same time each year, and different types of flowers differ in the times of day they open and close their blossoms. Many birds, butterflies, and other animals follow annual cycles of migration. And other creatures, including bears and chipmunks, hibernate according to a yearly pattern.

Until the 1700's, people believed that these rhythms represented a passive response to light and darkness. In 1729, however, French physicist and mathematician Jean Baptiste Dortous de Mairan showed that such rhythms persist even in constant darkness. He performed an experiment on a heliotrope plant that opens its leaves during the day and folds them at night. De Mairan kept the plant in total darkness and discovered that it still opened its leaves during the day and folded them at night. He concluded that an internal timing mechanism controlled the plant's movements. Since then, experimenters have reported finding such *free-running rhythms*—biological rhythms that continue in the absence of external indicators of time—in other plants and animals, including human beings.

To study free-running rhythms in people, scientists have conducted many experiments in which people have lived alone for long periods in caves and other controlled environments with no windows, clocks, radios, or other indicators of time. In such settings, the people continued to spend about two-thirds of their time awake and one-third asleep, but they usually developed free-running rhythms of rest and activity about 25 hours long. That is, they went to sleep and awakened about an hour later each day.

Because our free-running circadian rhythm of rest and activity normally tends to run on 25-hour days, we have to adjust our inner biological clocks by about an hour each day in order to remain in

The Timekeepers Outside

Research has shown that the body's inner rhythms tend toward a 25-hour cycle if left alone. But a number of external cues work with our inner rhythms to keep our biological clocks set to a 24-hour day. These external time cues—such as alarm clocks, regular mealtimes, patterns of daylight and darkness, even daytime traffic noises—help to fine-tune our inner clocks.

harmony with the 24-hour period measured by our wrist watches and clocks. Time cues in the world around us help reset our inner clocks. Scientists have identified more than 200 possible time cues. These include sunrise and sunset, daylight and darkness, regular meals, the ring of the alarm clock, and traffic noise—which is usually heavier during the day.

Scientists have also made important discoveries about the possible location of the biological clocks that regulate these biological rhythms. In 1967, psychologist Curt P. Richter of the Johns Hopkins University School of Medicine in Baltimore published the results of experiments on rats showing that an important biological clock is located in a lower part of the *hypothalamus*, an area of the brain. Destruction of the hypothalamus in rats abolished their rhythms of physical activity, rest, feeding, and drinking.

In 1972, two groups of researchers found that a key biological clock in rats is located in an area of the hypothalamus known as the *suprachiasmatic nuclei* (SCN), a tiny cluster of nerve cells located just above the optic nerves. In 1981, chronobiologist Martin Moore-Ede and his colleagues at Harvard Medical School in Boston located the SCN in monkeys. They claim that destruction of the SCN abolishes the rhythms of rest and activity and of many hormonal processes.

The cycle of sleeping and waking

Our most obvious circadian rhythm is the regular alternation of sleep and wakefulness. Most people in industrialized societies follow a sleep-wake rhythm based on a single period of sleep every night. But sleep researcher David Dinges of the University of Pennsylvania in Philadelphia suggests that our sleep-wake system is better satisfied by two sleep periods a day. One-quarter to one-half of the world's population follows what is known as a biphasic schedule. In Mexican, Spanish, and certain other cultures, for example, shops and offices close during the early afternoon to permit workers to nap. The habit of getting all one's sleep at night may simply reflect the cultural norm of a nine-to-five workday rather than an inner clock.

In fact, a period of drowsiness in the middle of the day commonly occurs in people throughout the world. Studies conducted over several years by sleep researcher Mary Carskadon and her colleagues at Stanford University's Summer Sleep Camp in Stanford, Calif., have shown that a tendency toward midday sleepiness begins to develop in most people at about age 13. Midday drowsiness is often referred to in industrial societies as the "post-lunch dip," but it occurs regardless of when people eat lunch—and even if they skip lunch altogether. This period of sleepiness corresponds to a regular, predictable dip in body temperature. Body temperature does not remain constant at 37°C (98.6°F.). Instead, it varies over the course of a day by about 1 Celsius degree (about 2 Fahrenheit degrees) and follows a predictable cycle. Although there is a great deal of individual variability, people generally feel more alert when body temperature is high and more sleepy when it is low. In people who ordinarily sleep from about 11 P.M. to 7 A.M., body temperature is generally at its lowest between 4 and 6 A.M. It starts to rise before the customary time of awakening and reaches about 37°C by midmorning. Around 3 P.M., body temperature dips a fraction of a degree for about an hour. This is the post-lunch dip, and napping is easy during this time. Temperature then rises to about 37°C again and stays steady until about 7 or 8 P.M., when it starts to fall. The evening decline makes it easier to fall asleep.

A number of experiments have also shown that people perform certain kinds of tasks better at particular times in their daily body temperature cycle. Most people find that mornings—while body temperature is rising—are better for complex tasks that involve considerable thought and verbal reasoning, according to chronobiologist Timothy H. Monk of the Western Psychiatric Institute and Clinic at the University of Pittsburgh School of Medicine. People are better at planning, organizing, creating, and decision making in the morning, he suggests. Afternoons and early evenings—when body temperature is at a plateau—are better for simpler, repetitive tasks, particularly those involving manual dexterity.

Rhythms must be in harmony

A person's many biological rhythms normally work together harmoniously. They relate to one another like the different instruments in an orchestra. In addition, the environment provides time cues—such as sunrise and sunset—that help keep the rhythms coordinated, much as an orchestra conductor provides the tempo and other cues for the musicians. Sudden changes in a person's daily schedule, however, can upset this harmony, causing distress. The result resembles what would happen if, midway through a piece of music, a new conductor took over, leading the orchestra to a different tempo. It would take a while for the musicians to adjust to the new conductor, and the sound produced during this time would be jarring to the ears. Similarly, certain situations and demands of modern life can disrupt the harmony of our biological rhythms. These include rapidly traveling by jet airplane across several time zones or abruptly changing one's work shift. Fortunately, research in chronobiology has led to the development of ways of helping our circadian system readjust after such disruptions.

Jet travel over long distances triggers the feelings of fatigue and disorientation commonly known as *jet lag*. A woman who leaves Chicago on a late-afternoon flight to London, for example, arrives at her destination at around 8 A.M., just as Londoners are starting their morning. Because London time is six hours ahead of Chicago time and her biological clock is still running on Chicago time, she feels as if it is 2 A.M. Her biological clock is telling her body to sleep when all the daytime cues in London are encouraging her to stay awake. At midnight in London, her body will feel as it does at 6 P.M. She

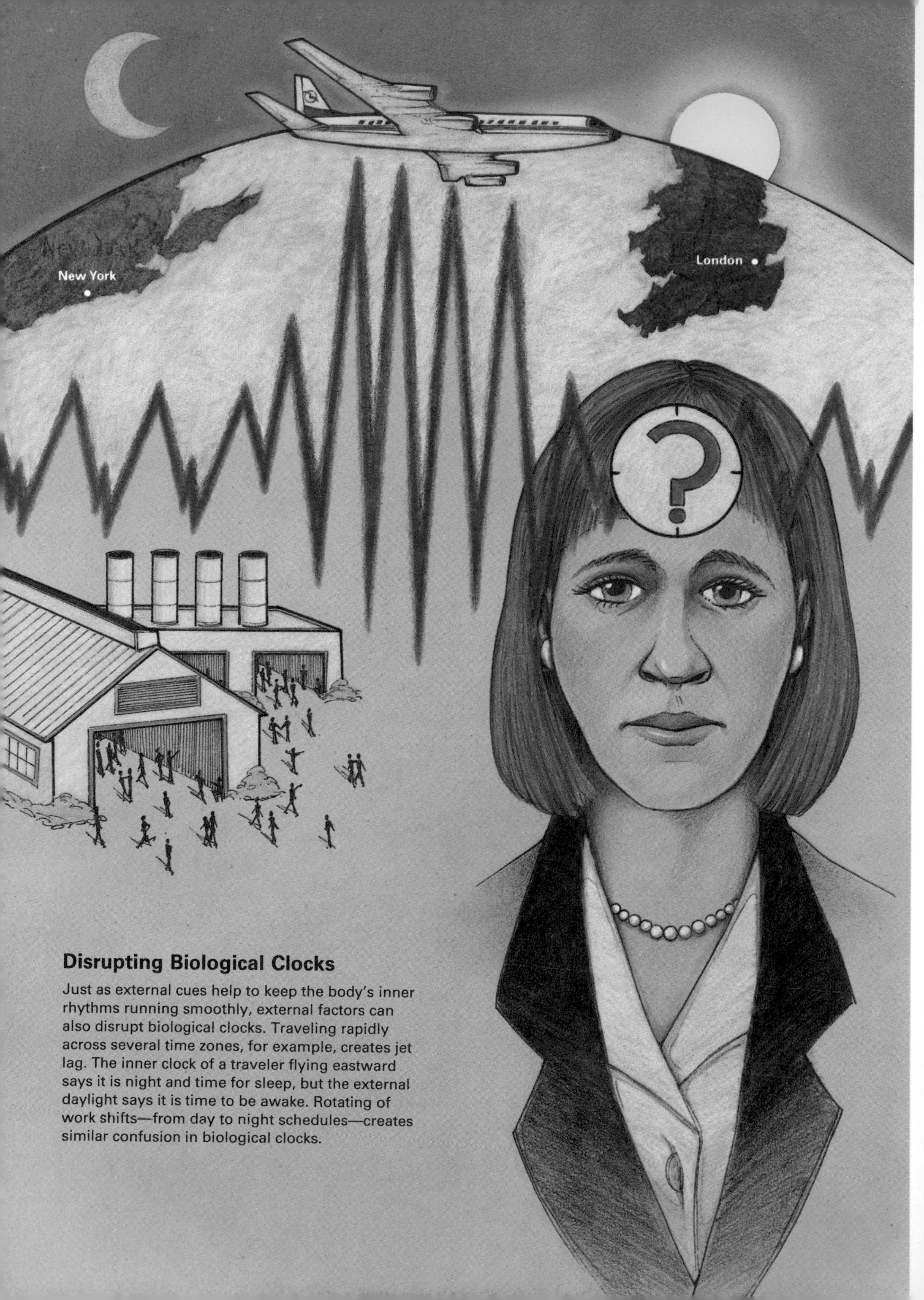

Disrupting Biological Clocks

Just as external cues help to keep the body's inner rhythms running smoothly, external factors can also disrupt biological clocks. Traveling rapidly across several time zones, for example, creates jet lag. The inner clock of a traveler flying eastward says it is night and time for sleep, but the external daylight says it is time to be awake. Rotating of work shifts—from day to night schedules—creates similar confusion in biological clocks.

may have trouble sleeping, at least on her first night in London. And her digestive hormones, still operating on Chicago time, may trigger feelings of hunger at unusual hours in London.

After such a switch in hours of sleeping and waking, it takes varying amounts of time for the body's biological rhythms to work together smoothly again. The sleep-wake cycle adjusts within a few days, for example, but the body temperature rhythm takes several days longer.

Readjustment usually is easier after travel in a westward direction. Moving westward, each time zone is an hour earlier, making the day longer. Because the body naturally prefers days longer than 24 hours, it is easier for people to stay up later, as westward travel demands, than to go to sleep at an earlier hour, as is the case when traveling east.

Overcoming jet lag

Scientists have suggested several ways to lessen jet lag. One such measure is to begin altering one's sleep-wake schedule several days before travel begins. Another method involves paying attention to the type of food consumed. Eating primarily such protein-rich foods as meat, fish, or cheese for breakfast or lunch, for example, may foster daytime alertness. Eating primarily carbohydrate-rich foods, such as potatoes or pasta, for the evening meal may promote drowsiness and help one fall asleep easier.

Exposure to very bright light—light that is brighter than ordinary indoor room light—also may hasten adjustment. Researchers found that pulses of bright light can reset biological clocks in animals, causing them, for example, to secrete increased amounts of the hormones necessary for reproduction just as if the seasons had changed from winter to spring. Psychiatrist Alfred Lewy of Oregon Health Sciences University in Portland reported in 1984 that increased exposure to very bright light in the morning appears to make the highs and lows of biological rhythms occur earlier in humans as well. Extra light in the late afternoon seems to make them occur later. Light exposure may thus be used to help minimize jet lag, Lewy says. For example, a traveler going from the West Coast to the East Coast of the United States, where time is three hours later, should spend at least 15 minutes in the sunshine outdoors in the morning. A traveler going in the opposite direction—where time is three hours earlier—should spend the same amount of time in the sunshine outdoors in the afternoon.

Shiftwork disrupts rhythms

Substantial numbers of the adult working population may experience disruption of their biological rhythms that amounts to a state of continuous jet lag. They are shiftworkers, and their jobs regularly require them to change the hours they work and sleep. In the

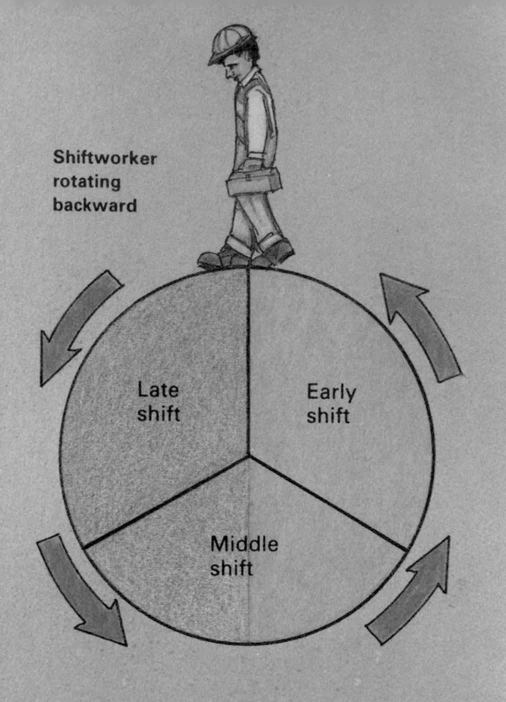

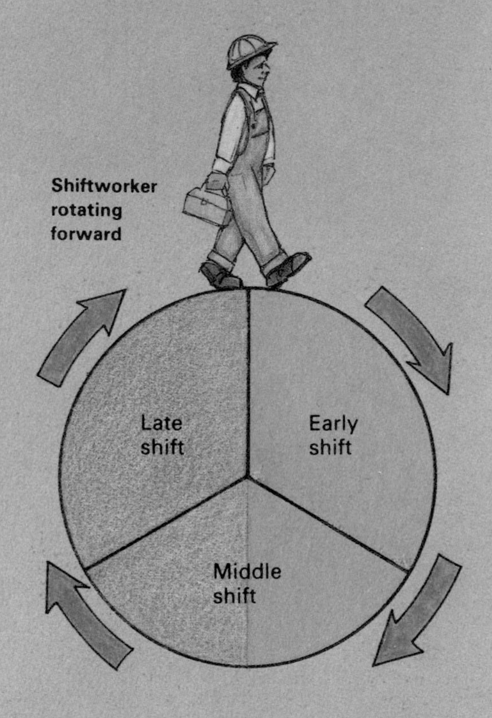

Resetting Biological Clocks

Disruptions to biological clocks in shiftworkers are greatest when schedules are rotated backward from morning to midnight shifts. The body's natural tendency to run forward on a 25-hour cycle produces more fatigue in workers who go from a day to a night shift. But rotating forward from a day shift to an evening shift, then to a night shift makes it easier for the workers' biological clocks to readjust.

United States, 1 in 4 working men and 1 in 6 working women—25 million people—are shiftworkers, and their number is steadily increasing. They include not only factory workers but also doctors, nurses, pilots, air-traffic controllers, police and other community service workers, military personnel, diplomats, sailors, athletes, and members of the news media. Because very few people are willing to work permanently at night, many employers find it necessary to rely on a rotating shift schedule. Such schedules vary widely. According to one such schedule, an employee may work from 7 A.M. to 3 P.M. for a week, then from 11 P.M. to 7 A.M. the next week, and then from 3 P.M. to 11 P.M. the following week.

The disruption of a worker's biological rhythms caused by such a schedule may lead to serious health and safety problems. Compared with workers who do not change shifts, shiftworkers complain more of lowered alertness and concentration; have a higher rate of digestive upsets, marital rifts, and divorce; and use more alcohol and sleeping pills to help them fall asleep and more caffeine to help them stay awake. Long-term shiftwork is also associated with an increased risk of heart disease.

In addition, researchers have found that workers who rotate from shift to shift make more mistakes on the job than other workers, particularly at night. Various studies have shown that between midnight and dawn, doctors make more mistakes reading electrocardiograms, nurses more frequently give the wrong medication, train engineers miss more warning signals, and truckers are more likely to run off the road.

Revising shiftwork schedules

Because studies of biological rhythms have demonstrated that more-alert, happier employees are also more productive, an increasing number of employers have become interested in work schedules that minimize the disruption of workers' biological rhythms. For example, in 1981, the Great Salt Lake Minerals and Chemicals Corporation in Ogden, Utah, sought expert help to improve the employees' work schedules. A team of researchers led by Charles A. Czeisler of Harvard Medical School in Boston designed a schedule that took the circadian rhythms into account.

For 10 years, many workers at this company had changed shifts every week, rotating from working nights, to evenings, to days. This direction of rotation forced employees to work and sleep at an earlier time with each shift change—counter to the body's natural preference. The researchers suggested that shifts be changed in the opposite direction. They also suggested that shift changes occur every three weeks rather than weekly—to give employees more time to adjust. After these changes were made, workers complained less about fatigue and poor sleep, and production rose dramatically.

Cycles of depression

Increased knowledge of human biological rhythms also holds profound implications for the practice of medicine. Scientists suspect that disruptions of biological clocks may be at the root of some illnesses. Biological rhythm abnormalities, for example, may be involved in some forms of depression. Depressed people often feel worse in the morning and better as the day progresses. Moreover, researchers have found that in some depressed patients the highs and lows of body temperature occur several hours earlier than in healthy people.

Speculating that circadian clocks may be running ahead of schedule in some depressed people, psychiatrist Thomas A. Wehr and his colleagues at the National Institute of Mental Health (NIMH) in Bethesda, Md., devised a method, first reported in 1979, to reset these clocks. The researchers persuaded depressed patients to go to bed and get up six hours earlier than usual. All the patients improved, though the improvement lasted only one to three weeks. This led the researchers to conclude that the patients' abnormal timing mechanism eventually reasserted itself.

In another study, reported in 1985, the NIMH team asked depressed patients to both advance their sleep time and take antidepressant drugs. With this combined treatment, the patients got better and stayed better, even after they returned to their previous nighttime sleeping hours.

People who suffer from another cyclic form of depression, known as seasonal affective disorder

Biological Clocks and Medicine

By studying biological clocks, doctors have learned that there are times of the day when medical treatments are more effective. They have found, for example, that certain cancer treatments, such as drugs or radiation, work better when given at times that coincide with peaks or dips in such rhythms as those of body temperature or cell division.

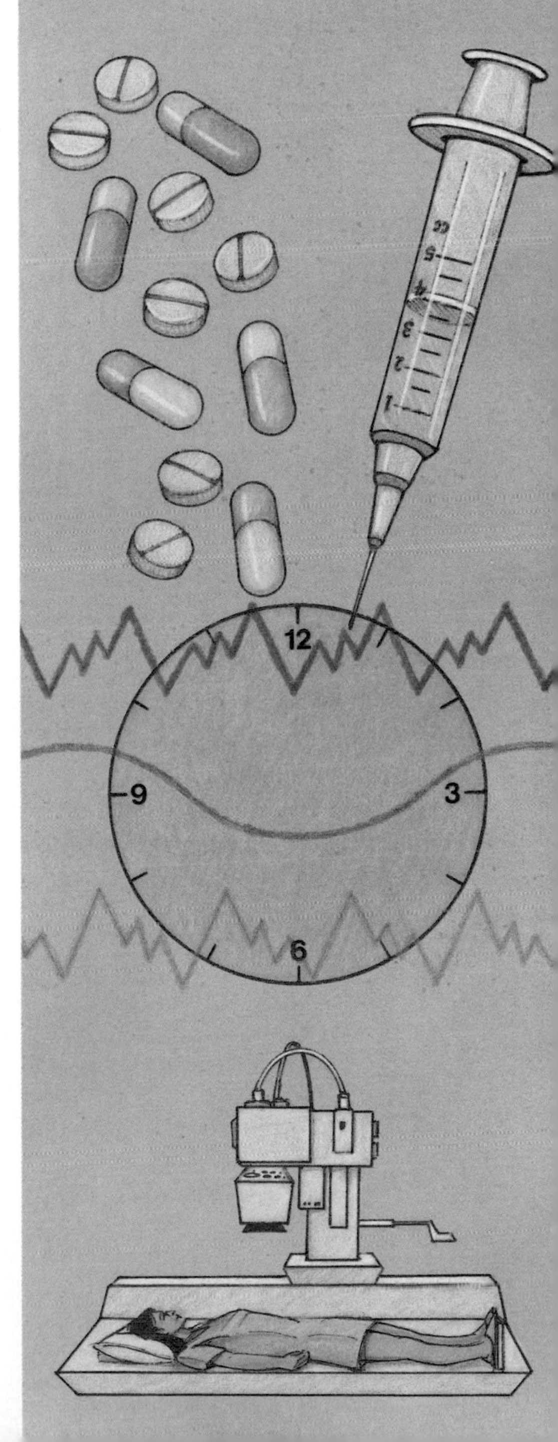

(SAD), become depressed during the fall and winter months and improve during the spring and summer. The reduced period of sunlight during fall and winter days is thought to trigger the symptoms, according to psychiatrist Norman E. Rosenthal of the NIMH. Exposure to extremely bright artificial light can, however, keep the depression from surfacing—in effect, making "spring" come early. In studies at the NIMH, underway since the early 1980's, both adults and children with SAD improved after sitting in front of very bright lights for several hours a day.

Using rhythms to fight disease

Awareness of variations in the rhythms of body chemistry and functions is helping physicians interpret laboratory test results and other measurements, leading to more accurate diagnosis of disease. Chronobiologists are also studying variations in the effects of medications when taken at different times of day. The results of such studies are helping doctors treat certain illnesses, including cancer, more effectively.

Internist William Hrushesky at the University of Minnesota in Minneapolis studied the effects of different treatment schedules of two commonly used anticancer drugs—adriamycin and cisplatin—on 58 women suffering from advanced cancer of the ovaries. These drugs are used to kill cancer cells, but they can also kill or severely damage normal body cells. Hrushesky and his colleagues evaluated various treatment schedules to determine which schedule allowed the drugs to kill the greatest number of cancer cells while causing the least damage to normal body cells. They reported in 1985 that giving adriamycin in the morning and cisplatin in the evening greatly reduced side effects. Most important, long-term follow-up of these women showed that timing treatment in accordance with the patients' biological rhythms of cell division, drug absorption, and hormone secretion resulted in greater shrinkage of tumors and an increase in the patients' survival rate.

A person's vulnerability to certain disorders of the heart and blood vessels may also be tied to biological rhythms. Merrill Mitler and his colleagues at the Scripps Clinic and Research Foundation in La Jolla, Calif., examined 4,619 randomly selected death certificates and reported in 1985 their finding that the peak time of death from heart disease for people over age 65 was around 8 A.M.

In another study reported in 1985, cardiologist James E. Muller and his colleagues at Harvard Medical School determined the exact time that heart attacks occurred in 2,999 patients by measuring the first appearance in the blood of a chemical known to show up four hours after a heart attack. The team found that more heart attacks occurred between 6 A.M. and noon than at any other time of the day, and that they occurred three times more often at 9 A.M. than at 11 P.M. The researchers believe the 9 A.M. peak may be due to stress associated with waking up. Indeed, around the time of waking and shortly thereafter, there is a rapid rise in heart rate, blood pressure, temperature, and the tendency of the blood to form clots. The Harvard researchers reported in 1987 that a temporary shortage of blood supply to the heart muscle is most likely to occur in the first two hours after waking. This finding implies the need to modify drug dosages or timing to assure adequate protection at this time.

Searching for the clockworks

Scientists are not only charting biological rhythms and devising practical applications for this knowledge but they also are beginning to discover the inner workings of the biological clocks that regulate these rhythms. They have, for example, identified genes that control biological rhythms in fruit flies and certain other organisms. Genes are the units of heredity present in every cell. They guide the production of proteins that make up cells and regulate their chemical processes. Scientists have found that certain *mutations* (changes in the chemical structure of genes) cause changes in the fruit fly's circadian sleep-wake cycle and also in the one-minute cycle of the courtship song of the male fruit fly. Different mutations lengthen or shorten both cycles or make them irregular. Moreover, researchers have found that transmitting these genes from fruit flies with normal rhythms to fruit flies that lack the genes partially restores the cycles in the deficient fruit flies.

Scientists have not yet identified genes in human beings that control biological rhythms, but the prospect of finding such genes may not be far off. This knowledge might help scientists develop drugs capable of quickly resetting our inner clocks—a comforting prospect for jet travelers, shiftworkers, and people with certain sleep disorders and forms of depression. Today's biological clock watchers may become the biological clock repairers—and perhaps even clockmakers—of tomorrow.

For further reading:

Arendt, Josephine. "The Pineal: A Gland That Measures Time?" *New Scientist,* July 25, 1985.

"Helping Workers Stay Awake at the Switch." *Business Week,* Dec. 8, 1986.

Hilts, Len. "Clocks That Make Us Run." *Omni,* September 1984.

"Jet lag: east vs. west." *Psychology Today,* July 1986.

Restak, Richard. "Master Clock of the Brain and Body." *Science Digest,* November 1984.

"Setting the Clock." *Scientific American,* July 1986.

Silberner, Joanne. "Timing Key in Cancer Chemotherapy." *Science News,* April 20, 1985.

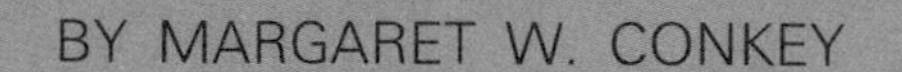

BY MARGARET W. CONKEY

Images from the Ice Age

Magnificent and mysterious, the cave art of prehistoric Europe has revealed tantalizing clues about its creators.

Preceding pages: Ice Age horses appear to race across a wall in Lascaux Cave, in southwestern France. The cave is decorated with almost 200 prehistoric paintings and 1,500 engravings.

It was a warm summer day, but we were unaware of the sun and the blue sky. We were at least 150 meters (500 feet) deep inside Enlène, a cave in the Pyrenees region of southern France, near Toulouse. About 12,000 years ago, near the end of the last Ice Age, prehistoric people called Cro-Magnons lived there. In 1986, scientists were digging up the things these ancient people left behind.

Along with many tools made of stone, bone, or antler uncovered in the layers of soil that had accumulated in the cave over the centuries, we discovered some chunks of red ocher and other pigments—colored minerals that may be found in geologic deposits near the cave. Such pigments are common at prehistoric sites like this, and scientists have proposed many theories to explain their presence. One use is certain: These pigments were often used by prehistoric peoples to paint and draw animals, human beings, geometric signs such as dots, and other images on cave walls.

We saw proof of this in another cave, called *Les Trois Frères* (The Three Brothers), down a narrow passageway from Enlène. The walls of this cave are covered with thousands of engravings, most of them of animals. But the cave's most striking image is a crouching figure called *The Sorcerer* or *Horned God.* Outlined in black pigment, the humanlike figure has antlers, a tail, and tufted ears.

The excavations at Enlène, led by a team of French archaeologists, have been going on for more than 10 years. As a scholar of prehistoric cave art, I am interested in such excavations for the light

Red dots, an example of geometric signs sometimes painted on cave walls and ceilings by Ice Age artists, mark the entrance to a decorated side chamber at Pech-Merle Cave in France.

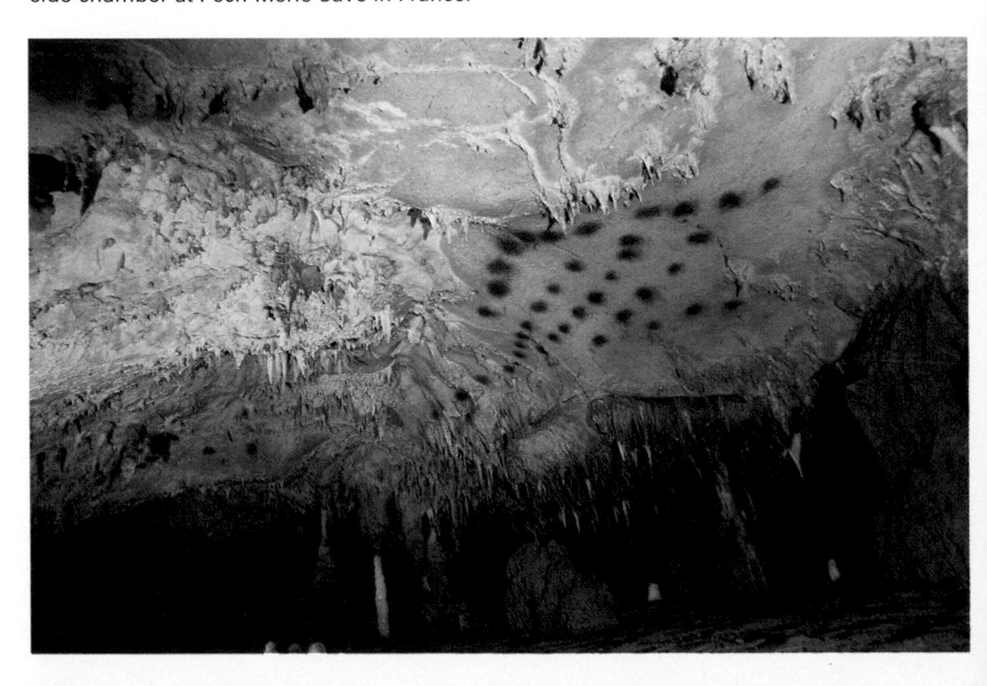

The author:
Margaret W. Conkey is an associate professor of anthropology at the State University of New York in Binghamton and the author of several books on Stone Age art.

Ice Age Artists and Their World

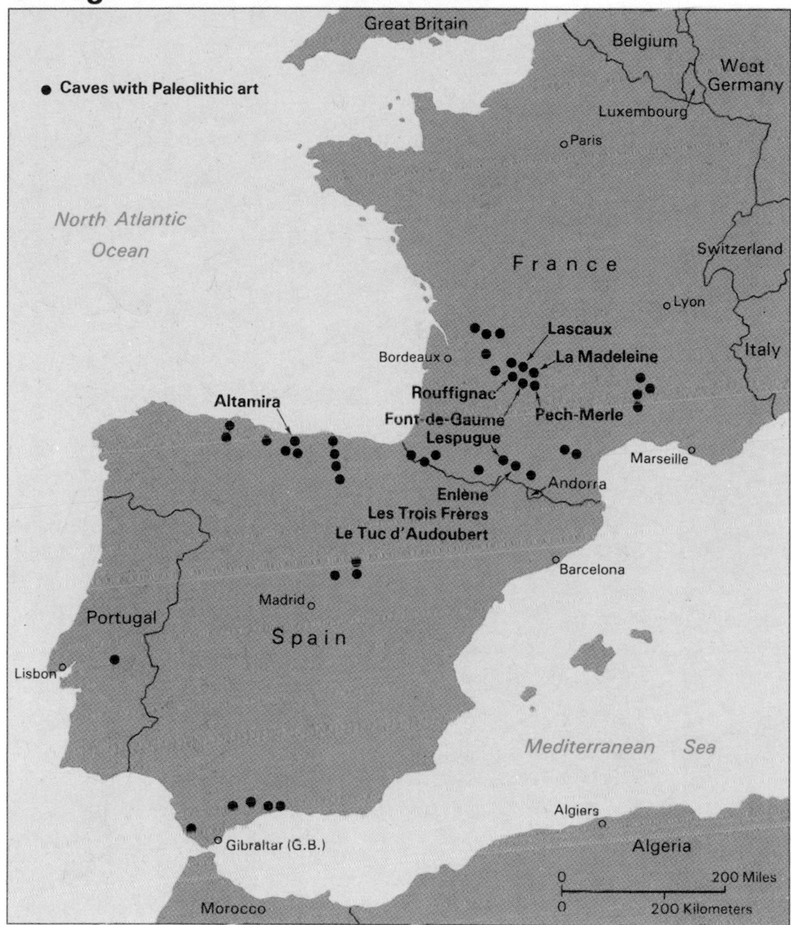

Nearly all caves with Paleolithic paintings and engravings are in southern France and northern Spain. Ice Age people elsewhere in Europe did not decorate cave walls even when suitable caves existed nearby.

they may shed on the meaning of the paintings and engravings. So periodically I have joined the workers in their exploration.

The tools, bones, and other objects found at Enlène and the drawings and photographs of the engravings and wall paintings from Les Trois Frères are only part of the very rich material record of life in Ice Age Europe that archaeologists have uncovered during the past 100 years. Of all these discoveries, however, cave paintings seem the most fascinating. In addition to black outlining, the paintings show subtle uses of reds, browns, and yellows; the colors are often still vivid after the passage of thousands of years.

Ice Age paintings and other such objects may be so intriguing because they are so puzzling. There are many images of animals but relatively few of people. In most cases, neither the animals nor the

Paleolithic Period
3,000,000 B.C.
First tools

2,000,000 B.C.

1,000,000 B.C.

Simple shelters

Ability to make fire

Burial of the dead

Upper Paleolithic Period
38,000 B.C.

Sewing

Cave paintings

Bow and arrow

Pottery

Mesolithic and Neolithic Period
8000 B.C.
Domestication of plants and animals
Farming

Small cities
Wheel Writing

Bronze Age
3000 B.C.

Iron Age
1500 B.C.

0

1988

The first cave painters lived about 21,000 years ago, during the Upper Paleolithic. The time line shows other major steps in human cultural development, though scientists disagree about exactly when some of them occurred.

Tools uncovered in archaeological excavations have helped archaeologists learn how Ice Age artists created cave paintings. A stone lamp, *top,* found at Lascaux Cave and decorated with abstract designs, was filled with animal fat, which burned to provide light for the artists. An artist's palette, *above,* found at La Madeleine Cave, is stained red from being used to grind up red ocher for pigment. The largest of the three ocher crayons on the palette is scratched, probably from being used on rough cave walls.

people appear to be doing anything. The meaning of the images remains elusive, but that is not to say the paintings and other art objects tell us nothing. From them, we have learned not only something about the Cro-Magnons' technological skills and incredible artistic sense but also something about the way they thought.

The cave art was created during the Upper Paleolithic, a 30,000-year period at the end of the last Ice Age, a time when huge glaciers covered much of northern Europe as well as parts of North America and Asia. The Upper Paleolithic, which lasted from 40,000 to 10,000 years ago, was the last phase of the Paleolithic or Old Stone Age. (The term *Paleolithic* comes from two Greek words—*paleo* meaning *old* and *lithos* meaning *stone.*) During this period, the people in Europe lived by hunting animals and gathering plants and made their tools and weapons from stone.

Archaeologists have found nearly 200 caves with Ice Age wall decorations. Nearly all these caves are in France and Spain, particularly southwest France and north coastal Spain. Archaeologists also have found what they call *portable art,* or art that can be moved—such as engraved bones and antlers, raised sculptures carved from stone blocks, and small statues made of ivory, stone, and antler—at Paleolithic sites across Europe. But it seems that people outside France and Spain almost never painted or engraved cave walls, even when suitable caves—with walls that weren't too wet, bumpy, or covered with a chalky mineral—existed nearby.

Even in France and Spain, not all suitable caves were painted. And in those caves that *were* painted, some parts were decorated and other parts left alone. In some painted caves, archaeologists have found tools, weapons, and other evidence that people lived in them. Generally, however, decorated caves such as Les Trois Frères apparently were uninhabited.

The caves used and painted by Paleolithic people can be mysterious and eerie places. While some caves are fairly open and easy to reach, most, like Enlène, are deep underground. Dark and damp, they have dead ends, side chambers, and winding passageways, some of which are so narrow that visitors must crawl or even slide along on their stomachs.

Until the late 1800's, Ice Age art remained hidden. Then one day in 1879, a young Spanish girl named Maria accompanied her father to a cave on the family's estate in northern Spain that he was exploring for prehistoric remains. Because she was short, Maria had a view of the cave's low ceiling that her taller father did not, and there she saw a fantastic display of animals painted in reds, browns, and black. Some of the animals were particularly lifelike because the painters had incorporated the bumps and bulges in the cave ceiling into the images. There were bull-like animals, for example, that had three-dimensional rumps.

Maria's discovery at the cave, called Altamira, was as controversial

A painting made in 1987 shows how Ice Age artists might have worked. By the light of a stone lamp, an artist paints a bull with a *tamping pad,* perhaps made of moss and used to "pat" pigment onto the cave wall. Another artist grinds pigment on a stone palette.

as it was exciting. Scientists had only recently begun to accept the idea that human history stretched back more than 4,000 years, a figure based on the number of generations mentioned in the Bible. This change resulted partly from the increasing number of stone tools and weapons being discovered in ancient layers of sediment. Also, in 1868, workers building a railroad near Les Eyzies-de-Tayac near Périgueux, France, made the first discovery of Cro-Magnon skeletons. The skeletons were like modern skeletons. But they lay in the same sediments as animal bones and stone tools now dated at more than 30,000 years old.

Accepting that human beings may have existed this long ago was one thing. Accepting that prehistoric people—widely believed to have been primitive and brutish—could have painted such dynamic and naturalistic images was quite another. There were accusations of forgery and hoax at Altamira.

The debate raged for almost 25 years. Finally, scientists were convinced by five lines of evidence. First, prehistorians began finding many more painted caves at other sites in Spain and France. Second, some paintings were buried beneath *geologic deposits*—ancient layers of sediment that had taken thousands of years to build up on cave floors and walls, or dirt and rock from collapsed cave entrances and walls. Third, prehistorians who believed the paintings were the

Ice Age images of people are rare. A sticklike figure, *above,* in a painting from Lascaux Cave is typical of the sketchy way in which people are represented in cave art, in contrast to the naturalistic images of animals. Handprints, such as that from Pech-Merle Cave, *right,* are the most common human image. They were made by blowing pigment from a bird-bone tube or by spitting pigment around a hand held against a cave wall.

work of Ice Age people also noted that many of the animals in the paintings, such as the mammoth, had become extinct or—like the bison—had disappeared from Europe. Given the relatively scanty knowledge of prehistoric animals at the time, how would a 19th-century forger have known how to paint these animals with such accuracy and attention to anatomical detail?

Fourth, prehistorians began finding more bones engraved and carved with the same animals and in the same styles as the cave art. These bones often were found in ancient geologic deposits and with stone tools known to have been made by prehistoric people.

Finally, during the late 1800's, anthropologists began learning more about the rich artistic and cultural life of modern hunter-gatherers, people in isolated areas of the world who had never developed farming skills but lived by hunting animals and gathering wild plants, much as Cro-Magnons did. Some skeptical prehistorians had argued that Paleolithic hunter-gatherers would have had the time to create such beautiful paintings only if they lived in a rich and bountiful environment that easily supplied their needs. But the climate of southwestern Europe during the Ice Age, with huge glaciers only a few hundred kilometers to the north, was much harsher than today's climate. The studies of modern Australian Aborigines and other hunter-gatherers in remote areas, however, revealed that people living in a harsh environment—for example, a desert—could have a rich artistic life. For example, the Aborigines paint cave walls and sheets of tree bark.

By the beginning of the 1900's, the evidence had persuaded most scientists that the cave paintings and engravings in Europe dated from the Upper Paleolithic. Determining exactly how old they are, however, is a tricky business. So far, scientists have not been able to date the paintings because there is, as yet, no way to calculate the age of minerals. (The pigments used to create the paintings are composed chiefly of minerals, such as manganese and ocher.)

All dates for the cave paintings, therefore, are *relative dates*, calculated by dating the paintings in relation to something else whose age is known—for example, the geologic deposits burying some paintings. By dating these deposits, scientists can determine the latest possible date at which the paintings were made. Using this process, however, scientists would not be able to determine a painting's earliest date—that is, how old the painting was when the deposits began to accumulate or the roof collapsed.

Some figurines and other forms of portable art, however, have been dated on the basis of the geologic deposits in which they were found. This became a basis for another method of dating cave paintings. Operating on the theory that paintings and portable art that look similar were created at about the same time, scholars have attempted to date cave paintings and engravings according to the ages of similar, archaeologically dated portable art objects. For ex-

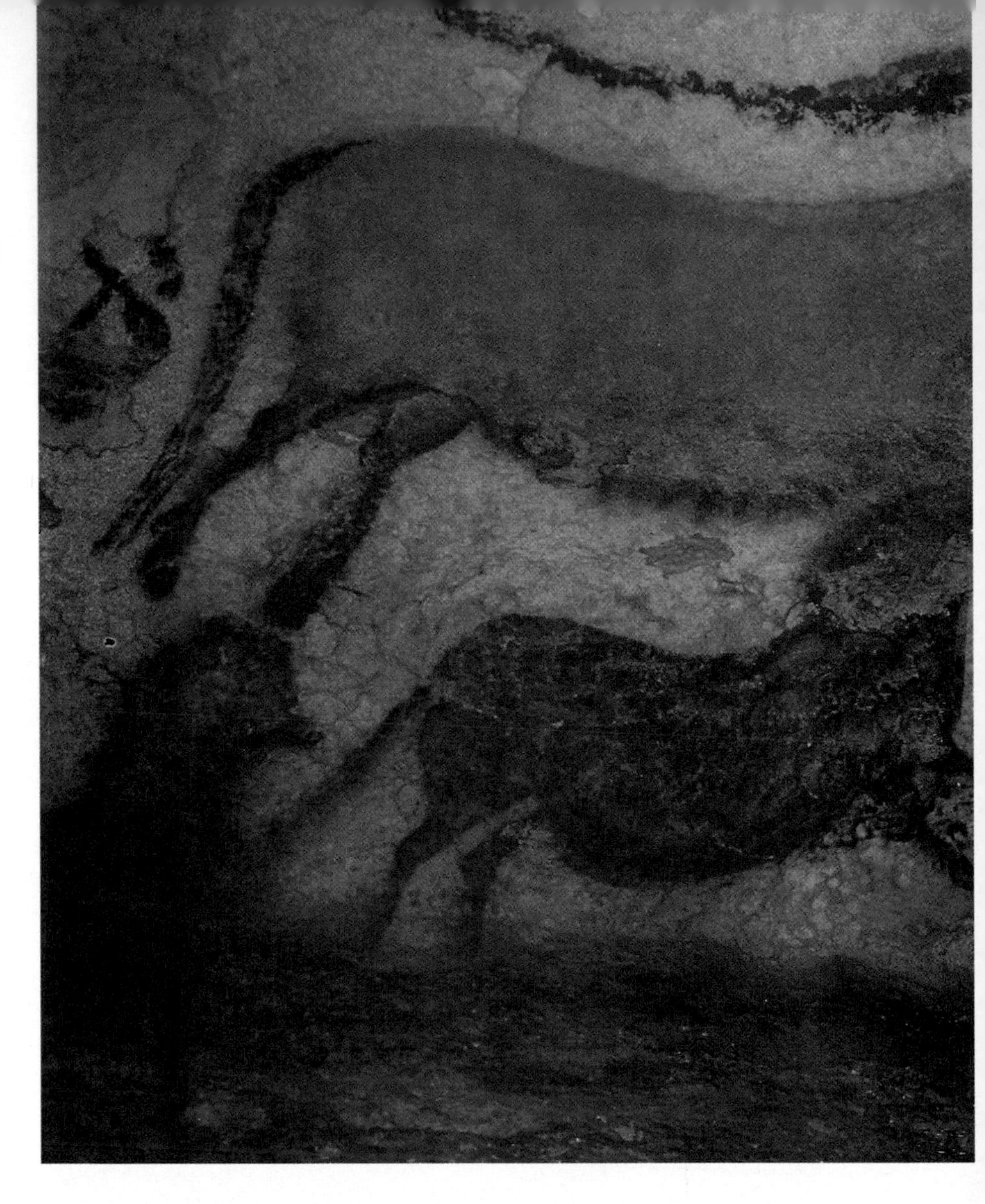

Larger-than-life-sized images of wild horses and a long-horned bull on a wall at Lascaux Cave still appear in vivid colors after thousands of years.

ample, they have compared the ways in which animals were portrayed and the relative simplicity or complexity of the image.

Using such indirect evidence, cave art scholars have concluded that not all of the different kinds of cave art were made throughout the Upper Paleolithic. For example, about 80 per cent of the cave paintings and engravings appear to have been created between 20,000 and 10,000 years ago. Many of the small female statues found at Paleolithic sites date from about 23,000 to 28,000 years ago. Also probably dating from this period are outlines of human handprints—made by blowing powdered pigment through a tube made from a bird bone or spitting the pigment around a hand held against a cave wall or rock.

What knowledge we have of the Cro-Magnon people who created this art comes from archaeological excavations such as we made at

the cave in France. From the bones, tools, and other objects found at these sites, scientists know that the Cro-Magnon hunter-gatherers lived in bands that were probably made up of related people. The bands traveled with the seasons to find different sources of food. Cro-Magnons knew how to make string and rope and how to sew.

And from the paintings, engravings, and statues they left behind, we know that the Cro-Magnons were also creative. Even some of their everyday objects, such as tools and weapons, are often decorated beautifully.

Since the discovery of the cave paintings and engravings, scholars have speculated on the artists. Were there specialists within each band? Were there traveling artists who moved from cave to cave? Were the cave painters men, the sculptors women? Unfortunately, neither archaeological excavations nor the art itself can answer these questions as yet. There are striking similarities—the shape of a horse's hoof, the way an animal is shaded—in some paintings and engravings, even in different caves. But there is no way to prove these images were created by the same individual or even by a group of artists trained in the same style.

Although the identity of the cave artists is lost in time, archaeologists have learned a great deal about the way in which they created the paintings. We know that in some cases, the artists collected their pigments from the countryside near the caves. For example, geologic studies at Lascaux Cave revealed mineral deposits that were likely to yield pigments. Chemical studies confirmed that these pigments were the same as those used in the paintings.

Sometimes, chunks of pigment, especially black manganese pigment, were used alone as a kind of crayon. Many of the black outline images, such as the bulls at Lascaux, are technically not paintings at all but drawings made with pigment sticks.

To make pigment sticks, the artists first ground the pigments to powder, mixed them with water and sometimes with clay, then formed the mix into a stick. Sometimes, the prehistoric artists mixed several pigments together to create a pigment stick of a new color. Archaeologists have found stone slabs that were used as palettes for this mixing process. At times, ground pigment—mixed with water and ground-up, baked animal bones—was applied to the walls with brushes made of animal hair. More often, the color was applied with *tamping pads*—pads, perhaps made of moss, used to "pat" the pigment onto the wall.

Studies by French archaeologist André Leroi-Gourhan of the University of Paris have shown that cave artists had definite preferences about which animals they liked to portray. Horses and bison are depicted in 60 per cent of the cave paintings. Deer, mammoths, ibexes, and reindeer make up another 30 per cent. The remaining 10 per cent are bears, lions, and rhinoceroses.

Only a few of the animals in the cave images appear to be doing something together. One painting of a row of *stag* (male deer) heads at Lascaux may depict a herd swimming. In another painting at Font-de-Gaume Cave in France, a male reindeer appears to be licking the head of a female reindeer facing him. Many of the animals appear to be in motion, but background images, such as lines representing the horizon or other ways we might use to orient the animals, are rare.

Human images, though not uncommon, are certainly less common than those of animals. Paintings and engravings of human figures also are much less naturalistic. In contrast to bison depicted so vividly that the animals almost seem to leap from the walls, the human images are usually sketchy, sticklike figures. Human faces are rare. The most common painted human image is a handprint.

As far as we can tell, almost none of the wall images or carvings

An engraving of a woolly mammoth, found at Rouffignac Cave, shows the animal's shaggy coat and humped back in great detail. Such detail in images of now-extinct animals helped convince scientists that cave paintings and engravings were created during the Ice Age, when the animals still lived.

on the portable art objects depict either such everyday human activities as hunting or gathering or such special occasions as rituals or ceremonies. As a result, the cave art cannot be used as a way to read about the Cro-Magnons' daily life. The information revealed by the imagery requires more *inferences* (conclusions reached by reasoning). For example, by examining the detail of the images we can infer that Cro-Magnons were keen observers of nature. Some of the animal paintings appear so realistic that we can tell whether the animals are wearing their summer or winter coats.

From similarities in the cave art images, we can conclude that there must have been extensive contacts among Cro-Magnon groups. The differences in the art, however, may indicate that the bands had their own customs and considered themselves at least somewhat separate from other groups.

Evidence discovered in the caves indicates that cave painting was often a social activity. At Lascaux, for example, the many pigments and pigment sticks, palettes, and stone lamps uncovered lead archaeologists to believe that several artists were working together.

We can take our deductions a step further: If it was a social project, cave painting must have involved planning and an exchange of information. Since most of the painted caves are by no means readily visible or easy to find, someone must have scouted for suitable locations. Expeditions to decorate the cave walls may have been highly organized affairs. Certainly, the artists had to anticipate the items they would need, including pigments and brushes, implements for sharpening dulled engraving tools, enough animal fat for lamp fuel, and food and water.

Some Ice Age sculptures, statues, and other artifacts can be dated, and thus provide clues to the age of art on the walls of caves in which they were found. The Venus of Lespugue, *above,* a 15-centimeter (6-inch) female statue carved of mammoth ivory, was dated from 25,000 to 27,000 years ago. A 10.5-centimeter (4-inch) spear throwing device, *above right,* made from reindeer antler carved in the shape of a bison licking its flank, was found at La Madeleine and dated to 13,000 years old.

One of the most important questions about Paleolithic cave art—what purpose it served—remains unanswered. Until the early 1960's, most cave art scholars believed that Cro-Magnons used the caves exclusively for magic and ceremonies to ensure success in the hunt and the fertility of the animals that they hunted. This theory, proposed in 1903 by French archaeologist Salomon Reinach, was vigorously promoted by Henri E. Breuil, a French priest who became one of the world's greatest authorities on cave art.

Reinach and Breuil argued that the animals in the paintings and engravings were either those that Cro-Magnons liked to eat, such as bison and reindeer, or those they needed to kill to keep their caves safe, such as bears or lions. They maintained that certain symbols painted on the tops of animals represented arrows or wounds, depending on their shape.

Research by Leroi-Gourhan, begun in the late 1950's, however, undermined the hunting-magic theory. He found, for example, that the percentage of "wounded" animals in the paintings was actually very small. And he discovered that Cro-Magnons, in fact, preferred

Bison, modeled in clay, were crafted by an unknown artist working almost 1.6 kilometers (1 mile) underground at Le Tuc d'Audoubert Cave, 15,000 years ago. The bison in the foreground is about 60 centimeters (2 feet) long.

to draw animals they did *not* often eat. For example, archaeologists have found far more reindeer bones—animals they ate frequently—than horse or bison bones at later Upper Paleolithic campsites in or near painted caves. Yet images of horses and bison are far more numerous on cave walls than those of reindeer.

In 1965, Leroi-Gourhan proposed that—although there were variations from cave to cave—Cro-Magnon artists followed an underlying set of rules for decorating caves. He called this master pattern a *mythogram*. According to Leroi-Gourhan, decorated caves can be divided into sections, such as a *main panel*—the wall in the main chamber of a cave, where most of the images are found—and *side chambers*—smaller caves off the main cave. He found that horses and bison, the most commonly portrayed animals, almost always are the central figures not only in the main panel but also throughout the cave. Images of bears, lions, and rhinos—the least common animals—usually appear in the deepest, most remote parts of the cave.

Some cave art scholars believe that the images served as a way of transferring knowledge from generation to generation among these people, who had no written language. These scholars argue that there was an information explosion during the Upper Paleolithic in nearly every aspect of life—from the manufacture of tools and weapons to art and music.

In his book *The Creative Explosion* (1982), journalist John Pfeiffer suggested that decorated caves were used for ceremonies initiating young people into adulthood. He noted that many of the decorated caves can be reached only by following tortuous passageways. Many of the painted images appear in hidden, out-of-the-way places; many are upside down or in other strange positions. Pfeiffer noted that caves are frequently disorienting places because they lack familiar landmarks or a view of the sun to mark the passage of time. He theorized that the trip to the caves and simply being in the caves created a heightened sense of awareness among the initiates. The sudden illumination of vivid images by flickering lamp or torch created shock and amazement, attracting and focusing the initiate's attention. According to Pfeiffer, cave art was a "way of attaching emotion to information for memory's sake."

If there are stories in the cave paintings and engravings, they are not composed in a way that we can easily "read." And there are no living Cro-Magnons to tell us.

Some cave art scholars believe that if we could find an accurate way of dating the cave paintings and engravings, we could "crack the code" of Ice Age cave art. Such dates certainly would be useful. For example, they would tell us exactly when the caves were painted. And if we could connect specific paintings of a known age with specific groups of Cro-Magnons, we could get an idea of how mobile Cro-Magnons were, whether they stayed in the same areas for long periods or moved about frequently. Dates would tell us whether cave art style and technique evolved over time, and if so, how. But it is unlikely that simply dating the paintings and engravings would reveal much about their meaning.

In the past several years, many cave art scholars have begun to reject the idea that there is only one interpretation for all the thousands of images made over the 30,000 years of the Upper Paleolithic. Researchers are now beginning to study limited sets of images and make specific interpretations. For example, in 1974, rock art scholar Nancy Olsen of De Anza College in Cupertino, Calif., argued that some of the bison at Altamira may not be charging or dying bulls as previously suggested but may represent females giving birth. Bison females and their calves generally live apart from the males, except during breeding and calving seasons. So the Altamira paintings of bison females may be a metaphor for a similar situation among some Cro-Magnons. Among some modern hunter-gatherers, such as Australian Aborigines, women and children often live together while the men are gone on hunting expeditions.

It is unlikely that Ice Age cave paintings and engravings tell stories in a straightforward way. The wall art usually does not deal with individuals but with animals, sometimes in groups. Perhaps this indicates the importance some Cro-Magnons placed on clan membership and cooperation. It is likely that the animals are symbols for valued human qualities, such as courage and loyalty, or even power.

I believe there can never be one single answer to the question of what cave art means. Human behavior is complex and everchanging, and we have endless ways to express our feelings and experiences, especially through making visual images. Perhaps Paleolithic cave art served several purposes at once. In addition to being used in rituals and ceremonies, images may also have been appreciated simply for their beauty. It is very likely that their meanings changed over time or according to the ritual in which they were used.

Archaeological excavations and cave art studies during the past 100 years have revealed a great deal about the prehistoric hunter-gatherers who were the first modern humans in southwestern Europe. In the future, studies of cave art will surely reveal more about the way in which it was created. Excavations in the caves should also tell us more about how the caves were used, which, in turn, may provide insights into the purpose of the art. The next century of research may reveal things that little Maria at Altamira could not possibly have imagined.

For further reading:

Leroi-Gourhan, André. *Treasures of Prehistoric Art.* Abrams, 1980.

———. *The Dawn of European Art: An Introduction to Paleolithic Cave Painting.* Cambridge, 1982.

Pfeiffer, John E. *The Creative Explosion: An Inquiry into the Origins of Art and Religion.* Harper, 1982.

White, Randall. *Dark Caves, Bright Visions: Life in Ice Age Europe.* American Museum of Natural History and Norton, 1986.

Because the camel can go for days without food or water, it may be the ideal farm animal for drought-stricken areas of Africa.

New Uses for "The Ship of the Desert"

BY NOEL D. VIETMEYER

At first glance, the scene is typical of many universities in warm climates—students in constant motion, heading from one sun-washed building to another. But from time to time, strange bellows, roars, bleats, snarls, grunts, and growls echo across the campus.

A raw barnyard smell permeates the area just behind the medical center. There, inside holding pens, gangly yellowish beasts peer out over steel bars, their lips upturned, their noses in the air. This campus is unmistakably in the Middle East. And these vocal but placid-looking animals, with gentle eyes and mouths that slant like smiles, are camels.

The campus is that of Ben-Gurion University in Israel's southern desert city of Beersheba. On its camel farm, veterinary professor Reuven Yagil is studying how the camel functions and what its feeding requirements are. In particular, he is trying to improve milk yields and speed up reproduction.

The South African-born Yagil is small, cheery, and usually dressed in a white lab coat and jeans. He loves all 12 of his camels, but his particular favorite is a short, combative female called Golda. The attraction is not mutual. Yagil takes so many blood samples that when Golda sees him she hollers and hoots in fear that he has come to jab needles into her again.

Camel-research programs similar to Yagil's are going on at institutions in Egypt, Pakistan, Saudi Arabia, and Syria. Sudan hosted an international conference on camels in 1979. Even Sweden and the United States support camel-study projects in various parts of Africa. The Soviet Union's central Asian republic of Kazakh has taken its research programs a step further, breeding thousands of camels for food and wool on its government-run farms.

All of these ventures are part of an international effort to develop alternative livestock and crops to feed the world, especially the developing nations of Africa, Asia, and Latin America. In the past, agricultural scientists focused most of their attention on improving the yields of a limited number of domesticated animals and crops rather than attempting to develop the productivity of a wide range of alternative species. Many researchers have studied cattle, corn, wheat, and rice. Few have investigated unfashionable—but potentially valuable—resources such as the camel or water buffalo and the hundreds of nutritious fruits and legumes of tropic regions.

The author:
Noel D. Vietmeyer is a professional associate of the U.S. National Academy of Sciences and an expert on underused animals and plants.

In the 1980's, however, agricultural researchers have begun to realize that dozens of neglected livestock animals and thousands of overlooked plants might help solve some of the world's food problems, especially in harsh, dry environments where many plant and animal species cannot exist. The camel, which is well known for its ability to survive long periods without water, is a classic example. "Camels are the livestock that are most adapted to dry conditions," Yagil points out. "They do well in many areas of the world where starvation continues to be a big problem."

This is especially important in the chronically drought-stricken areas of Africa, where 70 million people are subject to famine. It is crucial that African nations such as Ethiopia, Mali, Niger, Senegal, Somalia, and Sudan learn to feed their populations during drought as well as during plentiful years. Otherwise, some observers warn, 200 million Africans could be on the brink of starvation by the year 2000.

Yagil believes the camel can alter those grim projections. In recent years, he and anthropologist Scarlett Epstein of the University of Sussex in England have been traveling the world trying to win international support for this vision: "The people of drought-stressed Africa should herd camels," Yagil says forcefully. "Camels are exquisitely adapted to arid rangelands, the very regions where Africa's tragedy is occurring."

Residents of Mauritania in western Africa leave their drought-plagued homeland, where few plants or animals can survive on the barren land. Researchers are studying the use of camels as livestock in such areas.

Until recently, scientists have generally ignored the camel, and so it remains one of the least understood—and least appreciated—of all domestic animals. Many scientists and farmers believe that in comparison with cattle and sheep, camels are too big, too slow in maturing and breeding, and too ill-tempered to become productive domestic animals. Skeptics say that camels are "obsolete" and could become extinct without causing significant loss to the world. As a result, camel populations are dwindling, even in the Middle East and Africa. But pioneering researchers such as Yagil are learning exactly how the camel is able to survive on little water and seemingly inadequate feed. Their work is also beginning to dispel myths about this unusual creature and raise new interest in its use in the modern world.

Unfortunately, scientists cannot change the fact that the camel looks funny. An old joke says that the camel is a horse designed by a committee: gnarled neck, balloon back, barrel belly, leather lips, knock-knees, and spindle shanks. Adding to the animal's ungainly look is its height. Camels can be 2.1 meters (7 feet) tall at the shoulder and weigh up to 680 kilograms (1,500 pounds).

It might look misshapen, but scientists are intrigued—and sometimes awed—by the camel's body. Not only can it travel for days without water but it is also amazingly well adapted to its harsh desert environment. Thick eyebrows and a double fringe of eyelashes shield its eyes from the sun and sand. Flaps of skin over its nostrils and thick hair around and within its ears keep out windblown sand. The feet splay out from skinny shins so that even the heaviest camels do not sink into soft sand. The soles of a camel's feet are padded with fibrous tissue that allows it to tread painlessly across burning sand dunes. (Camels find rocky terrain difficult, however, because they cannot jump.) What appear to be bald spots on the chest, knees, and legs are actually calloused cushions that protect the camel when it rests on stony ground. When grazing, the camel uses its vertically divided upper lip to feel all around for its food. This conserves moisture by shielding its tongue from the dry air.

"Their foraging habits are the least destructive to the environment," Yagil says. "Cattle denude an area, but camels feed on trees and shrubs that have deep roots, so there is less chance they will destroy the vegetation. And because camels can be taken out to forage far beyond watering holes, their grazing is spread out and its damaging effects lessened."

Desert landscapes are typically composed of scattered trees with thorny branches that cattle find indigestible and mostly out of reach. Camels, however, stretch their necks and lift their heads up into the trees. Somehow a camel manages to chew on thorns sharp enough to pierce the sole of a shoe. Indeed, if hungry enough, camels will dine on anything in sight—thatched roofing, plastic sheeting, even copper wire. People who have left cars with fiberglass bodies unattended in the desert have returned to find neat chunks taken out of the fenders.

Thriving on desert vegetation that most animals shun, a camel grazes contentedly on the prickly spines of a thorn tree.

Zoologists have identified two modern camel species—the Bactrian and the Arabian. Bactrian camels have stocky, shaggy bodies with two humps. They inhabit the harsh deserts of Mongolia, China, and the Soviet Union. Arabian camels are taller, faster, and have only one hump. They generally live among the nomadic people of Africa and the Arabian Peninsula. A breed of Arabian camel, known as the *dromedary*, is raised for riding and racing.

Bactrian and Arabian camels can interbreed to produce vigorous, one-humped offspring that are fertile. Because of this, some scientists suggest that the two present-day species should really be classified as just one species.

Except for about 1,000 Bactrian camels that roam wild in Mongolia, almost all camels are now domesticated. Arabian camels have not existed in the wild for about 2,000 years.

We do not know exactly where or when people began taming and breeding camels, but Arabian camels were probably domesticated in what is now Saudi Arabia a little more than 4,000 years ago. Although we tend to associate camels with nomadic herders who live in the Sahara, camels are relative newcomers to that area. They were introduced during the A.D. 100's but were not used extensively there until after the Arabs invaded northern Africa in the 600's.

Millions of people came to rely on the camel for survival. Desert nomads—the Bedouins of the Middle East, the Tuaregs of the Sahara, and the Somalis of Somalia in eastern Africa—entrusted their lives to it. The camel was their only mode of transportation. In addition, camel milk and camel meat provided them with food. They made their clothes and rope from camel wool, camel skins gave them shelter, and camel droppings were used as fuel. Today, more than 15 million camels are still used for labor, transportation, and food throughout central and northern Africa, the Middle East, and parts of Asia.

Because the camel can carry burdens across vast wildernesses, it came to be called "the ship of the desert." The name is well de-

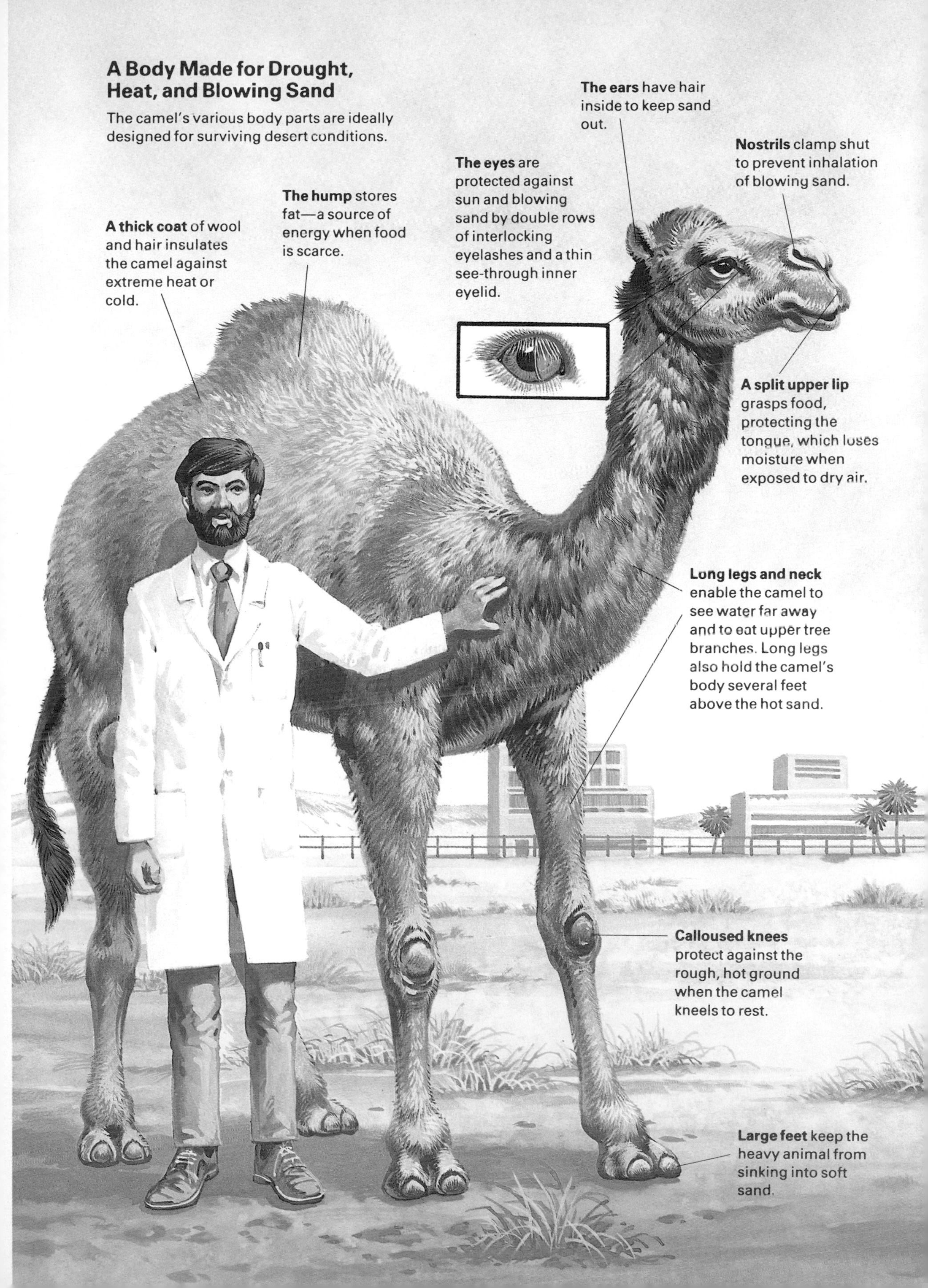
A Body Made for Drought, Heat, and Blowing Sand
The camel's various body parts are ideally designed for surviving desert conditions.
The ears have hair inside to keep sand out.
Nostrils clamp shut to prevent inhalation of blowing sand.
The eyes are protected against sun and blowing sand by double rows of interlocking eyelashes and a thin see-through inner eyelid.
The hump stores fat—a source of energy when food is scarce.
A thick coat of wool and hair insulates the camel against extreme heat or cold.
A split upper lip grasps food, protecting the tongue, which loses moisture when exposed to dry air.
Long legs and neck enable the camel to see water far away and to eat upper tree branches. Long legs also hold the camel's body several feet above the hot sand.
Calloused knees protect against the rough, hot ground when the camel kneels to rest.
Large feet keep the heavy animal from sinking into soft sand.

One hump or two?
There are two species of camel, Arabian and Bactrian. Arabian camels, *above left,* have only one hump and are taller and leaner than Bactrians, *above right,* which have two humps. Bactrian camels live in colder climates than Arabians do and are covered with a warm layer of long, shaggy hair.

served. The camel has been used as a cargo ship, a battleship, and a passenger ship. Its rolling gait—swinging both right legs forward, then both left legs—can even make its riders "seasick."

A camel can travel faster and longer than a horse and carry crushing loads. Weighed down with 225 kilograms (500 pounds), the camel can trudge 40 kilometers (25 miles) per day for three days. Without a load, a camel can pad along at 16 kilometers (10 miles) per hour for 18 hours straight—long past the point when a horse would collapse from exhaustion. Often going without water for days and sustaining itself on little but thorns and shriveled, bitter leaves, a camel can easily travel 48 kilometers (30 miles) per day for weeks.

Although the camel has been a beast of burden for centuries, it remains an animal with a mind of its own, often balking and whining when forced to perform its tasks. A mistreated camel can be a formidable enemy. The animal will seek revenge, no matter how long it must wait. An angry camel may attack without provocation—kicking backward and forward—and inflict serious injuries. When infuriated, it roars and bellows and bites anyone or anything in reach. Given careful handling, however, camels are gentle and well behaved.

The camel belongs to a group of animals—including cows, goats, and sheep—that *ruminate.* These animals swallow their food after chewing it only slightly and have either three or four stomach chambers. The camel, which has three chambers, stores partially chewed food in the first section of its stomach. Later, the animal draws the wet, sloppy food, called *cud,* back into its mouth for more chewing. As a camel plods silently across the desert, it is often busy chewing its cud. Finally the camel swallows the cud, which then goes to the other stomach chambers to be completely digested.

The camel's most amazing characteristic is its ability to endure long periods without drinking water. A United Nations survey revealed that in 1973, during Niger's worst drought in more than 50 years, all of the cattle and half of the sheep and goats in that African country died, but 80 per cent of the camels survived. During the annual dry season in northern Africa, cattle must be watered every 2 to 3 days and sheep and goats every 3 to 8 days. But camels are watered only once every 10 to 20 days. When grazing on fresh vegetation, which contains moisture, camels often show no interest in water for months. And although lambs and calves must always have drinking water available, young camels can live solely on their mother's milk, which actually increases in water content as the mother becomes more dehydrated.

Unlike other domesticated animals, camels are not particular about the salinity of the water they drink. They can thrive on water salty enough to cause death in other animals. This is a great advantage for the camel, because much of the water found in desert regions has a high salt concentration.

One popular misconception about camels is that they store water in their humps. A camel's hump actually stores fat when food is plentiful, enabling the camel to survive when forage is scarce. Because most of its fat stores are located in the hump instead of in a uniform layer under the skin, any excess body heat in the camel can quickly escape. Scientists once speculated that a "dry" camel produced water within its body by breaking down the fat in its hump, but research shows this would be an inefficient process, using as much water as it produced. The approximately 45 kilograms (100 pounds) of fat in a hump would yield only 49 liters (13 gallons) of water. The camel's body would need a great deal of oxygen as it broke down the fat, and the increased breathing would cause at least 49 liters of water to be lost by evaporation through its lungs.

Products from the Camel

Milk and meat (even ground-up hooves) provide food.

Hair and wool are made into rope and fabric for blankets and clothing.

Hides provide leather for whips, saddles, and water bags.

Dung, which burns readily, is used as fuel.

Since the 1950's, physiologist Knut Schmidt-Nielsen of Duke University in Durham, N.C., has studied how camels survive without water. In one experiment in northern Africa, he deprived a camel of water for eight days in heat so intense that human beings in the area needed to drink fluids every hour or risk death. The camel became thin and listless as it lost 100 kilograms (220 pounds) of water—20 per cent of its weight. Such a loss of body fluids would be fatal to most other animals. The camel, however, was never seriously endangered.

Schmidt-Nielsen learned that the camel relied on several physiological features to conserve water. All camels' body temperatures fluctuate several degrees throughout each day, and the camel in the test did not need to sweat—a process that cools the body through evaporation—until its temperature had risen from its normal low of 34°C (94°F.) to 40°C (105°F.). By insulating its body from outside heat, the camel's coat of wool actually helped the animal stay cool. (Schmidt-Nielsen also found that a camel sheared of its insulating

Studying the Camel

At Ben-Gurion University in Beersheba, Israel, veterinary professor Reuven Yagil uses modern research methods to learn more about the long-misunderstood camel.

In the laboratory, *right,* Yagil examines microscopic camel *embryos* (fertilized eggs) that have been flushed out of a female camel's womb. Some embryos will be implanted in other female camels as part of a camel-breeding program.

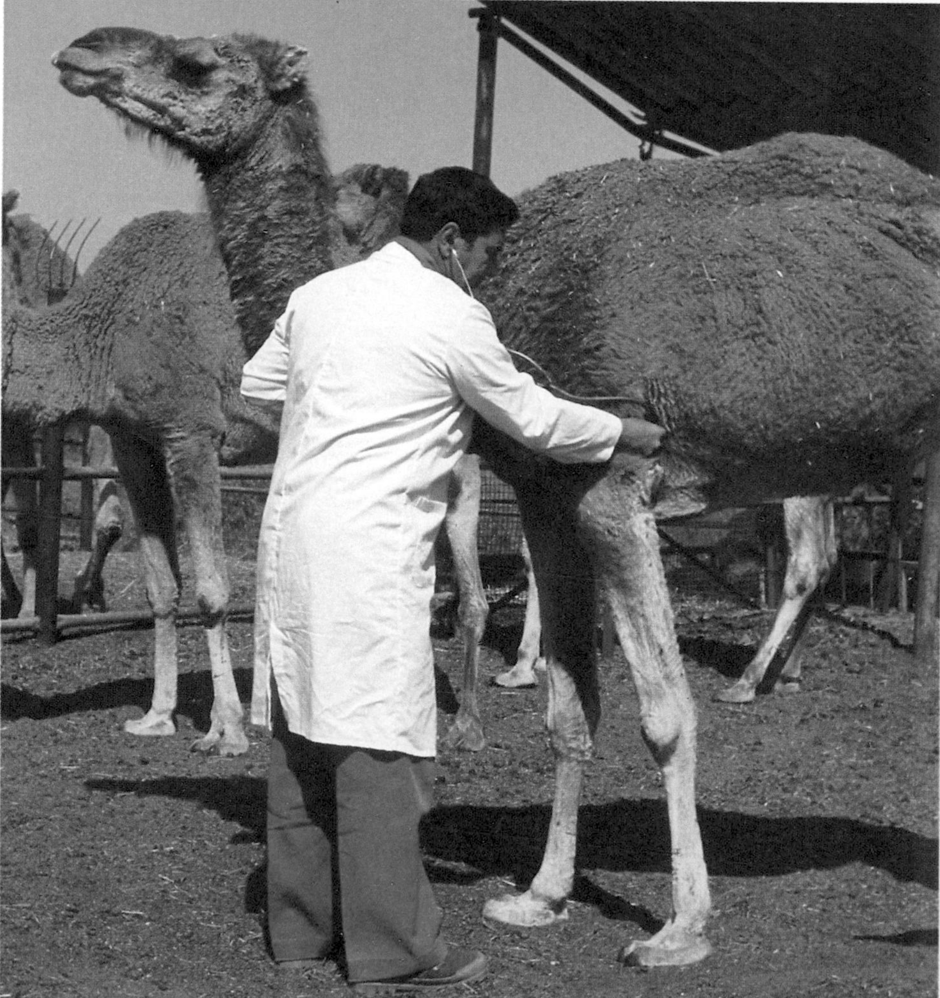

Using a stethoscope, *right,* Yagil listens to the heartbeat of Golda, his favorite camel.

An unusually docile camel allows Yagil to draw a blood sample, *left.* Frequent blood analyses have shown how the camel's body handles both the lack of water and its sudden replacement.

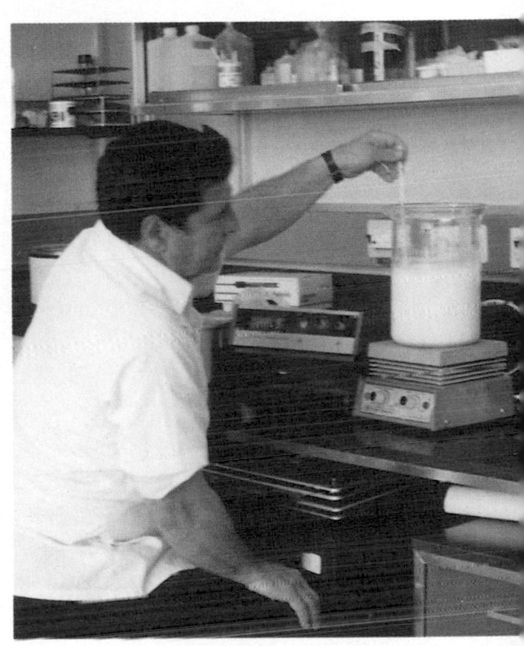

Yagil experiments with heating camel milk to condense it, *above.* His primary goal is to find a way to use camel milk for feeding starving people in Africa.

wool lost 60 per cent more water through sweating than a camel with a full coat.)

When it did sweat, Schmidt-Nielsen's test camel drew moisture from its body tissues—rather than from its blood as other mammals do. This ensured that the camel's blood remained thin enough to circulate freely and dissipate body heat.

The camel also excreted little water in its waste matter. The feces were dry as cardboard, and the highly concentrated urine was thick as syrup. In addition, the animal recycled its waste *urea*, a nitrogen substance found in urine, back into its digestive system to form protein. This process both saved moisture and provided more energy.

When thirsty camels finally get water, they drink huge amounts, and within minutes they can *rehydrate*, or replenish any fluid loss. Schmidt-Nielsen's camel downed buckets of water at the end of the experiment and regained its former appearance and condition within an hour.

At Ben-Gurion University, Yagil, who has also studied dehydration in the camel, once watched with amazement as a camel drank 57 liters (15 gallons) of water in three minutes. In human beings and most other animals, such a sudden increase in fluid volume

Dairy camels in a feed lot in Saudi Arabia, *top,* are kept in pens and eat from troughs so that they do not waste energy making long treks looking for food. These animals produce large quantities of milk, which is packaged in cartons, *above,* and sold in Saudi stores for about twice the price of cow's milk.

would drastically thin the blood. This dilution decreases pressure outside the blood cells, causing them to burst. This results in instant death. Only a few animals besides the camel can safely drink large quantities of water in a short time. A desert goat, for example, survives sudden rehydration because swallowed water remains in the goat's stomach and is released into the bloodstream very slowly.

To find out if a similar process occurs in camels, Yagil "marked" a container of water by adding a small amount of a radioactive substance to it. After a dehydrated camel drank the water, Yagil drew several samples of its blood at different time intervals and measured the blood's radioactivity. The measurements showed that the water did not remain in the camel's stomach but entered the animal's bloodstream almost instantly. How could an animal survive the sudden dilution of its blood? Further blood tests showed that the camel's body had stepped up production of a hormone called *aldosterone*, which increases the salt content of the blood to counteract the effects of dilution. Rehydrated camels also produce little *antidiuretic hormone* (a substance that inhibits the elimination of urine). Because of this, any excess water can be quickly excreted. Finally, Yagil discovered that the camel's red blood cells, which are oval instead of round as in other mammals, can swell considerably without bursting.

Since 1969, Yagil has studied the best way to raise camels as livestock animals. Regardless of locale—an Israeli experimental farm or the plains of Kenya—Yagil believes that female camels should be raised in stalls and small feed lots. Other scientists believe this method fails to exploit the camel's advantage as a wide-ranging forager that eats plants useless to other animals. But Yagil argues that

Keeping the camels in a protected environment spares them from long treks searching for food, and thus conserves the animal's energy resources for milk production.

Although camels can be farmed for meat and labor, Yagil is most enthusiastic about camel milk. At Ben-Gurion University, one of Yagil's camels—the feisty Golda—produces enough milk each day to feed several dozen children. And Yagil believes that selectively breeding camels that produce large amounts of milk and giving them nutritious feed can produce "supercamels" that will each provide enough milk for 100 children.

Yagil says that camel milk can provide most of the essential nutrients for children in drought-ravaged Africa. Laboratory analysis has found that camel milk provides three times as much vitamin C as cow's milk and greater percentages of vitamins A, B_1, B_2, and B_{12}. In addition, about a liter (1 quart) of camel milk can satisfy all of a small child's daily calorie requirements.

One drawback to the use of the camel as a farm animal is its low birth rate. Most female camels do not begin to reproduce until age 5, and then they bear a single calf only once every two or three years. By injecting hormones into 18-month-old female camels, Yagil has induced sexual maturity so that they start calving before the age of 3. With good feeding, these camels then produce calves annually. This birth rate is comparable to that of cattle.

Yagil's research has focused entirely on the Arabian camel. But in the Soviet Union's republic of Kazakh, biologist and camel specialist Vladimir Belokobylenko oversees farm management of the Bactrian camel. Kazakh is a land of vast *steppes* (plains) and large deserts. Although the region is sparsely populated, its cities are growing. This increases the need for fresh meat and milk—a serious problem in such a harsh environment. Belokobylenko reports that neither dairy cows nor beef cattle can stand the extreme conditions of this cold desert, but camels can. Domesticated Bactrian camels have long thrived on the arid lands found in Kazakh.

On the Kazakh farms, Bactrian camels are raised for meat, milk, and wool. Belokobylenko reports that meat from 1-year-old camels is more tender than veal and that camel sausages taste like those made from beef. Milking machines are used on 400 female camels in one farm's mechanized milking complex. Each animal also yields wool—anywhere from 1 to 5 kilograms (2.2 to 11 pounds)—when it sheds its coat in spring.

People who have never depended on the camel may think of it as an unpleasant beast that stinks, bites, spits, hisses—and has outlived its usefulness in today's world. Those who have studied the camel know otherwise. Someday, this animal may assume a place in farm fields and corrals. And if the camel fulfills the expectations that visionary scientists have set for it, this animal may go a long way toward relieving famine in the underdeveloped nations of the world.

Scientists are learning why some massive sheets of ice race over the land rather than creep along slowly.

Glaciers On the Go

BY MARK F. MEIER

As our helicopter circled above the tiny figure standing on the broken, jagged surface of Alaska's Columbia Glacier, I was uneasy. If we had to make an emergency landing, where could we set down on this shattered expanse? If we had to leave the area for some reason, would we be able to make our way back to that lonely person nearly lost on a sea of deep cracks and chasms?

Since 1983, Columbia Glacier, which sweeps 65 kilometers (40 miles) down a broad valley from Alaska's Chugach Mountains to Prince William Sound in the Bay of Alaska, has been on the move. So many icebergs are breaking off the front of the glacier that Columbia has *retreated* (shrunk in length) 2 kilometers (1.2 miles). Water from the sound has followed the glacier as it retreats, creating a deep inlet.

One of the most fascinating aspects of this retreat is that the speed at which the ice flows down the glacier from its *head* (point of origin) in the mountains to the sound has also increased dramatically. At times, the glacier races more than 15 meters (50 feet) per day, a gallop compared with a glacier's usual leisurely pace of about 30 centimeters (1 foot) per day. Our goal was to understand how this 1,165-square-kilometer (450-square-mile) river of ice moves so rapidly over its rocky bed.

Opposite page: Margerie Glacier sweeps majestically into Glacier Bay in southeastern Alaska.

Our helicopter flight that day was part of our attempt to "weigh"

the glacier. We had to determine the weight of the glacier so that we could calculate how much pressure this huge sheet of ice 760 meters (2,500 feet) thick exerts on its bed. The problem with weighing glaciers is that they are not solid blocks of ice whose weight can be calculated from measurements of their size and density. They are riddled with open *crevasses* (deep cracks), chasms, channels, and caverns. We had to find out just how much of Columbia Glacier is ice and how much is air. To do that, we made several helicopter trips to photograph the surface of the glacier so that we could map the crevasses. For scale, we used a member of the team—whose height we had carefully determined. Before the photo session, our helicopter would descend close enough to the glacier's surface for him to jump out. Once on the ice, he served as a measuring stick in the photographs. After the photo session, we would hoist our human yardstick back into the helicopter and head for our camp on Heather Island in Prince William Sound just below the glacier.

Heather Island is a beautiful place, with scores of bald eagles and an abundance of gulls, sea otters, porpoises, and even an occasional whale. Heather Island can also be a very noisy place due to the glacier's movements. From our camp we had a spectacular view of the front of the glacier, a wall of ice 6.5 kilometers (4 miles) wide that towers 60 meters (200 feet) above the waters of the sound. We could watch, awe-struck, as gigantic chunks of ice, some weighing millions of tons, broke away from the glacier and slammed thunderously into the water amid a fountain of spray.

Working on a glacier is difficult, sometimes frightening, but always exciting. In the 400 years since scientists first realized that glaciers *advance* (grow and move forward) and retreat, we have learned a great deal about how glaciers grind slowly over the land. But when it comes to understanding how these majestic rivers of ice can at times race along at such amazing speeds, it sometimes seems that the more questions we answer, the more questions arise.

Glaciers spread their icy fingers over part of every continent except Australia. They range in size from relatively tiny masses of ice in shaded mountain alcoves to huge sheets of ice as big as continents. There are three main types of glaciers: mountain glaciers, icecaps, and ice sheets. Mountain glaciers form in bowl-shaped hollows near high peaks, then spill down mountain valleys or slopes. Columbia Glacier is a mountain glacier. Icecaps are huge sheets of ice that flow out in all directions from a thick center, or *dome*. The Arctic Islands of Canada have many of the world's icecaps. Ice sheets are mammoth versions of icecaps. The Antarctic Ice Sheet, the world's largest ice sheet, covers 14 million square kilometers (5.4 million square miles), an area larger than the United States, including Alaska.

The ice in a glacier is constantly moving forward from the head, replacing ice lost by melting, evaporation, or the breakoff of ice-

The author:
Mark F. Meier is the director of the Institute of Arctic and Alpine Research and professor of geological sciences at the University of Colorado in Boulder.

An explorer investigates an ice cave in Muir Glacier. The cave is only one of many caverns and channels carved out inside the glacier by melted snow and ice seeping down from the surface.

bergs at the *terminus* (lower end). If the amount of snow and ice accumulating on the glacier equals the amount of ice that it loses, the glacier neither advances nor retreats. Its length stays the same. If the snowfall increases or melting decreases, the glacier advances. If the snowfall decreases or melting increases, the glacier retreats. A glacier is like a bank account with money constantly being deposited and withdrawn. The balance on hand—the size of the glacier—depends on the difference between these deposits and withdrawals.

Today, glaciers cover about 11 per cent of Earth's land surface. In the past, however, during the so-called ice ages, huge glaciers crept down from the polar regions and high mountains, turning vast areas of the planet into frozen wilderness. During the past 2.5 million years, ice waxed and waned at least 20 times over Canada and the Northern United States, as well as over much of Europe and Asia. During the last Ice Age, which ended about 10,000 years ago, glaciers covered nearly 25 per cent of Earth's land surface, including vast areas of Europe, North America, and Asia.

The movement of such massive sheets of ice has long fascinated scientists. Naturalists of the 1700's and 1800's who hiked or climbed Alpine glaciers observed that the ice underfoot was obviously hard and brittle, yet the ice was flowing down the valley, snaking around

Anatomy of a Glacier

A glacier is a moving mass of snow and ice. It flows from the *head,* where it begins, to the *terminus,* where it ends. Deep cracks called *crevasses* often mark its surface. Rocky material scraped from the bedrock by the glacier and deposited at the terminus forms a ridge called a *terminal moraine.* Some glaciers sit on a thin layer of *meltwater* (melted snow and ice) that forms a river at the terminus.

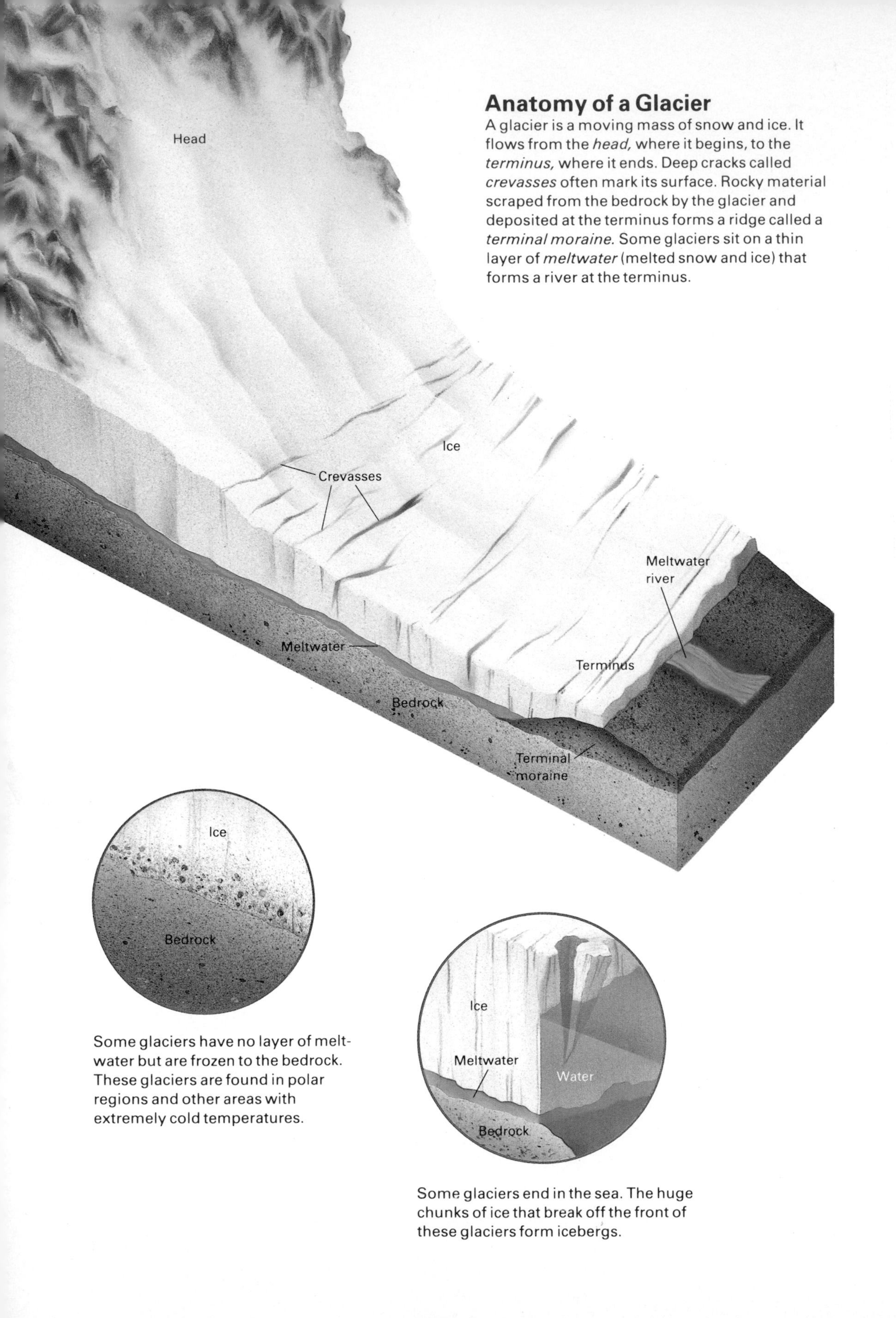

Some glaciers have no layer of meltwater but are frozen to the bedrock. These glaciers are found in polar regions and other areas with extremely cold temperatures.

Some glaciers end in the sea. The huge chunks of ice that break off the front of these glaciers form icebergs.

the curves like a great river. How could a material be hard and brittle and, at the same time, flow like a fluid?

For many years, scientists believed that glaciers flowed like an especially thick liquid, such as tar or honey. Then in the late 1940's and early 1950's, a group of British physicists and metallurgists discovered that ice flows in a manner similar to that of metals heated to near their melting point. This discovery resulted from unusual research undertaken during World War II to determine whether icebergs reinforced with fibers could be used as aircraft carriers. The British never built an iceberg ship, but they did learn a great deal about ice and how it flows.

Like metals, ice is made of crystals—particles whose atoms are arranged in a rigid regular pattern. The ice crystals in a glacier form from snow crystals that pile up over the years and become very dense. For a while, the weight of the ice pressed against the rocky ground keeps the ice in place. But eventually the ice at the head becomes thick and steep enough so that the layers of ice crystals and the atoms within them begin to slide over one another.

The combined movement of trillions of crystals causes the ice to *deform* (change shape). Tugged by gravity, the glacier moves downhill. Icecaps and ice sheets also move downhill. The caps and sheets spread out in all directions from the highest central point, where the ice is thickest, to lower areas, where the ice thins.

Glaciers that flow only by this process of *internal deformation* are called *cold glaciers*. They creep along very slowly, just a few meters per year. They are found in the polar regions and other areas with extremely cold temperatures, and their bases are frozen to the underlying bedrock.

Other glaciers flow faster, their movement measured in kilometers per year. In addition to moving by internal deformation, these glaciers move by sliding over the bedrock on a thin layer of water. Glaciers that move in this way—a process called *basal sliding*—generally are found in temperate regions where warmer temperatures keep the ice near the melting point. As a result, such glaciers, especially in the summer, hold a great

Slow-Moving Glaciers

Most glaciers move very slowly. They do so in two ways—by internal deformation and by basal sliding.

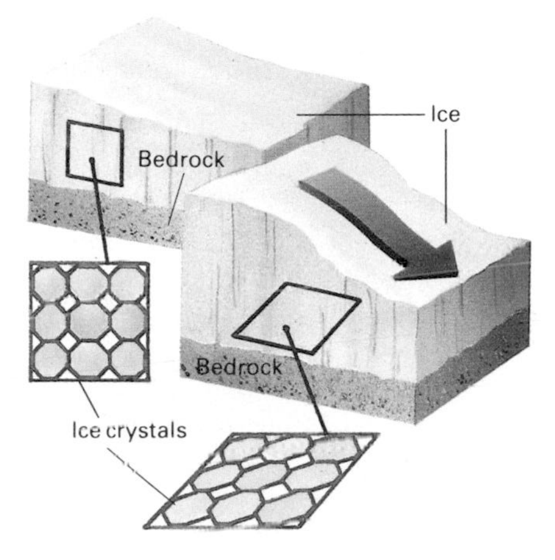

Internal deformation occurs when the head of a glacier reaches a certain thickness and steepness. This causes the layers of ice crystals to *deform* (change shape). Tugged by gravity, they slide over one another and move downhill. The upper layers of the glacier move faster than the lower layers, which are slowed by the friction between the base of the glacier and the bedrock.

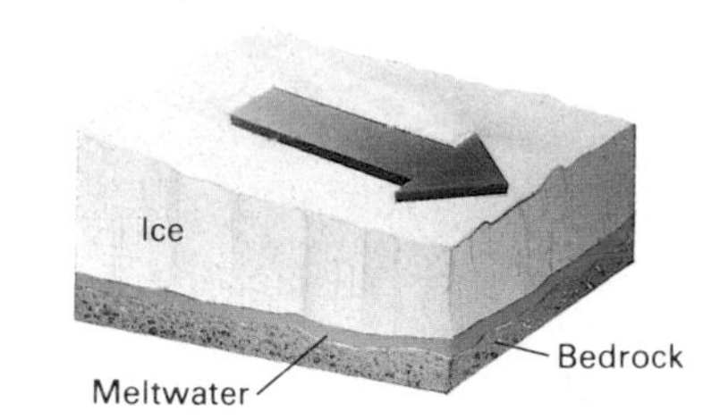

Basal sliding occurs when a layer of meltwater partially lifts the glacier from its rocky bed. This allows the glacier to slide along more easily. Because this reduces friction, the upper and lower parts of the glacier flow at about the same speed. The glacier may also be moving by internal deformation at the same time.

deal of *meltwater* (melted snow and ice). The meltwater seeps down from the glacier's surface through cracks in the ice to the base, where it escapes, creating a river at the terminus. Heat from the ground and from friction created as the glacier creeps along also cause melting at the base of the glacier. The meltwater is under high pressure because of the weight of the ice above. Confined to channels between the base and bedrock, the water exerts enough force to partially lift the glacier off the ground.

There are three types of fast-moving glaciers: ice streams, surging glaciers, and tidewater glaciers—though tidewater glaciers may move fast or slowly. These swift-flowing masses of ice have long puzzled scientists, but recent studies have revealed tantalizing clues about how they move.

There are hundreds of ice streams in the Antarctic Ice Sheet. These relatively fast-moving strips of ice—some 10 to 100 kilometers (6 to 60 miles) wide—are bordered by slow-moving ice.

In 1984 and 1985, scientists made the first intensive study of a major ice stream to find out why these ice streams flow so fast. Geophysicist Charles R. Bentley and his colleagues at the University of Wisconsin in Madison discovered that the answer seems to be internal deformation, but with a twist. Bentley and his group studied Ice Stream B, an ice stream in western Antarctica that is 50 kilometers (30 miles) wide and 500 kilometers (300 feet) long.

For their research, the scientists developed a sophisticated sonar device that sends sound waves into the ice and then records the

A helicopter hovers over huge crevasses that score the surface of Hubbard Glacier in Alaska. The crevasses were created by the rapid movement of the ice.

echoes received from below. Using this device, they found that between the glacier's bottom layer of ice and the bedrock is a 6-meter (20-foot) layer of wet rock particles of various sizes. This rock layer also contains meltwater under high pressure. The scientists deduced that the rapid motion of the ice stream is due to this water, which acts as a lubricant for the particles of rock, allowing them to slide readily over one another.

Research on Ice Stream C, a slower-moving ice stream nearby, however, raised more questions than it answered. In 1979, geophysicists Keith Rose and Gordon Robin of the Scott Polar Research Institute in Cambridge, England, noticed that the surface of Ice Stream C was smooth, with few visible crevasses. In contrast, the surface of Ice Stream B was heavily marked with crevasses. Using a special kind of radar that penetrates through ice, Rose and Robin found that 20 meters (65 feet) below the surface of Ice Stream C is a very rugged layer of ice. They concluded that this rugged layer was once the surface of the glacier. Evidently, the surface of Ice Stream C once looked like the surface of Ice Stream B. Did Ice Stream C once flow rapidly only to "turn off" several hundred years ago? If so, why? How ice streams form remains a mystery.

A scientist drills into a glacier on Mount Olympus in Washington to obtain an ice core for study.

Scientists still have a great deal to learn about surging glaciers, too, though a recent study of one such glacier, Variegated Glacier in Alaska, has provided a few answers. Surging glaciers are also called galloping glaciers. A typical surging glacier flows slowly, thinning for several decades. But then, mysteriously, the glacier starts to move faster, maintaining speeds of up to 6 to 9 meters (20 to 30 feet) *per hour* for a few months or even a year. Then the glacier abruptly slows, leaving the ice stretched and crisscrossed with crevasses. Several decades later, the glacier will surge again.

Surging glaciers are not rare. But a detailed study of a surge was not carried out until 1982, when three teams of scientists examined Variegated Glacier. The scientists chose Variegated Glacier because it has a history of surging about every 17 to 20 years, the last time in 1964-1965. So a new surge was expected between 1981 and 1985.

In 1973, the scientists began conducting tests on the glacier. To measure the glacier's *flow rate*—the rate at which ice moves from the glacier's head to its terminus—the scientists drove stakes into the glacier at various points. They then compared the position of the stakes over time with reference points off the glacier. To measure the speed of the meltwater flowing beneath the glacier, they drilled holes in the ice and pumped dyes into the water. Then they measured how long it took for the dyed water to flow out of the glacier at the terminus.

The scientists also dropped devices into holes in the ice to measure the pressure of the water at the base. Finally, using radar, they measured the thickness of the ice at various points along the glacier's length. They repeated these tests when the glacier began to

Fast-Moving Glaciers

Some glaciers, including ice streams and surging glaciers, speed over the landscape at rates up to 100 times faster than a glacier's normally slow pace. They do so by variations of basal sliding.

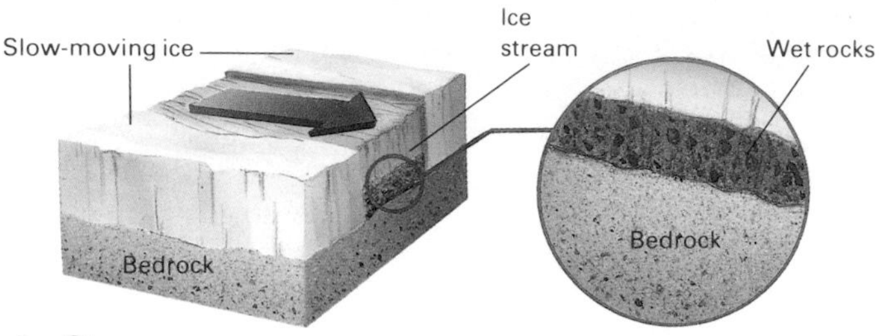

Ice Streams

Ice streams, which are found only in Antarctica, *left,* are fast-moving rivers of ice flanked by slow-moving ice. Ice streams appear to ride on a thick layer of wet rocks of various sizes. Water in this layer lubricates the rocks, allowing them to slide easily over one another. This movement of the rock layer causes the ice stream to flow rapidly.

Surging Glaciers

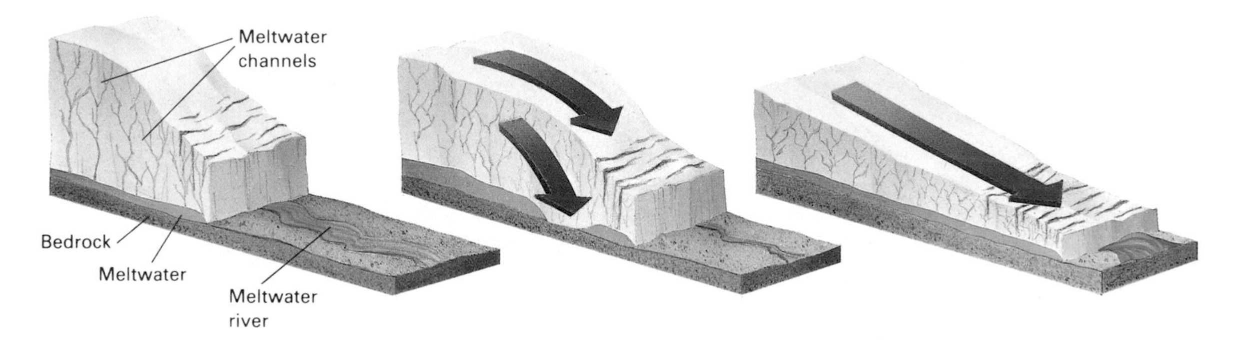

Surging glaciers, also called galloping glaciers, for some unknown reason begin to flow rapidly after moving slowly for decades. Before a surge, the head of the glacier becomes thicker and steeper because of a build-up of ice that normally flows slowly down the glacier. The surge begins when the head becomes so steep that the ice begins to flow rapidly downhill.

The force exerted by the weight of the moving ice smashes the channels that normally carry meltwater to the terminus. As a result, the water builds up at the base, partially lifting the glacier off its rocky bed and making it slide along rapidly. Like a wave moving downhill, more and more of the glacier begins to surge as channels farther toward the glacier's terminus collapse.

When the wave of motion nears the terminus, where the ice is thin, the trapped water escapes in a flood and the surge stops abruptly, leaving the glacier stretched and crisscrossed with crevasses. The glacier looks as though someone had grabbed its terminus and pulled hard.

surge in January 1982. By the time it stopped on July 4, 1983, they had learned a great deal about this puzzling phenomenon.

The scientists discovered that the surge was definitely the result of basal sliding, not internal deformation. They had drilled a hole in the ice straight down to the bed 385 meters (1,300 feet) below. When they checked the hole several days later, they found that its shape had not changed very much, indicating that ice near the surface was flowing at about the same speed as the ice near the base. If the glacier had been moving chiefly by internal deformation, the surface ice would have flowed much faster than the ice lower down and the hole would have been bent like a curved tunnel.

The scientists found that between surges, the head of the glacier became thicker and the terminus got thinner. This happened because the snow and ice that normally would have flowed down the glacier stayed put at the head. Eventually, however, the head of the glacier became so steep that ice began to flow downhill. The force exerted by the weight of such a heavy mass of moving ice partially smashed the meltwater channel under the head of the glacier. This channel, which runs along the base of the glacier much like a river, carries meltwater to the terminus. With the channel blocked, the depth and pressure of the meltwater under the head of the glacier increased, lifting the glacier higher off its bed and causing it to slide along rapidly. As the ice mass moved downhill, more and more of the channel was blocked, and more water accumulated under the glacier. At one point, the water layer was 1 meter (3 feet) thick, compared with 1 millimeter (1/30 inch) before the surge. Like a wave moving downhill, more of the glacier began to surge until the whole glacier was sliding along rapidly, at one point reaching a speed of 60 meters (200 feet) per day. Finally, when the wave of motion had nearly reached the terminus of the glacier, the water gushed from the glacier in a sudden flood, and the surge abruptly stopped.

Despite thorough probing by the scientists and their instruments, Variegated Glacier did not reveal all its secrets. For example, scientists still don't know why the glacier thickened at the head in the first place. Nor have they answered the most important question of all—why some glaciers surge while neighboring ones do not.

The third type of fast-moving glacier—tidewater glaciers—are in many ways even more mysterious than surging glaciers. Tidewater glaciers, such as Columbia Glacier, end in the sea and *calve* (break off) to form icebergs. Some tidewater glaciers flow slowly. Some, however, flow rapidly, like surging glaciers. Unlike surging glaciers, however, they do this continuously.

I began studying Columbia Glacier in 1973 because of an observation made that year by Austin Post, a glaciologist with the United States Geological Survey (USGS). While studying tidewater glaciers in Alaska, Post had noticed that all the tidewater glaciers that were slowly advancing or retreating ended in relatively shallow water less

Tidewater Glaciers

Tidewater glaciers end in the sea, where huge pieces of ice break off to form icebergs. The ice of a tidewater glacier can flow slowly or fast, depending on the pressure of water at the terminus.

The speed at which ice flows down a tidewater glacier depends on the relative pressures of the meltwater and the seawater at the terminus. If the meltwater pressure, *right* (small arrow), is less than that of the seawater (large arrow), the seawater prevents the meltwater from flowing out into the sea, and the ice flows slowly. This blockage, however, causes the meltwater to get deeper and its pressure to rise. Eventually, the pressure of the meltwater exceeds that of the seawater and the meltwater begins to flow into the sea, *far right.* The glacier, partially lifted off its base by the highly pressurized water, slides rapidly forward.

than 90 meters (300 feet) deep. A few glaciers were retreating very rapidly—virtually disintegrating because of massive calving. All of these ended in deep water.

Post theorized that the rate at which icebergs calved from tidewater glaciers depended on the depth of the water. Glaciers ending in shallow water calved few icebergs, so the glacier would retreat slowly—or even advance—compensating for the loss of ice with a similar amount of ice moving down from the head. But glaciers ending in deep water produced huge numbers of icebergs, much more than normal ice flow could keep up with. So these glaciers rapidly retreated.

Scientists are not sure why this should occur. One theory is that the action of the waves and tides in deep water causes notches to form in the ice at the water's edge. Eventually, these notches develop into large cracks and the ice breaks off.

Post's theory seemed to fit all of Alaska's tidewater glaciers except one—Columbia Glacier. Maps of Prince William Sound, where the glacier ends, showed deep water at the face of the glacier. But Columbia had neither advanced nor retreated much since at least 1899.

To determine whether the maps were correct, we equipped a small boat with a device that measures the depth of water. We found that the maps were wrong and that the water in front of the glacier was shallow, not deep. So Post's theory held up.

During the study, however, we noticed several signs that the glacier might be preparing for a retreat. For example, icebergs were breaking off rapidly in certain places, forming deep bays.

Using the specially designed radar that could penetrate ice, we

A massive chunk of ice breaks off from Columbia Glacier and crashes thunderously into Alaska's Prince William Sound, forming an iceberg.

Few icebergs are produced by glaciers that end in water 90 meters (300 feet) deep or less. The length of such a glacier varies little because the amount of ice flowing from the head compensates for the amount breaking off to form icebergs.

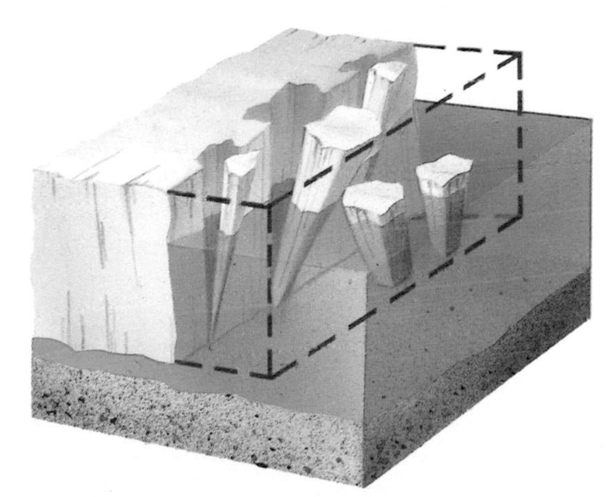

Many icebergs are produced by glaciers that end in deeper water. Such glaciers *retreat* (shrink in length) rapidly because the flow of ice from the head cannot keep up with the amount of ice breaking off as icebergs.

found that the terminus of the glacier rested on a relatively narrow underwater ridge. We also found that only 2 kilometers (1¼ miles) in from this ridge, or moraine, the land on which the bottom of the glacier rested was as much as 300 meters (1,000 feet) below sea level. If the end of the glacier moved back beyond the ridge, water from the sound would flood over the moraine, leaving the end of the glacier in very deep water. This would cause much more calving of icebergs.

Such a retreat to deeper water would have important practical consequences. Columbia Glacier is only 40 kilometers (25 miles) from Valdez, Alaska, the port for tankers carrying oil from the Alaskan Pipeline. If Prince William Sound were crowded with more icebergs, the risk of tanker-iceberg collisions and resulting oil spills would be much greater. So in 1974, the USGS asked us to study Columbia Glacier and predict whether it was about to retreat. This had never been done before.

In 1977 and 1978, we measured the amount of water—from rain and meltwater—flowing into the glacier and the speed of its flow. We thoroughly mapped the depth of the water at the face of the glacier and compared it with the water depth in front of other calving glaciers. Then in June 1980, we predicted that Columbia would begin to retreat in two to three years. By the summer of 1984, it was obvious that we were right. The glacier was calving four times as much ice and had retreated 1.6 kilometers (1 mile). We expect that within 30 to 50 years, the glacier will be 30 kilometers (19 miles) in from the point in the sea where it ended in 1987. Fortunately, the icebergs have not seriously interfered with the tankers' schedules so far, nor have there been any collisions.

What about the other side of the coin: How does a tidewater glacier move so far out into the sea in the first place? Scientists have found that the secret lies in the forward movement of its submerged ridge. The glacier creates this ridge as it scrapes along on bedrock. As the glacier moves forward, it slides up the face of the ridge, scraping off some of the rocky material. The glacier carries this material forward and dumps it on the other side of the ridge, in effect moving the ridge forward. This is a slow process, with the ridge moving at most about 30 meters (100 feet) per year.

This is currently happening at Hubbard Glacier, near Yakutat, Alaska. In 1986, the slow advance of Hubbard Glacier dammed Russell Fiord—a narrow inlet of the sea—temporarily making it a lake and trapping scores of seals, porpoises, and other marine animals. In October 1986, the ice dam broke and the water swept out to sea, freeing the animals. But Hubbard Glacier continues its relentless advance. Eventually the ice will become thick enough to seal off Russell Fiord permanently.

One fascinating characteristic of retreating tidewater glaciers is that while they are growing shorter, the ice is also flowing rapidly from the head toward the sea. Columbia Glacier flowed toward the sea at the rate of about 4.5 meters (15 feet) per day in 1978. By 1984, it was galloping along at about 15 meters (50 feet) per day.

Our recent studies at Columbia Glacier suggest that the speed at which a tidewater glacier flows is controlled—just as with surging glaciers—by the pressure of the meltwater at the base of the glacier. If the meltwater pressure is less than that of the seawater at the glacier's terminus, the seawater acts as a plug and prevents the meltwater from flowing from the terminus. This causes the meltwater to build up under the glacier and its pressure to rise. When the meltwater pressure exceeds that of the seawater, the meltwater can force its way out into the sea. And the glacier, partially lifted from its rocky bed by the highly pressurized meltwater, begins to move.

When a tidewater glacier retreats, it gets thinner as well as shorter and the ice exerts less pressure on the meltwater layer. This allows the meltwater to lift the glacier even higher off its bed and to move along even more rapidly.

Research on glaciers is valuable for reasons other than helping guard shipping lanes. It has helped us learn how glaciers have shaped the landscape of vast areas of Earth. And in their frozen depths, glaciers hold evidence of past climate changes—changes they themselves helped influence.

Studying glaciers also may be able to help us predict future climate shifts. Of particular interest is the *greenhouse effect* on global sea levels. The greenhouse effect is the warming of Earth's atmosphere that occurs when carbon dioxide and other gases trap heat from the sun. Some scientists believe that an increase in the greenhouse effect would—probably over a period of hundreds of years—melt enough glacial ice to raise sea levels by as much as 6 meters (20 feet), flooding low-lying areas such as Florida and coastal cities such as London, New York City, Tokyo, and Venice, Italy.

Such flooding would depend on a complex chain of events taking place, however, and scientists do not yet have a complete understanding of all the elements in this chain. Critical to any understanding is more knowledge of how ice streams, tidewater glaciers, and surging glaciers respond to changes in their external conditions, such as a warming climate that would cause increased melting.

During the past 10 years, scientists have penetrated some of the mysteries surrounding glaciers. More answers should unfold in the years to come as more and more attention is given to these intriguing and beautiful rivers of ice.

For further reading:

Eliot, John L. "Glaciers on the Move." *National Geographic*, January 1987.

Heacox, Kim. "Giants That Shape the Earth." *National Wildlife,* December 1985.

Matthews, Samuel W. "Ice on the World." *National Geographic,* January 1987.

BY BARRY CUNLIFFE

Excavating Roman Bath

Working in dark cellars and hot mud, archaeologists have learned much about the ancient Romans who built a temple and bathing complex in England almost 2,000 years ago.

A medieval abbey rises above the Great Bath, *facing page,* built by the Romans after they invaded England. Treasures found at the site include coins, *above,* and a bronze head of the goddess Sulis-Minerva, *right.*

Most people picture archaeologists as working outdoors in the healthy open air under a baking sun, carefully brushing the dust off some priceless ancient object. Sometimes it can be like that, but not very often. In Bath, a city located in southwestern England, we archaeologists work under very different conditions—in dark cellars under buildings, and sometimes even in sewers. But this makes our work no less exciting.

One afternoon my colleagues and I were digging in the cellar of a building constructed in the center of Bath in the 1790's. The cellar, 6 meters (20 feet) below the city, was narrow, dripping with water, and pitch-dark except for one lamp that hung from the ceiling. We had dug a trench 1.5 meters (5 feet) deep in the cellar floor, and were working in foul-smelling mud under a leaking sewer pipe. We were looking for some indication that a Roman temple once stood on this site. By the end of the day, our digging had yielded nothing, and we were fast losing our energy and enthusiasm.

Then came the discovery. Sticking out of the mud in the bottom of the trench was a large block of stone. At first glance, it looked like all the other stones we had dug up that day. We were tempted to ignore it, pack up, and go home. But something about this stone looked odd. So we cleared away some of the mud, sponged down the exposed surface with water, and found that it had been carved. After another hour of digging and sponging, we realized that we were looking at one of the corners of a Roman sacrificial altar carved with the figures of Bacchus, the Roman god of wine, and an unidentified water goddess. This was part of the altar that originally stood in the center of a temple courtyard built here nearly 1,900 years before. The altar had been dismantled around A.D. 400, and a marsh had eventually enveloped the courtyard and temple. We were the first people to see these fine carvings in 1,500 years.

But we were not the first people to try to unravel Bath's past. For 250 years, people had been finding pieces of the Roman settlement that once stood on the site of Bath. Since the accidental discovery of the first Roman artifacts in the early 1700's, *antiquarians* (people who study ancient relics)—and later, archaeologists—have worked to unearth the Roman temple, as well as a large bathing complex, built around a natural hot spring. This complex, made up of a series of bathing pools and exercising rooms, had been covered up by the debris of later civilizations. By carefully researching the records of earlier finds before we began our search, we had a good idea of where the remains of the great temple lay, about 6 meters below the modern city.

The author:
Barry Cunliffe is professor of European archaeology at the University of Oxford in England and author of *The City of Bath.*

For me, this search in the cellar was the first step in fulfilling a childhood dream. One hot summer afternoon, when I was a boy of 14, I visited the section of the Roman baths that had by then been uncovered. Looking at the map in the guidebook, I saw that a large area beyond the exposed spring was marked "unexcavated" and that the excavated Roman walls disappeared into this unexplored area. I was tantalized and fascinated by the idea that there might be

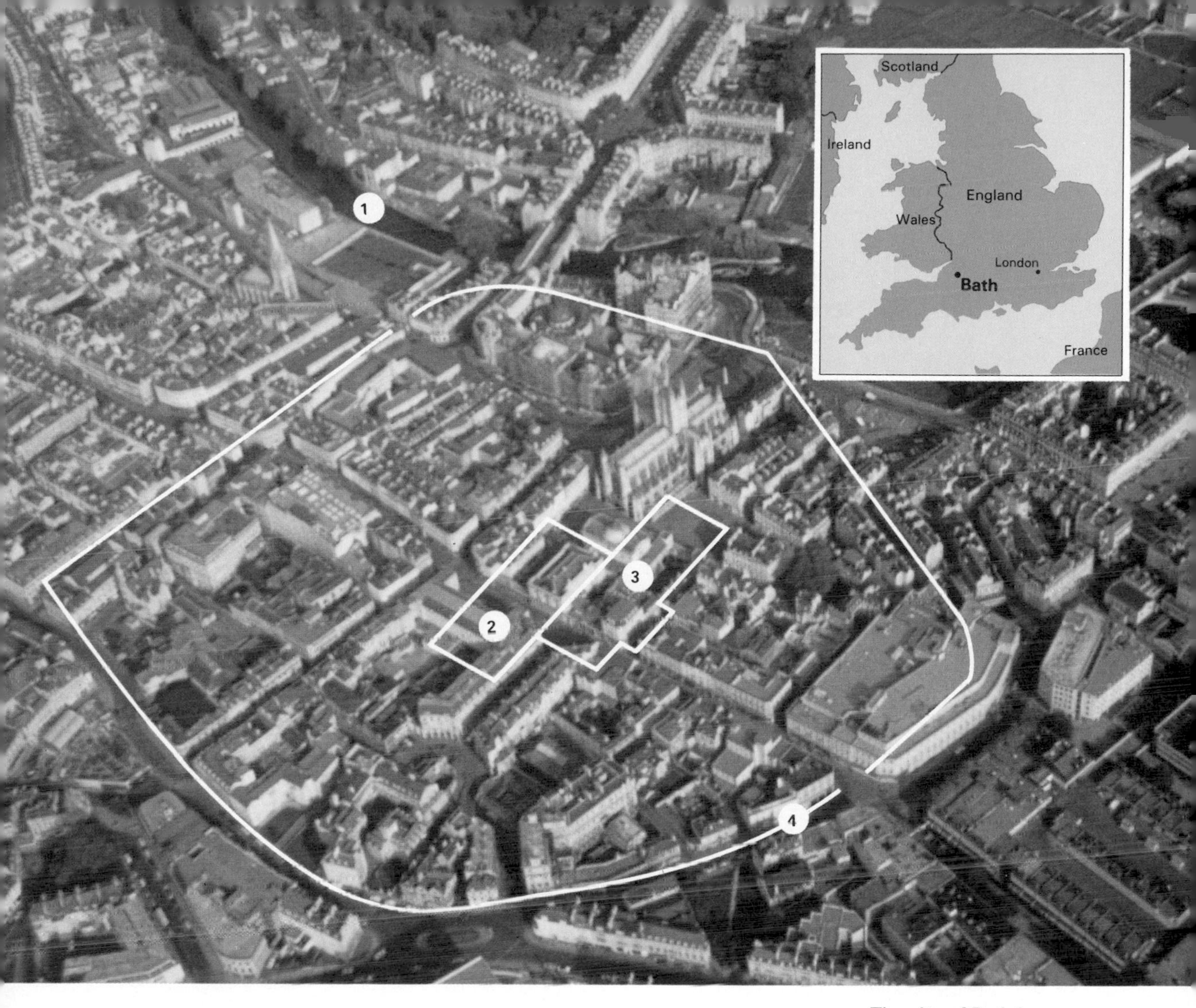

The city of Bath in southwest England (map inset) was the site of a settlement called *Aquae Sulis* by the ancient Romans, who occupied England from A.D. 43 to 410. The Romans transformed a marshland of hot springs close to the River Avon (1) into a walled settlement. At the heart of the settlement lay the temple (2) and the baths complex (3), now located in the center of modern Bath. The city has long since outgrown the original Roman walls (4), which no longer exist.

other Roman remains underneath the buildings of Bath. But where were these ruins and what secrets might they contain about the Roman people? These questions grabbed my teen-age imagination. Ten years later, in 1963, my curiosity led me back to Bath as director of excavations for The Bath Archaeological Trust. That was when I first uncovered the altar fragment in the dark and damp cellar and began to plan some major excavations.

From 1978 to 1984, my colleagues and I conducted two excavations in Bath. The first project was in the Romans' Sacred Spring, beginning in 1978. We planned to discover how the Roman engineers controlled the water of the spring and also to excavate from the sediment in the spring the offerings that Roman visitors might have thrown into the waters to appease the gods. The second project was the excavation of the temple. This took place from 1979 to 1984 in a large portion of that mysterious unexplored area on the map—which was located in the cellar where we first discovered the altar fragment. These were the most fascinating excavations I have

A Layer Cake of History

Under a piece of land in the center of the modern city of Bath are layers of civilizations dating back thousands of years. The Celts were the first people drawn to this marshland, where they worshiped at the Sacred Spring. When the Romans arrived, they built a temple and bath complex around the Sacred Spring. During the Saxon period, the baths fell into ruins, until medieval monks renovated the spring and built an abbey. A famous lounge called the Pump Room was built above the Roman baths during the Georgian period. Georgian architecture around the excavated baths can still be seen today.

Saxon (410-1066)

Roman (A.D. 43-410)

Celtic (before A.D. 43)

ever done. From our work, we gained a new understanding of the Roman way of life and a new appreciation of their engineering skills. We were also able to preserve much of the temple we exposed as a permanent underground addition to the Roman Baths and Museum, which was first opened to the public around 1900 and now attracts approximately a million visitors each year.

While Bath's Roman ruins have fascinated archaeologists for years, the city has something else that has been drawing people, including the Romans, since ancient times—three hot mineral springs. In this location next to the River Avon, water gushes up from the earth where it has been heated by underground rocks to a temperature of about 47°C (116°F.). The water is so hot that it is just bearable to touch, as we found out during our excavation of the spring. Only 9 meters (30 feet) from where we discovered the corner of the sacrificial altar in the cellar was the large spring that fed the Roman baths. This spring produces more than 1.1 million liters (250,000 imperial gallons) of water a day.

Excavations and ancient Greek and Latin documents have given archaeologists a picture of the people who were attracted to these springs. Before A.D. 43, Britain was inhabited by the Celts, an ancient people who lived on the hills and in the valleys surrounding the spring. These people believed that goddesses presided over all springs. We are now certain that they worshiped at the spring in what was to become the center of Bath. The local goddess was called

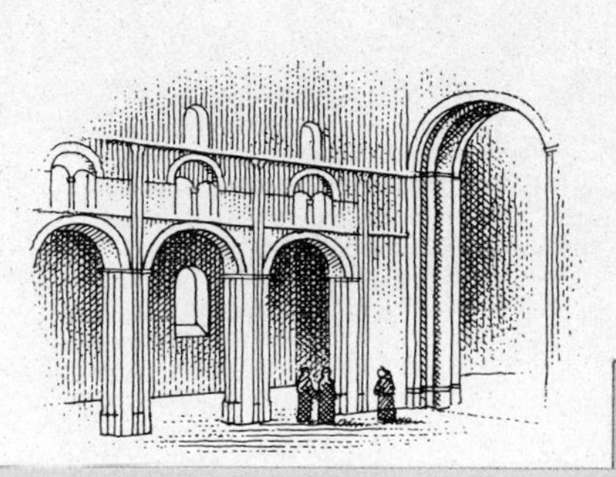

Modern
(1830-present)

Georgian
(1714-1830)

Medieval to Stuart
(1066-1714)

Sulis, and to gain her favor, they probably threw offerings, such as coins, into the spring.

In the summer of A.D. 43, the way of life for the Celts who lived in Britain changed forever. During that summer, the Romans invaded the island by boat, landing in what is now the southeast county of Kent. The Roman conquest of Britain was rapid. By the end of the first year, Roman troops had reached a ridge of limestone hills that ran across the country between the modern cities of Lincoln and Exeter. Along these hills they built a frontier road called the *Fosse Way* as the main supply route connecting their forts. The Fosse Way, parts of which survive today as a modern road, crossed the River Avon near Bath. Such an important river crossing would almost certainly have been guarded by a fort.

It was the springs, however, that attracted the Romans. Around the spring dedicated to Sulis, they built a temple and a huge bathing establishment. The baths were fed with spring water from a *reservoir*, or holding pool, that the Romans built above the spring. Bath, or *Aquae Sulis* (waters of Sulis) as the Romans called it, became a very fashionable place to visit.

The Romans expanded and improved the baths until sometime after A.D. 350, when the decline of Aquae Sulis began. The baths fell into disrepair, and sand and rubbish accumulated in the temple courtyard. As the engineering skills of the Romans were lost, the drainage system of the baths became blocked, creating a marshland

Water bubbles into the reservoir, *above,* that the Romans constructed on top of the Sacred Spring to control the flow of water through the baths. The walls surrounding the reservoir were built during medieval times, when the pool was transformed into the King's Bath. The Roman Circular Bath, *right,* was one of several cold baths.

around the spring. The ground level rose, covering parts of the temple and baths with mud. Archaeologists don't know exactly how long this period of decay lasted, but sometime during the 500's or 600's, the entire temple and bath complex was demolished by the Saxons—a Germanic tribe that had invaded England after the end of the Roman occupation. The Saxons built a cemetery and several buildings on the site of the baths and temple.

In the early 1100's, the Saxon buildings in the center of Bath were torn down, this time to make room for a magnificent medieval monastery. Medieval monks were able to refurbish the spring next to the monastery, turning it into what was called the King's Bath. Hundreds of years later, in the 1500's, the monastery was pulled down and replaced with a smaller abbey that still stands today.

It wasn't until the 1700's, however, that Bath regained its place as the most elegant spa in England. During this era, called the Georgian period, royalty and aristocratic families came to Bath to bathe, gamble, dance, and, most of all, to socialize and to drink the spring's water in a famous lounge called the Pump Room. Tourists still visit the Pump Room today.

Untangling the complicated history of a site such as Bath is an extremely difficult task for archaeologists, because over the centuries, Bath was a vigorous place, constantly growing and changing. New buildings were erected on the leveled remains of old buildings. Slicing down through this patch of ground in the center of Bath is like slicing through a layer cake. Each layer represents a period of the city's history. These historical layers are very precious and must be examined with great care by archaeologists looking for clues that will help us reconstruct Bath's history.

The most complete historical layer under the center of Bath belongs to the Roman period. This has been well preserved, because the rubble and mud that accumulated above it during the Saxon and medieval periods protected the Roman walls and floors from being destroyed by the foundations and cellars of buildings constructed later.

The first archaeological evidence of Bath's Roman past was discovered during the Georgian period when the city was installing a system of sewers. Workmen digging a sewer trench along Stall Street in the center of Bath in 1727 discovered a spectacular, life-sized bronze head of the Roman goddess Minerva. The discovery caused great excitement in Georgian Bath. It proved that there had been a temple dedicated to Minerva in the center of Bath. A reference to such a temple was recorded in a text written in the A.D. 300's. Archaeologists today are almost certain that this head came from a statue that once stood in the Temple of Sulis-Minerva.

Although the bronze head was a great find, no one attempted to launch an excavation of the site. Workmen continued to find Roman artifacts below the city, but only by chance. The slow archaeo-

A lead pipe, *top,* used to transport water, still sits in a slot the Romans cut into the stone floor north of the Great Bath. Stacks of bricks, *above,* are all that remain of a *hypocaust,* a chamber located below a sweatbath through which hot air circulated. The bricks were used to support the sweatbath floor.

logical progress was due, in part, to the location of the ruins underneath a thriving and densely populated city. Then, in the 1880's, workmen discovered the steps of the Roman Great Bath just south of the Pump Room. This was a large rectangular pool about 1.6 meters (5 feet) deep and 23 meters (75 feet) long. Caught up in a spirit of enthusiasm over the find, city authorities purchased many of the buildings that covered the baths and tore them down so that the baths could be unearthed and put on public display. City officials put city engineer Charles E. Davis in charge of removing the dirt and rubble that covered the Roman ruins. Davis had been excavating the spring beneath the King's Bath in 1878 and 1879. During the 1880's, he went on to excavate the Roman baths to the south of the spring. As he exposed the ruins, however, Davis destroyed the entire history of the site after A.D. 350. Gone, without any record, were all traces of the Saxon and medieval layers in this part of Bath.

An archaeologist descends into the Roman Circular Bath during the first systematic excavation of the site, in the 1880's.

When my colleagues and I started to plan our excavation projects in 1963, we were only too aware that by excavating the site we were, in fact, destroying it. It was essential, therefore, for us to use every archaeological and engineering technique available so that we could carefully record the progress of our work and analyze everything we found—from animal bones and pottery pieces to the soil and building foundations. In addition, untangling a site like Bath—with its layers of history—was made even more difficult by the fact that it is covered by buildings and streets that are still in use.

The first step in our temple project was to research old city records and newspaper accounts of the baths. These records showed that the site of the Temple of Sulis-Minerva was probably under the Pump Room. The only way to find out for sure was to dig a series of trial trenches in the Pump Room cellar. It was in the first of these trenches that we discovered the carved fragment from the altar.

Unfortunately, there was not enough room in the cellar for us to begin a major excavation. The only way we could dig up the entire temple area was to remove all the walls of the Pump Room's cellar. But we had to do this without destroying this beautiful Georgian building.

Working with engineers, we designed a way to underpin the Pump Room by placing massive steel girders in the cellar. The engineers also went down into the Sacred Spring. They found that the foundation holding up the south wall of the Pump Room, which had been built across the Sacred Spring, had, over the past 100 years, begun to sink into the mud that had built up during Roman times. This presented us with another problem: The underpinning to preserve the Georgian Pump Room could destroy some of the Roman structures. To avoid such a catastrophe, my colleagues and I first excavated only a small part of the debris that had filled the spring so that the engineers could underpin the Pump Room's

Using pumps to divert the flow of water from the spring, *above,* modern archaeologists sift through hot sediment at the bottom of the reservoir in search of coins and other artifacts. Waterproof clothing protects them from the contaminated spring water. Under the Pump Room, *left,* other archaeologists carefully uncover steps that once led to the Roman temple next to the baths.

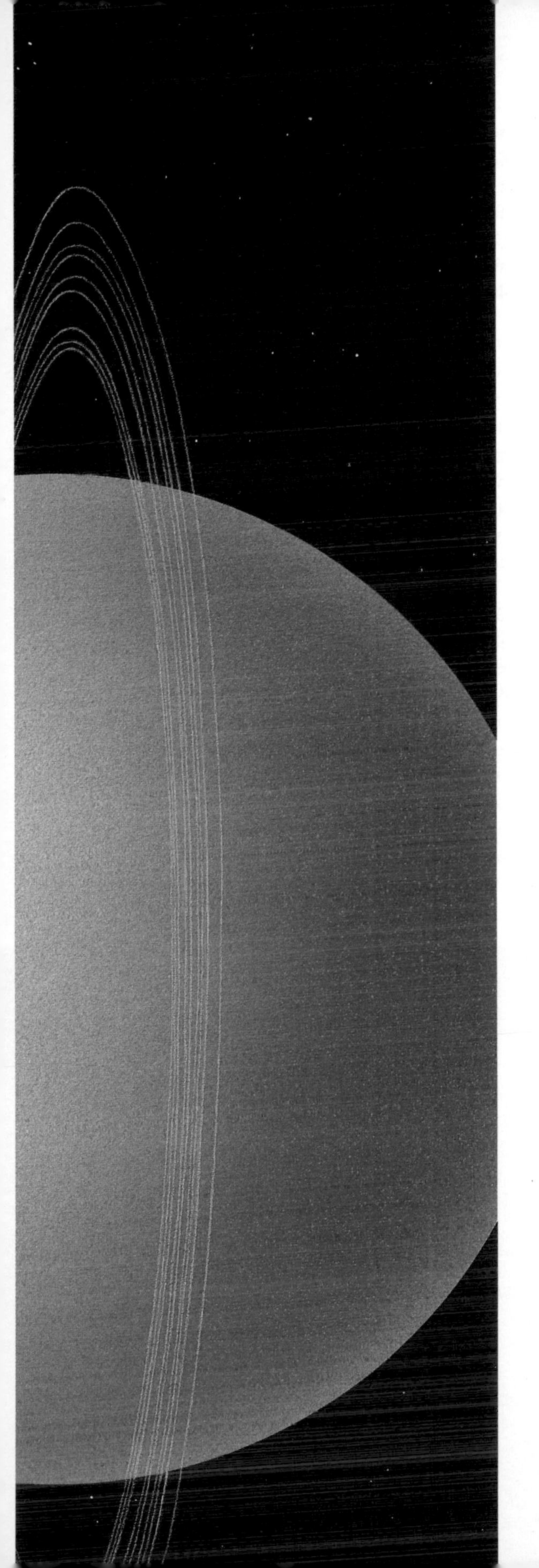

A small spacecraft's encounter with Uranus has changed our thinking about this planet—and has raised as many questions as it answered.

Voyager's Close Look at Uranus

BY LAURENCE A. SODERBLOM

A few minutes before 2 P.M. that Saturday afternoon I joined an anxious group of engineers and scientists watching a bank of television monitors in the Jet Propulsion Laboratory (JPL) of the California Institute of Technology in Pasadena. It was Jan. 25, 1986, and we were gathered at the mission control center for the *Voyager 2* spacecraft. Tension was high. The previous day, *Voyager 2* had hurtled past Uranus, the seventh planet from the sun.

On this day, *Voyager 2* was supposed to come within 28,600 kilometers (17,800 miles) of Miranda, the innermost of five large moons that orbit Uranus. We were worried about whether the spacecraft's cameras would obtain clear images of Miranda or even if the cameras would be pointed in the right direction. What we had attempted in order to capture images of Miranda verged on the impossible. Just navigating the spacecraft to arrive at precisely the right time and position near

Flight engineers at Jet Propulsion Laboratory in Pasadena, Calif., examine data from the *Voyager 2* spacecraft during its flight past Uranus in January 1986.

British astronomer, William Lassell, discovered two other moons—Umbriel and Ariel—orbiting closer to Uranus. Almost 100 years later, in 1948, U.S. astronomer Gerard P. Kuiper discovered Miranda, the fifth large moon, which orbits closest to Uranus.

The rings were not discovered until 1977, when several groups of astronomers observed a rare partial eclipse of a relatively bright star by Uranus. To their surprise, they found the star's brightness became less intense in a region outlying Uranus, indicating that something was present there to partially obscure the starlight. They concluded that the variations in brightness were caused by a ring system. Altogether, the astronomers discovered nine narrow rings. Most of the rings were only a few kilometers wide. The outermost ring, known as the Epsilon ring, was the exception, ranging from about 20 to 100 kilometers (12 to 62 miles) in width.

By analyzing the sunlight reflected from Uranus—a technique known as *spectroscopy*—scientists on Earth had determined by the 1940's some of the chemical elements that make up the planet's upper atmosphere. Astronomers learned that the planet's blue-green color was due to the presence of a small amount of methane gas in the upper atmosphere. Using spectroscopy, they also determined the presence of hydrogen gas. Other data indicated that helium, too, was probably abundant in the upper atmosphere.

By observing the orbits of Uranus' moons, astronomers determined that Uranus' *axis of rotation*—an imaginary line through its center around which the planet spins—was peculiar. All of the other planets in the solar system have axes that are almost perpendicular to their orbits around the sun, so that their equators point toward the sun. But Uranus' axis is tilted so far that it is almost level with its path around the sun. This tilt means that one or the other polar regions on Uranus points toward the sun and thus receives more sunlight than the region around the equator. It takes 84 years for Uranus to complete an orbit. For 42 of these years, the south pole faces the sun directly while the north pole is in perpetual darkness. The positions of the poles are then reversed for the next 42 years.

New findings about Uranus

Voyager 2 furnished much new information about Uranus' atmosphere and interior. Before the *Voyager 2* encounter, one of the major questions scientists had was whether Uranus' atmosphere had any clouds. Clouds could give astronomers clues about wind velocity and atmospheric circulation patterns.

Through careful computer processing of the camera images as *Voyager 2* grew closer to Uranus, the scientists were able to see faint patterns in the planet's atmosphere. The images showed cloud patterns on Uranus organized in bands parallel with the equator. Around the south pole—the only pole we could see—the atmospheric bands formed concentric circles, giving the planet the eerie appearance of a ghoulish eyeball peering out into space. We think

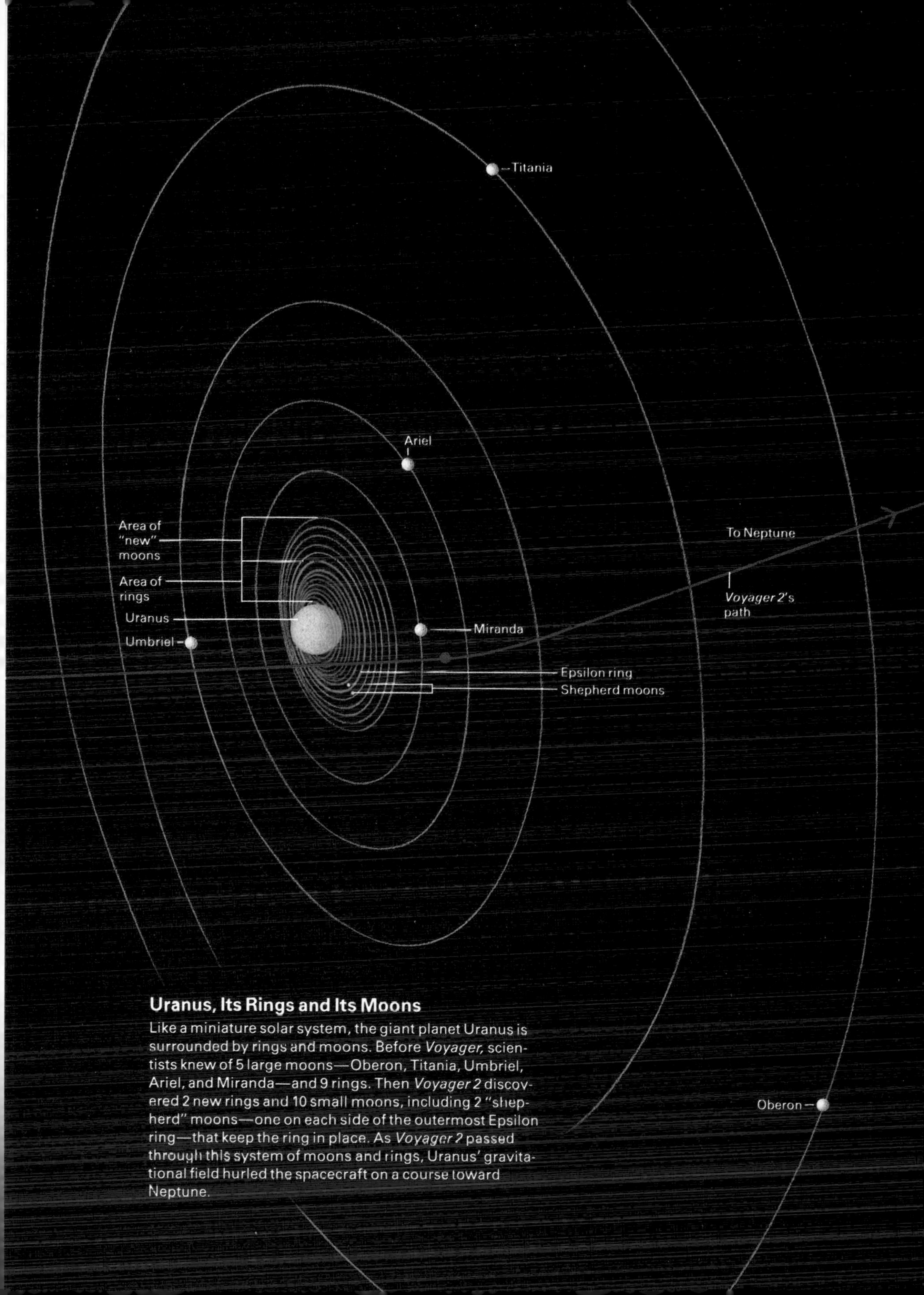

Uranus, Its Rings and Its Moons

Like a miniature solar system, the giant planet Uranus is surrounded by rings and moons. Before *Voyager,* scientists knew of 5 large moons—Oberon, Titania, Umbriel, Ariel, and Miranda—and 9 rings. Then *Voyager 2* discovered 2 new rings and 10 small moons, including 2 "shepherd" moons—one on each side of the outermost Epsilon ring—that keep the ring in place. As *Voyager 2* passed through this system of moons and rings, Uranus' gravitational field hurled the spacecraft on a course toward Neptune.

The Five Major Moons of Uranus

Voyager 2's photographs revealed that the major moons of Uranus look vastly different from one another.

Oberon, the outermost of the five major moons, has a surface marked by enormous craters resulting from meteorite impacts.

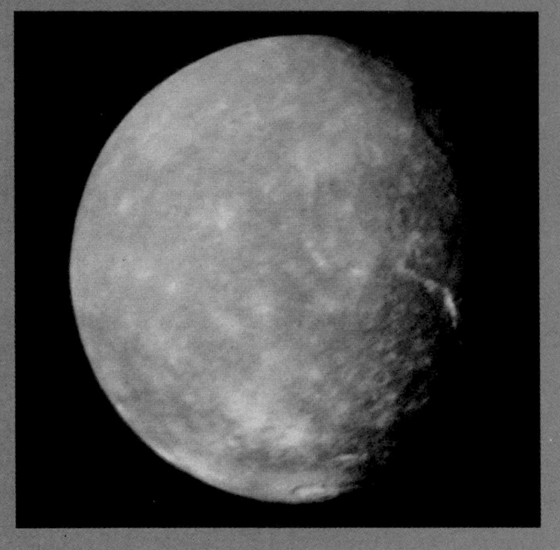

Titania, the fourth major moon from Uranus, has evidence of volcanic activity that buried most of its large impact craters.

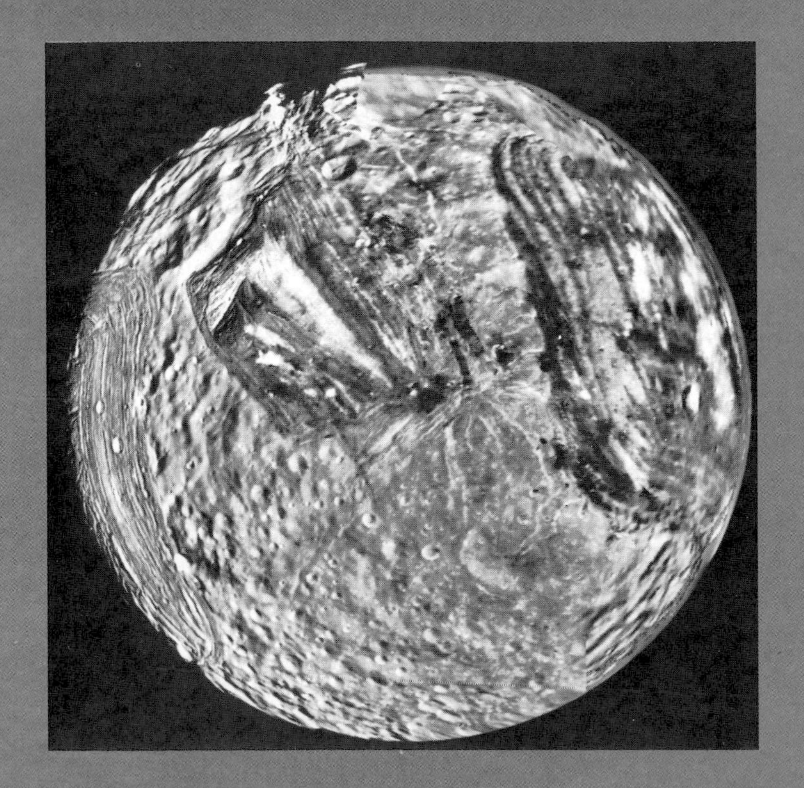

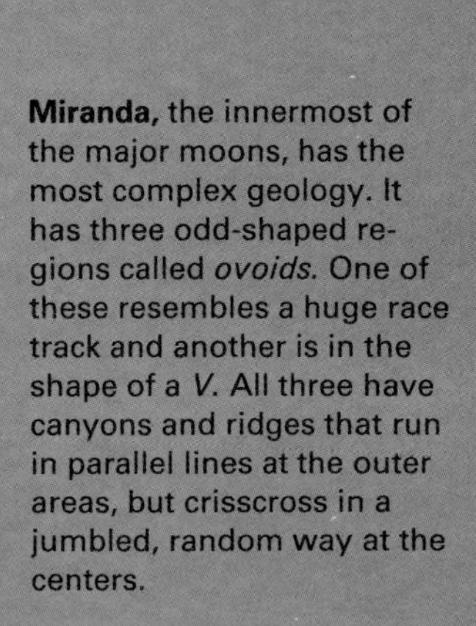

Miranda, the innermost of the major moons, has the most complex geology. It has three odd-shaped regions called *ovoids.* One of these resembles a huge race track and another is in the shape of a *V.* All three have canyons and ridges that run in parallel lines at the outer areas, but crisscross in a jumbled, random way at the centers.

to have been volcanic activity, however, because the craters themselves had not been erased. Umbriel remains a mystery.

Ariel, the second major moon from Uranus, is almost Umbriel's twin in size. Ariel is 1,170 kilometers (725 miles) in diameter. But here the similarity ends. While Umbriel is the darkest of the five large moons, Ariel is the brightest, reflecting about 40 per cent of the sunlight it receives. And while Umbriel's surface has many large craters left unchanged from the heavy bombardment of meteorites early in its history, Ariel's surface has few large craters. How two moons so similar in size and formed in the same vicinity could turn out so differently is a puzzle.

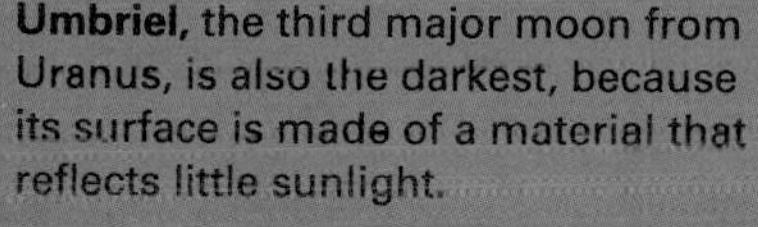

Umbriel, the third major moon from Uranus, is also the darkest, because its surface is made of a material that reflects little sunlight.

In many ways Ariel's geology resembles that of Titania. On both moons, large impact craters 100 kilometers (62 miles) wide are rare. A variety of evidence suggests that on Ariel, however, internally driven geologic processes were more intense and longer-lasting. For instance, Ariel has only about 5 to 10 per cent the amount of small craters that cover the surface of Titania. Both moons have systems of faults that range across their entire surface. But Ariel's network is much denser, and the faults have produced deep straight-walled canyons that intersect each other. Unlike canyon systems on Earth, which are usually produced by river erosion and which curve as a river does, the canyons on Ariel form straight lines. It is almost as if they were made with a ruler and a knife.

Ariel, the second major moon from Uranus, is the brightest and is free of large impact craters but has enormous *faults* (cracks).

Mysteries of Ariel and Miranda

Ariel showed dramatic evidence that some kind of volcanic activity occurred late in its geologic evolution, after the period of meteorite bombardment. A smooth geological feature—resembling a kind of volcanic lava flow—forms a vast plain several hundred kilometers in length and width that has buried older impact craters and canyon walls. This feature ends in a steep cliff several kilometers high. The material that makes up this feature must have been extremely thick and syrupy, judging from the steep slopes that mark the border.

Miranda, innermost of the five major moons, is also the smallest, only about 480 kilometers (300 miles) in diameter. The *Voyager* images revealed that Miranda has a very bizarre surface. Scientists

The Mysterious Moon Miranda

Voyager 2 obtained detailed pictures of Miranda's unusual surface. The pictures include a close-up view of a canyon wall (arrow), *right,* nearly 20 kilometers (12 miles) deep. A mosaic of *Voyager 2*'s photographs, *above,* reveals three ovoid regions (arrows)—bizarre features never found on any other object in the solar system.

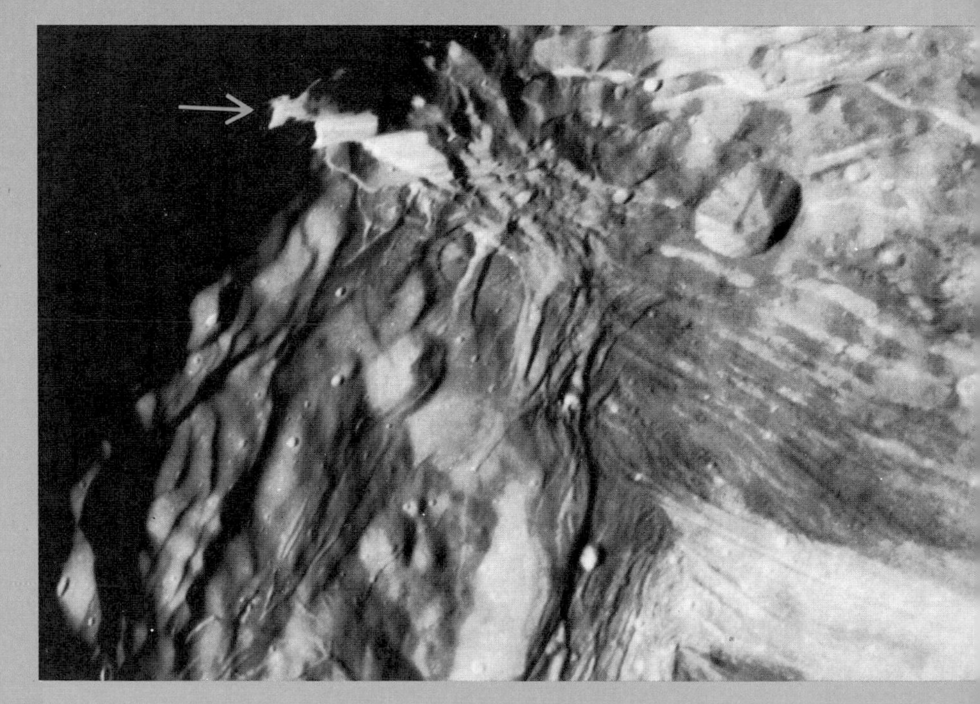

Two Theories of How Miranda Formed

Miranda's geological features are unlike anything seen on other moons and planets in the solar system. Scientists have proposed two theories to explain how a moon with such unusual features could have formed.

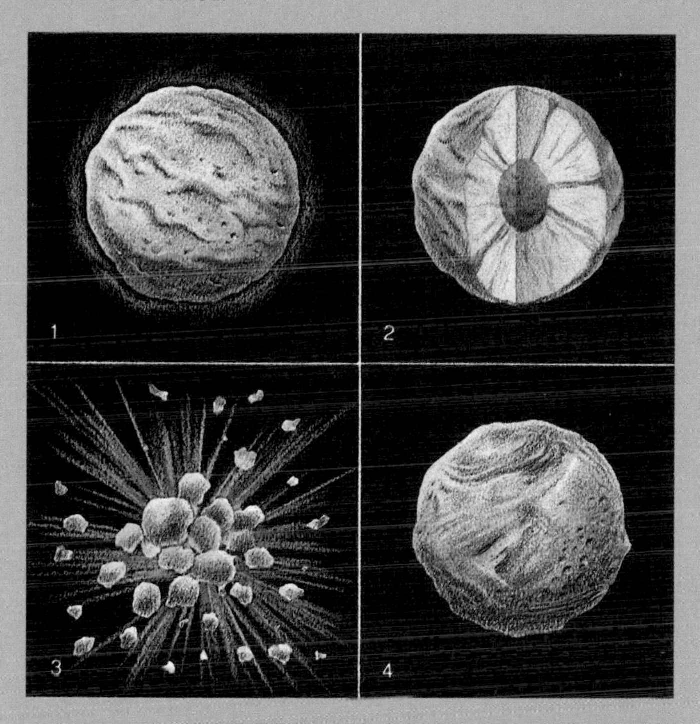

According to one theory, Miranda formed when a mixture of light and heavy materials orbiting Uranus came together (1). Gravity caused the heavy materials to sink to the core of the moon, while light materials rose to the surface (2). Miranda then experienced meteorite bombardment so intense that the moon broke apart (3). But the pieces stayed close enough together that they could be reassembled by the force of gravity. This created an odd mosaic with large chunks of both heavy and light material from the previous core lodged near the surface, forming the shapes called ovoids (4).

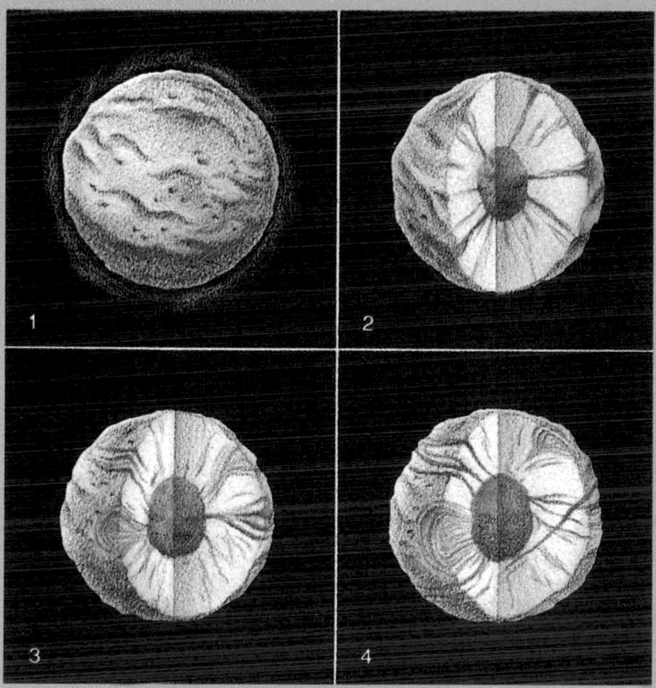

The second theory also maintains that Miranda formed as heavy and light materials came together (1) and (2). But this theory states that the process by which heavy materials sank to the core and light materials rose to the surface was not completed because Miranda lost its internal heat very quickly due to its small size. Light materials trapped deep in the interior erupted onto the surface (3). But not all the heavy materials on the surface had time to sink before Miranda became frozen completely (4).

coined a new geological term—*ovoids*—to describe three unusual oval-to-rectangular shapes that are prominent on Miranda's surface. Each of these ovoids is 200 to 300 kilometers (120 to 180 miles) across, and they are unlike anything ever seen in the solar system. The outer parts of these ovoids resemble gigantic race tracks in which parallel ridges and canyons wrap neatly around the centers. In the centers, however, ridges and canyons randomly crisscross one another in jumbled and chaotic patterns. Some violent process must have occurred to create such chaotic terrain.

From an analysis of the *Voyager* images, the material in the ovoids appears to be lighter in weight and less dense than the material in other regions of Miranda. And because there are few impact craters, scientists think the ovoids formed after the period of meteorite bombardment ended.

How Miranda may have formed

Scientists have proposed two theories to explain Miranda's curious geologic evolution and, specifically, why the ovoids are made of lighter, probably icy, material. Both theories involve *differentiation*, a geologic process that scientists believe occurred in all the moons and planets in the solar system soon after they formed. Differentiation causes heavy materials to sink and lighter materials to rise. Through differentiation, the surfaces and the interiors of the moons and planets began to change due to the force of their own gravity and the heat in their interior.

When Miranda first formed, for example, it was a mixture of light and heavy materials at all depths. Its interior was hot. Scientists believe this heat came from three basic sources. When orbiting materials left over from the formation of Uranus collided and stuck together to form the moon, their energy of motion was converted to heat. Then, as gravity compressed the material, this also created heat, and finally, the decay of radioactive elements in the materials gave off more heat. This internal heat set up currents that moved light and heavy elements. Lighter, less dense material, such as ice, rose toward the surface. Heavier, more dense material—aided by the force of gravity—sank toward the moon's interior.

According to one theory, this process left Miranda with a core of rocky materials, perhaps 300 kilometers in diameter, surrounded by a lighter icy covering, perhaps 100 kilometers thick. But Miranda must have undergone powerful meteorite bombardment because it is the closest major moon to Uranus, and Uranus' gravity would have increased the speed and power of the meteorites. Some scientists have suggested that the bombardment was so intense that Miranda could easily have been blasted into fragments during this period. If Miranda was blasted apart after its heavy and light materials had differentiated, large chunks of mostly ice or mostly rock could have been scattered along Miranda's orbit. The pieces were not so far removed from one another, however, that they escaped gravitational attraction. Gravity might then have pulled the pieces of the moon back together to form a jumbled lump with large chunks of what had been the rocky core lodged in the surface. These large heavy pieces eventually sank inward. Lighter icy materials then rose and broke to the surface to form the ovoids. The ovoids are large, so the heavy pieces must have been enormous. The problem with this theory is that it is difficult to explain how such enormous pieces could have remained intact, rather than breaking into small pieces, during the bombardment that blasted Miranda apart.

The other theory also contends that when Miranda formed, it was a uniform mixture of light and heavy materials. Then, as heavier materials sank deep into the interior, lighter materials began to collect on the surface, forming the ovoids. But the differentiation process was not completed because Miranda had little internal heat due to its small size. Instead, this process came to a halt, leaving Miranda frozen in a state where some surface regions—the ovoids—were composed of lighter material, and other surface regions were composed of heavier material too frozen to continue sinking to the interior. The problem with both theories lies in determining whether Miranda ever had enough internal heat for this process even to begin.

Unanswered questions

Detailed analysis of the new *Voyager 2* information will occupy scientists for decades. The new data have already profoundly altered our thinking about the Uranian system, and we can now pose far more advanced scientific questions about Uranus, its rings, and its moons. For instance, what is the dark material found throughout the rings and moons? Is it material from the formation of the solar system like that frozen in comets and meteorites deep in outer space? We know that certain types of meteorites contain amino acids, the building blocks of life on Earth. Could the dark material in the moons and rings contain amino acids? Or is it recently formed material created by the sun's radiation altering compounds such as methane? And what were the energy sources that apparently caused volcanic activity on such small moons in such a cold environment? These questions could not even have been imagined prior to *Voyager 2*, reminding us that no matter how sophisticated our scientific theories are, we will never be able to predict the true complexity of the universe. We simply have to go and look—at least with the "eyes" and scientific instruments of our unmanned spacecraft.

For further reading:

Ingersoll, Andrew P. "Uranus." *Scientific American*, January 1987.
Laeser, Richard P., McLaughlin, William I., and Wolff, Donna M. "Engineering *Voyager 2*'s Encounter with Uranus." *Scientific American*, November 1986.
Planetary Report (*Voyager 2* at Uranus issue), November/December 1986.

By understanding how heroin and cocaine affect chemical processes in the brain, scientists may discover a way to cure drug addiction.

Addiction and the Brain

BY ELIZABETH PENNISI

At age 18, Daisy is one of the lucky ones. After four years as a drug addict, she made the difficult decision one year ago to get professional help for her drug problem. Her voice quivers when she describes the loneliness and desperation she felt as an addict.

Like many teen-agers, Daisy first encountered mind-altering drugs at a party. Daisy was 13 years old at the time and experiencing the confusing emotions of adolescence. She was dissatisfied with her life, angry with her family, and fearful of the future. The marijuana joint her friends gave her at the party seemed to ease all her problems. Daisy liked the intense good feeling it gave her so much that she began smoking marijuana regularly.

Daisy also began using other substances that could make her feel good. She began drinking alcohol in the mornings before school. During the next four years, she snorted cocaine, popped barbiturates, and experimented with whatever drugs were available. Then one day, a drug dealer introduced her to heroin.

The more drugs Daisy used, however, the more she found herself unable to enjoy their effects. As she started getting high more often on heroin, she found she could not stop. Her schoolwork suffered. She began stealing money from her parents and borrowing money

Glossary

Axon: A branchlike extension of a nerve cell that carries signals to other nerve cells.

Dendrite: A branchlike extension of a nerve cell that receives signals from other nerve cells.

Endogenous opioids: Chemicals produced by the body that reduce the sensation of pain or stress and create a mild feeling of well-being.

Neurotransmitter: A chemical that carries messages from one nerve cell to another.

Opiate drugs (also called narcotics): A class of drugs, including heroin and morphine, that are made from opium. These drugs mimic the effects of the body's endogenous opioids.

Psychoactive drugs: Any drug, such as cocaine or heroin, that affects or alters the mind.

Receptor: A specially shaped site on a cell to which a biochemical, such as a neurotransmitter, binds.

Synapse: The space between an axon and another nerve cell body or dendrite.

from her friends to support her drug habit. When she was not high on drugs, she became irritable, jittery, anxious, and nauseous. Her moods swung wildly. Finally, unable to deal with the emotional turmoil that boiled over inside her when she was not high, Daisy checked herself into a drug rehabilitation center.

Daisy did not know it, but she was more than psychologically dependent on drugs. Her body had become chemically addicted to them. The substances in alcohol, cocaine, and heroin that made her feel so good at first had actually altered chemical processes within her brain.

Daisy's story is not an isolated case. Drug and alcohol addiction are major problems in the United States. The National Institute on Drug Abuse (NIDA) in Rockville, Md., estimates that 18 million Americans used marijuana on a regular basis in 1985. Cocaine—fast becoming the most popular illegal drug of the 1980's—has been tried by as many as 22 million people, according to NIDA's 1985 National Household Survey on Drug Abuse. An estimated 1.9 million people have used heroin.

More than 50 per cent of Daisy's peers in the United States have used an illegal drug at least once before they finish high school, according to a 1986 survey of high school students conducted by the University of Michigan in Ann Arbor. More than 17 per cent of the high school students reported using cocaine at least once in 1986. And almost 5 per cent—or 1 out of every 20—of all U.S. high school seniors drink alcohol daily, the survey found.

The human toll caused by drug and alcohol abuse is great. Half of all teen-agers who die each year in car crashes have been drinking, according to experts, and drugs or alcohol play a part in most suicides involving people under the age of 30.

In response to the problem, the U.S. government in 1986 declared war on illegal drugs. Scientists are also doing their part by studying how drugs affect the body. Their search has taken them deep into the brain. It is there that physical dependence—or addiction—occurs.

Scientists have long known that addictive substances—from cocaine to alcohol—share three important characteristics: They cause feelings of pleasure and changes in emotion; the body develops tolerance to the drug so that addicts must take ever-increasing doses to get the same effect; and the absence of the drug in the body causes painful withdrawal symptoms. Tolerance and withdrawal occur because addicting drugs upset the natural chemical balance in the brain.

Most addiction involves *psychoactive* substances, drugs that affect a person's emotions and energy level. Most of these powerful drugs are also painkillers. In fact, many of today's most addictive and abused drugs were first used in the United States as legitimate medicines. Physicians in the early 1800's prescribed morphine, opium,

The author:
Elizabeth Pennisi is a science writer based in Tucson, Ariz.

marijuana, and cocaine as "miracle drugs." They were believed to cure all sorts of ills, from seizures to indigestion. Many people became addicted to these drugs, however, and physicians realized that while the drugs had some beneficial qualities, they could be very harmful if used improperly. As a result, the U.S. government outlawed opium and marijuana in the early 1900's and subjected cocaine, morphine, and other drugs to strict controls.

Most addictive drugs are classed as either *depressants* or *stimulants*. Heroin, morphine, barbiturates, tranquilizers, and alcohol are depressants. They slow down processes in the brain and the central nervous system. Cocaine, amphetamines, nicotine, and marijuana are stimulants that excite the brain and the central nervous system.

Heroin and morphine belong to a class of drugs called *opiates* or narcotics. Most opiates are made from opium, a bitter-tasting substance extracted from the milky sap of opium poppies. Greeks as long as 2,000 years ago used opium to dull pain. In 1806, chemists discovered how to extract the powerful painkiller morphine from opium and, later, how to extract the more potent heroin from morphine. Opiates not only relieve pain but also create a dreamlike feeling of euphoria or happiness. Opiate drugs are highly addictive, however. If a person who is addicted to heroin or morphine suddenly stops taking the drug, withdrawal symptoms may occur, including severe vomiting, diarrhea, tremors, and chills.

Barbiturates and tranquilizers, although depressants, are not derived from opium. Barbiturates, which help people fall asleep, are made from a chemical, barbituric acid, and were first used in the early 1900's. Tranquilizers are more sophisticated chemical compounds and were widely prescribed by physicians during the 1960's to relieve anxiety. Although barbiturates and tranquilizers are useful in helping many people cope with a short-term crisis, the abuse of these so-called *downers* can become a serious drug habit for some people.

One of the most powerful stimulants is cocaine, a white powder made from the coca plant. South American Indians have chewed its leaves for thousands of years as a way to reduce fatigue and hunger. In the 1800's, scientists learned how to extract cocaine from coca leaves, and the drug became the rage of Europe as a cure for exhaustion, alcoholism, and morphine addiction. Cocaine withdrawal does not produce the violent symptoms that other addicting drugs do, so it was not until the 1950's that physicians realized cocaine was physically—as well as psychologically—addictive.

Cocaine has become one of the most abused illegal substances of the 1980's. The drug causes intense euphoria, alertness, and excitement followed by intense depression and irritability when the effects of the drug wear off. Cocaine is usually inhaled, or "snorted," through the nose. When snorted, cocaine passes into the bloodstream through the thin mucous linings in the nose. It takes the drug only about three minutes to reach the brain. Chronic snorting of cocaine not only overstimulates the brain and nervous system but it also destroys the mucous linings in the nose. Cocaine can also be smoked in an extremely pure form called *crack*. The effect of crack is 10 times more intense than cocaine, and so crack is even more addicting—and dangerous. Even infrequent or first-time crack users can die from an accidental overdose, because cocaine constricts the blood vessels and causes the heart to beat too fast or—in some cases—to stop altogether.

Amphetamines, sometimes called *uppers,* are stimulants made from synthetic chemical compounds. If used under a physician's supervision, these drugs can help people lose weight, work long hours, or feel less depressed. But "uppers" that are abused can be addictive. An overdose can lead to *paranoia* (irrational fear and distrust), violent behavior, convulsions, and death. People who are dependent on amphetamines experience apathy, irritability, depression, and disorientation when they stop using these drugs.

Scientists are studying nerve cells in the brain to discover how these different kinds of drugs cause addiction. The brain, about 1.4 kilograms (3 pounds) of tissue, contains 100 billion nerve cells. Heroin, cocaine, and other drugs enter this network of nerve cells from the bloodstream and travel to specific areas within the brain, including those that control muscle movements and thinking. Addicting drugs cause "highs" when they reach nerve cells in the *limbic system*, the pleasure center of the brain where emotions and physical sensations are controlled. Drugs stimulate or depress processes in the nerve cells in these areas by interfering with how the cells communicate with one another.

During the early 1900's, Spanish biologist Santiago Ramon y Cajal discovered how nerve cells communicate. He found that nerve cell bodies have fine branches that reach out toward the ends of branches from other nerve cells. The space between one nerve's branch and the next nerve cell or one of its branches is called the *synapse*. Electrical signals carrying messages travel through a nerve cell's body and down a long branch called an *axon*. When the signals reach the end of the axon, they cause the release of chemical messengers called *neurotransmitters* into the synapse. These chemicals drift across the synapse to the cell body of a neighboring nerve or to a type of branch called a *dendrite*.

Scientists have discovered that the surfaces of nerve cells, axons, and dendrites contain many kinds of *receptors* that pick up these chemical messengers. Each type of receptor is specially shaped to receive only one kind of neurotransmitter. After a neurotransmitter binds to its receptor site, the signal travels on to the next nerve cell, where the process is repeated over again. One nerve cell may "talk" to thousands of other nerve cells as the message is passed through thousands of synapses. A nerve cell is able to reabsorb excess neu-

Nerve Cells in the Brain

Nerve cells communicate by sending and receiving electrical impulses or signals. Drugs, such as heroin and cocaine, disrupt the normal functioning of this network.

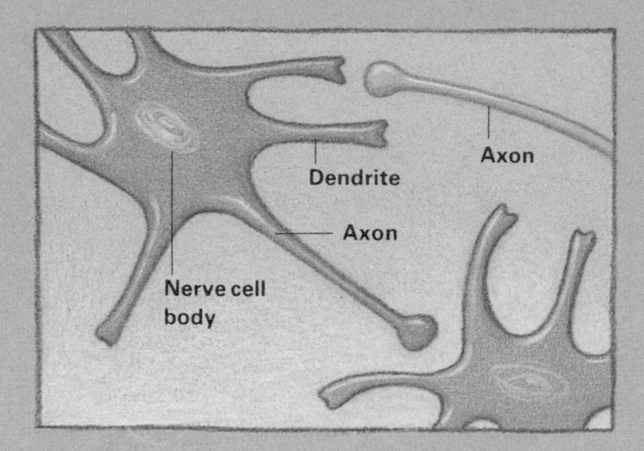

A nerve cell consists of three parts: a *cell body* that processes and receives electrical signals; an *axon* that carries signals to other nerve cells; and *dendrites,* branches that also receive signals from other nerve cells. Nerve cells send signals, or nerve impulses, from one part of the body to another through this network of connecting cell bodies, axons, and dendrites.

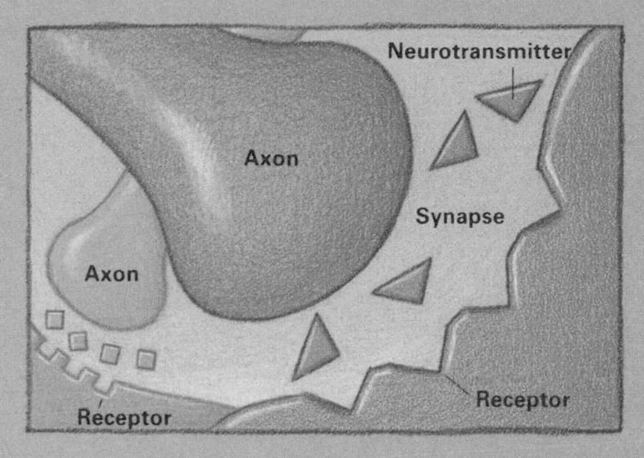

Chemical messengers, called *neurotransmitters,* transmit a signal from one nerve cell to another through a space between the axon and the cell body or dendrite called the *synapse.* Each neurotransmitter binds to its own *receptor,* a special molecule on a nerve cell that is shaped to receive only one type of neurotransmitter.

rotransmitters back into its axon so they can be used again to transmit another impulse. This reabsorption process prevents too many neurotransmitters from floating in the synapse and occupying too many receptors. If too many receptors become occupied, the nerve can be overstimulated.

All addicting drugs alter the function of these chemical messengers. Some drugs block the reabsorption of neurotransmitters so there is an excess in the synapse. Other drugs reduce the number of neurotransmitters that are released into the synapse.

Scientists have identified at least eight neurotransmitters in the brain—such as norepinephrine, dopamine, and serotonin—as well as other substances that seem to function like neurotransmitters, such as a chemical called Substance P. Norepinephrine accelerates the heart and elevates the blood pressure. Dopamine heightens the response of the limbic system and creates a high energy level that covers up fatigue or depression. Serotonin induces sleep. Substance P conveys the sensation of certain types of pain, such as pain caused by extremely high or low temperatures.

Opiate drugs provided scientists with the first clear indication of how addictive drugs can upset the natural flow of neurotransmitters. In 1954, A. H. Beckett and A. F. Casy, pharmacologists at Chelsea Polytechnic in London, were among the first to suggest that opiate drugs fit into specific receptors on nerve cells. In 1973, pharmacologist and psychiatrist Solomon H. Snyder and then-graduate

The Effects of Heroin

Heroin and other opiate drugs produce their effects by interfering with the nerve cells' communication system. These drugs mimic a natural neurotransmitter in the body that is usually produced to lessen the sensation of pain. In ways that scientists do not fully understand, they create a dreamlike euphoric feeling. Addicts denied the drugs experience nausea, cramps, and other withdrawal symptoms. An overdose can cause convulsions or death.

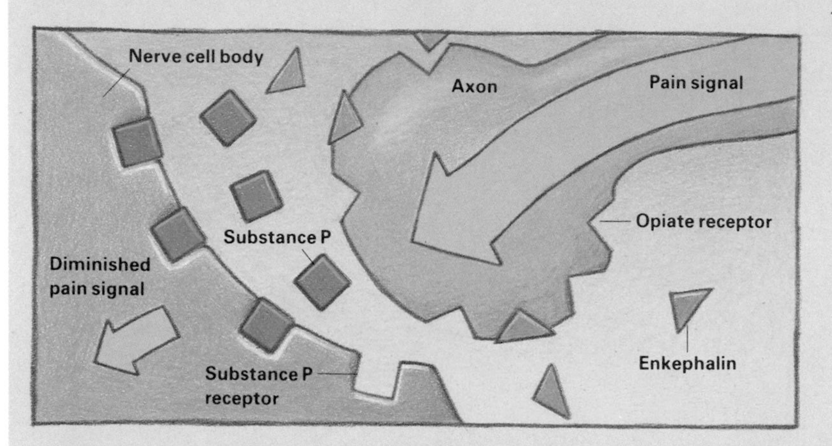

Natural opiates are produced by the body in response to pain or stress. *Enkephalin,* one of these opiates, regulates the release of *Substance P,* a chemical that relays the sensation of certain kinds of pain. Enkephalins bind to opiate receptors located on axons that produce Substance P, reducing the amount of Substance P that is released into the synapse. When less Substance P is released, pain is lessened.

student Candace Pert at Johns Hopkins University in Baltimore found the opium receptors. To trace the path of opiates in the brain, Snyder and Pert injected into tissue from rat brains compounds that would bind to opiate receptor sites but would not convey a message. These compounds were tagged with a radioactive substance that could easily be detected in a laboratory test. The scientists found that these compounds bound to specific receptors. They and other scientists later found these receptors on nerve cells in monkey and human brains.

If the brain had specific receptors for opiate drugs, the scientists reasoned that the body must produce natural substances that resemble and act like the drugs. In 1975, pharmacologists John Hughes and Hans W. Kosterlitz of the University of Aberdeen in Scotland identified two chemicals in the brain that bind neatly into the opiate receptors. They named these chemicals *enkephalins.* Hughes and Kosterlitz, along with other investigators, went on to find another group of chemicals in the brain that act like opiates, which they called *endorphins.* Like opiate drugs, enkephalins and endorphins appear to control the brain's response to pain and stress and pro-

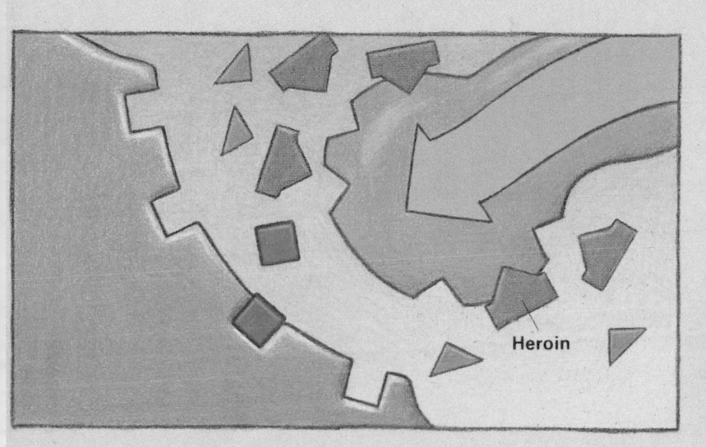

Opiate drugs have shapes that resemble the shape of enkephalins. This allows opiate drugs to bind to opiate receptors and mimic the effect of the enkephalins. They reduce the release of Substance P and trigger feelings of euphoria.

Addiction to heroin and other opiate drugs is thought to result when the opiate receptors on axons are constantly filled by opiate drugs. Scientists do not yet know why this causes addiction or the withdrawal symptoms that occur when people stop taking these drugs.

mote a feeling of well-being. Scientists call enkephalins and endorphins *endogenous opioids* to distinguish them from opiate drugs.

Researchers then found that the body's natural opiates are released from specific nerve cells in the brain. When the natural opiates bind to opiate receptors on other nerve cells, they may slow down the release of Substance P, dopamine, norepinephrine, or another neurotransmitter called acetylcholine. Heroin and other opiate drugs are enough like natural opiates that they can also bind to opiate receptors, producing effects that are similar to—but may be more intense than—those of the body's natural opiates. Scientists suspect that one of these effects may be the intense euphoria that heroin produces.

Frequent use of heroin or other narcotics keeps opiate receptors constantly full. Some scientists believe that this may cause the tolerance that develops to opiate drugs so that addicts must take larger doses to get high. Other scientists are investigating the possibility that tolerance develops because the body shuts down certain opiate receptors so that the nerve cells become less sensitive to the opiate drug. This would dull the effect of heroin.

Researchers have found that cocaine also interferes with the brain's chemical messenger network, but in a different way. In the brain, this drug affects certain nerve cells in the limbic system. During the 1960's and 1970's, several research groups performed experiments showing that cocaine blocks the reabsorption of excess serotonin, dopamine, and norepinephrine in the synapse. The excess amount of these neurotransmitters in the synapses results in more of their receptors being filled for longer periods of time. Some scientists believe that this overstimulates the nerve cells, making the cocaine user feel happy and excited. Cocaine may also cause these neurotransmitters to bind more tightly to their receptors, intensifying the stimulating message they carry.

Another theory about the cause of the cocaine high involves the overproduction of neurotransmitters. In 1960, L. G. Whitby, a British biochemist working at the National Institute of Mental Health in Rockville, Md., and his colleagues traced neurotransmitters tagged

The Cocaine High

Cocaine produces its effects by interfering with the absorption of several kinds of neurotransmitters. This causes a high—intense euphoria and excitement—followed by equally intense feelings of depression and irritability when the drug wears off. Researchers are trying to learn why cocaine causes such withdrawal symptoms, sometimes even convulsions and delirium, as well as a craving for the drug.

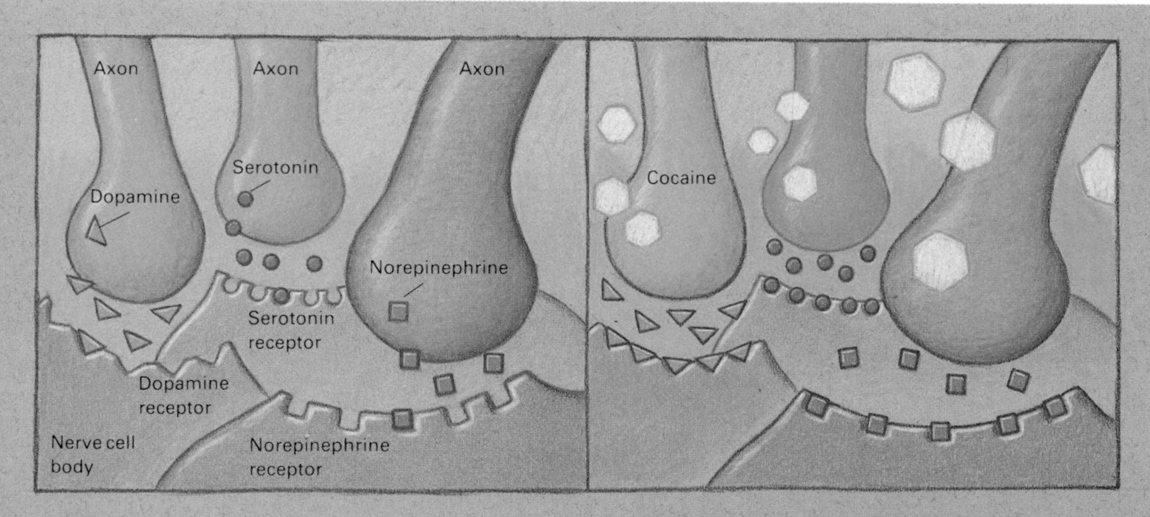

Nerve cells produce and release at least one type of neurotransmitter into the synapse. Three of these neurotransmitters—dopamine, serotonin, and norepinephrine—make the body more alert and alter a person's senses in various ways. Each nerve cell can pick up any excess neurotransmitter from the synapse and transport it back into the axon, for use at another time.

Cocaine prevents the axons from taking back their excess dopamine, serotonin, and norepinephrine, thus increasing the concentration of these neurotransmitters in the synapse. As a result, a greater number of receptors are filled by these neurotransmitters for a longer period of time, overstimulating nerve cells. Some scientists believe that this may be one cause of cocaine's intense high.

with a radioactive substance. They discovered that besides blocking the reabsorption of dopamine, cocaine stimulates the release of this neurotransmitter. This sudden excess of dopamine and its effect on the limbic system may be partly responsible for the intense high that cocaine produces.

Some scientists suspect that overstimulating nerve cells by heavy or prolonged use of cocaine depletes the brain's supply of serotonin, dopamine, and norepinephrine. Thus, when the drug wears off, the intense high is followed by equally intense feelings of depression, paranoia, and irritability. They also suspect that cocaine addicts crave the drug because they lack a natural reserve of neurotransmitters to stimulate the nerve cells in their brains. Addicts get caught up in a vicious cycle. They take more cocaine to relieve their withdrawal symptoms, but this only aggravates their brains' chemical imbalances.

Other addictive drugs also affect neurotransmitters, but in still other ways. Scientists believe that amphetamines may upset dopamine balance in the brain, while barbiturates and tranquilizers may interact with a neurotransmitter called gamma-amino-butyric acid (GABA). When GABA binds to a nerve cell, it prevents that nerve cell from releasing neurotransmitters. As a result, nerve cell communication is reduced.

Of all the addicting substances, tobacco and alcohol are probably the most commonly used. The American Heart Association in Dallas estimates that 350,000 Americans die annually from heart attacks, lung disease, and other illnesses caused by tobacco. More than 10 million adults in the United States are *alcoholics* (people who are physically dependent on alcohol), according to the National Clearinghouse of Alcohol and Drug Information in Rockville, Md.

The addictive chemical in tobacco is nicotine. High doses of nicotine stimulate the body to release endorphins, causing the calm feeling many people have when they smoke. Scientists know less about the effects of alcohol on nerve cells. Some scientists believe that alcohol, like sleeping pills and tranquilizers, acts on GABA in the brain.

The physical effects of one psychoactive drug—marijuana—still have scientists puzzled. Some researchers claim that marijuana does not produce the tolerance or the compulsive cravings associated with alcohol or cocaine. But other scientists, including researchers at the University of California in Los Angeles, say that marijuana is addicting and can cause flu-like withdrawal symptoms. They think that marijuana may reduce the levels of both the body's natural opiates and norepinephrine.

Scientists' understanding of the chemical imbalances caused by drugs has improved the chances of preventing or curing addiction. Researchers are working with a variety of drugs that may help correct imbalances of neurotransmitters.

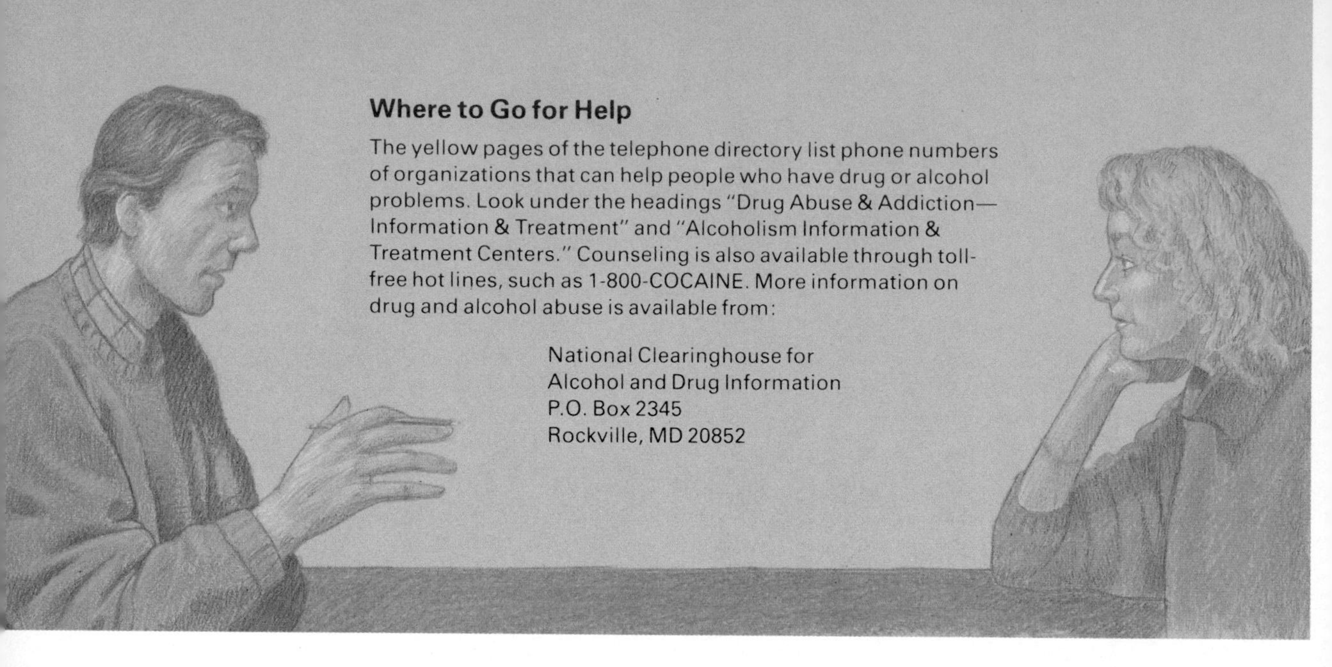

One promising drug is clonidine, a medication used to lower high blood pressure. In 1975, Liang-Fu Tseng, a pharmacologist at the University of California in San Francisco, found that clonidine could also be used to break the cycle of opiate addiction by relieving withdrawal symptoms. Tseng measured clonidine's effectiveness in rats addicted to morphine and found that the drug appears to help restore the normal flow of norepinephrine in the brain, reducing the body's craving for morphine more quickly. Clonidine is not addictive and does not create a high. The drug has some drawbacks, however. It can cause sleeping disorders and low blood pressure. Lofexidine, a close chemical cousin to clonidine, also appears useful in suppressing symptoms of morphine withdrawal. Another drug, called naltrexone, blocks the euphoria produced by heroin. Naltrexone prevents heroin from fitting into the opiate receptor sites, but it does not create a high or cause addiction.

Clonidine and naltrexone could be promising alternatives to methadone treatment for heroin addiction. Methadone, a synthetic opiate, has been used since the 1960's in programs designed to help people overcome narcotic addiction. When given in very high doses, methadone saturates opiate receptor sites so that addicts feel no effect when they inject heroin or morphine. This eliminates the addicts' craving for the drug. But although methadone does not cause the dreamlike state of other opiate drugs, it is addicting and can cause depression. Methadone treatment also does not work as quickly as clonidine or naltrexone treatments.

For cocaine addicts, two drugs look promising. One is bromocryptine, a medication used to treat infertility. Scientists do not know how bromocryptine breaks the cycle of cocaine addiction. The other drug is imipramine, which somehow dulls the nerve cells' sensitivity to cocaine.

But the development of new drugs is only part of scientists' efforts to counteract addiction. Some experts believe that heredity might play some role in substance abuse. Geneticists are developing special breeds of research animals that are particularly susceptible to abused substances, and the researchers are seeking markers in blood samples, such as certain enzymes, that would indicate whether the animal is susceptible to chemical dependency. Such markers, if found in human beings, might help identify in childhood people who run the greatest risk of becoming addicts. So far, the most significant clue to potential addiction seems to be a family history of alcoholism or drug abuse.

Several studies have already shown that alcoholism may be inherited. Psychiatrist Donald W. Goodwin of Kansas University Medical Center in Kansas City, along with researchers in Denmark, evaluated the risk of alcoholism in sons of alcoholic parents. The boys had been adopted and raised in families where drinking was not a problem. The researchers found that the adopted sons of alcoholic parents were three to four times more likely to become alcoholics than were adopted sons whose parents were not alcoholics. Another study reported in September 1985 by researchers at the State University of New York Health Science Center in New York City showed that young, nondrinking sons of alcoholics have the same type of a particular brain wave pattern as do their alcoholic fathers. This also seemed to indicate that a tendency toward alcoholism can be inherited.

Despite progress in finding drugs to treat addiction, psychological treatment remains a most important method of prevention and recovery. Recovering addicts need to change their life style and habits, and to stop associating with drug- or alcohol-using friends. Support groups help with the transition. To help prevent drug abuse and addiction, experts believe, people who are tempted to try drugs need to be taught how to say no. They must also be shown other ways of dealing with their emotional problems.

Psychological treatment—combined with new medications—can provide new hope for addicts. And as scientists continue to learn more about how these drugs work in the brain, other new solutions to the problem of drug abuse may be on the horizon.

For further reading:

Smart, Reginald. *Forbidden Highs: The Nature, Treatment, and Prevention of Illicit Drug Use.* Addiction Research Foundation of Ontario, 1985.

Snyder, Solomon. *Drugs and the Brain.* W. H. Freeman and Company, 1987.

Snyder, Solomon. "Opiate Receptors and Internal Opiates." *Scientific American,* March 1977.

By drilling into the ocean floor, the drill ship *JOIDES Resolution* is giving scientists a closer look at Earth and how it developed.

Drilling Under the Sea

BY PHILIP D. RABINOWITZ

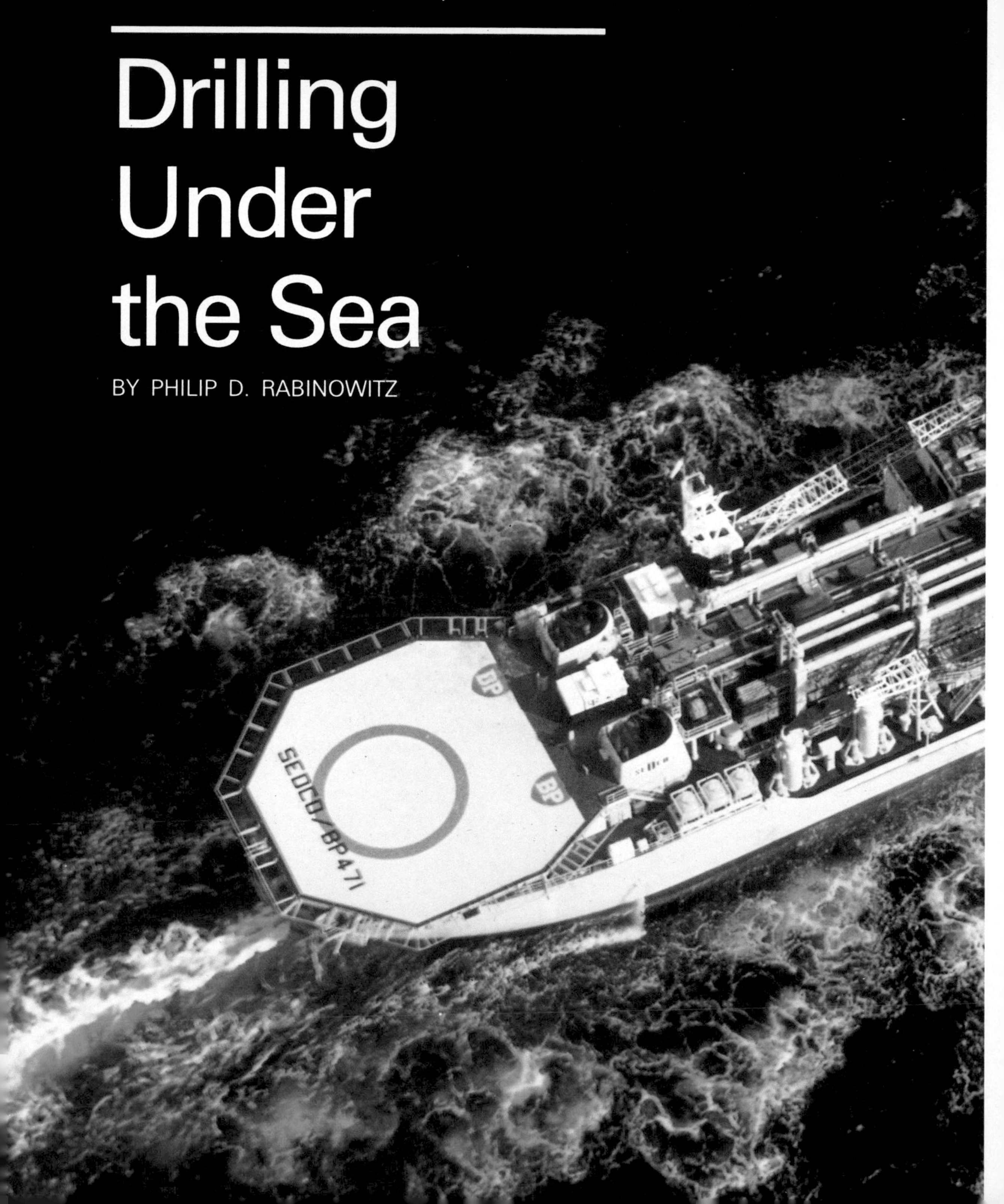

Preceding pages: The drill ship *JOIDES Resolution* is exploring the ocean floor, *inset,* as part of a 10-year scientific effort to learn what causes such geologic processes as the movement of gigantic plates that make up Earth's crust.

"Blast all scientists . . . curse the lot," Captain James Cook bellowed after his historic first voyage to Antarctic waters in 1773 on the ship *Resolution*. The famous British navigator complained that a small group of scientists allowed to accompany him on the voyage had been "cluttering up" the ship, "making excessive demands, never once getting into their thick skulls that the primary care of all shipmasters must be the ship herself and the safe preservation of her voyage."

More than 200 years later, another ship—named in honor of the first *Resolution*—explores the oceans. But this ship is dedicated solely to scientific discovery. And, instead of being tagalongs, scientists determine the ship's course and its purpose. Since its first cruise in January 1985, this ship—named *JOIDES Resolution*, after the *J*oint *O*ceanographic *I*nstitutions for *D*eep *E*arth *S*ampling—has been drilling into the rugged ocean floor and bringing up samples to help us learn more about Earth's development.

The Ocean Drilling Program (ODP) being carried out by *JOIDES Resolution* is expected to last 10 years. But it is not the first such project. The ODP is continuing the work begun by the Deep Sea Drilling Project, which lasted from 1968 to 1983. That project used the first drill ship dedicated to sea-floor exploration, *Glomar Challenger*. The ship made 96 voyages, drilled 1,092 holes in the ocean floor at 624 sites, and took 96,000 meters (315,000 feet) of core samples from some of the most remote regions of the oceans.

The cores retrieved during the *Glomar Challenger* expeditions revolutionized our understanding of geology, mainly by helping confirm the theory of *plate tectonics*. According to this theory, Earth's thin outer crust is divided into about 20 rigid plates upon which the continents and oceans rest. These plates drift on a subsurface layer of molten rock and move at an average rate of 1.3 to 10 centimeters (½ inch to 4 inches) per year. The theory explains the development of Earth's surface over millions of years, the growth of the oceans, the movement of the continents, and such phenomena as earthquakes and volcanoes.

Scientists hope that the cores being retrieved by the *JOIDES Resolution* will help explain the processes driving plate tectonics. Already, it has drilled into the region where Europe and North America originally separated 65 million years ago.

The late 1700's—when Cook explored the Pacific Ocean—was a golden age of discovery of remote lands. And for scientists studying the previously unknown ocean floor, the mid-1900's have been a golden age of discovery under the sea. One of the leading explorers of this latest golden age was oceanographer Maurice Ewing of Columbia University's Lamont-Doherty Geological Observatory in Palisades, N.Y. In the 1950's and 1960's, Ewing and a former student of his, Bruce C. Heezen, used sonar equipment on oceanographic research ships to map the sea floor. This equipment sent out sound waves, then recorded their echoes as the waves bounced off geologic features on the ocean floor. In the middle of the Atlantic, Indian,

The author:
Philip D. Rabinowitz, professor of oceanography at Texas A&M University in College Station, is director of the Ocean Drilling Program.

and Pacific oceans, Ewing and Heezen found that underwater mountain ridges form a continuous system encircling the globe like the seam on a softball. Formed by volcanoes, this system of mountain ridges and valleys extends along the ocean floors for more than 64,000 kilometers (40,000 miles). This finding paved the way for the theory of plate tectonics.

According to the theory, the midocean ridges are areas where two tectonic plates are separating. *Magma* (hot molten rock) from deep within Earth rises up through cracks at the midocean ridges, cools, and becomes new sea floor. The two plates continue to move apart, and this spreading of the sea floor causes the ocean basins to grow larger.

The theory also stated that tectonic plates could move toward each other. When they meet, they either collide head-on or the edge of one plate slides under the other. Since the late 1800's, oceanographers have known of deep trenches in the ocean floor. According to plate tectonic theory, these trenches are areas where one plate is sliding beneath another, a geologic process known as *subduction*. Friction caused by one plate scraping against another—along with the heat in Earth's interior that the edge of the plate encounters—melts part of the descending plate. This forms magma, which then rises to the surface of the plate above, producing volcanoes. This theory explains the famed *Ring of Fire*, a string of volcanoes that rim the Pacific Ocean from Mount Fuji in Japan on the west to the Cascade Range of North America on the east.

The theory of plate tectonics provided an excellent way of understanding how Earth's surface changes. But direct proof for this theory emerged only in the late 1960's, and drilling into the ocean floor helped provide that proof.

In gathering evidence to support or disprove the theory, scientists charted the age of the ocean floor in various locations. If the theory was correct, the ocean floor nearest the midocean ridges would be the most recently formed and therefore the youngest. The ocean floor farthest from the midocean ridges would have formed long ago and spread away from the ridges as new ocean floor was created. The *Glomar Challenger* drilled in regions both near and far from the midocean ridges. By determining the ages of the ocean floor samples recovered from the drilling, scientists were able to prove that the ocean floor is progressively older as it spreads out on either side of the ridges. This evidence was dramatic confirmation of the plate tectonics theory.

The *Glomar Challenger* expeditions also revealed that the entire ocean floor is relatively young. Even the oldest part of the ocean floor is only about 200 million years old, compared with 3½ billion years for the oldest rocks on the continents.

By taking careful measurements at many sites, scientists aboard the *Glomar Challenger* calculated the rate at which the sea floor is spreading. They found that new ocean floor in the Atlantic Ocean is being created at an average rate of 3 to 5 centimeters (1 to 2 inches) per year, causing Africa and Europe to move farther apart from North and South America. The average rate of sea-floor spreading in the Pacific Ocean is greater than that in the Atlantic, about 5 to 10 centimeters (2 to 4 inches) per year. But the east edge of the Pacific Ocean floor is disappearing as it slides underneath the continental plates of North and South America. It's conceivable that the Atlantic will be larger than the Pacific in 50 million years due to the westward movement of North and South America.

The *Glomar Challenger* expeditions made many other findings that revolutionized geological thinking. For example, in 1971, a drilling expedition revealed that the Mediterranean Sea dried up about 5.5 million years ago and was a huge "dead sea" full of gypsum and salt. Scientists think that the Strait of Gibraltar—a narrow body of water that connects the Mediterranean Sea and the Atlantic Ocean—must have closed about 5.5 million years ago due to the movement of plates. The Mediterranean became a stagnant basin of salt water isolated from the Atlantic Ocean. The seawater slowly evaporated over hundreds of thousands of years, leaving behind thick salt deposits. As the plates continued to move, the strait re-opened—perhaps hundreds of thousands of years later—and the dry basin was again filled with water from the Atlantic.

Glomar Challenger's greatest contribution, however, was that it gave scientists a glimpse of what the ocean floor looks like on a global scale. Until the drill ship began operating, no one had been able to examine the hard-rock ocean floor that is buried beneath layers of sediment. But scientists learned from cores recovered by the drill ship that the ocean crust is made up of various types of volcanic rock.

Even with such impressive accomplishments, *Glomar Challenger* barely scratched the surface of the ocean floor. The deepest penetration into the sea floor was 1.8 kilometers (1 mile)—less than one-fourth of the thickness of Earth's crust under the oceans. And the average depth of a drill hole was about 300 meters (980 feet).

When *Glomar Challenger* was retired in 1983, oceanographers decided that a new drilling program was needed to reach previously inaccessible areas in deeper waters and farther down into the

Proving the Theory of Plate Tectonics

Cores drilled from the ocean floor by *Glomar Challenger* in the 1960's showed younger rock near midocean ridges and older rock farther away. This finding helped confirm the theory of plate tectonics, which says that Earth's crust is made of huge plates that move on a layer of hot rock in the mantle. Cores of young rocks near midocean ridges showed that *magma* (molten rock) rises there to form new crust. The ocean floor spreads away from the ridge on both sides. In some cases, the edge of an oceanic plate slides under a continental plate, creating a trench in the ocean floor. As the edge of the plate melts, it also causes volcanoes to form on the continental plate.

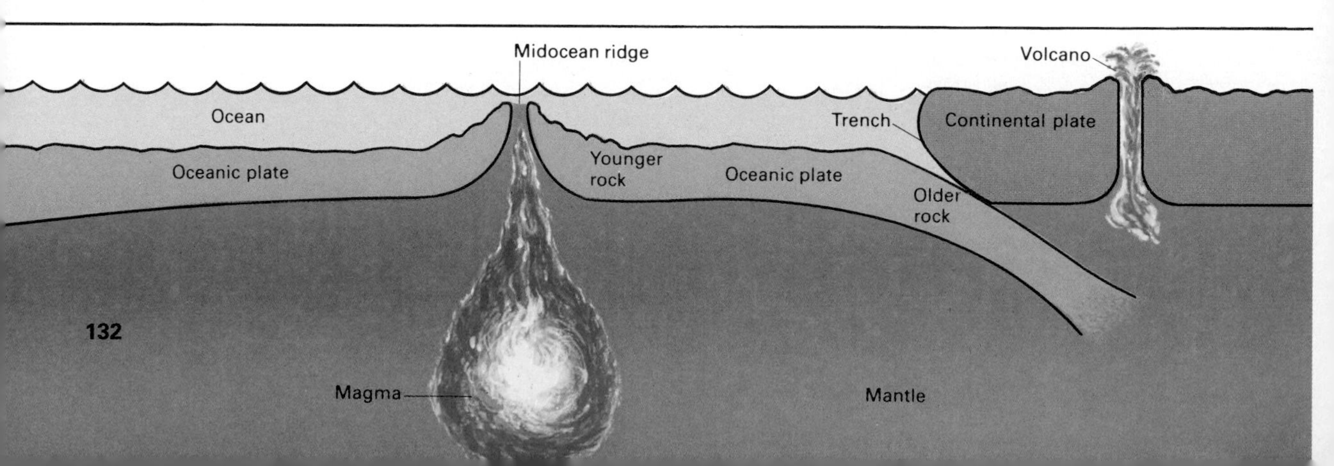

hard-rock crust. The ODP, of which I am the director, was established in 1984 to meet these goals. The program operates out of Texas A&M University in College Station, which is responsible for operating the drill ship and selecting its crew. The goals of JOIDES were set by an international group of earth scientists. JOIDES members include 10 institutions of higher learning in the United States and 6 institutions representing Canada, Japan, and 15 European nations. The program receives its funding from the U.S. National Science Foundation and other nations.

We want to study the origin and evolution of Earth's crust under the oceans—known as the *oceanic crust*—and we hope to learn what processes cause the motion of tectonic plates. While most scientists accept the theory of plate tectonics, there is disagreement about what causes the plates to move. Some scientists believe that the plates are being pushed apart by magma welling up at the midocean ridges, and other scientists believe they are being pulled apart due to subduction at the deep-sea trenches. Still other scientists believe that other forces contribute to this process. The *JOIDES Resolution* may help resolve this "push-or-pull" debate by such measures as drilling for the first time into the floors and steep walls of the trenches.

We also want to understand how sea-floor spreading has affected Earth's climate over the last 200 million years. We know that as the continents moved apart, the shapes of the ocean basins changed, and that this in turn changed the ocean currents. So we think we can learn how changes in ocean currents affected Earth's climate.

As ODP director, my first task was to find a drill ship large enough and well enough equipped to carry out the JOIDES research program. The search led to a ship registered as *Sedco/BP 471*, which was then drilling for oil in the Gulf of Mexico. The ship was 143 meters (470 feet) long. Its

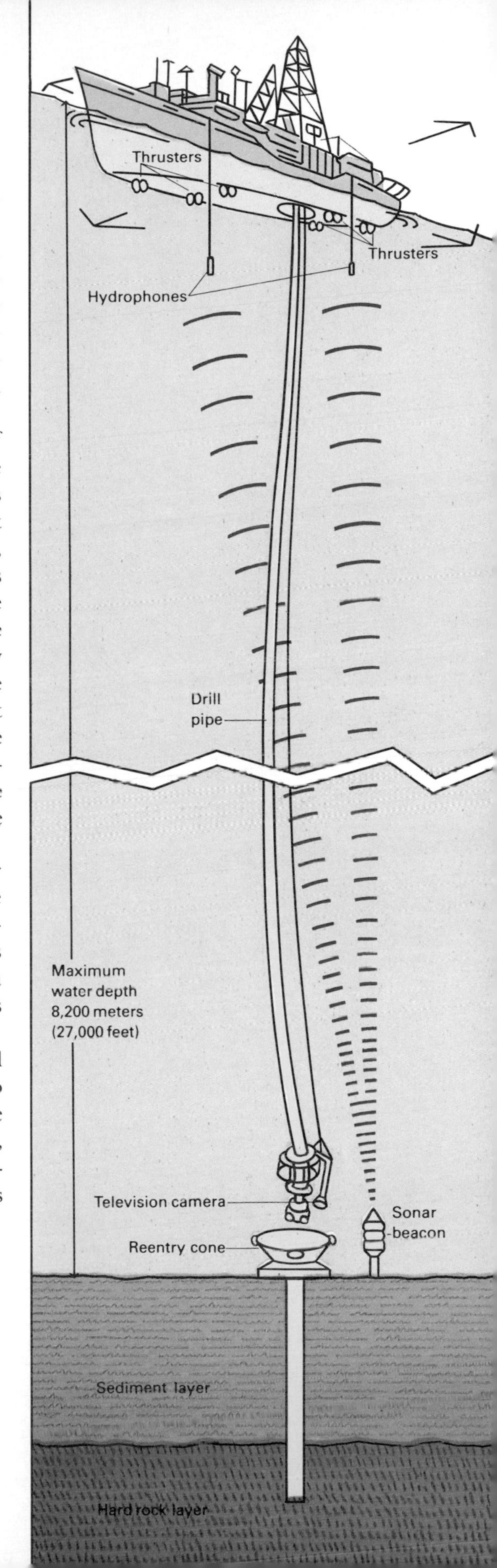

Drilling Under the Sea

Sound waves sent from a sonar beacon to the ship's hydrophones locate the drilling site on the ocean floor, while powerful thrusters hold the ship in place. A remote television camera aids in positioning the drill pipe into a metal cone at the top of the hole from which cores are being taken.

Taking a Core Sample

Crew members of the *JOIDES Resolution, right,* adjust the drill pipe that passes through a hole in the center of the ship to the ocean floor. After pulling a core barrel containing a rock and sediment sample up through the pipe and removing it, technicians cut the sample, *below,* into sections. A scientist, *below right,* then prepares part of a section for microscopic analysis to determine its age, while another scientist in the ship's chemistry laboratory, *opposite page,* examines a sample for signs of oil or gas.

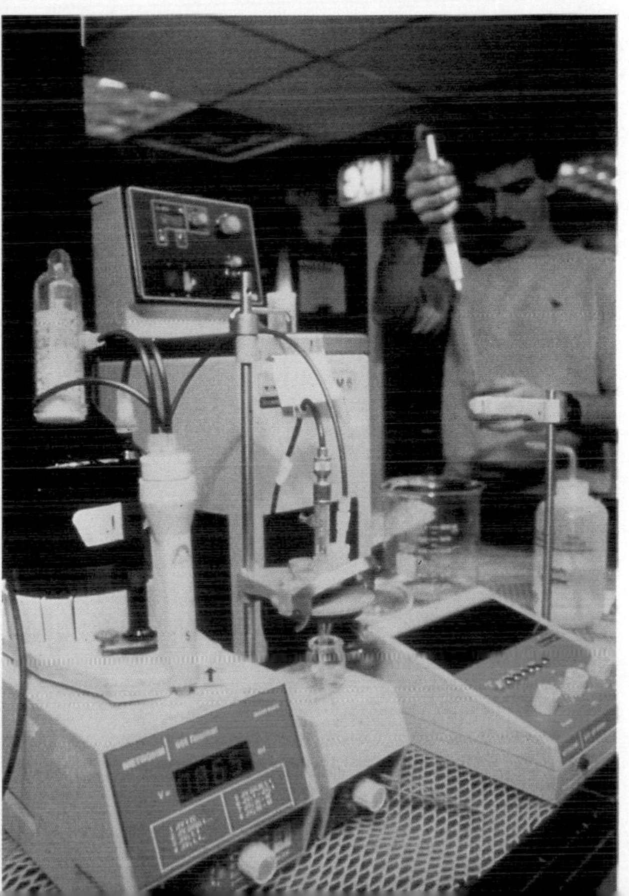

conversion to a research drill ship equipped with modern laboratories began in late April 1984. By January 1985, we had a floating laboratory that contains the world's most advanced research equipment in operation at sea.

The *JOIDES Resolution* has 12 laboratories on seven decks. The ship can carry drill-pipe lengths totaling about 15 kilometers (9 miles) and can drill in water about 8 kilometers (5 miles) deep. It has living quarters for up to 50 scientists and technicians and a crew of 65 and provides for leisure activities with a gymnasium, a library, lounges, and a movie theater.

A typical cruise of the *JOIDES Resolution* lasts about two months. We plan the expeditions so that the least amount of time is taken up sailing from one drill site to the next. For example, after an expedition in the Antarctic Ocean in January 1987, the next expedition was planned for the nearby South Atlantic Ocean. The scientists who take part in an expedition are flown to the port nearest each drill site.

Before each cruise, scientists from various oceanographic institutions around the world perform *site surveys* using other research ships. On these surveys, they collect information that determines where the *JOIDES Resolution* will drill. This information also tells them if oil or natural gas might be present at a possible drill site. We avoid drilling in such areas because, with our drilling system, we cannot use precautionary measures to prevent a dangerous oil or gas blowout that would endanger the people on the ship and damage the ocean environment. As a further precaution, while drilling we immediately check each core that is taken on board for the presence of oil or gas.

At the drill site, the crew holds the ship over the same spot on the ocean floor—even in the roughest seas—with a system known as *dynamic positioning*. First, we drop a *sonar beacon* (a device that sends out sound waves) from the ship to the site on the ocean floor where we want to drill. *Hydrophones* (instruments that detect sound waves underwater) mounted in the ship's hull pick up signals sent out from the beacon. A computer uses these signals to calculate the ship's position.

Clues in a Core
Scientists compare colors in a core sample with a special color chart, *right,* which gives them information about minerals and fossil matter found within the sediments. These data provide clues about the climatic conditions that existed on Earth when the sediments were deposited on the ocean floor. By analyzing core samples in a cryogenic magnetometer, *below,* scientists can trace the history of Earth's magnetic field reversals.

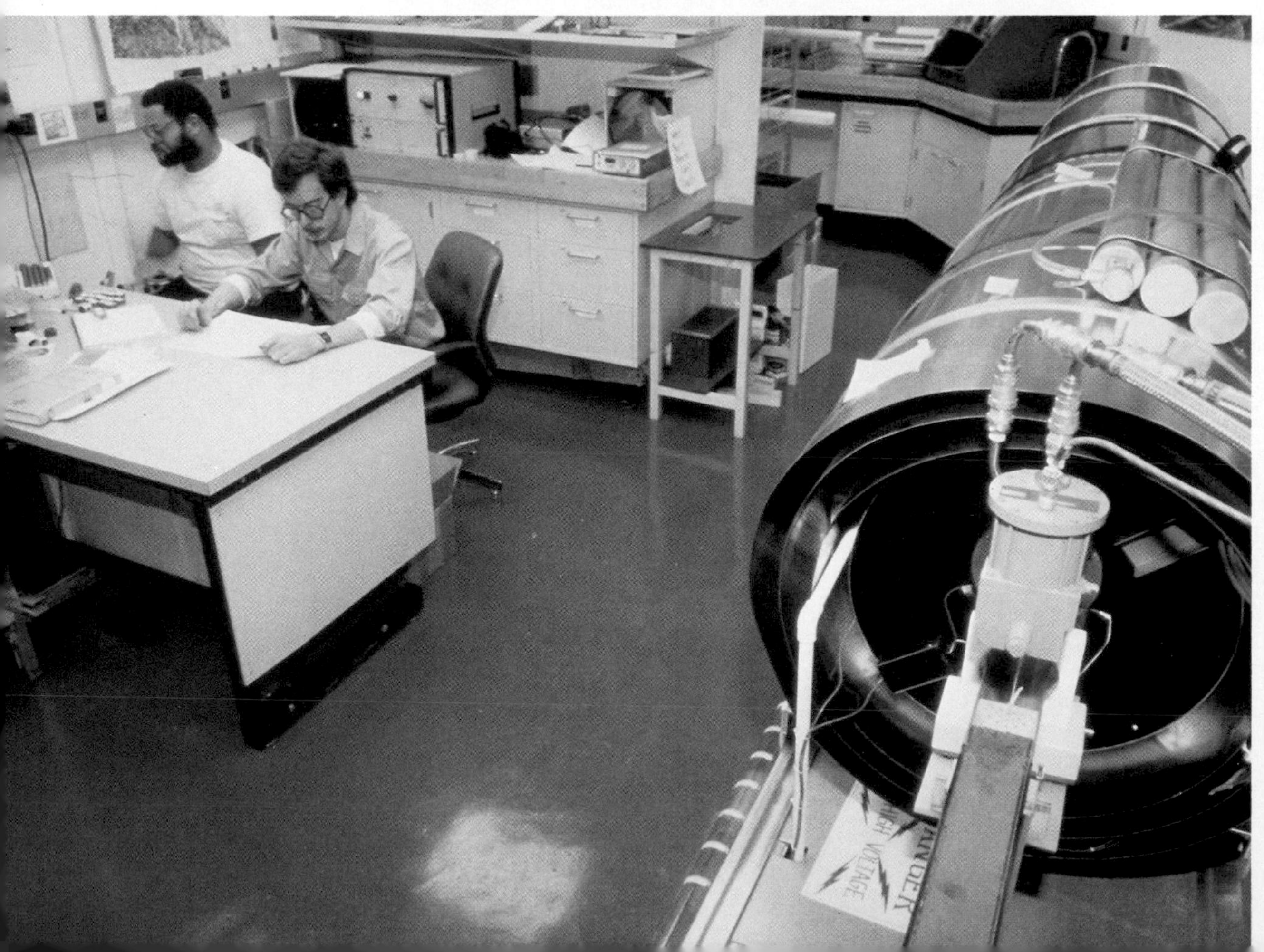

Storing the Cores
A scientist at the Lamont-Doherty Geological Observatory in Palisades, N.Y., files part of a core sample in a refrigerated archive for core samples from the *JOIDES Resolution.* Refrigeration preserves the moisture in the sample and prevents it from drying up and cracking. Three oceanographic institutions in the United States maintain permanent storage facilities for the core samples for possible future analysis.

If the ship begins to drift, the computer automatically turns on powerful *thrusters* (motor-driven propellers)—six on each side of the hull and one at the stern. The thrusters can move the ship sideways, forward, or backward to keep it directly above the drill site.

Once the ship is positioned, the crew often lowers a cone-shaped device called a *reentry cone* when a marker is needed where the hole is to be drilled. If the drill pipe later has to be pulled out of the hole—to replace a worn drill bit, for example—the crew can easily relocate the hole by using sonar. The crew attaches a sonar scanner to the drill bit at the end of the drill pipe. The scanner sends out sound waves that bounce off sonar reflectors on the reentry cone and guide the scanner to the cone. Sometimes we use a remote-controlled television camera to help position the drill pipe, making the sonar scanner unnecessary. Once the drill pipe is above the cone, the pipe is lowered into the hole, and the camera is removed.

The *JOIDES Resolution* method of drilling is much the same as that which was used on the *Glomar Challenger.* We drill until a *core barrel,* which fits inside a 9.5-meter (31-foot) joint of pipe, is filled with sediments or rock. The core barrel rests on top of a circular drill bit, which has a hole in the middle. As the drill bit's "teeth" dig into the ocean floor, the material it digs around goes up through the hole in the bit and into the core barrel above. When the core barrel is filled, the crew sends a wire down through the drill pipe,

An underwater television camera mounted on the drill pipe, at right in the picture, helps the crew of the *JOIDES Resolution* guide the pipe into a drill reentry cone on the ocean floor, at the left, which marks the spot where a drill hole was begun.

latching onto the core barrel and bringing it up onto the deck of the ship for study. A new core barrel is then lowered.

Drilling is a tedious and time-consuming process, but the area of the drill site can often furnish excitement. In January 1987, for example, the *JOIDES Resolution* dodged dangerous icebergs in the Weddell Sea off Antarctica. A patrol ship hired to accompany the expedition actually "lassoed" icebergs and towed them away to prevent a collision with the drill ship.

Not only can we drill in the roughest and most dangerous seas but a new drilling method also enables us for the first time to obtain samples of soft sediment. The uppermost layer of sediment on the ocean floor is often so soft that a drill bit simply churns up the sediments until they become unrecognizable and lose much of their value for analysis. But in our new method, we lower the core barrel first, ahead of the drill bit. A device using pressurized water then drives the core barrel into the soft sediments. Since the drill bit never touches the sediments, they remain intact inside the core barrel, and so the fossils can be analyzed.

With the *JOIDES Resolution,* we can also drill into areas where there is just a thin layer of sediment. Previously, drilling in deep-ocean basins required thick layers of sediment about 100 meters (330 feet) deep to "anchor" the drill. And since sediments are virtually nonexistent at midocean ridge areas, scientists could not drill at these important sites where new ocean floor is being created. Using an extremely heavy concrete guide base, however, scientists on a *JOIDES Resolution* cruise in the fall of 1985 were able to drill for the first time in this key region.

They first used a remote-controlled underwater camera—similar to the one that discovered the *Titanic* in 1985—to locate a volcano almost 3 kilometers (2 miles) below the ocean surface on the crest

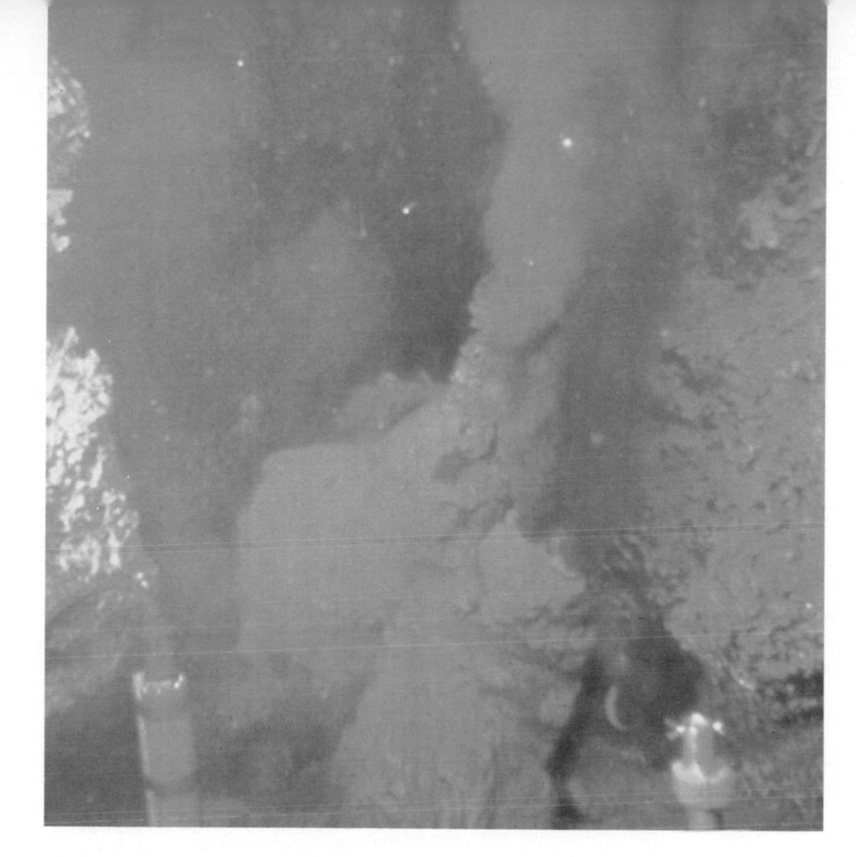

A black smoker in the Atlantic Ocean gives off plumes of hot, mineralized water. With the help of an underwater television camera, the crew of the *JOIDES Resolution* can position the drill precisely to take cores from around a black smoker, enabling scientists for the first time to investigate the source of the minerals.

of the Mid-Atlantic Ridge about 1,900 kilometers (1,200 miles) southeast of Bermuda. The crew then lowered an 18-metric-ton (20 short-ton) guide base to the site to hold the reentry cone. Next, they anchored the guide base with 45 metric tons (50 short tons) of concrete pumped through the drill pipe into bags on the guide base. This provided the support needed to drill through the reentry cone into the rocky, sediment-free surface. The researchers recovered the first samples of so-called *zero-age rock*, geologically young volcanic rock that rose to the ocean floor from deep within Earth about 100,000 years ago. Scientists are now studying the material to learn more about the creation of new ocean crust and the evolution of volcanic systems at midocean ridges.

The remote-controlled underwater camera has also made it possible to aim the drill very precisely. This proved extremely valuable for studies of the ocean floor near *black smokers*, volcanic formations that spew hot, mineral-rich water from within Earth's crust. Black smokers are relatively small mounds that rise from 1 to 5 meters (3 to 16 feet) from the ocean floor. The camera enables us to pinpoint their location and drill next to them so we can recover core samples that may tell us where the minerals originated.

Once a core arrives on deck, the shipboard scientists and technicians make use of sophisticated laboratory equipment to analyze the physical and chemical properties of the core material. Two samples are immediately taken. The sample that comes from the deepest—and thus the oldest—section of the core goes to a *paleontologist* (a scientist who studies prehistoric life). By examining fossils in the core, the paleontologist can determine about how old the core is. The other sample goes to scientists in the chemistry laboratory, who analyze it for signs of oil or gas.

Other scientists on board then cut the core into sections 1.5 meters (5 feet) long and label them with information about when and where the core was recovered. The core is then ready for detailed scientific analysis.

Analyzing samples of rocks from the ocean floor provides scientists with a variety of important information about conditions on Earth over the last 200 million years. For example, we can learn about past changes in Earth's magnetic field by analyzing core samples with an instrument known as a magnetometer. This instrument enables us to determine the orientation of Earth's magnetic field at the time the rock formed. For example, when magma erupts through a midocean rift and turns into solid rock as it cools on contact with ocean water, magnetic minerals within the rock become aligned in the direction of Earth's magnetic field at the time. Scientists know that Earth has a history of random but frequent reversals of its polarity—that is, the north magnetic pole becomes the south magnetic pole, or vice versa. These reversals have occurred at intervals ranging from 50,000 to 20 million years.

As we continue drilling and study more rock samples, we can obtain a detailed history of Earth's magnetic reversals. This is important for several reasons. For example, some scientists believe that during a magnetic reversal, Earth temporarily loses its magnetic field until the reversal is completed. Since the magnetic field shields living things on Earth from dangerous cosmic radiation, the loss of this shield may have been a factor in mass extinctions. Paleontologists are now comparing the fossil record of extinctions with magnetic reversal data to determine if this is in fact what happens.

After scientists analyze the core sections for their magnetic orientation and their ability to conduct heat, the cores are split in half lengthwise. One half is designated the "working half" and is set aside for further study. The other half—known as the "archive half"—is stored. Both halves are kept refrigerated. Refrigeration preserves the moisture in the core samples and prevents them from drying up and cracking. After each cruise, all of these cores are shipped in refrigerated containers to storage in one of three oceanographic institutes: Scripps Institution of Oceanography, Lamont-Doherty Geological Observatory, or Texas A&M University. If new methods of analyzing core samples are developed in the future, we will be able to use them on the stored cores instead of going back to previous drill sites for more samples.

Meanwhile, the working halves of the core samples aboard ship continue to be examined by a variety of scientific specialists for clues about Earth's history and environment. For example, *sedimentolo-*

gists—scientists who study sediments—examine the core samples to determine how the layers of sediment were deposited on the ocean floor. Sediments may build up due to the decay of plant and animal life that has fallen to the ocean floor over thousands of years. Or they may build up from the runoff of soil from continents. Fine-grained sand in a layer of sediment, for example, is evidence of continental runoff.

Petrologists—scientists who study the origin and structure of rocks—examine core samples that contain layers of volcanic rock. Because every volcanic flow has a slightly different chemistry, petrologists analyze the samples to learn how many separate volcanic eruptions occurred over a period of time, thus telling us how active a particular underwater volcano was.

Finally, paleontologists study the core samples for clues about Earth's past climate. For example, they look for certain fossils of animals they believe lived in a cold climate. If they find those fossils at a drill site near the equator, it may be evidence that the climate at the equator was much cooler millions of years ago.

So far, the expeditions of the *JOIDES Resolution* have told us much about Earth's past. We have, for example, made important advances in our studies of how the continents first began to separate. Scientists believe that about 200 million years ago all the continents formed a single land mass known as Pangaea. Gradually, this land mass broke apart and separated into the continents and oceans we know today.

Scientists believe that *continental margins*—the underwater edges of continents—represent the areas where the continents first began to break apart and the oceans began to form. In the summer of 1985, we learned more about this separation by drilling for the first time into a key region of a continental margin. The drill site, off the coast of Norway, is the region where scientists believe Norway and the rest of Europe began to split apart 65 million years ago from a land mass that included Greenland and North America. The core samples revealed that huge outpourings of volcanic lava erupted as the separation occurred. At that time, this coastal region off Norway was above sea level, but it gradually sank underwater as the continents drifted apart, and an ocean basin formed between Norway and Greenland. Because other continental margins appear to resemble the area off Norway, further analysis of the cores may help us understand more about how this separation took place.

The *JOIDES Resolution* appears destined to return results that will continue to alter our understanding of Earth. Since the 1960's, when oceanographers first began drilling into the ocean floor, our view of Earth's history has changed radically. By the year 2000, after another 20 years of drilling, we may understand exactly why and how those changes occurred—and what future changes may be in store for Earth.

Warning: This house may be hazardous to your health.

Polluted air indoors, rather than outdoors, is rapidly becoming a major health concern, with the greatest threat coming from radon, a natural radioactive gas.

Your House Can Make You Sick

BY JANE SAMZ

A steel-blue haze hangs in the air. Sharp, bitter smells sting your nose. Airborne dust and molds slowly swirl around you, irritating your eyes and throat. Harmful fumes from gases such as methane, butane, and propane make your head spin; carbon monoxide makes it ache. With every breath, you inhale radon, a radioactive gas that bombards your lung cells, threatening to slowly turn them cancerous. But this poisonous, cancer-causing air is not found inside a steel mill, a coal mine, or a chemical plant. Nor is it part of the cloud-shrouded, hostile atmosphere of another planet. Amazingly, air this dangerous is sometimes found in the seemingly harmless, homey environment you'd call a living room in an increasing number of houses found to have polluted indoor air.

Researchers have found that indoor air pollution can be worse—sometimes as much as 100 times worse—than pollution outside. Because many people, especially urban dwellers, spend 70 to 90 per cent of their lives inside houses and other buildings, the health consequences can be serious. Experts link high levels of indoor air pollutants to health problems ranging from flulike symptoms to serious allergic reactions, pneumonia, and an increased risk of lung cancer. Perhaps cause for greatest concern is radon, which is sometimes found in soil, building materials, and water and gas supplies. Since

1984, health officials have found a large number of radon-contaminated dwellings, and public concern has begun to mount. Federal officials estimate that radon may be responsible for 1 out of 25 lung cancer deaths. Fortunately, radon and most other air pollutants are dangerous only after long exposure, and they can be eliminated from the home.

Indoor air pollution became a major problem in the 1970's. Homeowners concerned about the rising cost of fuel and the dwindling supply of coal, oil, and natural gas began to weatherproof their houses by tightly plugging holes, installing storm windows and doors, caulking, weatherstripping, adding insulation, or hanging up plastic sheets or heavy drapes to eliminate drafts. These homeowners may have saved on heating and cooling bills but at a cost to the occupants' health unless they also installed ventilation and air-filtering devices. Energy conservation is a very worthwhile goal, but without a constant flow of air in and out of a building—which is measured in *air exchanges* per hour—pollutants become trapped inside. In older, "leaky" houses, the air exchange rate may be as high as 4 per hour, which means that all the air inside the house is replaced by outside air every 15 minutes. But in tightly sealed, energy-efficient houses without good ventilation systems, the air may be replaced as infrequently as once in 10 hours. Air pollutants inside "tight" houses rise to unhealthful levels.

Indoor air pollution is a problem for office workers, too, especially because many modern office buildings are built with windows that do not open. Such buildings may simply recirculate much of the inside air. This means, for example, that even those who work in "no smoking" areas may be subjected to tobacco smoke from smokers working in other parts of the building. In 1982, a Washington, D.C., office building had to be abandoned because of its polluted indoor air. Designers had created the office space by converting a portion of a parking garage. But harmful exhaust fumes seeped inside work areas, giving employees headaches and creating other health problems. Ironically, the office building belonged to the Environmental Protection Agency (EPA). Another government building in the same city was seriously polluted when slime growing in air-conditioner drain pans produced microscopic plants called *fungi* that were blown throughout the building. Technicians from the National Institute for Occupational Safety and Health (NIOSH) measured the office's airborne fungi level and said it was comparable to that of a chicken coop.

In 1979, scientific concern about the effect of restricted air exchange in homes and offices led the EPA to ask the National Academy of Sciences to evaluate the possible dangers of indoor pollution levels. Since 1980, the Centers for Disease Control, NIOSH, and many other health organizations have given serious attention to the problem. Today, environmental scientists are concerned with estab-

The author:
Jane Samz is a free-lance science writer and editor of *Science World* magazine.

lishing clean air standards for homes and offices. Researchers are also finding ways to eliminate or minimize indoor air pollution caused by tobacco smoke, chemical fumes, and airborne particles and gases produced by objects in the workplace and home.

The indoor air pollutant that many people are most concerned about is the naturally occurring radioactive gas radon. Radon is colorless and odorless. It is created from a chain of reactions that begins with the radioactive decay of uranium, which is found in tiny amounts throughout Earth's crust. In the first reaction, the uranium atom emits an *alpha particle*, a tight cluster of two *protons* and two *neutrons*—subatomic particles that make up an atom's nucleus. The atom that is left is no longer uranium, but another radioactive element called thorium, which decays into another element, radium, by emitting another alpha particle. The reactions involving radioactive decay change one element into another because the number of protons in an atom's nucleus determines what element that atom is. For example, an atom of uranium has 92 protons in its nucleus, thorium has 90, and radium has 88. The radium atom also decays, emitting an alpha particle to become radon.

Because it is a gas, radon can enter living spaces by seeping from uranium-containing soil or other substances into a house. In indoor air, radon creates a health hazard because, like uranium and radium, it decays into a number of radioactive elements. These elements (known as *radon daughters*) cling to airborne dust particles. When radioactive dust is inhaled, the smallest particles get trapped deep inside the lungs.

The kind of damage a radon daughter inflicts depends on its *half-life* (the time it takes for half the atoms in any radioactive element to decay). Long-lived daughters such as lead 210, which has a half-life of 22 years, may accumulate in body tissues or bones. Short-lived radon daughters include polonium 218, which has a half-life of 3 minutes, and polonium 214, with a half-life of 0.00016 second. These elements break down so rapidly—and so soon after being inhaled—that the alpha particles they give off are absorbed by the lining of the lungs. Scientists theorize that the alpha particles damage lung cells, leading to cancer.

Experts have been studying the radon-cancer link for many years. As early as 1879, researchers in Germany noticed that the cause of death among German uranium miners was often lung disease. By the 1930's, scientists had pinpointed radioactivity in the air in uranium mines as a direct cause of lung cancer. Since then, several major studies have linked radon to lung cancer.

For years, experts thought that uranium miners were the only people exposed to the dangers of radon. But in the 1960's, health officials learned that many houses and other buildings in towns such as Grand Junction, Colo., had been built on uranium-mine waste and that the residents were at risk of lung cancer because of radon

Understanding the Radon Threat

Scientists estimate that up to 12 per cent of U.S. houses have high indoor levels of radon—a colorless, odorless radioactive gas that has been linked to lung cancer.

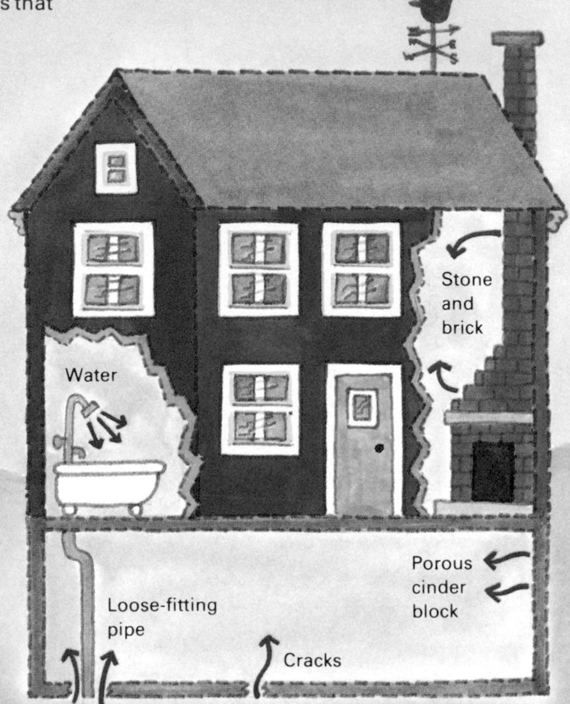

How Radon Enters a House

Radon gas exists naturally in some soils, water, and building materials. It can seep from the soil through cracks in a basement floor or a building's foundation. Radon can also be released from radon-containing water or from radon-containing bricks, stone, or other building materials. Tightly sealed windows and doors prevent the gas from escaping, causing indoor radon levels to rise dangerously.

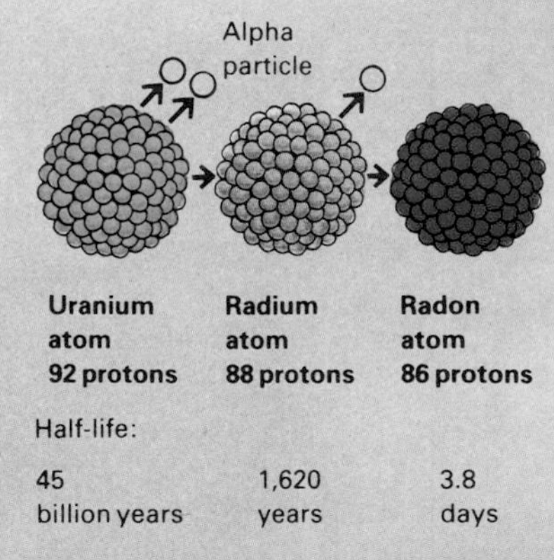

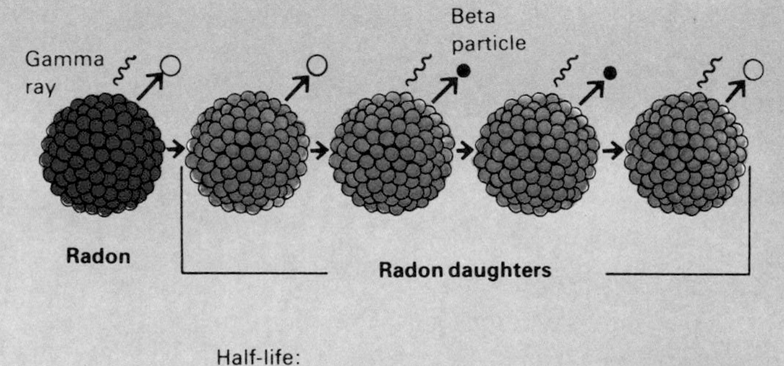

Where Radon Comes From

Radon is formed by the radioactive decay of uranium. Through radioactive decay, an atom can change from one element to another by emitting alpha particles, each of which contains two protons and two neutrons. The number of protons in an atom's nucleus determines what element that atom is. The emission of two alpha particles—a total of four protons—changes a uranium atom into a radium atom, which then decays into radon.

Radon and Its Deadly Daughters

Because it is a gas, radon can move out of the soil and into the air we breathe. Radon atoms then decay into a series of radon daughters—by emitting alpha particles and other forms of radiation called beta particles and gamma rays. Radon and most radon daughters are dangerous because they have a very short *half-life* (the time it takes half the atoms in a radioactive substance to decay). Due to this short half-life, radon and its deadly daughters rapidly release their radiation. If these atoms are inhaled, this barrage of radiation can damage lung cells, leading to cancer.

exposure. Still, experts believed radon was a threat only in such unusual situations. Their thinking changed after December 1984, when Stanley Watras, a nuclear power plant worker, set off radiation monitors as he reported for work at the Philadelphia Electric Company's Limerick plant. Watras was contaminated with high levels of radiation that could not have come from the plant, which was still under construction. Radiation levels in the Watras home in Boyertown, Pa., were monitored, with alarming results. Technicians found that the radon level in the Watras house was 2,500 times higher than what was considered average—the highest ever found in the United States. Scientists say that people living in such a house would receive radiation exposure equivalent to 455,000 chest X rays per year. Long-term residents of such a house are virtually assured of developing lung cancer.

The Watras home had been built along a uranium-rich granite formation known as the *Reading Prong*, which extends from Reading, Pa., through northern New Jersey and into New York. Soon after the Watras incident, environmental officials began monitoring radon levels in many other houses built over this geologic formation. Pennsylvania officials announced in early 1986 that 60 per cent of the 22,000 houses tested in Pennsylvania's Reading Prong area had radon levels above the EPA's recommended maximum.

Government officials in other areas also began radon testing programs after the Watras incident. Because there are millions of houses in the United States, testing throughout the nation may take years to complete. The undertaking is especially difficult because radon levels vary greatly from neighborhood to neighborhood and even from house to house. The house next door to the Watras home, for example, had average radon levels.

Using data collected from 17 states and every major geographic region in the United States, physicist Anthony V. Nero and other researchers in the Indoor Environment Program of the Lawrence Berkeley Laboratory (LBL) in Berkeley, Calif., calculated average indoor radon concentrations. The scientists in 1986 estimated that 1 million of the 70 million single-family homes in the United States may have annual radiation exposure levels equal to or greater than those received by uranium miners.

What such findings mean in terms of public health is under dispute. Some scientists believe that the EPA has overstated the lung cancer danger of relatively low indoor levels of radon. Radiation expert Naomi H. Harley of New York University, for example, notes that of a group of miners who had the highest known radon exposures, fewer than half died of lung cancer. In her view, risks associated with the relatively low-level exposure found in the majority of contaminated houses may be minimal.

Scientists will probably continue to disagree until direct evidence links lung cancer among the general population to the presence of

Where Radon Risk May Be Highest

Using the locations of uranium deposits as a guide, the U.S. Environmental Protection Agency has identified areas that may have high radon levels in soil. Houses in any region may become contaminated, however, because of other risk factors in houses such as poor ventilation or the presence of radon-containing building materials.

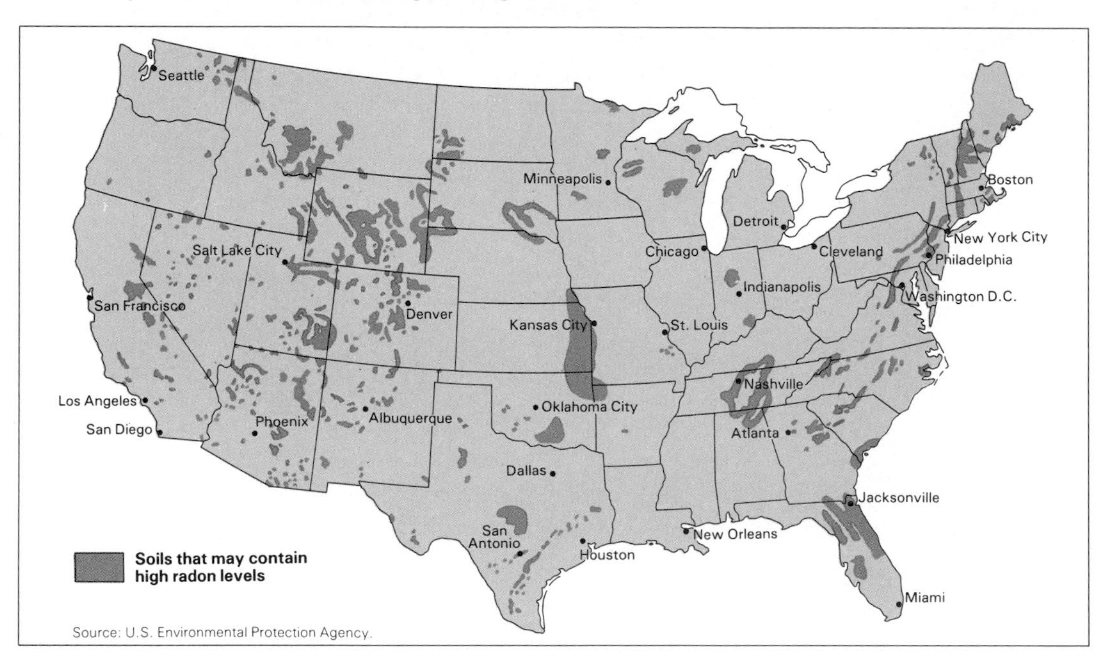

radon in buildings. Researchers in Maine, New Jersey, and Pennsylvania are looking for such evidence by comparing radon levels found in lung cancer victims' homes with radon levels in houses of people who are free of the disease.

How radon enters buildings is also being researched. It is clear that radon in some uranium-rich areas can move up through the soil and into the air. Most of the radon that enters a building in these areas seeps through cracks in basement floors and walls, through air spaces around pipes, or through floor drains. For this reason, radon levels tend to be lower on higher floors. Radon levels in a basement are often higher than on the first floor, for example.

Researchers have found that houses in areas not considered high risk may also be contaminated by radon seeping from building materials—such as concrete, brick, gypsum, or rock—that contain radon. Most building materials, however, give off less radon than soil does because the radon is often trapped inside the solid material. Paint and wallpaper may also help keep radon from escaping from building walls. Other sources of radon pollution are water and natural gas that pick up radon from uranium deposits in the soil or rock they travel through. If radon exists in local water supplies, it can be released into the air when the water is heated or agitated, as it is when used for showers, washing clothes, or washing dishes. Radon may also be released by the burning of natural gas in ovens, stoves, or furnaces.

Houses that are free of radon-contaminated building materials and are in areas with low radon levels in the soil may still have a radon pollution problem. High indoor radon levels can occur in airtight buildings that do not have adequate ventilation. Over time, even small amounts of radon leaking into such a house can rise to unsafe levels. A 1986 study by LBL's Nero reports another major cause of high indoor radon levels—lower air pressure inside than outside. Heating a house, for example, causes air to flow upward through the building, sucking gases from the soil through the basement floor or foundation walls. Wind causes similar pressure differences, as do some vent fans.

Technicians in Montclair, N.J., remove radioactive soil from a neighborhood where 200 houses had been built on landfill that contained factory waste contaminated with radon. Radon contamination requiring such drastic treatment is rare.

Radon, however, is not the only substance that can cause indoor air pollution. Anyone who has ever spent time in a smoke-filled room knows that ordinary tobacco smoke can sting and burn eyes, make throats feel scratchy and sore, irritate noses, and make people cough. Medical researchers suggest that tobacco smoke can also cause serious damage, including an increased risk of heart disease and lung cancer. In 1984, the U.S. surgeon general reported that smoking is clearly a health hazard for nonsmokers as well as for smokers. In December 1986, he said that the effects of *passive smoking* (when nonsmokers breathe smoke-filled air) was so severe that smoking should be limited in workplaces. Research chemist Alfred H. Lowrey and environmental protection specialist James L. Repace estimated in 1985 that tobacco smoke in indoor air causes 100 times more deaths than are caused by *carcinogens* (cancer-causing substances) in outdoor air pollution. They also estimated that about 5,000 of the 130,000 Americans who die of lung cancer each year are passive smokers.

Another lung cancer hazard and component of indoor pollution is *asbestos,* a number of fiberlike minerals used to make insulating materials and fireproof clothing. Asbestos can be found in many older buildings. Products containing the material may include ceiling and floor tiles. Asbestos may also be packed around water pipes. When asbestos-containing materials are damaged and crack or break, microscopic asbestos fibers are released and can be inhaled into the lungs. Physicians have known since the 1920's that asbestos fibers can cause a severe lung disease called asbestosis. By the late 1940's, British scientists found that asbestosis victims had a 50 per cent chance of dying of lung cancer. In the mid-1960's, researchers at Mount Sinai Medical Center in New York City showed a definite link between asbestos inhalation and lung cancer. The health effects are so pronounced that pathologists at the University of Vermont College of Medicine reported in 1982 that no level of exposure to absestos fiber is without risk.

Dangerous Fumes That Get Trapped Indoors
Many items and products used in the home may release gases and chemical fumes that can rise to dangerous levels when used or stored in an airtight house.
N
E
W
S
Urea-formaldehyde insulation
Paint and paint thinner
Airplane glue
Felt-tip markers
Furniture polish
Newly dry-cleaned clothing
Moth balls
Hair spray
Drain cleaner
Bug spray
Carbon monoxide
Oven cleaner
Nitrogen dioxide
Tobacco smoke
Carbon monoxide
Asbestos insulation
Chlorine bleach
Spot remover
Exhaust fumes
Asbestos floor tile

Health Hazards of Indoor Air Pollutants

Pollutant	Sources in the home	Possible health effects
Asbestos	Floor and ceiling tile, pipe insulation, cement, plaster, wallboard	Lung diseases including cancer; cancer of the linings of the chest cavity and abdomen
Benzene	Cleaning fluid, spot removers	Nausea, fatigue
Carbon monoxide	Poorly vented furnaces, stoves, fireplaces, and attached garages	Headaches, nausea, dizziness; high levels can cause death
Formaldehyde	Foam insulation, new furnishings, plywood, and particle board	Shortness of breath, nosebleeds, asthmatic reactions, headaches, fatigue, nausea
Nitrogen dioxide	Poorly vented furnaces, stoves, and fireplaces	Respiratory disease and lung problems in children and those suffering from bronchitis, asthma, or emphysema
Radon gas	Radioactive soil, building materials (such as brick and stone), water, and natural gas	Lung cancer
Styrene	Plastics	Liver and kidney damage
Tobacco smoke	Burning cigarettes, cigars, and pipes	Allergies, shortness of breath, increased heart rate and blood pressure, emphysema, lung cancer

Source: Consumer Federation of America.

Burning fuel in poorly ventilated areas produces another dangerous indoor pollutant—carbon monoxide. When inhaled, this colorless, odorless gas prevents blood from carrying oxygen to the body's cells, and exposure to high levels can result in death. The most obvious source of carbon monoxide in the home is car exhaust fumes that leak into a house when someone runs a car in an attached garage. Carbon monoxide is also given off by burning cigarettes and by poorly ventilated coal and gas furnaces, heaters, and stoves. Gas stoves and heaters also produce nitrogen dioxide, which can cause lung problems.

Common products often found in homes—bug sprays, paint and paint thinners, moth balls, chemicals for cleaning, recently dry-cleaned clothes, even spray deodorants and hair sprays—can release *toxic* (poisonous) fumes. Such fumes can cause headaches, dizziness, nausea, blurred vision, burning or itchy eyes, allergic reactions, shortness of breath, tightness in the chest, and skin irritation.

One strong substance that few people recognize as a household chemical is *formaldehyde*, which has been used in some types of insulation and as a binder in building materials such as plywood and particle board. In large concentrations, formaldehyde's pungent

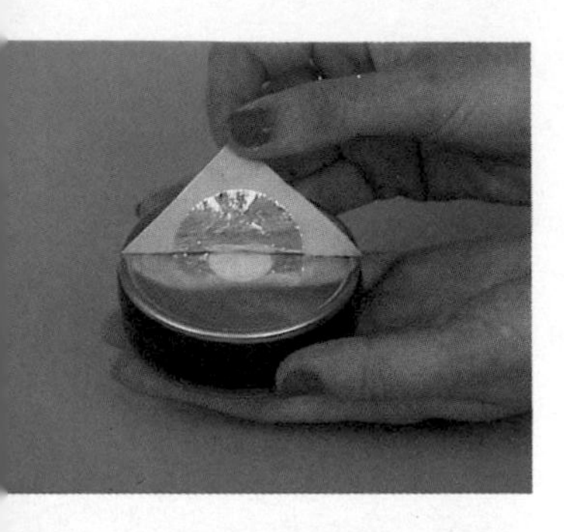

Radon detection
The Environmental Protection Agency recommends two types of radon detectors. The charcoal canister, *top,* is exposed to the air in a house for about one week. A more precise, long-term monitor, such as the alpha track device, *above,* is left in place for up to one year. After the test periods, both devices are sent to laboratories for analysis.

odor may make you run for an open window, but even at levels too low to be detected by smell, formaldehyde in the air can be dangerous. In 1979, the Chemical Industry Institute for Toxicology Laboratory announced that rats subjected to large amounts of formaldehyde fumes developed cancerous tumors in their noses. On the basis of this and other tests using animals, some health organizations have concluded that formaldehyde may also cause cancer in human beings. Although scientists do not agree on the human cancer danger, it is clear that the chemical—especially when in the form of urea formaldehyde insulation—sometimes causes nosebleeds, dizzy spells, skin rashes, and eye and lung irritation.

How does a homeowner know whether his or her house contains high levels of radon or other dangerous indoor air pollutants? There are two types of do-it-yourself radon monitors recommended by the EPA. Both cost less than $50. One, called a diffusion barrier charcoal adsorption (DBCA) collector, records radon levels over short periods—a few days to a week—and is used to quickly screen houses for potential radon problems. The DBCA collector is a canister containing about 25 grams (0.8 ounce) of charcoal. The charcoal absorbs water vapor in the air and, with it, radon. The canister is then returned to the manufacturer, who determines the radon content based on the amount of gamma rays coming from the charcoal and reports the finding to the customer.

Whether or not the results show a high radon level, homeowners may want to follow up with a long-term measurement using the second type of home monitor, called an alpha track device. Long-term measurements are more accurate than short-term ones because radon levels can vary from month to month and season to season. (In warm weather, for example, people open windows, allowing pollutants to escape, which results in lower radon levels.) Alpha track monitors are made of radiation-sensitive film and are taped to walls, floors, or ceilings. When alpha particles from radon or radon daughters strike the film, they leave "tracks." As with the DBCA, the consumer sends the device back to the manufacturer for analysis. Using microscopes, technicians count the tracks to calculate the amount of radon in the house and notify the consumer.

Testing for other pollutants, including asbestos, is not so easy. Because the procedure can be very expensive, many companies that test office buildings will not bother with private homes. As more studies of indoor air pollution are conducted, however, it may be possible to volunteer a house to be part of a local study. Regional EPA offices and local medical centers or universities may have information about future studies.

If analysis of a monitoring device reveals high indoor radon levels, a homeowner should contact the EPA for a list of contractors qualified to determine the radon source and the best way to eliminate it. In general, reducing dangerous radon levels can cost from

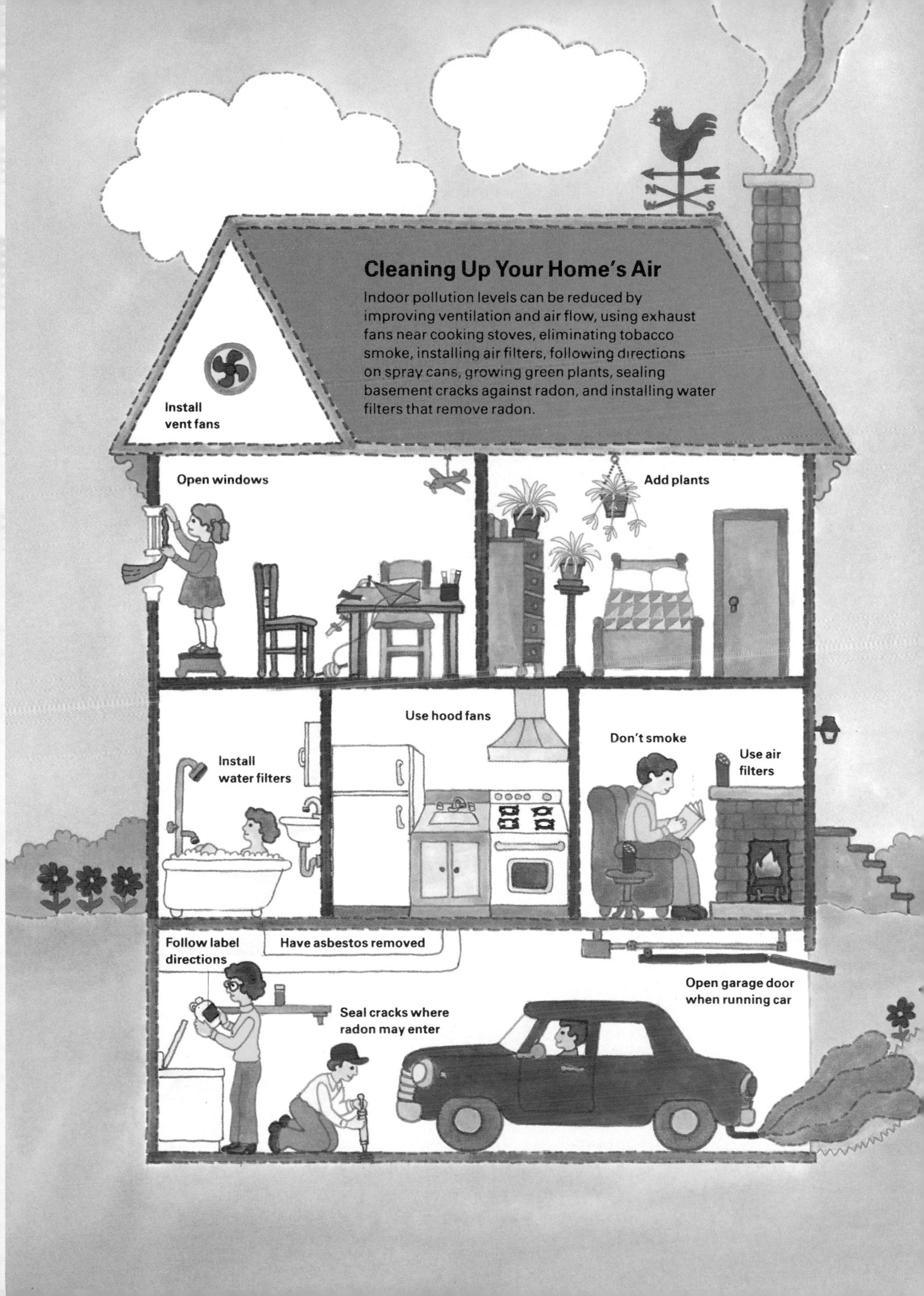
Cleaning Up Your Home's Air
Indoor pollution levels can be reduced by improving ventilation and air flow, using exhaust fans near cooking stoves, eliminating tobacco smoke, installing air filters, following directions on spray cans, growing green plants, sealing basement cracks against radon, and installing water filters that remove radon.
Install vent fans
Open windows
Add plants
Use hood fans
Install water filters
Don't smoke
Use air filters
Follow label directions
Have asbestos removed
Open garage door when running car
Seal cracks where radon may enter
N
E
W
S

$50 to $5,000 or more, depending on the severity of the problem and its source.

The EPA recommends a number of radon-reduction methods. If radon is leaking into the basement, a contractor may suggest simply filling cracks in basement walls and floors with flexible sealants or concrete. A sometimes necessary but more expensive solution is to install a special ventilation system under the basement floor to pull radon outside. The reverse type of system, which blows fresh air into the basement, is also used.

If an expert determines that the radon source is a house's building materials—radioactive brick or stone in a fireplace, for example—the substance may need to be removed. If the amount of radiation is small, the EPA advises covering the polluting material with concrete, epoxy, or other sealants that prevent radon from escaping into the air.

If the source of airborne radon is the water supply, special charcoal filters may be able to remove the gas. Although water softeners can remove radium from water, they may not remove radon.

Finally, there are some extreme cases of unusually high levels of radon outdoors—when, for example, houses have been built on top of radioactive waste from factories or from uranium or phosphate mines. In such cases, the radioactive soil may have to be dug up and stored at hazardous waste sites under the direction of government officials.

Asbestos is another pollutant whose removal should not be tackled by the typical homeowner. People who suspect there is an asbestos hazard in their home should not even touch the material. It must be handled with extreme care to avoid spreading the fibers. The EPA should be contacted for assistance.

If there is no serious radon or asbestos problem, homeowners can take some simple precautions to eliminate or minimize indoor air pollution. One of the best ways to lower indoor air pollutant levels is to increase the amount of fresh air inside a house—assuming, of course, that the air quality outdoors is better than indoors. Increasing fresh air can be as simple as opening windows. This can also help avoid the low inside air pressure that "pulls" radon inside. Homeowners improving the energy efficiency of their houses should also consider the need for fresh air and install air filtration systems.

Another simple way to fight indoor pollution is to prevent pollutants from contaminating the air in the first place. This can be accomplished by following directions on product containers. *Volatile* (quickly evaporating) chemicals should be used only in well-ventilated areas. If possible, people should work outdoors when using sprays such as waterproofers and adhesives and should paint indoors only when windows can be kept open.

Mechanical devices can also help clean up indoor air at relatively low cost. Filtering devices called room air ionizers and centralized electrostatic precipitators electrically charge airborne dust particles. The charged particles, which radon daughters sometimes stick to, then settle on walls and floors or are attracted to an electrified collecting plate. No longer airborne, the particles cannot be inhaled. Fans that push dust particles to the walls and floors accomplish the same thing. Exhaust fans and range hoods are also helpful for eliminating some of the carbon monoxide, nitrogen dioxide, and radon produced by gas stoves.

Sophisticated models of air filters are also available. Very fine air filters were first used in computer manufacturing plants, at the National Aeronautics and Space Administration (NASA), and in other places that require dust-free environments. These machines, known as high efficiency particulate absolute filters, are now available for home use to remove dust and other pollutants.

Finally, researchers have learned that gases such as carbon monoxide and formaldehyde can be cleaned up the old-fashioned way—with plants. Environmental scientist Bill C. Wolverton at the National Space Technologies Laboratory in Bay St. Louis, Miss., reported this discovery in 1984 after conducting research for NASA designed to find plants that could clean the air inside spacecraft. Plants that seem to be especially good at removing formaldehyde, carbon monoxide, and nitrogen dioxide, Wolverton found, include the Chinese evergreen, the golden pothos, the peace lily, and the spider plant.

According to Wolverton, one spider plant filtered out 85 per cent of the formaldehyde in a sealed container the size of a dishwasher in a single day. This indicates that about 15 plants would be needed to clean the air in an average-sized house. The plants must be kept clean and free from bugs and disease, because mildew, fungi, and plant pests can themselves be sources of indoor air pollution.

Governmental agencies continue to study the problem of radon and other indoor pollutants, searching for innovative remedies and formulating clean air standards for the home. Meanwhile, families should do whatever they can to minimize the invisible hazards of an environment that once seemed so safe.

For further information:

The following pamphlets are available free or at a small charge.

Air Pollution in Your Home and *Home Indoor Air Quality Checklist.* Contact your local lung association or American Lung Association, Box 596SY, New York, NY 10001.

A Citizen's Guide to Radon and *Radon Reduction Methods.* Contact your state health department or Radon Division (ANR-460), Environmental Protection Agency, 401 M Street SW, Washington, DC 20460.

Formaldehyde: Everything You Wanted to Know But Were Afraid to Ask. Contact the Consumer Federation of America, 1424 16th Street NW, Washington, DC 20036. Enclose 25¢ and a self-addressed, stamped envelope.

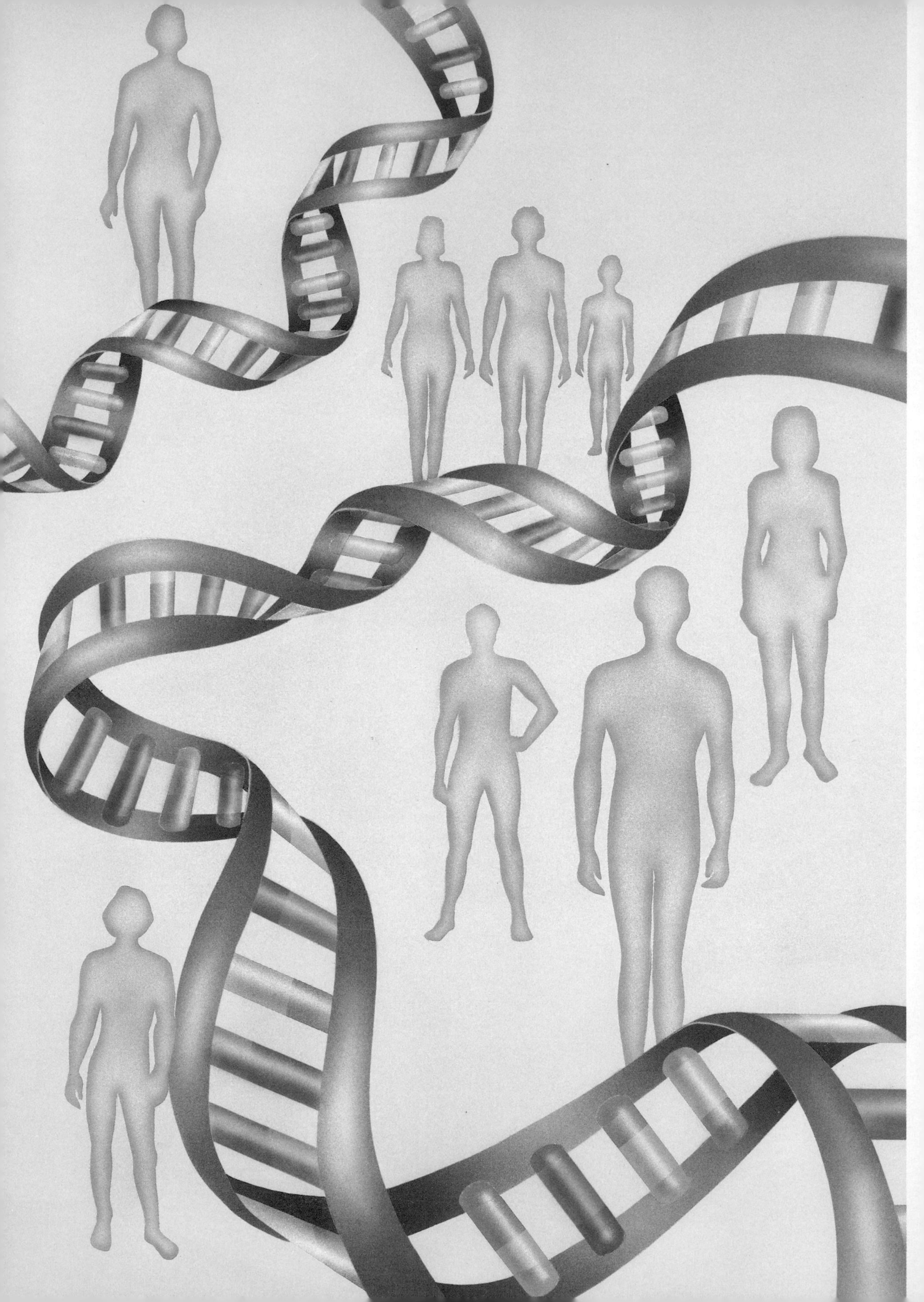

Scientists are exploring the mystery of how a normal cell turns cancerous and the role genes may play in this deadly process.

Cancer: A Genetic Time Bomb?

BY JOANN ELLISON RODGERS

For nearly a century, medical experts have been intrigued by "cancer families," families whose members are prone to the deadly disease.

- In New York City, a young woman, her mother, her grandmother, and two aunts—all on the same side of the family—have breast cancer.
- In Alabama, five different kinds of cancer have threatened the lives of six brothers and sisters.
- Each year in the United States alone, hundreds of couples face a 50-50 chance of bearing a child who is likely to develop a deadly form of inherited eye cancer.

The word *cancer* is used to describe more than 100 diseases, but all forms of cancer have one thing in common—wild, uncontrolled growth of cells. Somehow, normal cells are transformed into cancerous ones that eventually overwhelm the body.

Cancer families have long fascinated researchers because they seem to support a promising theory—that cancer is the "enemy within," an enemy that is somehow carried inside each body cell. The very existence of cancer families suggested that cancer may erupt when *genes*, which control the cell and carry hereditary material from generation to generation, are somehow damaged or disrupted. Since the 1970's, scientists have gathered substantial evidence that certain genes in each cell have the potential to play a role in the development of cancer. These genes became known as *oncogenes*—a name taken from the Greek word *onkos*, meaning *tumor*. These cancer genes—which have odd-sounding names such as *myc, sis, ras, yes,* and *ros*—are rare among the approximately 50,000 other genes present in each human cell. But by the mid-1980's, scientists had linked them to at least 14 human cancers, including some of the most common, such as lung and breast cancer.

Scientists who study cancer genes suspect they lead a Jekyll-and-Hyde existence. It appears that—like Dr. Jekyll, the unfortunate physician who turned into the evil Mr. Hyde in Robert Louis Stevenson's novel—a cancer gene normally is not dangerous but plays an important role by instructing the cell to make useful substances. When these precancerous or potentially cancer-causing genes are performing their normal roles in the body, they are called *proto-oncogenes*. Scientists believe the proto-oncogenes are transformed into true cancer genes when something—perhaps a virus, a chemical, or radiation—alters the genetic material of the cell.

Scientists now believe that each of us possesses this handful of vulnerable genes—described by one scientist as "little genetic time bombs." The fact that these genes have the potential to sabotage normal cell growth sounds like bad news. But there is a good side to this story. The presence of cancer genes in cells strongly suggests that cancer is not some mysterious, random process or a matter of bad luck that leaves us helpless. Their presence means that cancer develops according to biological rules that we can identify and manipulate. Their existence in everyone also means that most cancers are *not* inherited in the traditional sense, but are the result of a lifetime of harmful changes to the genetic material after birth.

Many scientists believe that cancer genes offer the key to learning how to diagnose, prevent, and even cure the disease. Cancer gene research is teaching us how normal cells guide their own growth and why cancer cells cannot.

The author:
Joann Ellison Rodgers is a free-lance writer specializing in science and medicine.

How genes work

Understanding how certain genes play a role in the development of cancer requires some basic information about their normal role in the cell. Genes are composed of a compound called *deoxyribonucleic acid* (DNA), a long molecule that resembles a twisted ladder. DNA contains the blueprint for all of an organism's characteristics, from the color of our eyes to the length of our bones. All instructions are contained in this biological library, and copies of it are passed on to each new cell that reproduces.

In each of our body's 100 trillion cells, the DNA also provides the information necessary to sustain life. Every cell gets its orders—heart cells pump, nerve cells relay electrical signals, and red blood cells carry life-sustaining oxygen.

The genes reside, like a vast array of beads on a string, on *chromosomes*, tiny threadlike structures found in the nucleus of the cell. Most human cells have 46 chromosomes—23 pairs.

The information carried by the genes is written in a code that directs the body's production of thousands of *proteins*—substances such as enzymes, hormones, and material that make up the body's tissues. Each "rung" of the DNA ladder consists of a pair of simple molecules called *bases*. There are four different kinds of bases, and the order of these bases in a particular gene carries the code for that gene's corresponding protein, just as an ordered sequence of dots and dashes in Morse code conveys a specific message. Each of the thousands of proteins has a particular function in the body.

When the body needs to replace old or injured cells, the DNA blueprint orders cells to grow and divide. With the aid of various enzymes and other cell chemicals, the chromosomes *replicate*—duplicate themselves so that each new cell has a copy of the genetic code. The genes then direct the cell to make the necessary proteins and divide itself into two *daughter cells*. Normally, when enough new replacement cells are made, the process of growth and division stops.

Glossary

Cancer: A group of diseases in which cells multiply without control and invade and destroy healthy tissues in the body.

Chromosomes: Structures in the cell nucleus that contain the genes.

DNA: Deoxyribonucleic acid, the substance of which genes are made.

Gene: A piece of DNA that contains coded instructions for making a protein.

Metastasis: The spread of cancer from the original tumor to one or more body sites.

Mutation: A change in a gene.

Oncogene: A cancer gene—a gene with the potential for causing a cell to become cancerous.

Proteins: Complex molecules that are found in all living things, and which perform many growth and repair functions.

Tumor: A mass of cells arising from a single cell.

How cancer develops

Occasionally the cell's genetic material is *mutated*, or changed. This could be caused by a virus, a chemical, radiation, or some accidental rearrangement of the chromosomes. Cells usually detect and repair these changes. But sometimes they fail to do so, and the cell divides, passing the mistake along to the daughter cells. Scientists believe that if the error disables the cell's ability to stop reproducing, the cell will divide again and again, forming a tumor.

Some tumors are noncancerous, or *benign*. These growths do not invade surrounding tissue and are usually not dangerous unless they press on vital organs or block passages, such as the bowel. Cancerous tumors are *malignant*. Cells from these tumors have the ability to break off, invade surrounding tissue, and even get into the bloodstream or *lymph* (fluid in body tissues), which carries them to other sites in the body. This process is called *metastasis*. Some tumors, such as colon cancer, grow slowly. Others, such as a common form of lung cancer, grow and metastasize quickly. The type of tissue in which the cancer arises, along with many other factors, determines how quickly a tumor grows and spreads.

Sometimes the body's immune system detects the deadly cells and attacks them. But if a cancer cell eludes the body's defenses, it can take root in new areas and create new tumors. When important organs such as the lungs or liver are invaded and damaged, the body is overwhelmed. What began as a disorder of one cell becomes a full-fledged killer cancer.

Although the belief that genes play a role in cancer began with the observation that some families seem particularly prone to the disease, researchers uncovered other clues that pointed them in the same direction. Beginning in the early 1900's, some researchers noted that cancer cells commonly contain damaged chromosomes. Later, scientists suggested that cells that have trouble repairing DNA—a routine function of a normal cell—tend to become cancerous. For example, patients with a rare skin disease called *xeroderma pigmentosum* tend to develop skin cancer. Skin cells of these patients cannot repair damage to DNA caused by ultraviolet light, a component of sunlight.

Cancer genes and animal viruses

Although such circumstantial evidence had pointed to a link between genes and cancer, no one had demonstrated that a specific gene could transform normal cells into cancerous cells. The first successful work of this kind came not from human cancer studies but from studies of animal *tumor viruses* that rapidly produce cancer in newborn rats, chickens, cats, and mice. These viruses have an unusual—and dangerous—ability. They can insert their genes into the genetic material of the cells they infect and alter part of the cell's genetic blueprint.

By the early 1970's, some crucial tools for genetic research had been developed—such as enzymes that make it possible to isolate specific genes. Armed with these new techniques, scientists were able to study animal tumor viruses in a more detailed fashion.

Researchers found that in each of the animal tumor viruses they studied, a single gene could transform healthy cells into malignant ones. It became clear that many animals, such as rats, mice, birds, cats, and monkeys, develop cancer after being infected by viruses that carry deadly cancer genes.

It also became clear that cancer genes are not really viral genes at all—even though they were carried by viruses. In the early 1970's, scientists discovered that healthy, uninfected cells from a variety of animals—from fruit flies to human beings—contain genes that are nearly identical to the cancer genes found in tumor viruses. Researchers believe that, perhaps 600 million years ago, certain viruses infected animal cells and accidentally incorporated copies of the animal's genes into their own genetic material. These genes became "hitchhikers" that traveled with the virus. Unfortunately, when a hitchhiking gene was a cancer gene, the virus became a tumor virus, able to transform cells it infected into cancer cells.

Because there is very little evidence to link particular viruses to particular human cancers, many scientists believe the role of viruses in human cancers is limited. So researchers searched for a link be-

What Cancer Is

Cancer is the uncontrolled growth of cells. It begins with a single renegade cell that does not know when to stop dividing. The new cells that it produces form tumors, which can cause death by destroying vital organs and sapping the body's resources.

A normal cell grows and divides to replace cells that have died or to provide additional cells, such as those needed for building up muscle tissue or fighting infection. When enough new cells are made, the process of growth and division stops.

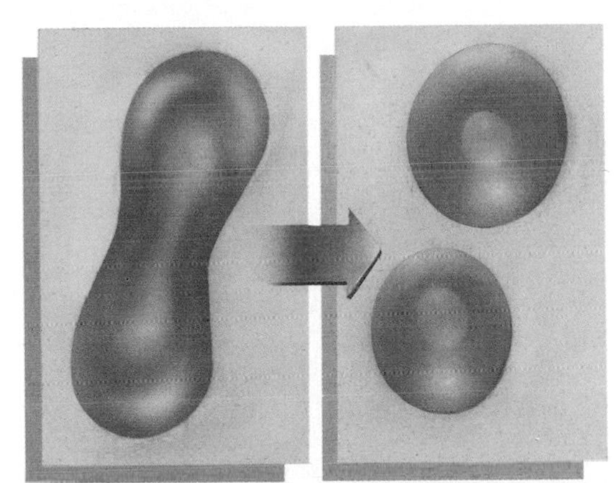

A cancerous cell does not have the normal controls on cell growth and development. It divides endlessly, forming a tumor.

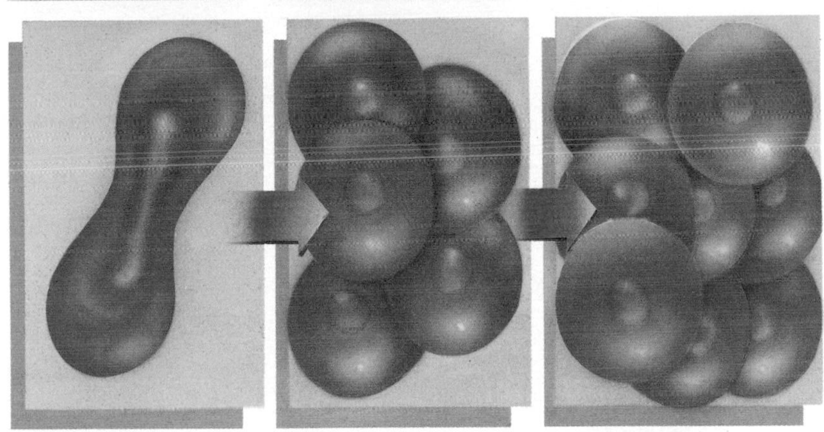

A benign tumor is made of cells that do not invade surrounding tissue.

A malignant tumor is made of cells that are able to invade surrounding tissue, enter the bloodstream, and spread the disease to other parts of the body.

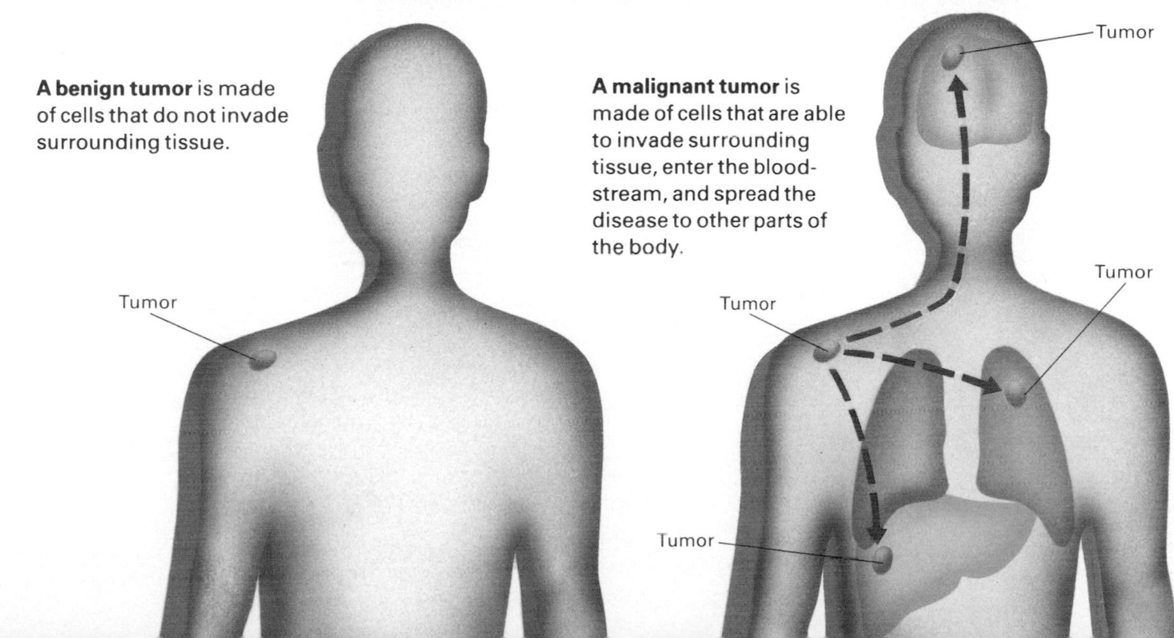

What Controls Cells
The instructions for every cell's functions and characteristics are contained in its genes. Genes are located on structures called chromosomes in the cell nucleus. Genes are made of deoxyribonucleic acid (DNA)—a molecule that resembles a twisted ladder.

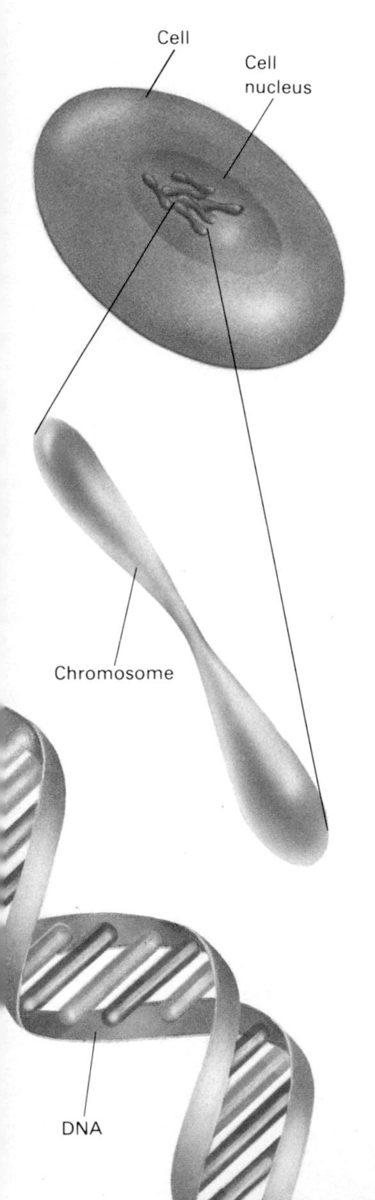

tween cancer and genes that did not involve viruses—and they found one. In 1982, scientists studying malignant and normal cells *grown in culture* (grown in a special solution in laboratory dishes) discovered that a single cancer gene, plucked out of a malignant cell culture and put into a healthy cell in culture, could transform that normal cell into a cancerous one. Something in the DNA alone—with no assistance from viruses, radiation, or cancer-causing chemicals—could make a healthy cell malignant.

Scientists have since found that cells from many human cancers—including cancer of the lung, colon, blood, and breast—contain genes that can be inserted into laboratory cultures of healthy cells and cause them to become cancerous. Moreover, most of these cancer-inducing genes turned out to be similar or identical to previously identified cancer-inducing genes from animal viruses.

When "good" genes go "bad"

Where do such human cancer genes come from if viruses are not responsible for introducing them into human cells? The most likely possibility is that there are normal versions of the cancer genes—the proto-oncogenes—present and playing normal roles in healthy cells. After all, experts reasoned, it is unlikely that these genes would have been retained throughout evolutionary history if their sole purpose in the cell was to cause cancer. And indeed, scientists found that healthy cells contain versions of cancer genes that are not identical to the cancer-causing genes but are very similar in structure. Significantly, studies showed that a human bladder cancer gene could make mouse cells in culture grow like cancerous cells—but the normal version of the gene could not.

Since the early 1980's, scientists have found important links between normal genes and cancer genes and are learning how both fit into the cancer process. All cancer genes that scientists have studied closely have something to do with how cells grow and *differentiate*. Differentiation is the process by which cells take on specialized forms and functions in the body—becoming blood cells or brain cells or liver cells, for example. Cancer genes usually remain in their normal precancerous state, making the correct protein when it is needed and remaining "turned off" when the protein is not needed. Precancerous genes must be activated in some abnormal way before they become full-fledged cancer genes.

Scientists suspected that changes in the DNA of a cell must be at least partly responsible for activating a cancer gene. To date, scientists have proposed three different processes that alter DNA in ways that may transform a normal gene into a cancer gene.

The first of these processes involves a mutation that changes just a tiny portion of DNA—a single base pair of a single gene. In 1982, molecular biologist Robert A. Weinberg and his colleagues at Massachusetts Institute of Technology (M.I.T.) in Cambridge discovered that such a mutation—called a *point mutation*—may have activated a cancer gene. They compared a cancer gene in bladder cancer cells to that gene's normal counterpart in healthy tissue and found that the two genes differed by just one base pair out of thousands. This would be the equivalent of a single letter out of hundreds of words in a book. Such mutations can occur when DNA is exposed to radiation or to certain chemicals—such as those found in cigarette smoke. And this tiny change in the DNA made the difference between a normal gene and its cancer-causing version.

Unlucky combinations

But scientists believe that it usually takes more than a single cancer gene to turn a normal human cell into a cancer cell. Studies such as Weinberg's generally involve inserting a single cancer gene into mouse cells grown in laboratory culture dishes. Many researchers think that the single cancer gene is able to turn these mouse cells cancerous because they are already in a precancerous state. They suspect that a normal human cell reaches this state only after other genetic and biochemical changes have taken place.

The second way in which cancer genes may be activated is by genes moving from one chromosome to another. For example, in 1982, scientists discovered evidence that such a genetic accident, called a *chromosomal translocation*, can activate a cancer gene. During a translocation, two chromosomes swap pieces of DNA. As a result, some genes are relocated to a different chromosome. This can affect their behavior in the cell. The researchers found that one particular translocation is found in the cells of patients afflicted with *Burkitt's lymphoma*, a cancer of the lymph system.

What activates a cancer gene when it is relocated? In the case of Burkitt's lymphoma, the gene swap involved a precancerous gene that scientists call *myc*. Scientists believe that the relocation may have moved a *myc* gene away from portions of the DNA that act as "on" or "off" switches telling the gene when to make its protein product. And, in fact, the scientists found that the translocated *myc* gene and the *myc* gene on its normal chromosome make the same protein. But the cancer gene makes the protein in huge amounts, as if the "off" switch for the *myc* gene were absent.

Human Genes and Cancer

All human beings have genes that can play a role in the development of cancer. Although these genes have normal roles in cell growth and development, they can be activated in some way to become cancer genes. Radiation, cancer-causing chemicals, or genetic changes—such as a rearrangement of genes on chromosomes—can cause one of these genes to become an activated cancer gene. More than one activated cancer gene may be required to produce cancer in a human being.

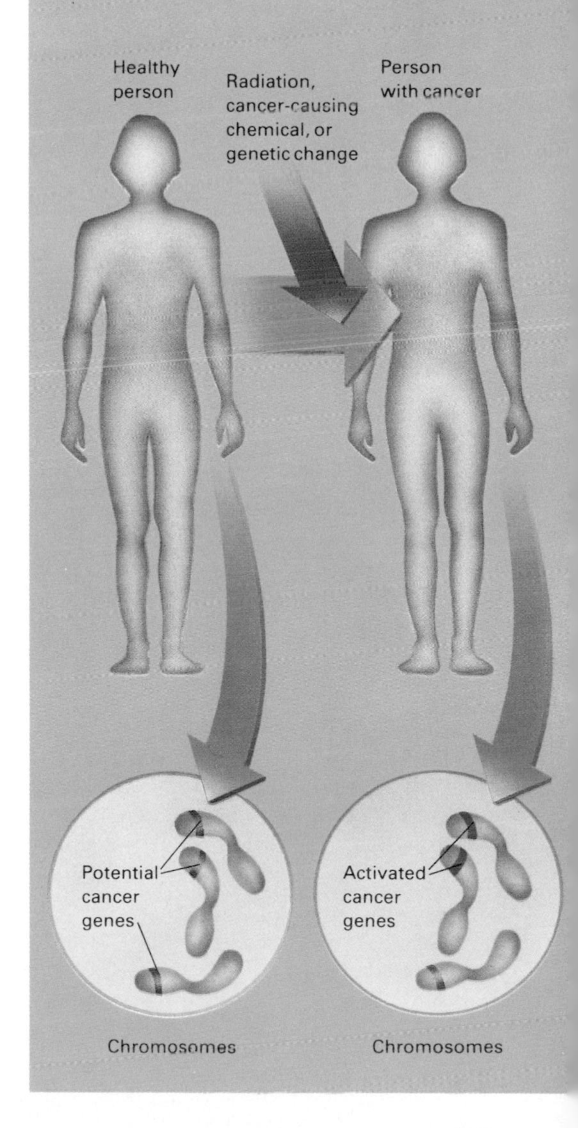

When "Good" Genes "Go Bad"

Scientists have identified three types of changes in genes that can transform a normal gene into a cancer gene. Both normal genes and cancer genes *code for* (contain instructions for making) proteins involved in cell growth and division. Normal genes code for normal proteins. A cancer gene, on the other hand, causes the cell to make too much of a protein, or a protein that does not work correctly.

A Point Mutation—a tiny change in a gene—can cause a cell to make an abnormal protein. For example, the normal version of one cancer gene instructs the cell to make a protein that plays a role in controlling cell growth. But when a point mutation transforms this gene into an activated cancer gene, the cell makes an abnormal version of the protein, which improperly regulates cell growth.

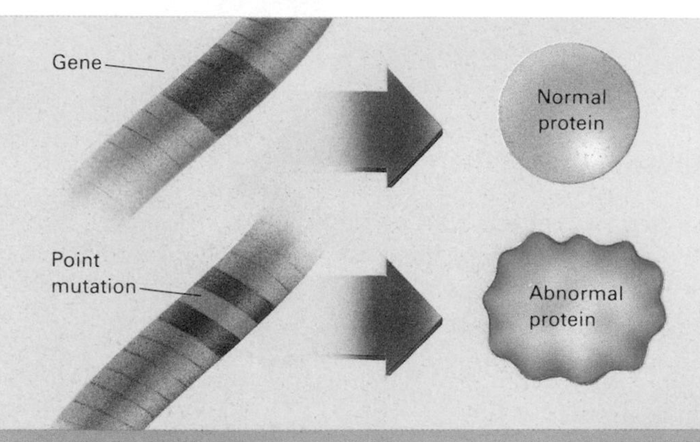

A Chromosomal Rearrangement—an accidental swapping of genes between chromosomes—can alter a gene's behavior by placing the gene near a piece of DNA that acts as an "on" switch. As a result, the gene is constantly turned on, instructing the cell to make excessive amounts of a protein. For example, one form of a human blood cancer appears to be triggered by moving one potentially cancer-causing gene to another chromosome. As a result, the gene instructs the blood cell to make excessive amounts of a protein involved in cell division, causing the cell to divide uncontrollably.

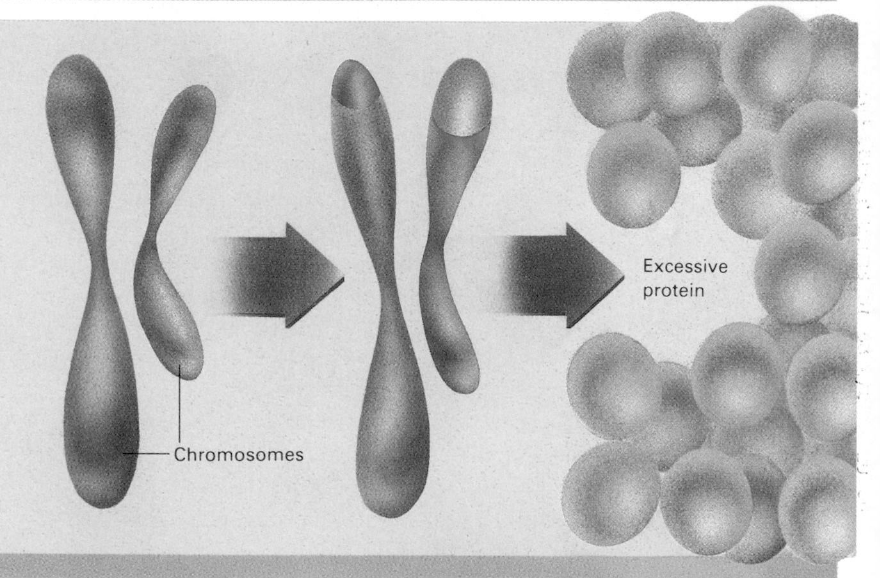

Gene Amplification—a process that causes extra copies of a gene to be made—can turn a cell cancerous. For example, cells taken from several kinds of human cancers contain extra copies of a gene that codes for a protein involved in cell division. The extra copies of this gene may cause the cell to make too much of the protein and cause the cell to constantly divide.

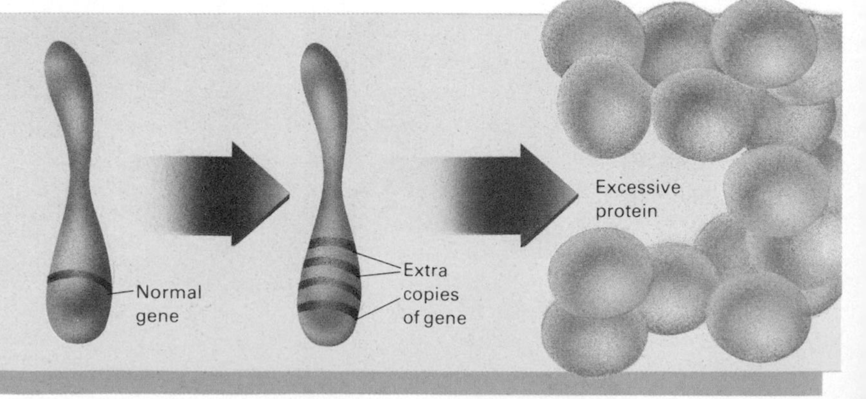

Scientists have since linked other human chromosomal translocations with two blood cell cancers.

Scientists know that chromosomes break, permitting genes to be rearranged or lost. But they do not yet know why this happens. Some researchers believe that chromosomes may have fragile areas that could make some individuals more prone to genetic changes that trigger cancer. If those fragile areas are inherited from generation to generation, this could help explain why certain cancers seem to run in some families.

A third process that may activate cancer genes is called *amplification*. Normally, only two copies of a particular gene are present in a cell—one on each chromosome of a pair. But some cells have a mysterious ability to *amplify* (make many copies of) particular genes. Researchers have found multiple copies of cancer genes in a variety of tumor cells, including colon, lung, and breast cancers. If all of these extra copies of the cancer gene are turned "on," they will instruct the cell to make excessive amounts of the protein for which they code. This extra protein may encourage unrestrained cell growth.

The role of protein

These findings led to the next step in unraveling the mysteries of the cancer process. Because genes exert their effect through the proteins they direct the cell to make, researchers are homing in on the identity and function of the proteins for which cancer genes code. And all such proteins examined so far resemble proteins that appear to control cell growth and development.

Some of these cancer gene proteins closely resemble substances called *growth factors*, protein molecules that enhance cell growth. In 1983, scientists found the first link between growth factors and cancer genes. They discovered that a protein produced by a cancer gene called *sis*—which is found in a virus that causes cancer in monkeys—closely resembles part of a growth factor found in blood *platelets*, disklike structures in the blood that play a role in blood clotting. This substance, called platelet-derived growth factor, triggers tissue repair in humans by stimulating cells to divide.

Molecular biologist Stuart A. Aaronson of the National Cancer Institute (NCI) in Bethesda, Md., then found a similar precancerous gene in human cells. Furthermore, he was able to insert the gene into cultures of normal cells and make them cancerous. Thus, it seems that protein produced by this particular cancer gene is a growth factor whose normal function can be perverted, so that it pushes the cell into uncontrolled growth.

Other growth factors are important during the development of a fetus. In 1984, researchers at the Imperial Research Fund in London examined a cancer gene called *erbB*, which is found in a tumor virus that causes a blood cancer in chickens. They discovered that this gene codes for a protein that closely resembles a human protein believed to play a role in cell division and differentiation during the

Cancer Caused by an Absent Gene

Not all cancers are caused by the presence of a bad gene. At least one cancer is caused by the absence of a protective gene. *Retinoblastoma*—a rare eye cancer of children—occurs when cells lack protective "anticancer" genes.

Most people are born with two chromosomes containing this anticancer gene, so the eye stays healthy.

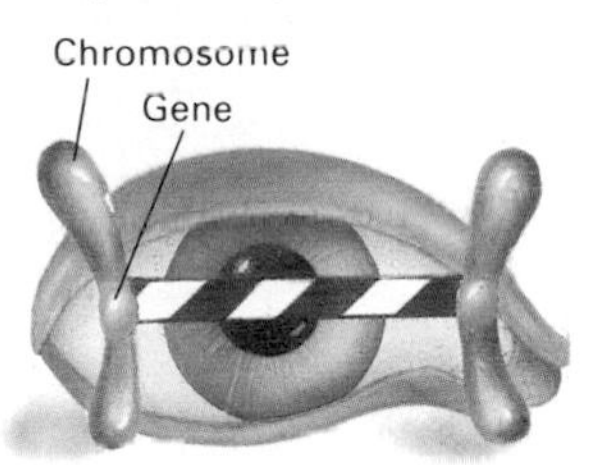

Some children inherit one defective chromosome that has a missing or disabled anticancer gene. If the gene on the other chromosome remains intact, the eye stays healthy.

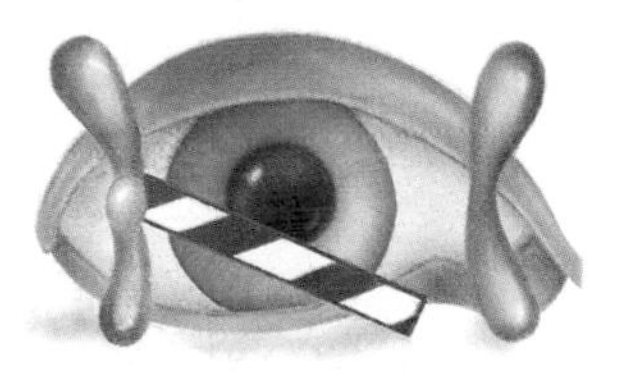

If the anticancer genes on both chromosomes are lost or disabled, cancer can arise in the eye.

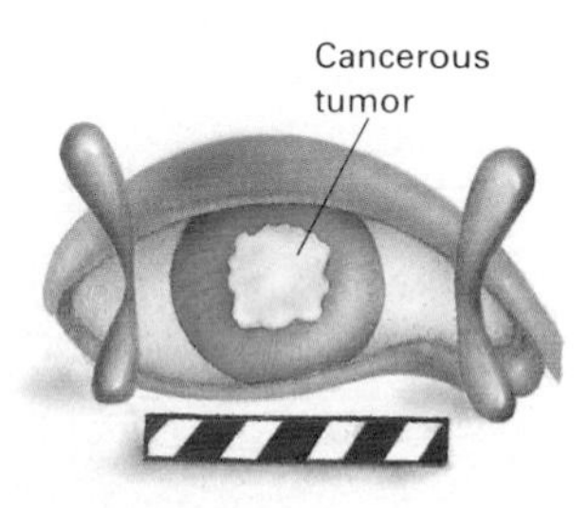

early development of a fetus. Biochemist Arthur B. Pardee of the Dana Farber Cancer Institute in Boston thinks that some cancer genes normally act as "switches," shutting off the production of growth factors after fetal development is complete. He thinks that somehow, in cancer cells, the genes tell the embryonic growth factors to turn on again.

Since the early 1980's, knowledge of cancer genes, the proteins they make, and their links to animal viruses and human cancer has rapidly advanced. Scientists have discovered about 40 animal and human cancer genes already, and many believe that at least 100 cancer genes will be found by 1990.

Cancer: a multistep process

One question that continues to intrigue scientists concerns the length of time it takes cancer to develop in the body—which could be years. Scientists believe one reason for this is that cancer is a multistep process, and it is not unusual for years to elapse between the individual steps. Also, cancer gene researchers are examining why it seems that a single activated cancer gene is not enough to cause cancer in human beings. For example, in the 1970's, geneticist Alfred Knudson, then of the M. D. Anderson Hospital and Tumor Institute in Houston, theorized that two genetic changes, or "hits," are needed for the development of three childhood cancers—*neuroblastoma*, a cancer of the nervous system; *retinoblastoma*, an eye cancer; and *Wilm's tumor*, a kidney cancer. The first hit may involve a mutated or missing gene or a chromosomal rearrangement, inherited from a parent. The second "hit"—perhaps an alteration of the DNA by chemicals or radiation—occurs after birth.

Although some people may inherit a higher risk of getting some cancers, most cancers appear to be the result of an accumulation of genetic hits, such as mutations caused by exposure to radiation or a chemical. Certain combinations of altered genes may then result in a cancer cell.

Scientists are exploring yet another aspect of the cancer puzzle. They want to know if the body has "anticancer" genes—genes whose normal function is to protect a cell from cancer. And in October 1986, researchers at three laboratories in the Boston area discovered a gene that seems to have such a protective function against retinoblastoma. Most people have two copies of this protective gene—one on each chromosome pair. But some children are born with only one copy of the protective gene and already have a strike against them. These children are at high risk for developing retinoblastoma. If their only copy of the protective gene is somehow lost, the cancer can develop. This discovery suggested to scientists that some genes have, as their normal role in the cell, anticancer duties. If so, some cancers, like retinoblastoma, may be caused when genes that serve as "off" switches—keeping cells from dividing endlessly—are lost or disabled.

Using Genes to Diagnose and Treat Cancer

Experts predict that special molecules called *probes,* which seek out cancer genes and the proteins for which they code, could be used in laboratory tests to diagnose cancer. Other molecules could be used to deliver cancer-killing drugs.

Biochemical probes can be designed to fit onto the abnormal proteins produced by some cancer genes, like a key fits in a lock. Such probes, linked with a chemical or radioactive tag, would be mixed with a blood or tissue sample taken from the patient. Laboratory tests to determine whether the probe binds to a protein would help diagnose cancer.

Cancer gene

Directs the production of

Abnormal protein

Abnormal protein

Abnormal protein

Biochemical probe

Biochemical probe

Biochemical probe

Gene probes, pieces of DNA designed to attach only to activated cancer genes, could help diagnose cancer in its earliest stages. A probe tagged with a special chemical or a radioactive molecule would be mixed with a sample of DNA from the patient. If tests indicated that the probe was binding to a place on the DNA, this would show that a cancer gene was activated.

Gene probes

Cancer gene

DNA

Powerful drugs attached to probes that seek out abnormal proteins on the cancer cell surface could be used to kill only cancerous cells.

Probe with drug attached

Probe with drug attached

Probe with drug attached

Abnormal protein on cell surface

Retinoblastoma is one of the few cancers that has a clear-cut inherited component. Scientists still do not know why some cancers, such as breast and colon cancer, tend to run in families—or how cancer genes might fit into the picture.

Cancer genes for diagnosis

In addition to broadening our understanding of how cancer develops, many scientists believe cancer gene research will yield payoffs in the diagnosis, treatment, and prevention of the disease. For example, experts predict that *molecular probes*—molecules specifically tailored to detect cancer genes or their protein products in blood and tissue samples—will help doctors diagnose cancer in its earliest stages and select treatment.

Today, some medical researchers employ such probes to count the number of copies of cancer genes in tumor cells and use this information as a way to assess the patient's future health. Scientists at the University of Texas Health Sciences Center in San Antonio and at Genentech, Incorporated, in South San Francisco, Calif., have found that breast cancer patients with multiple copies of a cancer gene called *neu* in their tumor cells are far more likely to suffer a recurrence of the disease than patients with the normal number of *neu* genes. The researchers believe that such information could help doctors determine which patients need more aggressive treatment for the disease and more frequent checkups for recurrences. Scientists are trying a similar approach to design better treatments for lung cancer and for neuroblastoma.

Molecular probes also are being used to detect the presence of chromosomal rearrangements that are linked to certain cancers. One biotechnology company is testing a probe that detects a specific chromosomal rearrangement linked to a form of blood cancer. Some cancer experts predict that a laboratory test using this probe will make it easier to diagnose this cancer.

Scientists are also using a gene-detecting probe to disclose the presence or absence of the anticancer gene that protects children against retinoblastoma. Before the anticancer gene was discovered in 1986, prospective parents with a family history of the disease knew that their children would have a 50-50 chance of inheriting the tendency to develop the cancer.

Researchers are now able to test family members—including unborn infants—to see if they lack the protective gene. For example, geneticist Webster K. Cavanee of the Ludwig Institute for Cancer Research in Montreal, Canada, was able to ease the fears of an expectant couple whose first son had died of retinoblastoma, and assure them that their unborn child would not be vulnerable to the cancer. Today, they are the grateful parents of a healthy boy.

Experts also predict that biochemical probes designed to detect the proteins produced by cancer genes will become important in diagnosing and treating cancer. A probe designed to detect differences between a cancer gene protein and the normal version of the protein could be used to diagnose whether a cancer gene has been activated. The probe will recognize and attach to a specific cancer gene protein, like a key fitting in a lock.

These probes can be tagged with a radioactive molecule or *fluorescent* (glowing) dye, to make it possible to detect the cancer proteins to which they attach. The probe is mixed with a blood or tissue sample, which is fixed to a microscope slide, a piece of filter paper, or some other surface. If the sample does not contain the cancer protein, the probe can be washed off the sample.

Probes designed to find cancer cells can also be tagged with radiation and injected directly into a patient. Research trials at the NCI show that these probes can be used to find cancers that do not show up on X rays. A special device called a gamma-ray camera is used to detect the radiation and thus produce an image that reveals the location of cancer cells. Experts predict that in the near future, all these tools will allow earlier diagnosis for leukemia, as well as for breast, lung, and colon cancers.

Better treatments for cancer

Cancer gene research is also beginning to yield new treatments. For example, biochemical probes could be joined with a radioactive molecule or a toxic anticancer drug. Like deadly guided missiles, such probes would search out cancerous cells and deliver the drug or radiation to destroy them while leaving healthy tissue unharmed.

Other scientists are looking for ways to block the action of cancer genes. Virologist George Todaro, scientific director of Oncogen in Seattle, has isolated proteins that stop the growth of cells after cancer genes are activated. He believes that a whole family of chemicals may exist that do the opposite of what cancer genes do—shut down the relentless cell division.

Researchers also hope that cancer gene studies will suggest ways to actually prevent cancers from developing in the first place. One preventive strategy might involve supplying a protective protein, such as providing children who lack the protective gene for retinoblastoma with the protein that gene codes for—once the protein is identified. Another strategy could involve using cancer genes and the proteins they make to prepare anticancer vaccines.

Experts estimate that more than half of all cancer patients today could be saved by early diagnosis and prompt treatment with surgery, radiation, or drugs. But only a more fundamental knowledge of how cancer develops is likely to produce ways to cure all individuals or prevent the disease altogether. Cancer genes appear to be a major route to that knowledge.

"Cancer used to be a very mysterious process," says M.I.T.'s Weinberg. "But now it's different. It's a whole new world, in which we've begun to precisely define the molecular changes by which cancer arises. It's very exciting."

Today's high-technology planes are a far cry from the Wright brothers' home-built craft. The planes of the future will be even more incredible.

Airplane Design Takes Off Toward Tomorrow

BY JERRY GREY

Even the most veteran skywatchers stared wide-eyed at the strange shape that flashed by overhead at Edwards Air Force Base in California's Mojave Desert—a jet airplane whose wings slanted forward instead of backward. The design of the experimental aircraft, the X-29, was so unconventional that many engineers considered it too risky to fly. Indeed, prior to the development of advanced new materials and controls, it could *not* have flown.

The X-29's maiden flight took place on Dec. 14, 1984. On Dec. 14, 1986, two years later to the day, another extraordinary airplane rose from the runway at Edwards—*Voyager*. When it returned nine days later, the oddly shaped craft had accomplished one of the few remaining "firsts" of airplane flight—a nonstop trip around the world without refueling. *Voyager*, too, was made with the latest materials and could not have gotten off the ground without them.

Like all experimental aircraft, the X-29 and *Voyager* can teach us a great deal about how to build planes that fly faster, farther, and more safely, maneuver more surely and quickly, and cost less to operate and maintain. The X-29 is just one example of what the future will bring. Built by the Grumman Aircraft Corporation, it is the forerunner of the next generation of jet fighters, scheduled to join the United States military arsenal in the 1990's. *Voyager*, on the other hand, may have no direct practical application, but it illustrates the innovative thinking going into today's aircraft design.

An exciting era is underway in the aircraft industry. It is an era that has already produced some remarkable airplanes—not only ex-

Opposite page: The experimental X-29 has unusual rear-mounted, forward-swept wings.

Glossary

Drag: Resistance to a plane's forward motion, caused by such factors as air friction and turbulent airflow.

Hypersonic: About five or more times the speed of sound.

Laminar airflow: Airflow over a plane's wing that is smooth and does not break up into turbulence.

Lift: The upward force that supports a plane's weight in the air. Lift is produced when air flows over the curved upper surface of a wing, resulting in lower air pressure above the wing than under it.

Mach number: A way of describing the speed of a plane flying near or above the speed of sound. The Mach number is obtained by dividing the plane's speed by the speed of sound at the plane's altitude. For example, a plane flying 2,650 kilometers per hour (kph)—1,650 miles per hour (mph)—at an altitude of 12,000 meters (40,000 feet) is moving at a speed of Mach 2.5 because the speed of sound at that altitude is about 1,060 kph (660 mph) (2,650 ÷ 1,060 = 2.5).

Speed of sound: The speed at which sound moves through air. The speed of sound is about 1,190 kph (740 mph) at sea level at 0°C (32°F.), decreasing at higher altitudes as the air gets colder.

Supersonic: Faster than the speed of sound. This term is used for speeds up to about Mach 5.

The author:
Jerry Grey, a former professor of aeronautical engineering at Princeton University in New Jersey, is publisher of the monthly magazine *Aerospace America.*

perimental craft but also several new commercial airliners and business planes. And taking shape on the drawing board—or already in development—is a dazzling variety of superhigh-technology aircraft that will appear in coming years: jet fighters that are faster, deadlier, and electronically "smarter" than those of today; bombers and other warplanes that are nearly invisible to radar; passenger airliners that will streak through the upper atmosphere at many times the speed of sound; airplane-helicopter hybrids; and other kinds of planes that can take off and land vertically.

Advances in aviation technology are making such aircraft possible. Aircraft design today involves a complex mix of many diverse disciplines and tools. The engineer's slide rule of yesteryear has given way to computers, and a variety of modern test facilities enable engineers to try out their designs. Those facilities include specially equipped tunnels, or chambers, in which models of new designs are exposed to the harsh conditions of flight.

Each major part of a new aircraft, such as engines and structural materials, must be designed to exacting standards and rigorously tested. New metals and alloys are developed in advanced chemical and metallurgical laboratories and tested by sophisticated inspection techniques that reveal hidden flaws. *Composites*—nonmetallic materials that are built up in thin layers—call for entirely different design, production, testing, and inspection facilities. And wings and other structures are designed with the help of computers.

Aviation has made incredible strides since Dec. 17, 1903, when Orville and Wilbur Wright's crude home-built airplane first rose into the air near Kitty Hawk, N.C., and flew 37 meters (120 feet). In proving once and for all that powered flight was possible, the Wright brothers confirmed the theories of European aviation pioneers of the 1800's. Those earlier experimenters had constructed gliders and explained the principle of *lift*, the aerodynamic force that makes flight possible. Lift is created when a wing—whether of a bird or an airplane—cuts through the air. The air that passes over the curved upper surface of the wing must travel farther and faster than the air that passes under the wing. The result of this difference in airflow is a decrease in air pressure above the wing. The higher pressure under the wing then pushes the wing upward.

The main problem faced by those early engineers in building a workable airplane was powering it. The plane had to move through the air fast enough for the wings to generate lift. The solution lay in developing an engine-driven propeller. The principle of the propeller was understood in the 1800's, but all of the engines capable of turning a propeller fast enough to generate the necessary speed, such as steam engines, were much too heavy to mount in an airplane. Then, along came the answer: the internal combustion engine, which produces a great deal of power by burning fuel inside a system of enclosed cylinders. The Wright brothers connected a

pair of propellers to a small gasoline engine, and aviation was born.

The airplane was largely a novelty for a few years, but its potential as both a means of transportation and a weapon of war soon became clear. During World War I (1914-1918), airplanes proved their worth in reconnaissance and combat. From then on, aircraft evolved at a rapid pace. The first planes needed two sets of wings to provide adequate lift. More efficient wings, engines, and propellers eliminated the need for the second set of wings, so these double-winged *biplanes* gave way to sleeker *monoplanes* with only one set of wings. New materials also came into use. The wood and cloth of early planes were replaced by metal alloys.

By the late 1920's, airplanes had become so dependable that pilots were routinely flying hundreds of miles without stopping. Less than a decade later, in 1936, the twin-engine Douglas DC-3 entered service. As the first efficient, dependable airliner, the 21-passenger DC-3 ushered in the era of low-cost passenger air travel. Many DC-3's are still flying.

In the 1940's, military fighters and bombers played a decisive role in World War II, and a new kind of airplane—the jet—made its appearance. Germany developed the first jet-powered plane, the experimental Heinkel He-178, in 1939.

In a jet engine, compressed air at a very high pressure is forced into a combustion chamber, where it mixes with fuel and burns. The gases produced by combustion shoot out the back of the engine at great speed. Pressure within the combustion chamber is thus high at the front and low at the back. This pressure difference drives the engine, and the plane on which it is mounted, forward.

Jet engines enabled airplanes to fly faster. Some planes came close

On Dec. 14, 1986, *Voyager, top,* an extremely lightweight plane designed by aeronautical engineer Burt Rutan, heads westward on its globe-circling mission. Nine days later, *Voyager*'s crew members—Jeana Yeager and Richard G. Rutan, the designer's brother, *above*—exult after completing the first nonstop, unrefueled flight around the world.

Milestones in Aviation

Aircraft designs developed rapidly after 1903, when the Wright brothers' *Flyer* achieved the age-old dream of powered flight. By the 1930's, commercial airliners were carrying passengers thousands of miles each year. In the 1940's and 1950's, jet planes came into use and the "sound barrier" was broken. In 1986, new lightweight materials enabled *Voyager* to make the first unrefueled flight around the world.

Wingspan—12.3 m (40 ft. 4 in.)
Length—6.4 m (21 ft. 1 in.)

The Wright *Flyer*, built by Orville and Wilbur Wright, made the world's first powered flight on Dec. 17, 1903, near Kitty Hawk, N.C. The plane's two propellers were driven by a lightweight gasoline engine.

Wingspan—7.8 m (25 ft. 7 in.)
Length—8 m (26 ft. 3 in.)

The Blériot XI was the first plane to fly across the English Channel—in 1909. It was built and flown by Louis Blériot of France, a self-taught pilot.

Wingspan—16.8 m (55 ft.)
Length—9 m (29 ft. 8 in.)

The Junkers J-1, built in Germany in 1915, was the first all-metal airplane. The J-1 was also the first plane with cantilever wings, which are supported by the fuselage rather than by braces and wires.

Wingspan—29 m (95 ft.)
Length—19.7 m (64 ft. 5.5 in.)

The Douglas DC-3, designed and manufactured in the United States, entered commercial service in 1936 and soon became the world's most widely used airliner.

Wingspan—7.2 m (23 ft. 7.5 in.)
Length—7.5 m (24 ft. 6.5 in.)

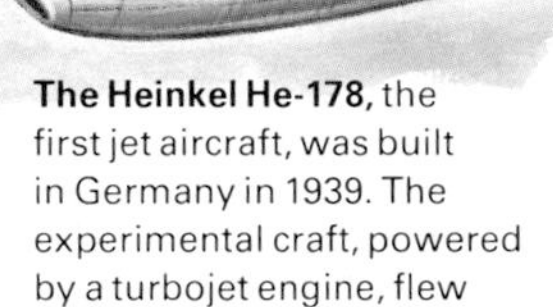

The Heinkel He-178, the first jet aircraft, was built in Germany in 1939. The experimental craft, powered by a turbojet engine, flew 580 kilometers (360 miles) per hour.

Wingspan—8.5 m (28 ft.)
Length—9.4 m (31 ft.)

The Bell X-1 in 1947 became the first plane to fly faster than the speed of sound. The experimental X-1 was powered by a rocket engine.

Wingspan—25.6 m (83 ft. 10 in.)
Length—62.1 m (203 ft. 9 in.)

The Concorde was the world's first supersonic passenger plane. Built as a joint venture by France and Great Britain, the Concorde began flying passengers in 1976.

Wingspan—33.7 m (110 ft. 9.5 in.)
Length of fuselage—7.7 m (25 ft. 4.8 in.)

Voyager, designed by aeronautical engineer Burt Rutan, made the first nonstop, unrefueled flight around the world in December 1986.

to *Mach 1*, the speed of sound—about 1,060 kilometers per hour (kph) or 660 miles per hour (mph) at an altitude of 12,000 meters (40,000 feet). (The speed of sound varies with altitude, decreasing the higher a plane flies.) But as they approached Mach 1, these planes encountered an unforeseen phenomenon—severe buffeting that caused the craft to shake and sometimes to spin out of control.

Wind tunnel tests showed that shock waves—sharp variations in air pressure—formed on a plane's surfaces at speeds close to Mach 1. Those tests explained what had been causing the bone-rattling flights reported by many jet pilots. Flying at high speed through the atmosphere, a plane created air-pressure disturbances that traveled away from the craft in all directions at the speed of sound, much like ripples in a pond. When the plane reached Mach 1, it caught up with the pressure disturbances ahead of it, which then piled up in a shock wave in front of the craft and on its wings. Aeronautical engineers speculated that the build-up of shock waves created a "sound barrier" that limited how fast a plane could fly.

In 1947, however, a United States Air Force test pilot, Captain Charles E. (Chuck) Yeager, laid that notion to rest by flying faster than sound in an experimental rocket plane, the Bell X-1. The X-1 was radically new in its design. It was highly streamlined, with a pointed nose, a *fuselage* (body) free of projections, and wings that looked almost knife-thin. Those features enabled the X-1 to slice through the air with ease, minimizing the effect of shock waves.

The shock waves that occur at Mach 1 are a type of *drag*, or resistance to a plane's forward motion. There are several other kinds of drag, including drag caused by friction between the plane and the air through which it is passing and drag produced by turbulent airflow around the wings and other surfaces.

All forms of drag increase with speed. But as the success of the X-1 showed, good design can significantly reduce drag. Engineers realized that *supersonic* (faster than sound) aircraft must be built much like the X-1, with a slender, tapering fuselage and thin wings. Those changes were made, and military aviation entered the supersonic age. Today's military fighters can make hairpin turns at *Mach 2* (twice the speed of sound), and most can climb almost straight up using their engine *afterburners*—devices that increase thrust by adding extra fuel to the jet exhaust and igniting it.

Great changes also took place in civilian aircraft design after World War II. Jet airliners, notably the Boeing 707, began replacing propeller-driven cargo and passenger planes in the 1950's. Wide-body jumbo jets that carry more than 400 passengers and the Concorde, a supersonic transport (SST), entered the scene in the 1970's.

The newest commercial airplanes, such as the Boeing 767, the French Airbus 320, and the McDonnell Douglas MD-80, incorporate the latest technology. In the cockpit, for example, the familiar rows of circular gauges are mostly gone, replaced by video screens

Reducing the Effects of Drag

The flow of air over an airplane's wing, *right,* creates lower pressure over the wing, thus producing the force of *lift.* But the wing's motion through the air also generates forces that try to slow it down. These forces—friction and turbulent airflows—are known collectively as *drag.* Designs that minimize drag enable planes to fly faster and use less fuel.

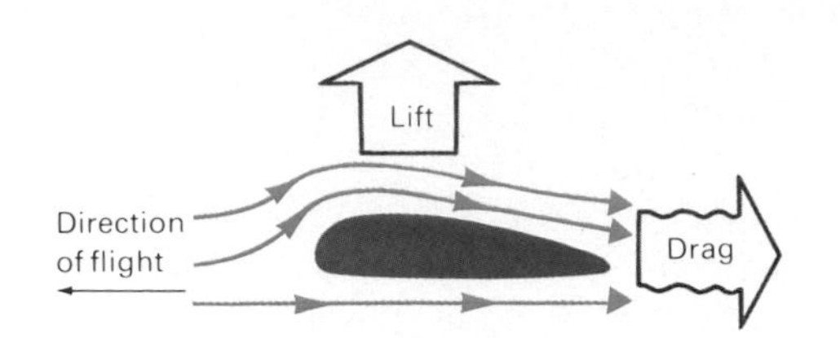

Supercritical Wings

When a plane flies at close to the speed of sound, air accelerating over a conventional wing, *below,* causes a shock wave, followed by turbulence, to form over the wing. Many newer airliners, however, have *supercritical wings, bottom,* which are slightly flatter on top and rounder underneath than conventional wings. That shape reduces the size of the shock wave and moves it, and the resulting turbulence, farther back on the wing, thereby reducing drag.

Conventional wing

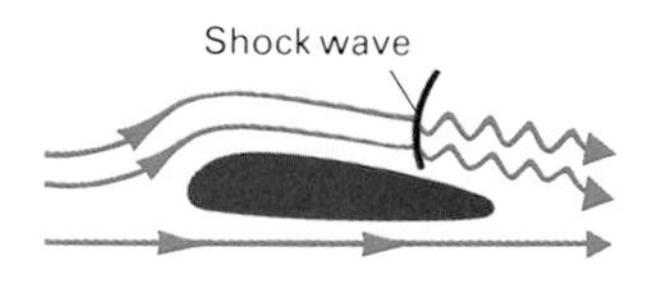

Supercritical wing

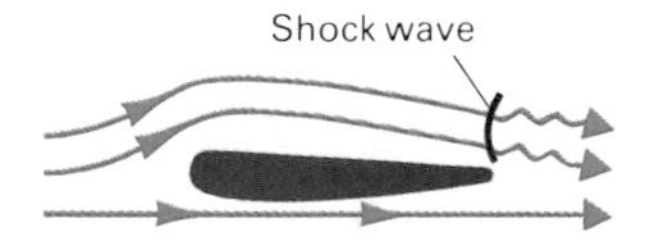

that present easy-to-read mechanical and flight information. Computerized control systems navigate the plane and warn the pilot of problems. The pilot still flies the plane, but the computer amplifies every touch on the controls and compensates for errors.

Advanced electronics systems are making airplanes safer and easier to fly. Making them cheaper to fly has been another major goal of engineers, especially since the mid-1970's, when fuel prices began to climb. Many of today's airliners can fly 30 to 40 per cent farther on each gallon of fuel than their predecessors could, thanks to redesigned wings and more efficient engines.

The main objective in redesigning wings is drag reduction. Engineers are constantly seeking new ways to make airplanes slip through the air more easily. One of their recent innovations is the so-called *supercritical wing*, which is slightly flatter on top and rounder on the bottom than ordinary wings. Supercritical wings are the latest solution to the problem of shock waves, a nuisance that designers have never been able to entirely get rid of. When an airliner flies close to Mach 1, air flowing over the wings reaches or exceeds the speed of sound, producing a small shock wave above each wing. The shape of supercritical wings reduces the shock wave and its resulting turbulence and pushes them farther toward the trailing edge of the wing, thereby cutting drag. Supercritical wings have become standard for new airliners.

Aeronautical engineers are also taking a new look at a still-experimental drag-reducing approach to wing design called *laminar flow control*, which has been studied off and on since the 1930's. The object of laminar flow control is to make air move more evenly over the wing. As any object, no matter how well streamlined, moves through the air, it is slowed by the friction of air molecules flowing around it. At the high speeds of modern aircraft, that friction is strong enough to stir the air in the *boundary layer*—a thin layer of air adjacent to the plane's skin—making it turbulent. If the stirring could be reduced, the airflow would remain *laminar*, or smooth. In wind-tunnel and flight tests, engineers have achieved laminar flow with experimental wings perforated with hundreds of thousands of tiny holes. Boundary-layer air is drawn into the holes and pumped out of the plane, usually through the engines. Test results indicate that laminar flow control could cut drag by up to 30 per cent.

In another effort to reduce fuel consumption, commercial-aircraft engineers have designed a new engine called the *propfan*. The propfan uses rotating turbines to turn two propellers, each with 8 to 10 blades. The new propellers, which spin in opposite directions, are aerodynamically superior to earlier propellers. The tips of ordinary propeller blades create disruptive shock waves if they move at supersonic speeds, a problem that has been overcome with the computer-designed blades of the propfan's propellers. Tests by Boeing and McDonnell Douglas indicate that the propfan can match the thrust of the best *turbofans*—the kind of engine used on almost all jet airliners since the late 1950's—while cutting fuel consumption by 30 per cent. Both companies hope to be installing propfans on many of their planes by the mid-1990's.

The development of the revolutionary new propellers for the propfan, like many breakthroughs in aviation today, could not have been accomplished without advances in computers and materials. Computerized simulators enabled engineers to electronically test and discard one design after another until they found the ideal shape for the propfan blades. And new composite materials, which can take great stress, enabled them to build the new propeller.

Composites consist of thin fibers of carbon, glass, or other substances embedded in tough epoxy resins. Composites are lighter and stronger than metal and thus are perfectly suited for many airplane parts. When used in the construction of wings or fuselages, composites are formed into panels that are glued together rather than riveted. Because they are expensive to manufacture and assemble, composite materials are still in limited use.

One plane built largely of composites is the X-29. The new materials made possible the X-29's most radical feature, its forward-swept wings. Unlike most aircraft, which are designed to be stable, the X-29 was deliberately designed to be *un*stable.

In a conventional airplane, the *center of pressure* (the point at which all aerodynamic forces balance) is well behind the *center of gravity* (the point about which the airplane would turn if you could balance it on a stick). In flight, if the airplane's nose deviates from a straight path, the aerodynamic forces at the center of pressure compensate for the movement. The plane thus resists being turned. With forward-swept wings, an airplane's center of pressure is in front of the center of gravity. This reversal of the plane's flight dynamics makes the craft unstable but highly maneuverable. It can change direction at a mere flick of the controls.

Engineers had known for years from theoretical studies that forward-swept wings offered tremendous advantages for fighter aircraft. In addition to making the plane more maneuverable, they improved lift and low-speed handling. But until the development of composite materials, engineers could not build a plane with such wings. The aerodynamic forces acting on a forward-slanting wing tend to twist it upward and backward. A metal wing would continue to twist until it ripped right off the plane. The fibers of a composite

Laminar Flow Control

Another—but still experimental—approach to reducing drag is *laminar flow control.* When air flows over a wing, friction between the wing and the air creates a thin, slow-moving *boundary layer* of air next to the plane's skin. With a conventional wing, *below,* fast-moving air above the boundary layer stirs boundary-layer air. Airflow over the wing thus changes from *laminar* (smooth) to turbulent. An experimental wing, *bottom,* maintains laminar flow by sucking most of the boundary-layer air through hundreds of thousands of tiny holes. The boundary-layer air is pumped to another part of the plane and exhausted to the outside.

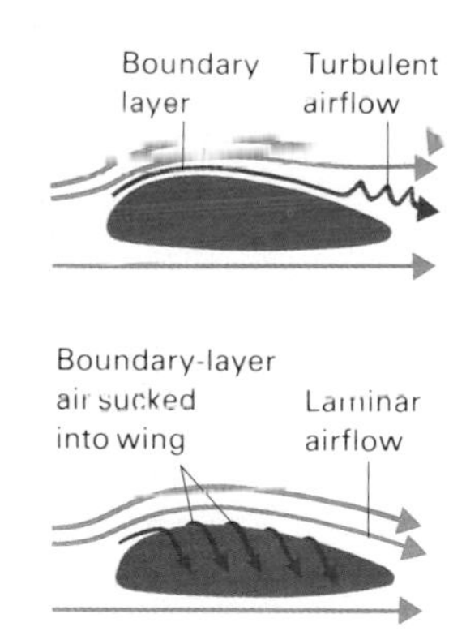

Designed To Be Unstable

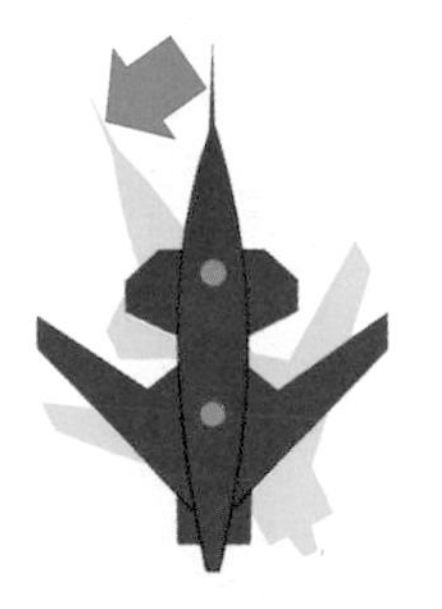

The X-29, *above right,* was deliberately designed to be unstable. Ordinarily, a plane is designed so the *center of pressure* (green dot)—the point where all aerodynamic forces balance—is behind the *center of gravity* (blue dot), the point around which the plane's weight is evenly distributed, *top.* This arrangement produces a corrective tendency (arrow) that resists any movement away from a straight flight path. In the X-29, the center of pressure is *in front* of the center of gravity. With such a design, the tendency (arrow) is for any flight deviation to increase, a highly unstable situation. The benefit is that the plane is extremely maneuverable.

wing, however, can be "tailored" to counteract the twisting motion and keep the wing at the correct angle.

The instability caused by the forward-swept-wing design presented another stumbling block. No pilot, even with the fastest of reflexes, would be able to keep an airplane such as the X-29 from tumbling out of control. It was the development of compact high-speed computers in the 1970's that finally made the X-29 possible. On board that remarkable aircraft are three computers—a main computer and two backups—that adjust the wing and tail surfaces 40 times a second to keep the plane firmly on course.

At the same time that the Grumman Corporation was designing and building the X-29, two other aircraft companies—the Lockheed Aircraft Corporation and the Northrop Corporation—were working with the Air Force on top-secret projects to develop "invisible" airplanes. Electronics experts think these so-called *stealth* airplanes will be almost undetectable to enemy radar.

Radar works by sending out pulses of high-frequency radio waves that bounce off distant objects. Reflected waves are picked up by a receiver and translated into an electronic image. There are three ways in which an airplane can be made to fool radar: by shaping the plane so it scatters radar waves instead of reflecting them; by coating the plane with a material that absorbs radar waves; or by fitting the plane with instruments that create false images on the enemy's radar screens. Stealth employs all three methods.

The Air Force is now ready to embark on its next big project: the development of a new fighter for *tactical* purposes—supporting battlefield maneuvers, making short-range attacks, and engaging enemy aircraft. Envisioned as the most formidable warplane that has ever taken to the skies, this Advanced Tactical Fighter (ATF) will incorporate stealth features and the technological know-how gained from experimental planes such as the X-29. The ATF will be able to cruise for long distances at Mach 2 and, with its afterburner, reach speeds up to Mach 2.5. The ATF's composite wings will contain computer-controlled "muscles" that alter the wing's curvature for maximum performance at different speeds. The on-board computer will, in fact, directly or indirectly control almost every function of the aircraft. Serving as an "electronic copilot," the computer will maintain stable flight, guide the plane around obstacles when flying at treetop level, track targets, and aim missiles and bombs with pinpoint accuracy. The computer may also have the ability to understand the spoken words of the pilot—whose role will become primarily that of decision maker—and to respond in a synthesized voice. The pilot will likely control the plane with the aid of an elaborate helmet, already under development, that displays a variety of computer-generated information on the inside of the visor.

The military is also interested in developing high-speed aircraft that can take off and land like helicopters. One experimental design currently being studied is called the X-wing. The X-wing has a four-bladed rotor powered by a jet engine. The rotor is designed to spin rapidly like a helicopter blade for vertical take-offs and landings. In flight, the rotor will be locked in place and perform like a conventional airplane wing. When it is no longer being used to power the rotor, the engine will provide jet thrust to propel the X-wing at speeds up to 800 kph (500 mph). Some aeronautical engineers also foresee the development of small airplanes that take off and land vertically that will be used for personal and business transportation. Such craft will probably be jet airplanes that move vertically by directing their jet exhaust downward.

Just as flying in personal airplanes may someday be routine, so may winging halfway around the world in less time than it takes to see a double feature at a movie theater. The U.S. government has committed itself to the development of an *aerospace plane* as a possible replacement for the space shuttle, and a modified version of the craft—a *hypersonic transport* (HST)—will probably be built for use by

High-speed computers and new kinds of materials have become indispensable in the design and construction of aircraft. A computer simulation of the U.S. space shuttle, *below,* showed designers how air would flow around the body of the craft. At the Grumman Corporation, *below left,* engineers assemble the wings of the X-29 by gluing together layers of *composite materials.* Such materials, composed of high-strength fibers embedded in an epoxy resin, are lighter and stronger than metal.

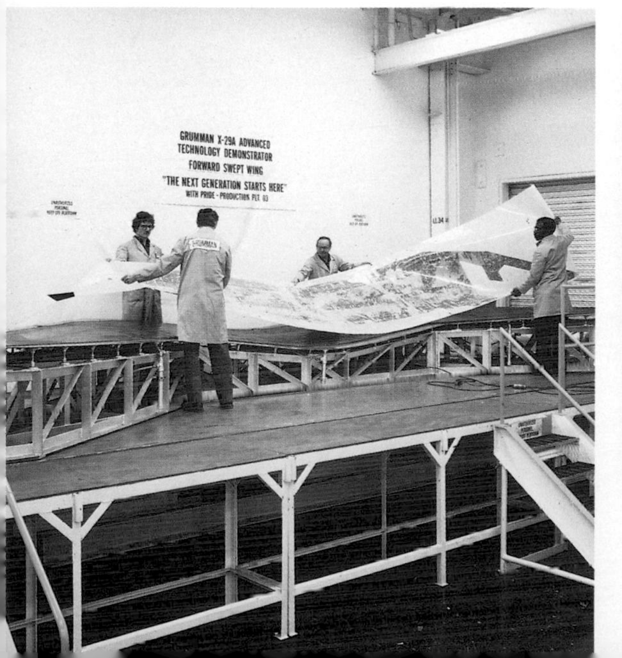

Jet Engines for the 1990's and Beyond

The *turbo-ramjet* and the *scramjet*—experimental variations of existing engines called the *ramjet* and the *turbojet*—will enable hypersonic flight. And with the *propfan, opposite page,* airliners will again have propellers.

In a ramjet, used primarily in guided missiles, incoming air is compressed as it "piles up" within the engine. The air then rushes into a combustion chamber, where it mixes with fuel and burns, producing thrust.

The turbojet, which powers most military fighters, is similar in operation but uses a fanlike compressor to squeeze incoming air. The burning gases in the combustion chamber spin a turbine that drives the compressor. But beyond Mach 3.5 (3½ times the speed of sound), the compression of the airflow in a turbojet engine would raise combustion temperatures high enough to destroy the turbine.

The turbo-ramjet will avoid the problem of excessive heat up to about Mach 6 by diverting air around the turbine, which is turned by a high-pressure stream of fuel heated to a constant temperature.

The scramjet (*s*upersonic *c*ombustion *ramjet*) will enable speeds up to Mach 25 without a destructive build-up of heat by allowing air to flow through without slowing it down as much as other kinds of engines would.

Ramjet

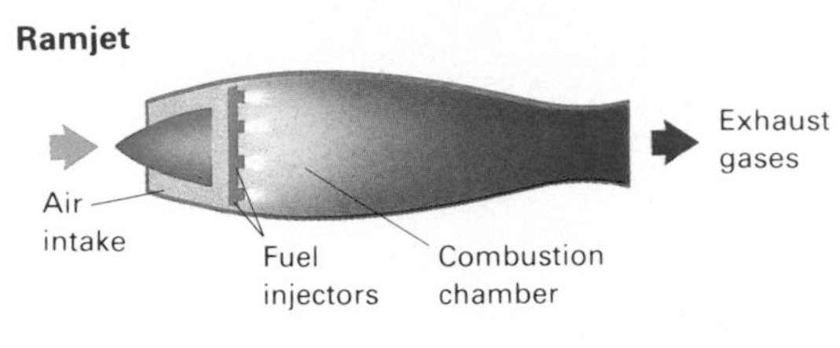

Turbojet

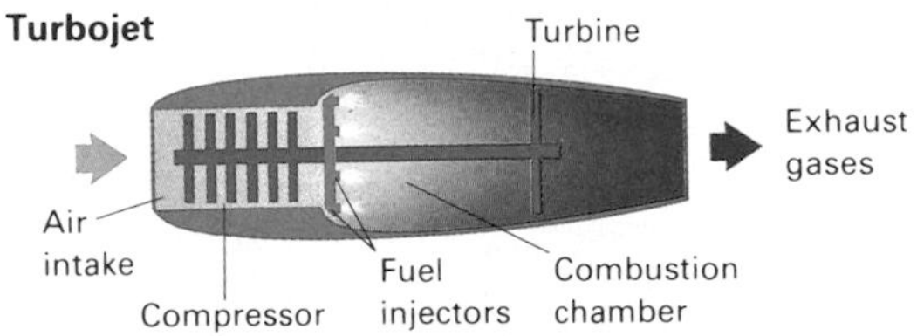

Turbo-ramjet

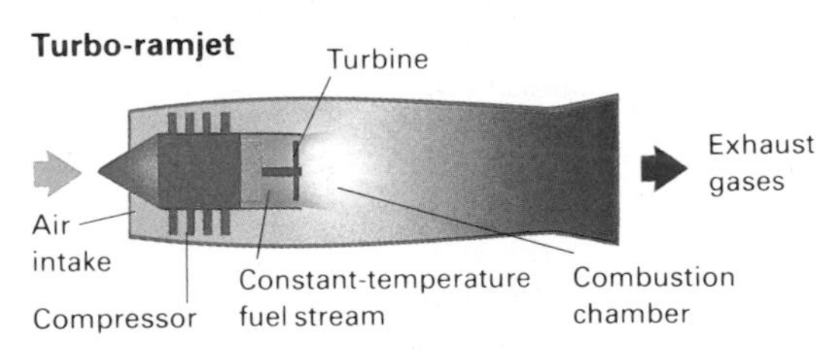

Scramjet

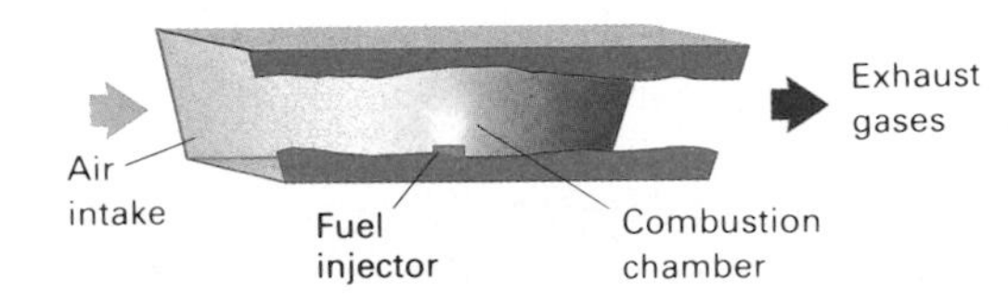

The proposed hypersonic transport, the "Orient Express," shown in an artist's conception, soars above Earth at five to six times the speed of sound. Such a plane would most likely be powered by turbo-ramjet engines.

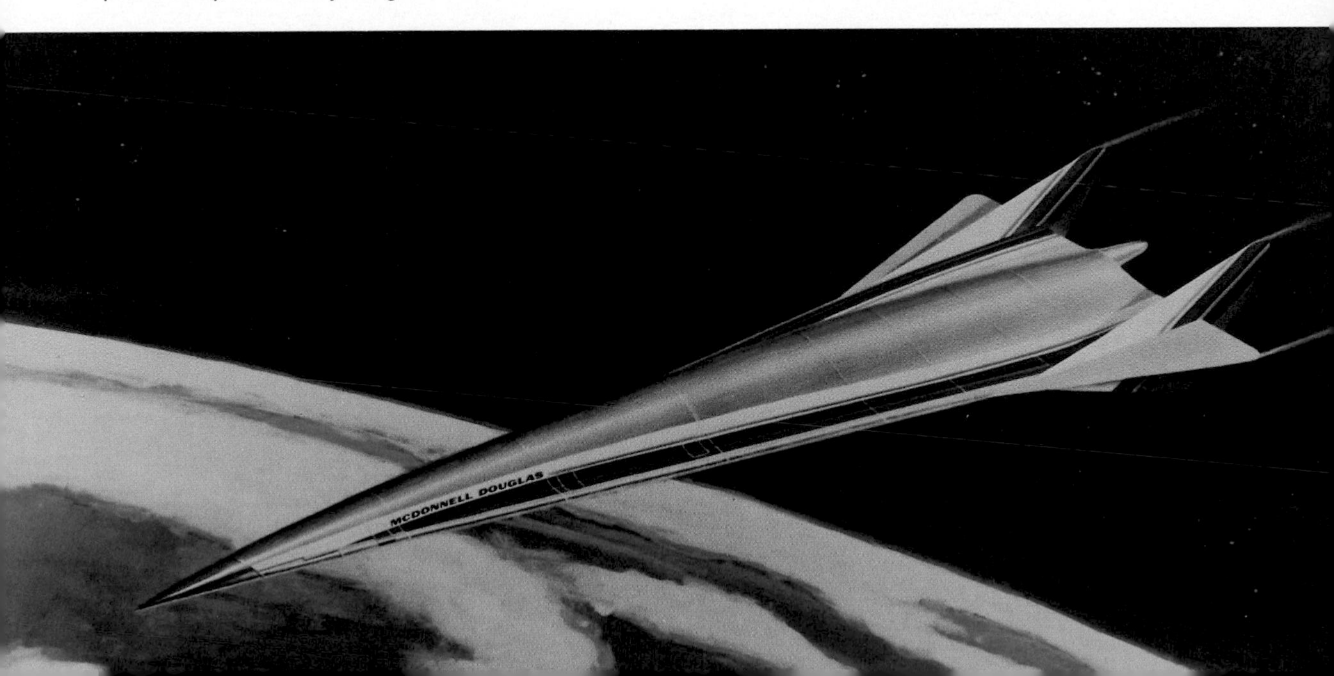

The Propfan

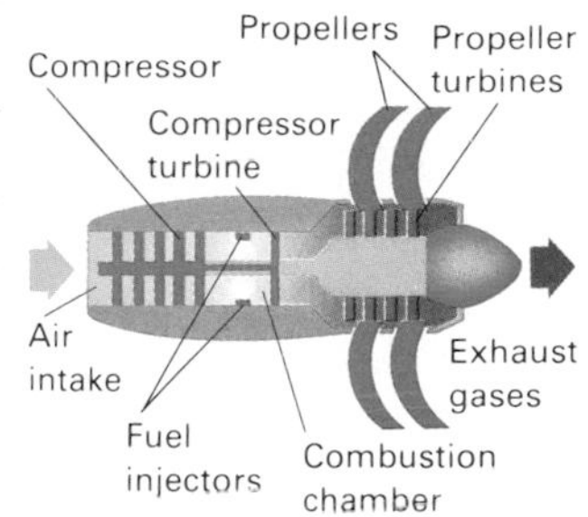

The propfan, *above,* is a return to an old idea: using a jet engine to turn a propeller. But the propfan's computer-designed propellers are a vast improvement over earlier propellers, whose tips created disruptive shock waves if they moved at supersonic speeds. Compressors in the propfan force air into a combustion chamber, where it mixes with fuel and burns. The hot gases produced by combustion turn a set of turbines. The first turbine drives the compressor; the other turbines power a pair of multibladed propellers that spin in opposite directions. The propfan produces as much thrust as the jet engines now being used on many passenger planes while using much less fuel, and thus promises to lower airlines' operating costs. Airliners powered by propfans, such as the Boeing-designed plane shown *above left* in model form, are scheduled to go into service in the 1990's.

the airlines. Flying at speeds up to Mach 6, or nearly 6,400 kph (4,000 mph), the HST will carry 250 to 300 passengers from New York City to Tokyo in about two hours. Such a trip today takes 14 hours on a wide-body jet. Because it would make the Far East so readily accessible to travelers, the HST already has been dubbed the "Orient Express."

Both the HST and the aerospace plane, whose *prototype* (original model) has been designated the X-30, would take off horizontally from a runway. But the X-30, which would fly into orbit around Earth, would travel much faster—an incredible Mach 25, or about 28,300 kph (17,000 mph). The plane would be used for a variety of missions, most of them military, including reconnaissance and perhaps protecting U.S. satellites.

Building an aircraft that can fly at 25 times the speed of sound will be no easy task. For one thing, the plane must have a skin capable of withstanding the intense heat that would be generated by atmospheric friction. Another serious problem is that no wind tunnel can duplicate flight conditions beyond about Mach 8 for more than a few thousandths of a second, so much of the X-30's performance prior to flight will have to be electronically simulated.

A major challenge in building both the X-30 and the hypersonic transport is propulsion. The *turbojet* engines that power present-day supersonic aircraft use a fanlike compressor to force incoming air into a combustion chamber, where it is mixed with fuel and ignited. As well as producing thrust, the burning gases power a turbine, which in turn drives the compressor. At about Mach 3.5, however, turbojets encounter a limit far more severe than the old sound barrier: the *thermal barrier.* When an airplane flies at high speed, air piles up in front of it, becoming very hot. When the air is slowed down in the engine and compressed, it gets even hotter. At Mach

3.5 and beyond, temperatures in the combustion chamber would exceed 1500°C (2700°F.)—high enough to destroy the turbine.

An ingenious variation on the turbojet called the *turbo-ramjet* can evade the thermal barrier up to approximately Mach 6 and thus would be an ideal engine for the HST. The still-experimental turbo-ramjet is a cross between the turbojet and the *ramjet*, a simple jet engine with no moving parts that is used chiefly in missiles. Like the turbojet, the turbo-ramjet has a compressor powered by a turbine. In a turbo-ramjet, though, the hot incoming air is diverted around the turbine to a combustion chamber at the rear of the engine. Gaseous fuel, preheated by partial combustion to a constant temperature, is forced through the turbine, causing it to spin. The fuel then mixes with the airstream in the combustion chamber and burns completely.

But what about the aerospace plane—what sort of engines will be able to push an aircraft to a speed of Mach 25? At velocities over Mach 6, air would be moving so fast into a plane's engines that com-

A model of what the U.S. Air Force stealth fighter might look like, *below,* was created by a model-airplane manufacturer in 1986. Stealth airplanes—including stealth bombers, also under development—are designed to be nearly invisible to radar. These top-secret planes have shapes that scatter radar waves, coatings of radar-absorbing materials, and radar-fooling instrumentation.

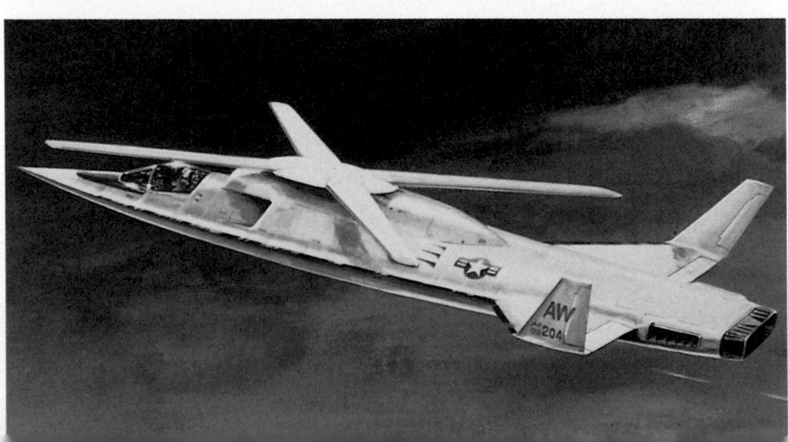

The jet-powered X-wing, *right,* in an artist's conception, an aircraft that is now in an early stage of development, will be part airplane, part helicopter. The four-bladed rotor will spin rapidly for take-offs and landings but will lock in place during forward flight.

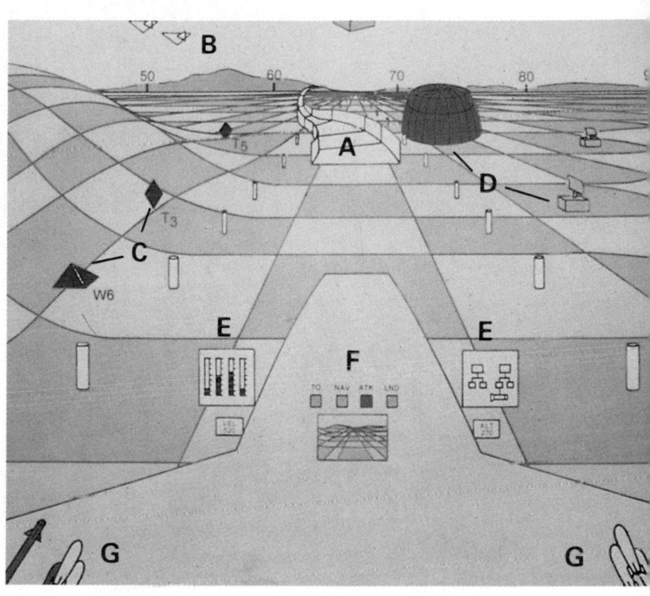

An experimental helmet for U.S. fighter pilots, *above left* projects information on the inside of the visor. The cartoonlike panorama, *above,* generated by the plane's onboard computers, shows the terrain and sky ahead of the aircraft as well as information about the plane's various systems. At a glance, the pilot can see the plane's expected flight path (A), other aircraft (B), and enemy missile installations (C) and radar stations (D). In the lower part of the display are instrument readouts (E), electronic controls (F), and pictures of the plane's available weapons (G).

bustion temperatures would rise to about 1930°C (3500°F.), at which point even nonmoving engine parts would begin to melt.

Aeronautical engineers think the answer lies with another experimental engine called the *scramjet,* which is short for supersonic combustion ramjet. As its name implies, the scramjet does not appreciably slow down and compress the incoming air—the major cause of soaring temperatures—but rather lets it flow through the engine at supersonic speeds. The scramjet would burn hydrogen, carried in liquid form at a very low temperature—about −255°C (−425°F.). The last leg of the journey into space orbit, where there is no air, would be accomplished with rocket engines.

Although the aerospace plane will undoubtedly be used exclusively in U.S. space and military programs, the HST will be for everyone. Hypersonic airliners could usher in an era of ultrahigh-speed transportation in which we would think nothing of meeting a friend or business associate on the other side of the world for lunch or dinner. By the turn of the century, according to some predictions, we might be able to take a short-hop aircraft, such as a civilian version of the X-wing, to a huge regional airport. There we would board an HST for a nonstop flight to Japan, Australia, or some other faraway destination. Estimated flying time from take-off to any point on Earth: less than 3½ hours. Elapsed time since Kitty Hawk: less than 100 years.

For further reading:

Grey, Jerry. "The New Orient Express," *Discover,* January 1986.
Kinnucan, Paul. "Superfighters," *High Technology,* April 1984.
Schefter, Jim. "Engineering Tomorrow's Airliners," *Popular Science,* April 1987.

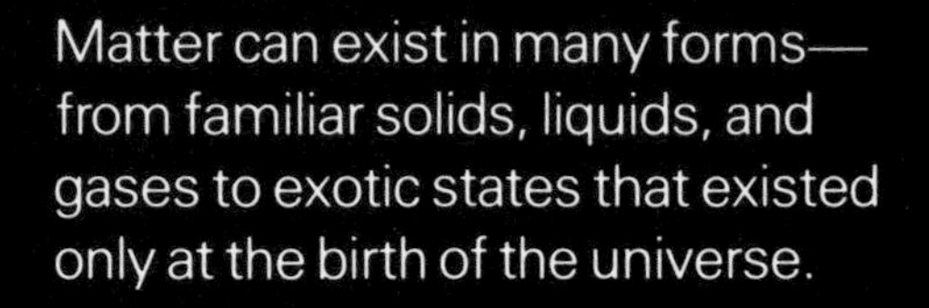

Matter can exist in many forms—from familiar solids, liquids, and gases to exotic states that existed only at the birth of the universe.

Exploring the States of Matter

BY ROBERT H. MARCH

ASSOCIATED STAT

Temperature has provided the route for exploring the world of matter, from the tropical realms of quarks to the frigid latitudes of superconductivity.

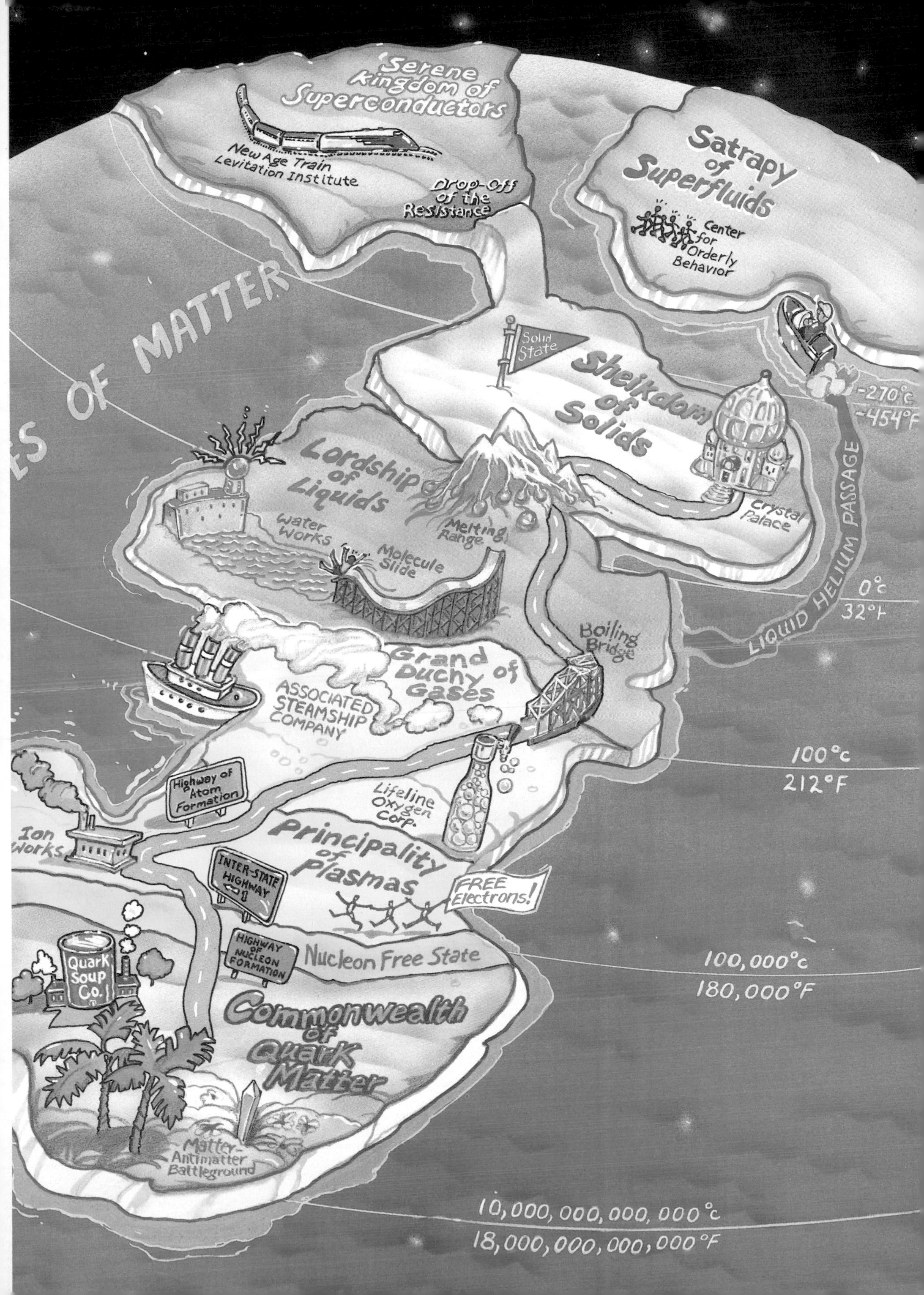

Serene kingdom of Superconductors

New Age Train Levitation Institute

Drop-Off of the Resistance

Satrapy of Superfluids

Center for Orderly Behavior

ES OF MATTER

Solid State

Sheikdom of Solids

Crystal Palace

-270°c
-454°F

LIQUID HELIUM PASSAGE

Lordship of Liquids

Water Works

Melting Range

Molecule Slide

0°c
32°F

Boiling Bridge

Grand Duchy of Gases

ASSOCIATED STEAMSHIP COMPANY

100°c
212°F

Highway of Atom Formation

Lifeline Oxygen Corp.

Ion Works

Principality of Plasmas

INTER-STATE HIGHWAY

FREE Electrons!

HIGHWAY OF NUCLEON FORMATION

Nucleon Free State

100,000°c
180,000°F

Quark Soup Co.

Commonwealth of Quark Matter

Matter-Antimatter Battleground

10,000,000,000,000°c
18,000,000,000,000°F

Glossary

Conductor: A substance such as copper that conducts, or carries, electric current easily.

Crystal lattice: An orderly geometric pattern of atoms or molecules linked together strongly in a solid.

Free nucleon state: A form of matter made up of free-flying protons, neutrons, and electrons.

Plasma: Matter composed of free-flying electrons and charged atoms.

Quark matter: A dense soup of free-flying quarks and electrons.

Superconductivity: The ability of a substance to conduct electric current without resistance.

Superfluidity: The ability of a liquid to flow without resistance.

Viscosity: A liquid's resistance to flow.

On a sunny winter day, water drips from an icicle. Before our eyes a solid—ice—changes into a liquid through the familiar process of melting. Soon the icicle is gone, leaving a puddle of water on the ground. If we wait long enough, that too will disappear. The liquid water of the puddle will evaporate, converted by the sun's heat into water vapor, an invisible gas.

When a substance such as water changes from its liquid form to a solid or to a gas, scientists say the substance changes its *state*. Everyone is familiar with the three most common *states of matter*—solid, liquid, and gas. But scientists have discovered other states that are very hard to imagine—including a state that existed at the birth of the universe and one that allows liquids to flow up walls. And discoveries in the mid-1980's about still another state—called superconductivity—promise to bring great changes to our daily lives through such developments as power lines that transmit electricity cheaply and trains that speed along above a track, held up in the air by a powerful magnetic force.

For thousands of years, people have observed the changing states of at least one form of matter—water. Water's ability to change states deeply impressed scientists of ancient Greece. In 600 B.C., Thales of Miletus wondered: If water can exist in three very different forms, why stop with just three? Thales proposed that all matter—every solid, every liquid, and every gas—might simply be different states of water.

Although this idea has long since been discarded, Thales was not entirely wide of the mark: Many substances do have more than one solid state. Diamonds and graphite, for example, are simply two different solid forms of the chemical element carbon.

Our understanding of the states of matter owes much to another Greek, Democritus of Abdera, who lived about two centuries after Thales. Democritus regarded matter as being made up of tiny *atoms*, solid particles that cannot be divided and are always in motion. This idea was not developed much further until the 1800's. By then, scientists realized that atoms did indeed exist and that certain attractions between atoms tend to link them together to form combinations called *molecules*. A water molecule, for example, is made of two hydrogen atoms and one oxygen atom—expressed in the chemical formula H_2O. But ice is also H_2O, as is steam. The concept of molecules made the differences among the *properties* (characteristics) of the three familiar states of matter easy to explain.

All states have properties by which we can recognize them. A solid is *rigid*. It holds its shape and stubbornly resists any attempt to compress it into a smaller volume. Ice, like most other solids, is rigid because its molecules are linked together strongly in an orderly geometric pattern called a *crystal lattice*.

A liquid is as difficult to compress as a solid because the molecules are in close contact with one another. But liquid is *fluid*; it takes the shape of its container and can flow out of its container. Water flows because the H_2O molecules are not strongly linked but are free to

The author:
Robert H. March is a professor of physics at the University of Wisconsin and the author of the Special Report MIND-BOGGLING MYSTERIES OF MATTER in the 1987 edition of *Science Year*.

bump, jostle, and slide past one another. Not all liquids, however, flow as freely as water. Honey, for example, oozes rather than pours out of a jar. In other words, honey resists flowing more than water does. Resistance to flow is called *viscosity*.

A gas is compressible. Although it can expand as far as the walls of any container, gas can also be squeezed into a smaller volume—within limits. Steam, like all other gases, is compressible because there is a lot of empty space between its molecules. A gas is compressed by pushing the molecules closer together.

The main driving force behind changes of state is temperature. For example, the changes that occurred when ice went from a solid to liquid water and then from a liquid to a gas were brought about by rising temperatures. These changes can be reversed by cooling. As air temperature drops, water vapor condenses into liquid water, and then the water turns to ice.

We are familiar with all three states of H_2O because, in everyday situations, we can encounter water's freezing and boiling points—0° and 100°C (32° and 212°F.) at sea level. (The boiling point of a liquid depends on the pressure of the atmosphere and varies with altitude.) In fact, H_2O is the only substance we normally encounter in all three of the common states of matter.

A stream in winter reveals the three most common states of matter—solid, liquid, and gas. Solid water—ice—lines the banks as water in liquid form runs past. The warm air causes some of the water to evaporate, to change from a liquid to a gas.

Understanding how temperature forces matter to change its state came from Democritus' idea that atoms are always in motion. By 1850, most scientists had recognized that heat is a measure of the motion of atoms and molecules: The higher the temperature, the faster they move.

Even when locked into solids, atoms and molecules vibrate. As heat is added, a few atoms or molecules acquire enough energy to overcome the attractions that hold them rigidly in place, and melting begins.

When the solid becomes a liquid, each of its molecules is still attracted by its nearest neighbors. But with neighbors on all sides, the attractions neutralize one another, so the liquid's molecules move in all directions. An exception to this occurs on the surface of the liquid. Almost all the neighbors of a surface molecule are on the inside and pull the surface molecule inward. This powerful tug is called *surface tension*. Sometimes, a few molecules on a liquid's surface have enough energy to overcome surface tension and evaporate, entering the gas state. The rate of evaporation depends upon the warmth of the surface.

The connection between heat and atomic motion gave rise to the idea of *absolute zero*. As a substance cools, its atoms and molecules move slower. Scientists in the 1800's expected that, when the atoms and molecules stopped moving, no more heat could be removed. The temperature at which they expected motion to cease—calculated to be about −273°C (−460°F.)—came to be called *absolute zero*.

In the late 1800's and early 1900's, scientists got caught up in a race to see who could get closest to this temperature. The major milestones in the race were the conversion of a succession of gases to liquids. The race began in 1877, when French physicist Louis P. Cailletet succeeded in liquefying oxygen at a temperature near −183°C (−297°F.). Nitrogen succumbed at −196°C (−321°F.).

In 1908, Dutch physicist Haike Kamerlingh Onnes passed the last milestone by making a liquid of helium, with the lowest boiling point of all, −269°C (−452°F.). Ironically, at this milestone was another starting point, the gateway to the exotic states of *superfluidity* and *superconductivity*.

Metals can exist not only as solids but also as liquids. Unlike metals that change from solids to liquids only at high temperatures, the metal mercury exists as a liquid at room temperature. Unlike water, liquid mercury is very heavy and dense—even able to support solid rocks on its surface.

A superfluid is a liquid that flows without resistance. In other words, it has no viscosity. As a result, it can form a thin film that flows right up the walls of its container. Soviet physicist Peter L. Kapitsa discovered superfluidity in 1938. He found that liquid helium abruptly became a superfluid when cooled to about 1 Celsius degree (1.8 Fahrenheit degrees) below its normal boiling point. Since then, scientists have not been able to make a superfluid of any other substance.

Superconductivity is the disappearance of resistance to electric current. Electric current is a flow of *electrons* (negatively charged subatomic particles) through wires. Normal wires are made of substances—such as copper—that are known as *conductors* because they conduct, or carry, current easily. Even the best conductors resist current, however, by absorbing energy from the flowing electrons. Electrons swarm through a normal conductor like an unruly crowd. They jostle the atoms in the solid conductor's crystal lattice, making the atoms vibrate more rapidly and thus heat up. At the same time, bumping into atoms makes the flowing electrons slow down or stop. But electric current flows through a superconducting wire without bumping into or being stopped by any of the wire's atoms.

Superconductivity was found in metals cooled by liquid helium to temperatures of −269°C (−452°F.) and below. It was discovered in 1911 by Kamerlingh Onnes while he was measuring the electrical resistance of frozen mercury. More than half the metallic chemical elements and a wide variety of mixtures and compounds become superconducting if cooled by liquid helium.

The discovery of superfluidity and superconductivity led to a new understanding of the meaning of absolute zero. No longer did it imply the disappearance of all atomic and molecular motion. Today, we realize that heat is simply the disorderly part of the motion. At or near absolute zero, atoms, molecules, and electrons may still move rapidly, but their movements are orderly.

In superfluid helium, the movement of the atoms involves none of the bumping, jostling, and sliding responsible for viscosity in a normal liquid. In fact, the atoms do not even touch.

In a superconductor, orderly motion involves both the electrons that make up the current and the nature of the superconducting material. All the electrons move in exactly the same direction and at exactly the same speed. And, because the superconducting material's crystal lattice is extremely cold, its atoms and molecules do not vibrate much. As a result, the electrons slip between atoms without contact. There is none of the jostling and loss of electron energy that is responsible for electrical resistance.

So far, researchers have found no practical uses for superfluids. Superconductors, however, are finding widespread application because in many cases, their ability to conduct electricity without absorbing energy makes them less expensive to use than ordinary conductors—even though it is costly to liquefy the helium that cools superconducting wires.

A gas expands to fill its container, and this property is what causes a hot air balloon to rise. The air—a mixture of gas molecules—expands as the heat drives the molecules farther apart. Some of the air escapes out of the balloon, leaving the remaining air inside lighter than the air outside. Any gas surrounded by a heavier gas will rise, so the balloon goes up.

Today, coils of superconducting wire are used in powerful electromagnets for a wide range of devices, including *particle accelerators* (atom smashers); magnetic resonance imaging machines, which provide pictures of the inside of the human body without the use of X rays; and experimental high-speed trains that do not run on wheels but float several centimeters or inches above their rails, held aloft by powerful magnetic forces.

In January 1987, scientists reported the development of materials that superconduct at temperatures as high as −180°C (−292°F.). Since this temperature is considerably above the boiling point of nitrogen, liquid nitrogen can be used to cool the new superconductors. Liquid nitrogen costs 6 cents per 1 liter (1 quart), while a liter of liquid helium sells for $3, so this development is good news to anyone working on practical uses of superconductors.

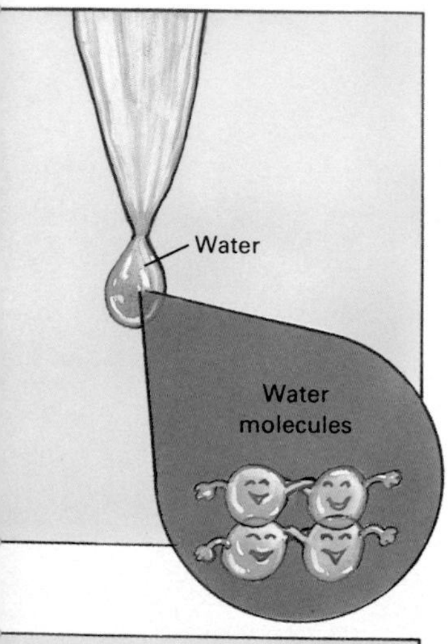

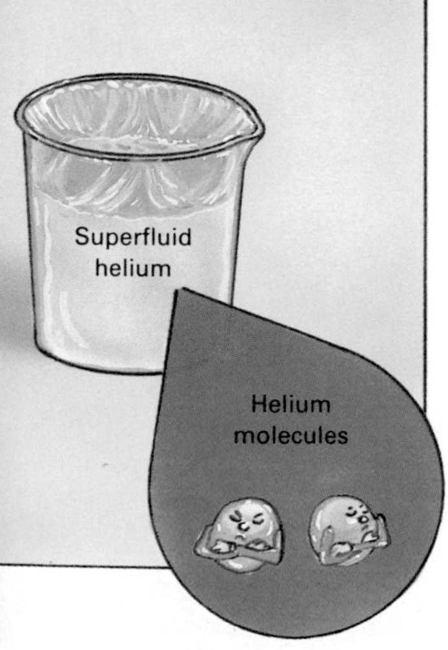

From Fluid to Superfluid

Water resists dripping, *top,* because its molecules—like those of all other ordinary liquids—are strongly attracted to one another. But helium molecules in an extremely cold state of matter called superfluidity, *above,* have no such attraction, so the helium flows without resistance, even flowing up the sides of a beaker—in much the same way that water flows upward when it soaks a napkin.

Not all the exotic states of matter exist at the cold end of the temperature scale, however. At the hot end, beyond gases, are several such states. The coolest of these, *plasma,* forms from gas when the temperature of the gas reaches 100,000°C (180,000°F.). Plasma is simply a gaslike collection of *free electrons* and *positive ions.* A free electron is an electron that is not in orbit around an atomic nucleus, and a positive ion is an atom with less than its normal number of electrons—in some cases, an atom with no electrons. (In a normal atom, one or more electrons orbit a nucleus.)

The change of state from gas to plasma is easily understood in terms of atomic motion. When gas atoms are moving swiftly enough, many of them collide so violently that electrons are knocked loose from them. These electrons become free electrons, while the atoms with the missing electrons become positive ions.

Although plasma is an exotic state, we have all seen at least one form of plasma—in a neon light. The glowing substance inside the glass tube is a plasma.

Electricity creates this plasma. The tube has two electrical connections, one at each end, and is filled with neon gas. When the tube is turned on, about 15,000 volts of electricity knock electrons out of their orbits around neon nuclei. The resulting plasma glows between the two connections. The tube remains cool because most of the energy is in the electrons rather than in the neon ions.

Nature also uses plasma as a "lighting device." Our sun and other stars that shine are composed almost entirely of plasma. At the sun's core, the temperature reaches 15,000,000°C (27,000,000°F.). The ions in the core are moving so rapidly that they would fly from one another in an instant if they were not confined by the rest of the matter in the sun. The tremendous weight of this matter presses in on the core so that the ions cannot escape. As a result, ions collide so violently that their nuclei join together. This process, called *nuclear fusion,* produces a tremendous amount of energy, which travels from the interior to the solar surface. Some of the energy causes the

Freezing Out Resistance

When chilled to extremely low temperatures, certain solids change their state, becoming *superconductors,* materials through which electricity can flow without any resistance.

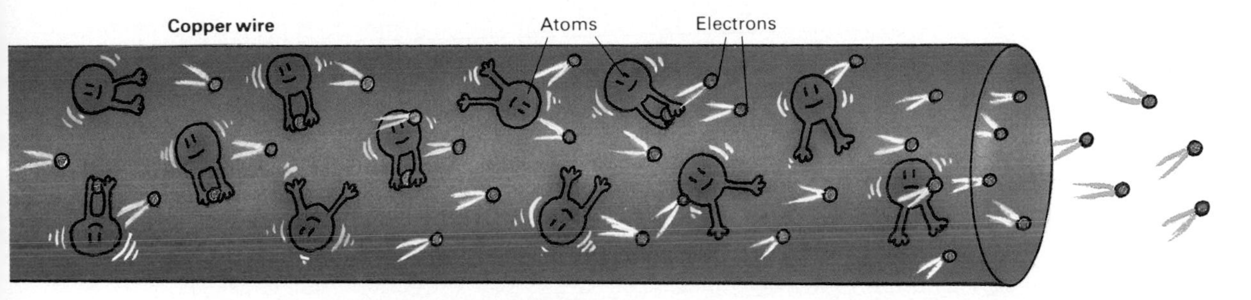

Current in an Ordinary Wire

An electric *current* (a flow of electrons) encounters resistance in an ordinary conductor such as a copper wire, *above,* because atoms in the wire vibrate strongly. This causes the electrons to flow in a disorderly way, and many are "caught" by vibrating atoms. Not only does this cause a loss of current, but these electrons transfer their energy of motion to the wire, heating it. Resistance can be useful in some appliances designed to generate heat, such as a hot plate, *left,* whose heating element is made of an *alloy* (metal mixture) designed to have high electrical resistance.

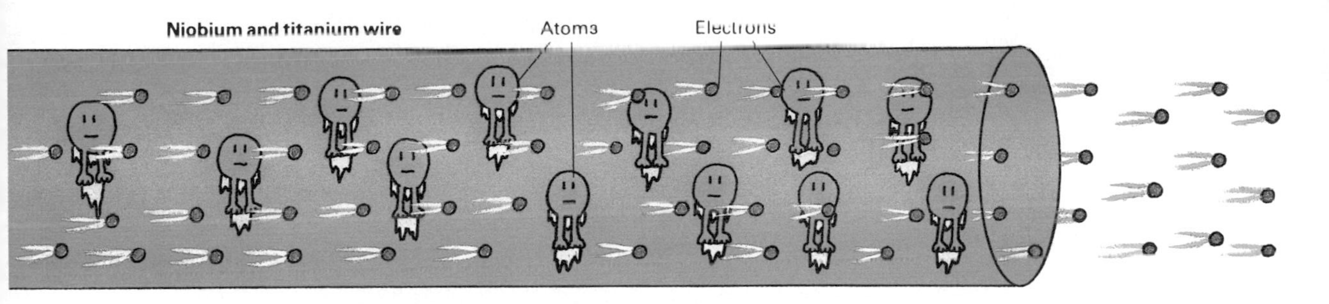

Current in a Superconducting Wire

Electric current flows without any resistance through a superconductor such as a wire made of a niobium and titanium mixture, *above.* All the electrons flow through the wire in an orderly way because the atoms are virtually motionless. Superconductors must be very cold to stop the motion of the atoms, but scientists are developing materials that are superconductors at close to room temperature. Used as wire coils in compact but extremely powerful electromagnets, superconductors mounted in an experimental high-speed train and its track, *left,* create fields of magnetism that hold the train up in the air so that it runs without friction.

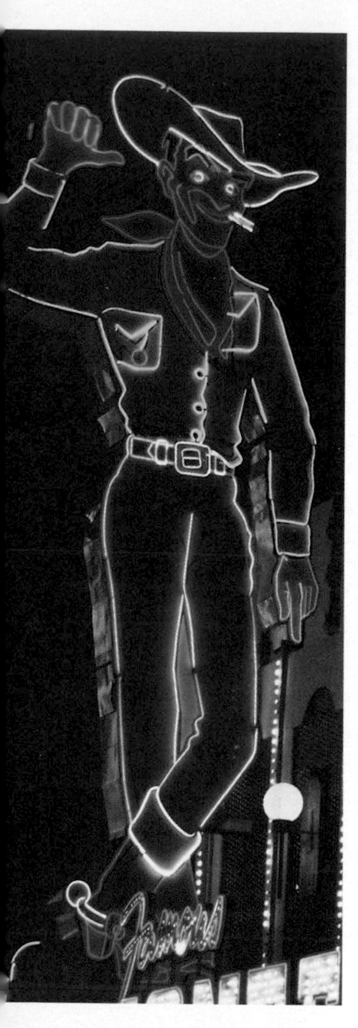

Neon signs contain a state of matter called *plasma*. Inside a sign's thin glass tubes, electric current strips electrons from gas atoms, creating a glowing plasma made up of charged atoms and free-flying electrons.

sun's atmosphere to radiate with a glow that we know as sunlight.

Beyond plasma at the hot end of the temperature scale is the *free nucleon state*, which forms from a plasma that is heated to a temperature of 10,000,000,000°C (18,000,000,000°F.). At this temperature of billions of degrees, *nucleons* (protons and neutrons) that make up the nuclei in the plasma move so rapidly that they fly apart from one another.

Nothing in the natural universe has a temperature above 10,000,000,000°C. Particle accelerators, however, can heat nuclei far beyond that temperature by accelerating two beams of nuclei to nearly the speed of light, then forcing the beams to collide head-on. On the rare occasions when two nuclei meet head-on, most of their energy of motion is transformed to heat. The temperature gets high enough to liberate nucleons, but, because only a few nuclei are involved, the collision does not heat the accelerator.

Beyond the free nucleon state is the state of *quark matter*, consisting of a soup of electrons and particles called quarks. Quarks are the building blocks of nucleons. A proton or a neutron is made up of three quarks.

The temperature needed to transform free nucleons into free quarks is in the trillions of degrees—10,000,000,000,000°C (18,000,000,000,000°F.). Machines now on the drawing board will one day force heavy nuclei to collide so violently that for a few instants they will behave like a mass of free quarks.

What lies beyond quark matter? In the 1960's, Soviet physicist Andrei D. Sakharov made a suggestion that still holds up in theory. Sakharov suggested that, at almost unimaginable temperatures—possibly as high as hundreds of trillions of trillions of degrees—quarks and electrons would turn into formless particles unlike any that exist today.

Such temperatures could never be generated in any conceivable device. *Cosmologists*—scientists who study the universe as a whole—say, however, that the temperature of the universe was even higher than this during the first instant of its existence.

Cosmologists are almost certain that the universe began 10 billion to 15 billion years ago with the *big bang*—the bursting into existence of a tiny, unimaginably hot, expanding blob of formless matter. As the universe expanded, it cooled, and the matter went through a sequence of changes of state. The variety that we see today—from subatomic particles to galaxies—arose from this sequence of changes.

Cosmologists have pieced together a fairly complete picture of when all the known states of matter first came into existence. With the cosmologists, we can "jump back in time"—in our imaginations—to the big bang, then travel forward to the present, stopping along the way to witness the various changes of state. We would see how each change left an imprint on today's universe.

It is hard to pin down an exact time for our first stop—the time when the formless particles turned into quark matter. Most cosmologists believe, however, that it was an amazingly short time—less than one-billionth of one-trillionth of one-trillionth of one second—after the big bang.

At this stop, we would see the formless particles slow down as the universe cooled. As the particles slowed down, they would turn into electrons, quarks, and *antimatter particles*. For every type of particle, there is an object called an *antiparticle*. The antiparticle has the same mass as the particle, but its other properties, such as electric charge, are the opposite of the particle properties. For example, the positively charged *positron* is the antiparticle of the negatively charged electron.

When a particle meets its antiparticle, they annihilate each other, converting their mass into energy. Most of the energy then turns right back into particles. Matter-antimatter annihilations—tremendous flashes—are followed in an instant by new materializations of matter and antimatter.

And perhaps we would learn at this stop why this change of state, or perhaps an even earlier one, has left a puzzling imprint on today's universe—the domination of the universe by matter, with only small traces of antimatter. This has baffled scientists because matter and antimatter are normally created and destroyed together in equal amounts.

Sakharov theorized that, during an early change of state, matter might have somehow got slightly ahead of antimatter. After a tremendous number of annihilations, some matter would have been left over—the matter that makes up everything that there is in the universe today.

At our second stop, one-thousandth of a second after the big bang, the universe would have cooled to a temperature of less than

Evidence of an exotic state called *quark matter* might be found only amid a shower of subatomic particles produced when nuclei collide inside an "atom smasher," *below*. Quarks, which make up protons and neutrons, have not existed by themselves since a fraction of a second after the beginning of the universe.

Matter Takes a Trip on the Inter-State

Temperature determines what state matter is in. At temperatures of trillions of degrees, matter is in an exotic state consisting of only indivisible particles called quarks and electrons. As this matter cools, quarks combine in groups of three known as *nucleons* (protons and neutrons). When protons and neutrons lose heat, they join together to form atomic nuclei, which, with free-flying electrons, make up a state called *plasma*. As a plasma's temperature drops, electrons go into orbit around nuclei to form atoms of *gas*. As a gas becomes colder, the atoms move closer together, eventually forming molecules that begin to slide over one another in the *liquid* state. Chilling a liquid forces its molecules to take up definite positions relative to one another, creating a rigid structure called a *solid*. Certain substances assume exotic states at extremely low temperatures. Helium becomes a *superfluid,* a liquid that flows without resistance, not even forming drops. Certain solids become *superconductors,* capable of carrying electric current without any resistance.

10,000,000,000,000°C (18,000,000,000,000°F.). We would see the soup of free-flying electrons and quarks gradually slowing down. The electrons would still be moving much too rapidly to change state, but the quarks would join together in sets of three to form protons and neutrons, creating the free nucleon state.

We would be astonished by the thoroughness of this change of state. Not a single quark in the entire universe escaped the process. The imprint on today's universe is that all quarks appear to be imprisoned in protons, neutrons, and other subatomic particles.

At our third stop, the universe would be about 3 minutes old and would have a temperature of about 10,000,000,000°C. (18,000,000,000°F.). Here we would witness the slowing of free-flying electrons, protons, and neutrons. Cooling nucleons would fuse to create the first atomic nuclei. Together, the free electrons and the nuclei would form plasma.

This change of state was far from complete. The imprint of this change is the tremendous amount of hydrogen in today's universe. This indicates that only a fraction of the nucleons fused, because the nucleus of the most common form of hydrogen is a single proton. All other nuclei are products of fusion. Today, even after all the fusion that has taken place in all the stars that ever existed, 75 per cent of the matter in the universe is still hydrogen.

At our next stop, the universe would be 300,000 years old, and its temperature would have fallen to 100,000°C (180,000°F.). The free

Principality of Plasmas
"Birthplace of Nuclei"
e-
Grand Duchy of Gases
"Where Atoms Become Complete"
1
Lordship of Liquids
"Famous for Flowing water"
1
H2O WATER COMPANY
to the Satrapy of Superfluids
Sheikdom of Solids
"Specializing in Crystal and Glasses"
to the Serene Kingdom of Superconductors
H2O Ice Company
VanSeveren

A History of the States

The universe began with the *big bang,* when a tiny, unimaginably hot blob of matter exploded into existence. As the universe expanded, the matter cooled and—at very irregular intervals—changed its state. The earliest known state, quark matter, lasted for only one-thousandth of a second. Plasma and gases coexisted for about 1 billion years, after which they were joined by liquids and solids. In the early 1900's, scientists created two previously unknown and very cold states—superconductivity and superfluidity—in the laboratory.

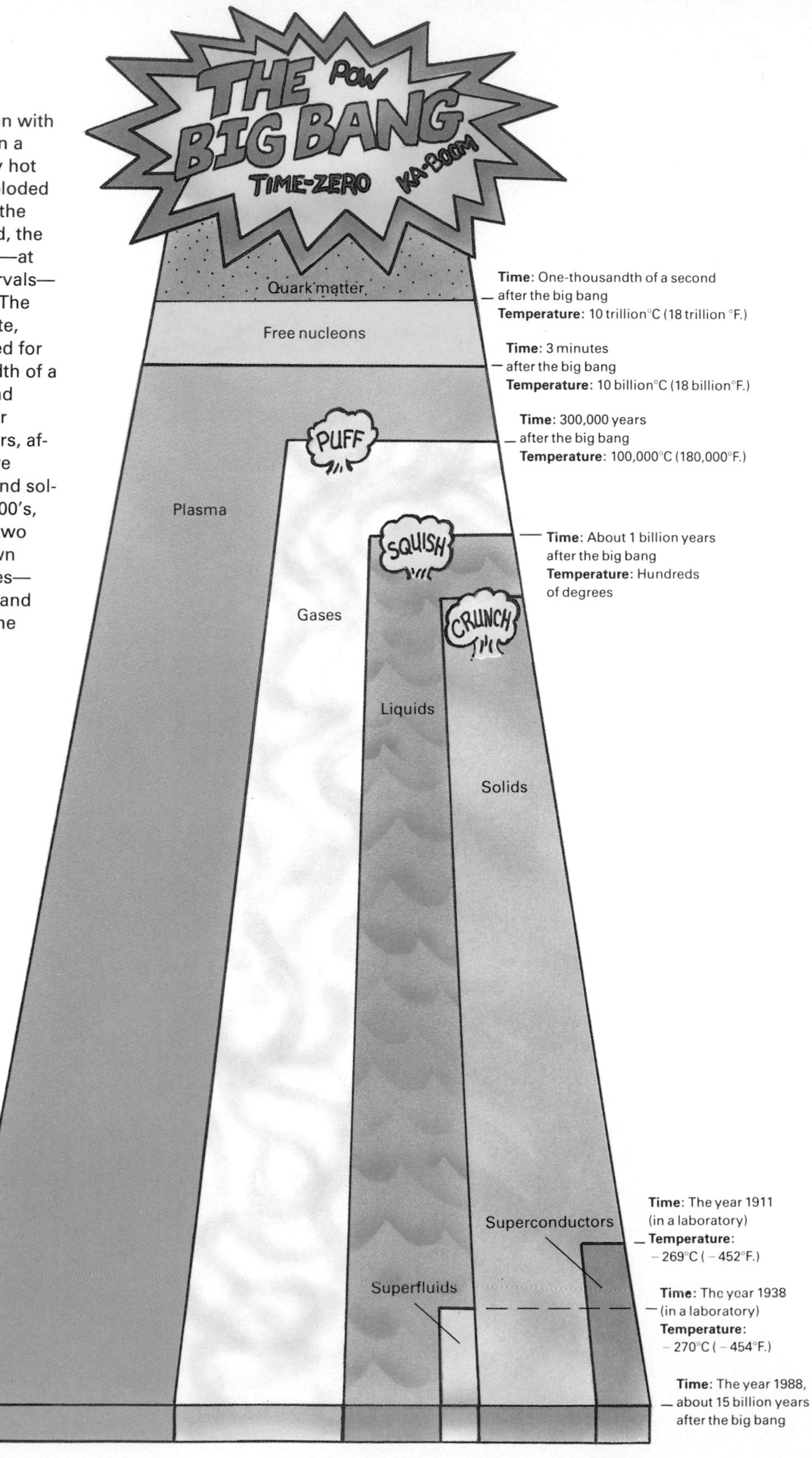

electrons would have slowed so much that they would begin to go into orbit around nuclei, creating the first complete atoms of gas.

At this stop, we would witness a great event. Orbiting electrons do not absorb light waves as readily as do free electrons. As a result, when the change to a gaseous state occurred, space became capable of conducting light. The universe "lit up."

The change from plasma to gas has left another imprint—the *cosmic background radiation*—the microwaves scattered throughout today's universe. Microwaves are a type of electromagnetic radiation, as are radio waves, infrared rays, visible light, ultraviolet rays, X rays, and gamma rays. Electromagnetic radiation has varying wavelengths, with light being shorter than microwaves. Nearly all the light that was present when the change from plasma to gas occurred is still with us, but not in the form of visible light. As the universe expanded, the light waves expanded with it, stretching into the cosmic background radiation. Physicists Arno Penzias and Robert Wilson of AT&T Bell Laboratories in Holmdel, N.J., discovered this radiation in 1965.

We would make two more stops before returning to the present. At the first one, the universe would be about 1 billion years old, and various parts of it would have different temperatures. We would stop at a place where the temperature was measured in the hundreds of degrees.

Here, we would observe two changes of state. Slowing gas atoms would move closer and closer together, unable to escape the attractions between them. Eventually, they would bump one another. Some of them would begin to slide over one another. Masses of them would draw together, forming the first liquid. Some atoms would cool so much that they would form orderly geometric patterns—the crystal lattices of the first solids. The resulting collection of gases, liquids, and solids would be well on its way to becoming the first planet. These particular changes left no imprint on our world. Billions of years later, however, identical changes led to the formation of Earth and all its living creatures.

Our final stop is in Miletus, around 600 B.C.—not to observe another change of state, but to honor Thales. It matters little that something as common as water did not prove to be the universal substance. One has to look deeper inside the atom, and further back in time, to find a more likely claimant to that throne—some sort of formless particle. But Thales' basic insight—that changes of state play a major role in creating the rich variety of the world around us—has stood the test of time.

For further reading:

Heppenheimer, T. A. *The Man-Made Sun: The Quest for Fusion Power.* Little, Brown, 1984.

Lemonick, Michael D. "Superconductors!" *Time,* May 11, 1987.

Morrison, Philip. *Powers of Ten.* Scientific American Books, 1982.

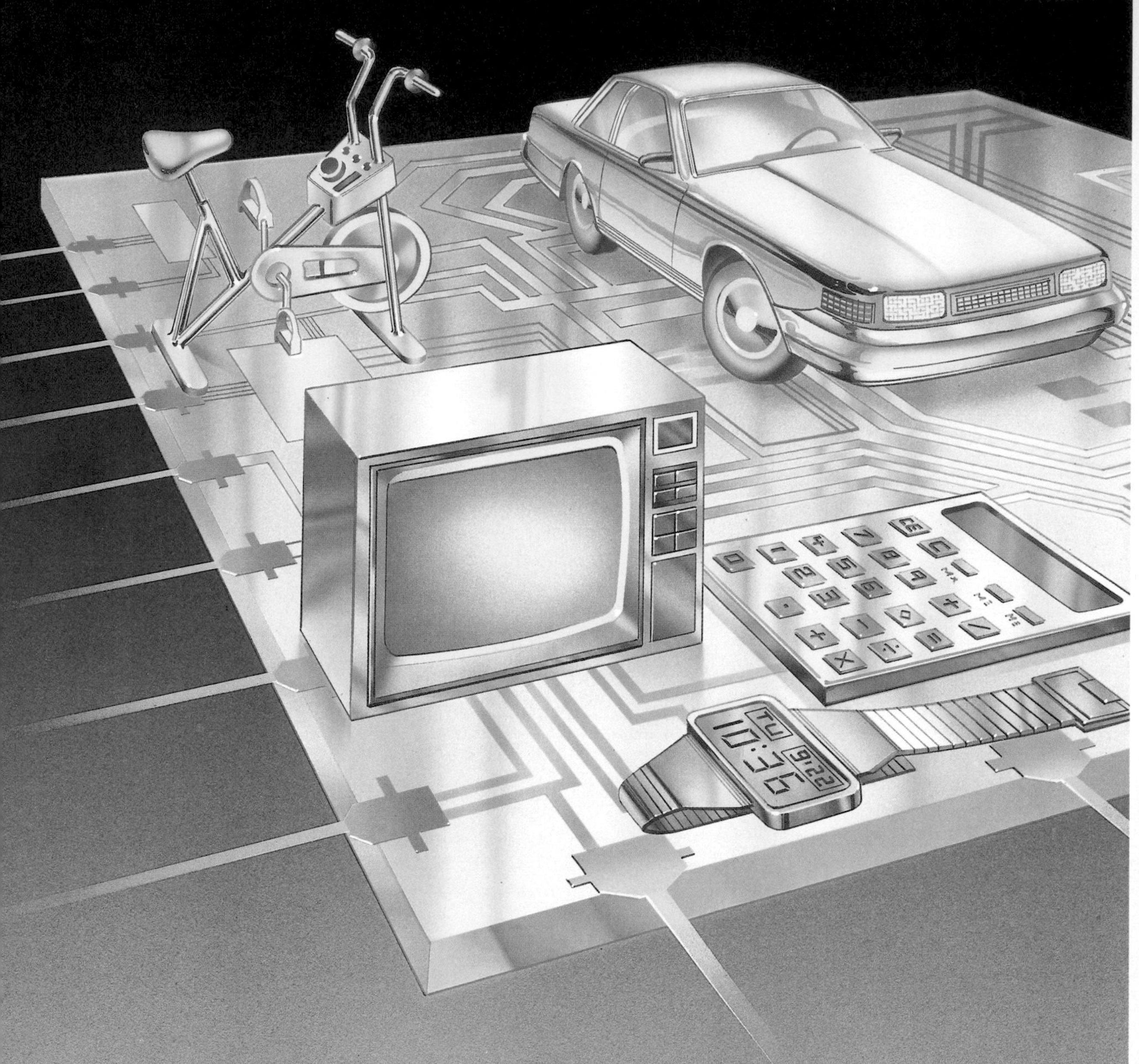

BY ARTHUR FISHER

The Microchip—A Miniature Marvel

Electronic circuits built into tiny slabs of silicon have revolutionized products ranging from automobiles and computers to cameras and wrist watches.

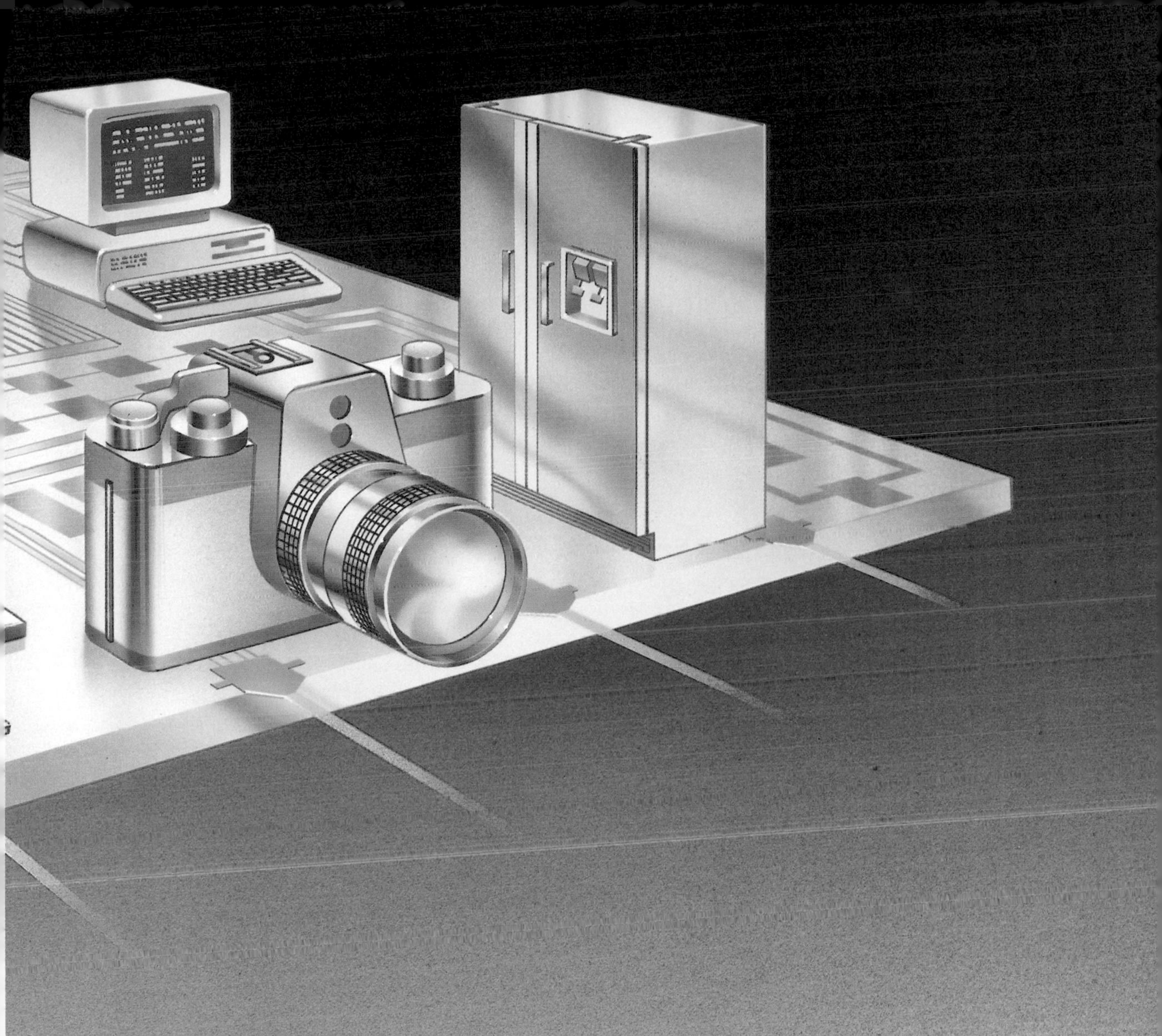

Here is a high-tech riddle: What can you find in a refrigerator, microwave oven, sewing machine, toaster, exercise bicycle, portable computer, automobile ignition system, hearing aid, video game, camera, digital watch, and home security system—in short, practically everywhere you look? The answer, as you have surely guessed by now, is a microchip—a miniature marvel that has transformed our lives in an amazingly short time.

This device—also known as an electronic chip, a microprocessor, or a computer chip—is a network of electronic circuits made up of components built into a flat crystal substance called silicon. A modern microchip may have 1 million components on a square only 6 millimeters (¼ inch) on a side—no bigger than a baby's thumbnail.

Tiny they may be, but some of these chips have enough computing power to outperform the world's first electronic computer—a

Glossary

Bit: A *bi*nary digi*t*—0 or 1—in the code used by electronic chips.

Byte: A group of eight bits.

Component: A circuit part with a specific task such as storing an electric charge or controlling current.

Integrated circuit: A circuit made up of components built into a surface that also acts as part of the circuit.

Microprocessor: A microchip that performs computer functions such as arithmetic and logic operations and memory storage.

Semiconductor: A material that conducts electricity poorly but, when impurities are added, can be used for electronic components.

Switch: Area in a chip that is either "on" or "off" to represent 0 or 1.

Transistor: A semiconductor component used to control the flow of electric current.

giant called ENIAC (*E*lectronic *N*umerical *I*ntegrator *A*nd *C*omputer). ENIAC weighed 27 metric tons (30 short tons) when it was completed in 1946. Today, small personal computers made possible by powerful microchips have endowed people from all walks of life with the means to perform feats of computation and memory that scientists and engineers could only dream of in 1946.

The chip is also behind the revolutionary development of the pocket calculator. Before the 1970's, calculators were heavy electromechanical machines about half the size of a typewriter, and they cost hundreds of dollars. Then, in 1970, came the first calculators using microchips. They were *four-function* machines—they could add, subtract, multiply, and divide—and were handheld. But they were heavy and cumbersome and cost between $300 and $400. The price fell as low as $29 within a year, as chips began to be produced in volume. Today, for less than $12, you can buy a small calculator that handles the four basic functions and performs such other tasks as computing square roots and trigonometric functions.

Chips have revolutionized the computer, the calculator, and a host of other products because chips process information with amazing speed and "remember" tremendous amounts of data. New products continue to be developed because chips are becoming faster and their memory capacity is growing by leaps and bounds.

Increases in chip speed and capacity are a direct result of dramatic reductions in the size of the components built into the chips. The components are much too small to be seen with the naked eye. By mid-1987, chip makers had shrunk components on experimental chips to an astonishing 0.7 micron. (One micron equals 0.001 millimeter, or 0.00004 inch.)

Today's manufacturing methods cannot make components much smaller than this. And even if they could, the circuits made of materials used today would not conduct electric current efficiently. So to achieve further reductions, researchers are experimenting with new materials and new ways to make chips. Some scientists are working on circuits that are not even electronic but instead use beams of light to compute, store information, and operate devices.

The rapid advance of technology, however, has not changed the way in which circuit components communicate with one another. Even the most sophisticated chips of today speak the same "language" that ENIAC used. The basis of this language is a very simple code: All data—words and numbers—are translated into strings of 1's and 0's. The code is called *binary* (consisting of two) because only two digits are used. Devices that use a digital code are described as *digital*, as in *digital computer* or *digital watch*.

To communicate with human users, digital devices translate the code back into words and numbers. A digital watch, for example, turns 1's and 0's into the numerals displayed on the watch face.

Why do circuit components use a code that seems fit for simple-

The author:
Arthur Fisher is science and technology editor of *Popular Science* magazine and author of the Special Report THE FABULOUS LASER in the 1987 edition of *Science Year*.

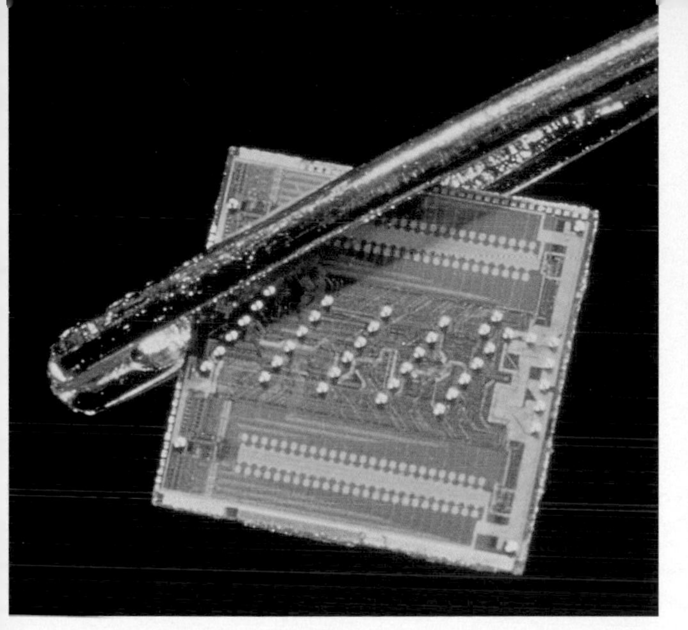

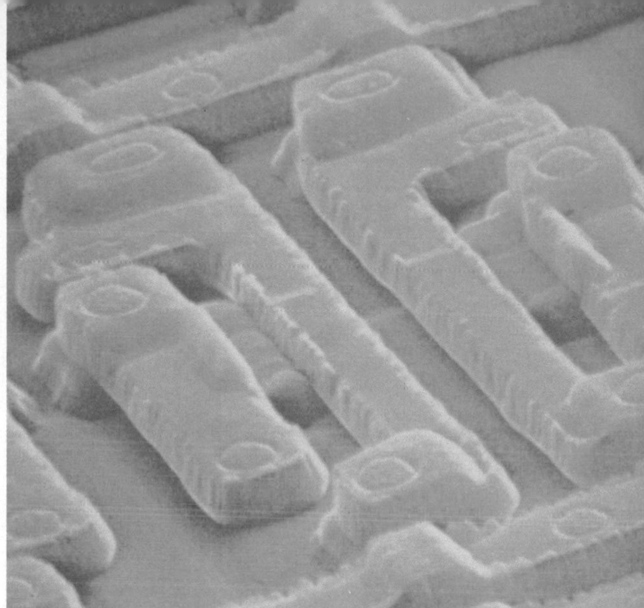

A state-of-the-art electronic chip is small enough to pass through the eye of a needle, *above left*. An experimental chip, *above,* contains components only 0.0005 of a millimeter wide.

tons? Because it allows chips to operate in a very simple way. A chip contains switches, wires, and connections that can exist in only two electrical states—one state corresponding to a 1 and the other to a 0 in the binary code. A switch can be on or off, a wire can be conducting or not conducting current, and a connection can have a high or low voltage. Coded messages travel throughout a digital device as strings of electric pulses. Each pulse is called a *bit*, a contraction of *binary digit*. Many devices handle these bits in groups of eight, called a *byte*, from *b*inar*y*, digi*t*, and *e*ight.

There is a disadvantage to the binary code, however. The total number of symbols that have to be manipulated is much greater than it would be in normal communications using the letters of the alphabet and the digits 0 through 9. For example, the number *1,000* in binary code is *1111101000*, and the letter *A* is *01000001*. More symbols mean that more switches, more wires, and more connections must be used.

In the case of ENIAC, more symbols meant more *vacuum tubes*, electronic devices used as switches in that computer. A vacuum tube—a technological ancestor of today's electronic chip—is a glass globe from which almost all the air has been pumped out, and which houses various combinations of wires and metal parts. The picture tube in your television set is a vacuum tube, but this is the only kind of vacuum tube in common use today.

At the time ENIAC was built, however, tubes were widely used in radios as well as the giant computer. A typical radio housed a cluster of vacuum tubes, each no larger than a light bulb. One kind of tube commonly used as a switch had three main parts separated from one another by empty space but connected by wires to the base of the tube. When electrons flowed through the space to one of the parts, called the plate, the vacuum tube—acting as a switch—was on. This represented a 1 in the language of computers. When electrical

How Chips Work

All microchips work the same way, whether they store information or operate devices. Thousands of switches built into a chip turn on and off with incredible speed, transmitting data and commands in the form of pulses of electric charge. In a computer chip, the switching operation represents symbols of computer language.

A 01000001

B 01000010

C 01000011

The Language of Chips
Computer language is made up of 0's and 1's. Ordinary letters such as A, B, and C must be translated into a series of 0's and 1's in order to be understood by a computer chip. The 0's and 1's are represented by the turning off (0) and on (1) of switches that are built into chips.

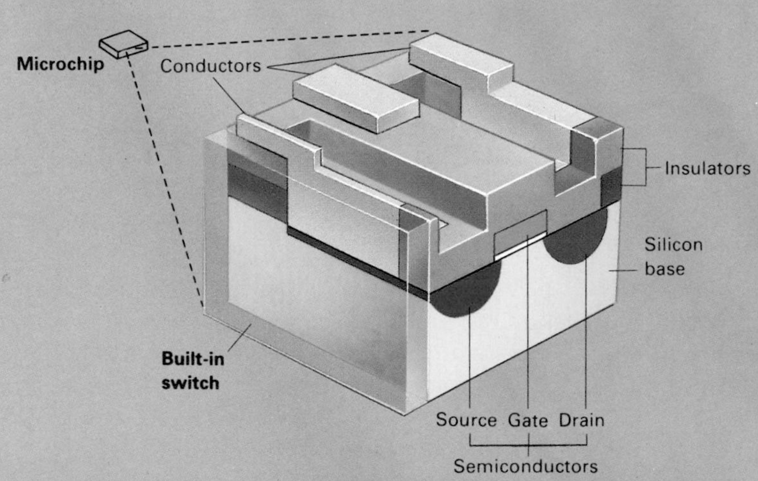

Anatomy of a Switch
A typical microchip has thousands of switches built into it, *above.* Each switch is built in layers, and each layer has a specific function. Conductors carry pulses of electric charge between switches. Insulators, which cannot conduct, prevent charges from wandering off their correct paths. Semiconductors, which are "reluctant" to conduct but can be forced to do so, perform the actual switching. The switch has three special semiconductor areas—the gate, the source, and the drain—each connected to a conductor. The gate controls the flow of electric charge pulses from the source to the drain, turning the switch on and off, *below.*

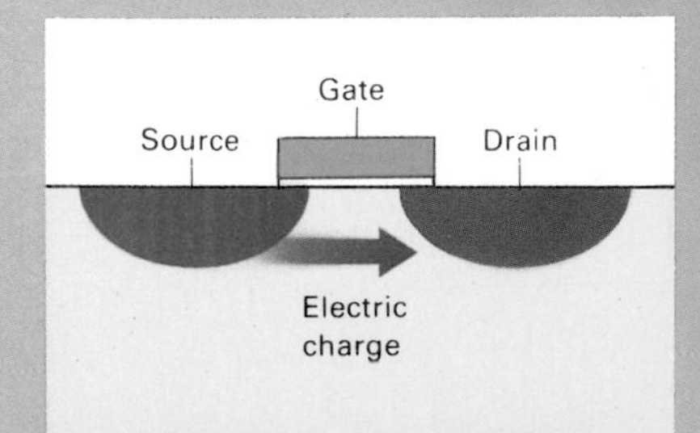

Switch On, Representing 1
When the gate is on, electric charge flows from the source to the drain, turning on the switch. This represents a 1 in the language of computers.

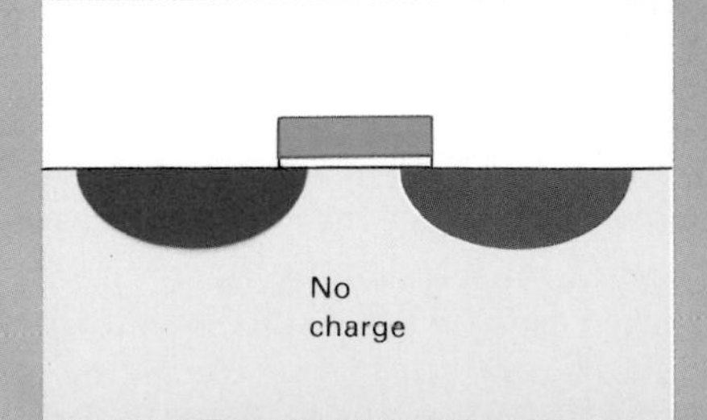

Switch Off, Representing 0
When the gate is off, no charge flows from the source to the drain, and the switch is off. This represents a 0 in the language of computers.

forces prevented electrons from flowing to the plate, the "switch" was off, representing a 0.

But vacuum tubes had drawbacks. They took up a great deal of space and consumed large amounts of power. ENIAC, with about 1,800 tubes, occupied a chamber measuring 10 by 17 meters (30 by 50 feet) and required about 140 kilowatts of electric power to perform calculations. This is enough power to run about 40 typical homes in the United States today. The glass tubes also were unreliable. Fifty of ENIAC's tubes burned out in its first month of operation.

Scientists wanted to replace the vacuum tube with something smaller, less power-hungry, and less delicate. This was the spur to the invention of the transistor, another ancestor of the microchip.

Physicists had known for years that a group of materials called *semiconductors* might control electric current. Semiconductors are crystalline materials such as silicon, the main component of sand, and germanium, a metallic element. If a small number of atoms of another material, such as phosphorus, could be "scattered" among semiconductor crystals, scientists believed, devices made of such crystals might perform the functions of vacuum tubes.

There are two basic types of microchips—*memory*, to store data; and *processing*, to manipulate data. A personal computer, *top*, has a bank of memory chips to store information, such as numbers, letters, and graphics. By contrast, a calculator, *above*, has no memory chips but instead stores small amounts of data in memory circuits on its single processing chip.

The transistor, invented in 1947, was this kind of semiconductor device. The inventors of the transistor—United States physicists John Bardeen, Walter H. Brattain, and William Shockley—shared the 1956 Nobel Prize in physics for their work.

The first transistor was tiny compared with a vacuum tube. It was a piece of germanium inside a metal case about 13 millimeters (½ inch) long. The transistor could do everything the vacuum tube could do. A few years passed, however, before transistors became inexpensive enough to compete with vacuum tubes—an achievement that launched a transistor revolution.

But the search for the ideal vacuum tube replacement was far from over. Electronic circuits containing transistors took a long time to manufacture. Workers had to mount transistors and other electronic components—such as *resistors*, which regulate current flow, and *capacitors*, which store electric charge—by hand on a flat piece of plastic called a *circuit board*. Then they connected the components by soldering wires between them. This was a slow process, and the soldered joints were prone to failure.

Moreover, many designers wanted to build even smaller circuits. One of them, a young engineer named Jack Kilby, was working in 1958 for Texas Instruments Incorporated (TI) of Dallas. Researchers there had developed the first commercially successful silicon transistors and were trying to find a way to miniaturize certain electronic components for military use. Kilby had a brainstorm: Make all the circuit components from a single semiconductor material so that a small piece of that material not only served as a circuit board but also acted as an integral part of the circuit. Because of this dual function, such a circuit is called an *integrated circuit* (IC).

How Chips Are Made

The creation of a microchip is a long and detailed process that begins with large circuit patterns and ends with microscopic layers on a silicon chip.

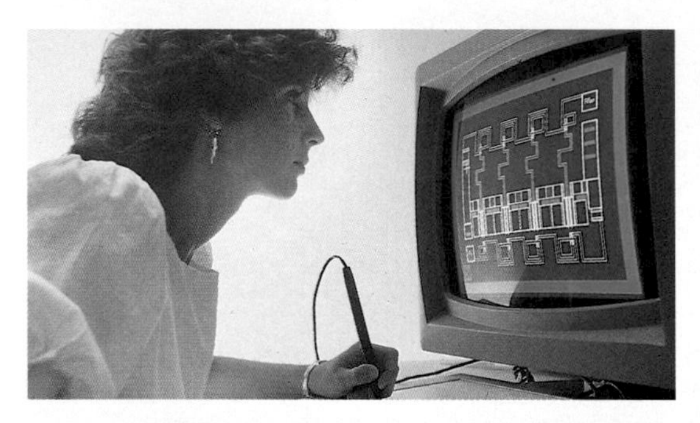

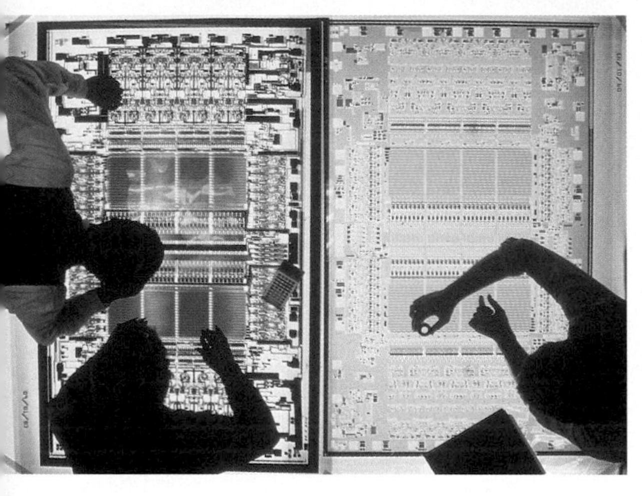

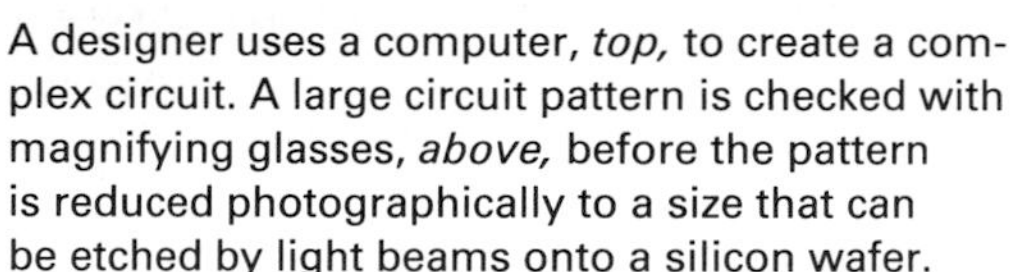

A designer uses a computer, *top,* to create a complex circuit. A large circuit pattern is checked with magnifying glasses, *above,* before the pattern is reduced photographically to a size that can be etched by light beams onto a silicon wafer.

A technician, *above,* holds gleaming silicon wafers that have been polished to a mirror finish to remove any imperfections on their surface that might interfere with the workings of the circuits to be etched on them.

Kilby built the first IC in 1958. It was a crudely assembled affair—a sliver of germanium bearing just five components—but it was the first electronic chip.

Kilby's chip evolved into today's microchip with breathtaking speed. In the early 1960's, a chip held 6 to 10 components; by the mid-1960's, a chip held 10 to 100 components; in the early 1970's, 100 to 10,000 components; and in the early 1980's, more than 100,000 components. Chips holding 2 million components are now in production.

Why squeeze so many components into such tiny spaces? There are four reasons. First, it saves space—a crucial consideration in everything from spacecraft to weapons to consumer appliances. Second, it saves money; it is much less expensive to make one chip with

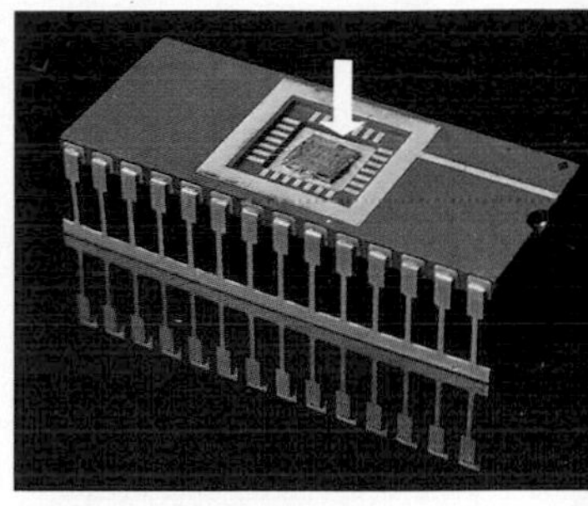

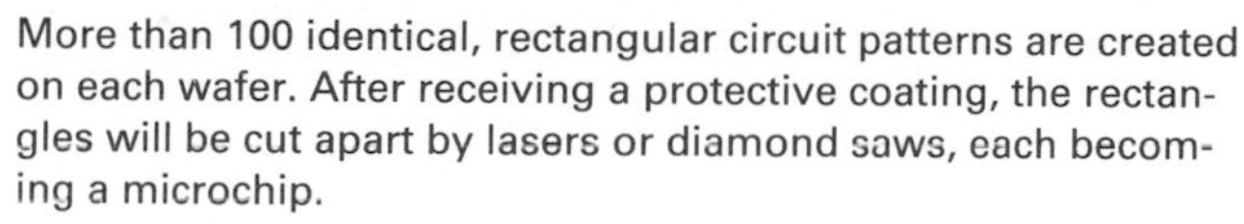

More than 100 identical, rectangular circuit patterns are created on each wafer. After receiving a protective coating, the rectangles will be cut apart by lasers or diamond saws, each becoming a microchip.

A finished chip (arrow) is mounted in a *package, top,* for easy installation in a product such as a microwave oven, *above.* Metal contacts on the sides of the package conduct electric current to and from the chip.

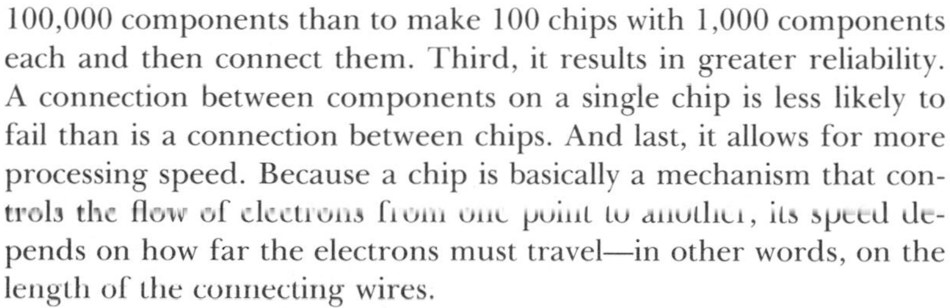

100,000 components than to make 100 chips with 1,000 components each and then connect them. Third, it results in greater reliability. A connection between components on a single chip is less likely to fail than is a connection between chips. And last, it allows for more processing speed. Because a chip is basically a mechanism that controls the flow of electrons from one point to another, its speed depends on how far the electrons must travel—in other words, on the length of the connecting wires.

Fast-acting chips have made today's swiftest computers more than 300,000 times as fast as ENIAC. That pioneering machine could do about 5,000 calculations per second. By contrast, a Cray 2 supercomputer made by Cray Research, Incorporated, of Minneapolis, Minn., can perform more than 1.7 billion calculations per second.

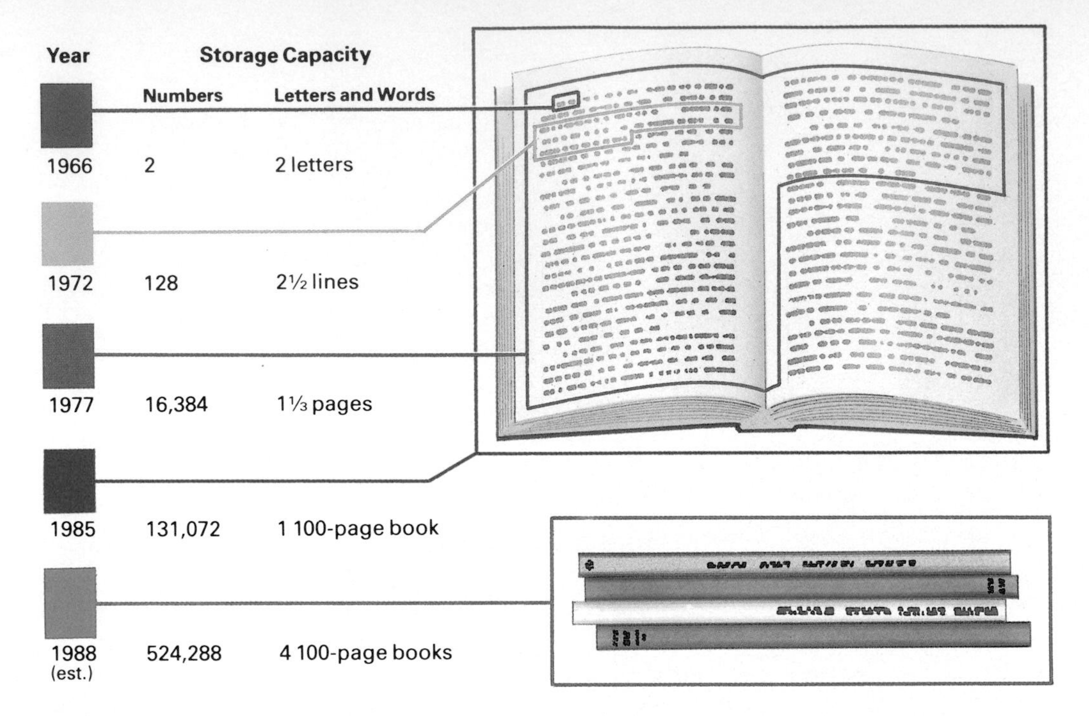

Surge in Storage
The amount of information that can be stored on a microchip has skyrocketed since chips were first developed. In 1966, a chip called a dynamic random access memory (DRAM) could store the equivalent of two numbers or two letters. Storage capacity grew steadily and, by 1985, the entire contents of a 100-page novel could fit on a DRAM chip. In mid-1987, researchers were in the process of developing chips that could hold four of these books.

Chips have become not only smaller and faster but also specialized to fulfill a variety of tasks. One special breed is the memory chip, which stores information in the form of electric charges. There are two basic kinds of memory chips—the *read-only memory* (ROM) chip and the *random-access memory* (RAM) chip.

The ROM chip permanently stores information put into it at the time of manufacture. This information can neither be erased nor added to and generally consists of a set of instructions for operating a device such as a computer, a calculator, or a video game.

On the other hand, information on a RAM chip can be erased or added to. A RAM does not store memory permanently. When the power is turned off, all the electric charges stored in its circuits are wiped out and the memory disappears. RAM chips are used in computers and programmable calculators, two devices that need to store large amounts of information for brief periods of time.

Manufacturers of personal computers have taken advantage of the tremendous surge in the storage capacity of RAM chips to beef up their machines' memories. For example, the Macintosh computer, introduced in January 1984 by Apple Computer, Incorporated, had a RAM of 128 kilobytes (128K), enough to store 100 pages of double-spaced typewritten text. (A kilobyte is about 1,000 bytes.) The Macintosh Plus, introduced in 1986, could store about 1 million bytes, or 1 megabyte, on RAM chips. One megabyte is enough for 800 typed pages. By 1987, permanent memory storage devices for personal computers could hold up to 40 megabytes.

Remarkable as they are, memory chips can neither add nor make decisions. Those functions are reserved for circuits called *arithmetic-logic units* (ALU's). When memory circuits, ALU's, and certain other circuits are combined on one chip, the result is a *microprocessor*—basically a tiny computer.

Microprocessors all work the same way in products that are as different as a computer and a camera. Both products have *input devices* that send pulses of electric charge representing something in the "outside world." In a computer, the input device is usually a keyboard. It transmits pulses representing numbers, letters, and other symbols typed by the individual using the computer. In a camera, the input device can be a light sensor that transmits information about the amount of light hitting the object to be photographed.

The microprocessors then use this information. In the computer, the microprocessor might multiply one number by another. In the camera, it might determine how the focus should be set.

Finally, the microprocessors emit pulses that operate *output devices*. The chip in the computer might command a televisionlike screen to display the result of a calculation. The camera's chip might operate a tiny motor that changes the position of the lens.

Product manufacturers want as many components as possible crowded onto each chip. So chip makers have borrowed a construction technique from architecture. Just as constructing a building with many stories conserves land, building components into the chip, and wires onto the chip, in layers conserves chip surface area. A complex microprocessor might have a dozen or more layers.

The process of producing a microchip begins with design. Chip designers use computers to help determine the intricate circuit patterns for each layer. Drafting machines then trace the patterns onto paper on a scale tens of thousands of times larger than chip size. Next, the patterns are photographed and reduced to chip size on glass plates. All areas of the plate except the circuit pattern are coated with a material through which light cannot pass. The plates are called *masks*, and they serve a function similar to that of a stencil.

Meanwhile, other workers prepare the silicon material on which the circuits will be "printed." First, they slice circular *wafers* from a cylinder of silicon about 13 centimeters (5 inches) in diameter and 75 to 100 centimeters (30 to 40 inches) long. Each centimeter of cylinder length provides about 10 wafers.

The surfaces of the wafers are rough, so the workers polish them with abrasive liquids. The workers then expose the wafer to oxygen in an oven, causing a hard layer of silicon dioxide to form on the surface. Silicon dioxide is an *insulator*, a material that does not conduct electricity. Finally, the surface receives a coating of a light-sensitive chemical.

A technician then puts a mask and a wafer into a machine that shines a light through the stencillike mask, imprinting an exact duplicate of the circuit patterns onto the wafer surface. Next, chemicals *etch* (eat away) the unexposed portion of the wafer surface, leaving a silicon dioxide replica of desired circuit patterns. Depending on the design of the chip, other substances are then deposited on the wafer in various ways. In one technique, called *diffusion*, a worker feeds the wafer into an oven filled with a gas of the substance to be deposited. Gas atoms enter the areas of exposed silicon. When the wafer cools, the atoms insert themselves into the silicon crystal, forming sectors that will act as transistor parts. Repeating such steps—coating, shining light through a mask, etching, and diffusion—creates multiple layers.

To make electrical connections from layer to layer and transistor to transistor, a layer of a metal such as aluminum, perhaps 1 micron thick, is deposited on the wafer, then etched to form strands 2 to 10 microns wide that function as electric wire. Finally, the hundreds of chips on the wafer are cut apart with a diamond saw or a laser.

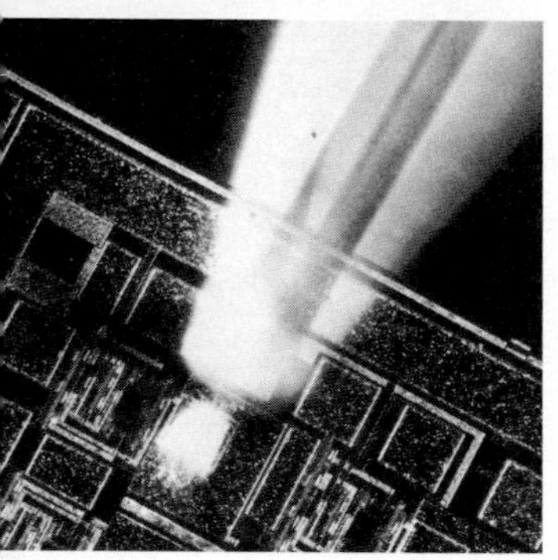

Because electronic chips are reaching their limits of speed in processing digital information, scientists have begun to investigate ways to send these signals on flashes of light, *top*. An experimental chip, *above,* contains components that translate light flashes into pulses of electric current. In the future, chips that both communicate and process data in the form of light will operate faster and overcome problems, such as heat build-up, that plague electronic chips.

Chip making is a big business. Economists forecast that the industry will have annual worldwide revenues of $60 billion by the year 2000. And international competition is intense, with Japan and the United States struggling for an edge. Because all modern technology, including military technology, relies on computers—and therefore on chips—U.S. government leaders are concerned that the United States will fall far behind in chip manufacturing.

The United States has also been concerned about trade practices of Japanese chip exporters. In the summer of 1986, the United States persuaded Japan to stop Japanese exporters from selling chips below cost in markets outside the United States and to grant U.S. chip makers a larger share of the Japanese chip market. In March 1987, the United States said that Japan had not lived up to the agreement and slapped retaliatory tariffs on a wide range of Japanese electronic products.

Spurred by intense competition, how many components will future chip makers be able to jam onto a sliver of silicon? No one is certain, but designers are already running into limits imposed by the laws of physics. One limit involves the wavelength of the light used to etch circuit patterns on wafers. Light that passes through the openings in a mask spreads out slightly. The amount of spreading depends upon the light's *wavelength*—the distance between successive wave crests. As a result, light cannot be used to print anything smaller than its own wavelength. The wavelength used today is about 0.5 micron, and with components and spacing shrinking rapidly, soon light will no longer be able to do the etching job. To overcome this limitation, experimenters have been etching wafers with X-ray beams and electron beams, both of which have much shorter wavelengths than does visible light.

Another limit to component size and spacing is heat. All chip components and connecting strands that carry electric current resist the flow of current by absorbing electric energy and converting this energy to heat. A build-up of heat can cause a chip to fail, so the chip must get rid of heat by radiating it to the surrounding air—in some cases with the help of a fan—or by transferring it to a liquid coolant. But the closer the circuits are packed together in the future, the harder it will be for them to get rid of heat.

One promising solution to the problem of heat is to use "wires" made of materials that become *superconductors* at extremely low temperatures. A superconductor loses all resistance to the flow of current and therefore generates no heat.

Superconductors introduced in 1986 and 1987 brought this solution closer to reality. Previously, the highest temperature at which a material would superconduct was −250°C (−418°F.). The only practical way to cool a material to that temperature is to bathe it in liquid helium, which costs about $3 per liter (1 quart). But the most advanced of the new materials becomes a superconductor at −180°C (−292°F.). Liquid nitrogen, which sells for 6 cents per liter, can cool this material to the superconducting state. In the Special Reports section, see EXPLORING THE STATES OF MATTER.

What if scientists and engineers handled the problems of wavelength and heat so well that manufacturers could build components with dimensions of less than 0.1 micron, about one-tenth of their present size? The resulting circuits might fail because layers of insulation would be too thin to prevent electrons from leaking from one circuit to another.

One solution to the leakage problem would be to develop better insulating materials. But as insulation layers became thinner and thinner, the demand for better materials would become increasingly difficult to meet.

A more radical solution would be to abandon electronic chips and use *photonic circuits* instead. These circuits use pulsed beams of light to transmit data and commands through hair-thin strands of glass called *optical fibers*, so they need no electrical insulation. In addition, using light eliminates the problem of heat build-up

Researchers are experimenting with a variety of photonic circuits. The most radical design is the so-called *biochip*, which would have circuit components made of *organic molecules* (carbon-based compounds that are the building blocks of living matter). Some molecules that could be used are so small that a three-dimensional biochip occupying 1 cubic centimeter (0.06 cubic inch) could be crowded with an astonishing million billion molecular switches.

Will such devices eventually be developed? It is hard to say. But who would have predicted—only 20 years ago—that we would have the kinds of tiny electronic microchips that we have today and that they would play such a large and varied part in our lives?

For further reading:

Heppenheimer, T. A. "Micromicromicrochips." *Popular Science*, December 1986.
Physics Today (special section), October 1986.
Shurkin, Joel. *Engines of the Mind: A History of the Computer*. Norton, 1984.

Science File

Science Year contributors report on the year's major developments in their respective fields. The articles in this section are arranged alphabetically.

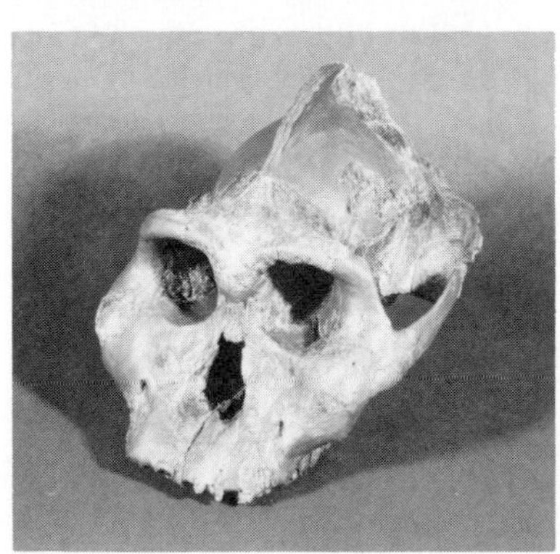

Environment

Genetic Science

Geology

Immunology

Medical Research

Medicine

Meteorology

Molecular Biology

Neuroscience

Nobel Prizes

Nutrition

Oceanography

Paleontology

Physics, Fluids and Solids

Physics, Subatomic

Psychology

Public Health

Science Education

Science Fair Awards

Space Technology

Zoology

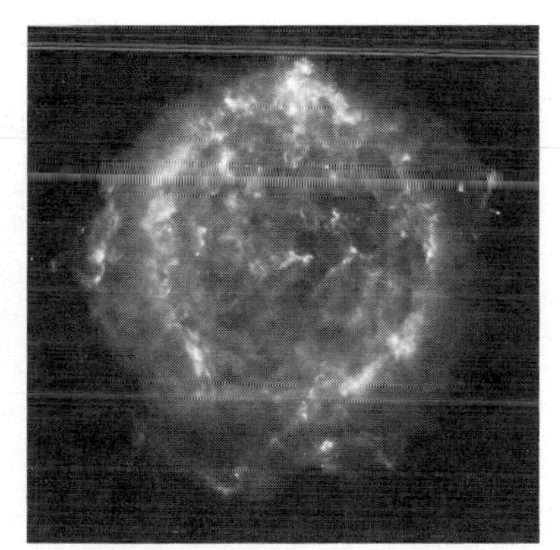

Agriculture

Researchers from Advanced Genetic Sciences, Incorporated, of Greenwich, Conn., sprayed genetically altered bacteria on a small field of strawberry plants in northern California in April 1987. The experimental spraying, the first authorized outdoor test of genetically altered bacteria in the United States, had been delayed by court challenges since 1984.

A judge of the California Superior Court gave the final go-ahead for the field test after concluding that the altered bacteria posed no threat to human beings or the environment.

The bacteria were a strain of *Pseudomonas syringae*, which ordinarily produces a protein that acts as a "seed" for the formation of ice crystals on plants. The gene carrying the genetic instructions for the production of the ice-forming protein was deleted in the experimental *P. syringae*. When sprayed on plants, the altered bacteria were expected to prevent ice damage in temperatures as low as −5° C (23° F.). Because temperatures in April were already much higher than that, the purpose of the first test was simply to see whether the engineered bacteria would displace naturally occurring *P. syringae*. Results of the test were being evaluated in late 1987 by Advanced Genetic Sciences.

Centennial marked. One of the leading systems of agricultural research in the United States—state agricultural experiment stations—marked its centennial in 1987. This system, which has greatly benefited U.S. society and influenced the development of agriculture elsewhere in the world, was established by the Congress of the United States on March 2, 1887. Combining research and education, the more than 50 experiment stations, most of them located at state colleges and universities, have made an important contribution to the world's agricultural knowledge.

Growth hormone. Experiments with genetically engineered *somatotropin*, or growth hormone, in 1986 and 1987 indicated that the substance greatly increases the milk production of dairy cows and causes cattle and hogs to

A researcher sprays genetically engineered bacteria on a field of strawberry plants in northern California in April 1987. This was the first authorized release in the United States of a genetically altered organism into the environment. The bacteria were altered so that they could not produce a protein that ordinarily promotes the formation of ice crystals on plants. Scientists predicted the altered bacteria would help protect crops against frost damage.

grow faster and to put on more lean meat.

Somatotropin is a protein that is ordinarily produced in tiny amounts within an animal's body to stimulate growth. The somatotropin being used in research is produced in large quantities in biotechnology laboratories and given to animals daily by injection.

Scientists at Cornell University in Ithaca, N.Y., reported in October 1986 that somatotropin may increase dairy cows' milk production by up to 30 per cent. And researchers at Cornell and several other institutions reported in October 1986 and March 1987 that somatotropin administered to young pigs increased the animals' muscle mass.

Whole plants from single cells. Researchers in France, Great Britain, and Japan reported in December 1986 and January 1987 that they had succeeded in growing mature plants from rice *protoplasts* (cells with their walls removed). This process, called *protoplast regeneration,* was tried in the early 1970's with tobacco, tomato, and petunia plants, but scientists had been unable to regenerate the protoplast of any major cereal plant.

Protoplast regeneration opens the way for introducing desirable new traits into rice by transferring genes from one variety of rice to another or by fusing two protoplasts from different strains. These procedures can be done only with protoplasts; they do not work with regular cells.

The ability to produce new rice hybrids by means of protoplast regeneration is important because rice is the world's number-one food crop. This advance removed the last barrier to the genetic engineering of rice.

The cultivated morel. The Neogen Corporation, of Lansing, Mich., a private biotechnology institute affiliated with Michigan State University, announced in July 1986 that it had developed the first workable method for cultivating the morel mushroom. The morel, which looks like a golfball on a stick, is one of the most highly prized edible *fungi* (nongreen plants) in the world.

In Neogen's patented process, morel spores are first germinated in culture dishes. The mature mushrooms are later grown in trays of soil under controlled temperature and humidity.

Neogen was licensing its growing method to commercial mushroom producers in 1987. Large-scale cultivation of morels was expected by 1990.

High-tech seeds. Increasing numbers of farmers and gardeners in 1986 and 1987 used *primed seeds* for planting. Primed seeds—also called conditioned, refined, or activated seeds—are seeds that have been coated with protective or growth-promoting substances, usually pesticides or fertilizer. Primed seeds develop into plants that grow vigorously and are more resistant to pests.

Often, in the production of primed seeds, seeds are partially germinated under controlled temperature, light, and moisture conditions. The seeds are then coated, and germination is arrested by drying. These primed seeds can be stored for later planting.

Researchers in 1986 and 1987 were looking for new ways to prime seeds. In March 1987, biologists at Macquarie University in New South Wales, Australia, reported that when they primed tomato and carrot seeds with *potassium nitrate* (saltpeter), the seeds sprouted sooner than usual, and all sprouted at about the same time.

Little-known plants. Many forgotten or underused plants around the world could be of great benefit to humanity, according to a June 1986 report by biologist Noel D. Vietmeyer of the National Research Council (NRC). The NRC is an independent group headquartered in Washington, D.C., that advises the federal government on matters of science and technology.

Vietmeyer said there are thousands of species of plants that could be used for food, fuel, fibers, building materials, and raw materials for industry. He said many of these plants have been neglected "merely because they are native to the tropics, a region generally neglected because the world's research resources are concentrated in the temperate zones."

Among noteworthy plants that have been overlooked, Vietmeyer said, is the pejibaye palm, or peach palm, native to Central America. The pejibaye's chestnutlike fruit contains a nutritionally balanced blend of carbohydrates,

A U.S. Department of Agriculture geologist studies a map showing soil density and underground channels at a test plot in Georgia. Such maps, produced by a radar device towed behind a tractor (background), will help scientists learn how pesticides move through soil and into ground water.

protein, oil, minerals, and vitamins. Another nutritious but largely ignored plant is amaranth, a grain that was grown by the Incas and Aztecs of Central America and Mexico before their conquest by Spanish invaders in the 1500's. According to Vietmeyer, underutilized plants such as these—as well as underutilized animals—could be of particular value in developing nations, where people often do not get enough to eat. In the Special Reports section, see NEW USES FOR "THE SHIP OF THE DESERT."

Disappearing plastic. Agricultural scientists in 1986 and 1987 tested several new types of plastic sheeting for *mulching*. Mulching involves covering the soil with a material that holds in moisture and helps keep weeds from growing. Traditionally, organic materials such as straw or plant residues have been used as mulches.

Although plastic makes a convenient and easy-to-use mulch, it is unsightly and contributes to the pollution of the environment because it is not *biodegradable* (capable of being decomposed). Researchers at Belland, a company in Switzerland, in 1987 were field-testing a plastic that breaks down when it comes in contact with water. The plastic is made of ammonia-based compounds that dissolve in water and fertilize the soil.

Researchers at the USDA's Northern Regional Laboratory in Peoria, Ill., were trying a different approach. They developed a type of polyethylene plastic containing starch molecules. Soil microorganisms devour the starch molecules in the plastic, causing the remaining, synthetic material to break up (see CHEMISTRY). Still another kind of plastic, developed by Eco Plastics Limited of Ontario, Canada, falls apart when exposed to sunlight.

Pesticide resistance. Increasing numbers of plants' natural enemies were becoming resistant to chemical pesticides, according to a report released in November 1986 by the National Research Council's Board on Agriculture. The report stated that 447 insect species had become resistant to insecticides; at least 100 species of fungus, to fungicides; and 55 weed species, to herbicides. [Sylvan H. Wittwer]

In WORLD BOOK, see AGRICULTURE.

Anthropology

The discovery of a 2.5-million-year-old fossil skull in northern Kenya, announced in August 1986, has challenged accepted theories about the development of australopithecines, a type of ancient *hominid* (modern human beings and their closest human and prehuman ancestors). The hominid skull was found west of Lake Turkana in 1985 by anatomist Alan Walker of Johns Hopkins University in Baltimore. Walker and anthropologist Richard E. Leakey of the National Museums of Kenya reported the discovery.

Walker and Leakey identified the fossil skull, known as WT 17000, as a member of a *robust* (ruggedly built) hominid species called *Australopithecus boisei*, a side branch on the human evolutionary tree. The new fossil—about 300,000 years older than the oldest previously known *A. boisei* fossil—suggests that australopithecines had a complicated evolutionary history in which extremely robust forms existed much earlier than had been previously thought.

Like other *A. boisei* fossils, the 2.5-million-year-old WT 17000 has huge cheekbones and rear teeth; a large, rugged skull; and a bony crest running along the top of the skull, to which large jaw muscles were attached. These are features thought to have evolved later in hominid history. But the skull also possesses a series of features found only in earlier, more primitive hominids. These include an apelike protruding face and a small brain.

The new fossil challenges one of the most commonly accepted theories of hominid evolution—that australopithecines evolved in an orderly fashion from *A. afarensis*, which lived from 3 million to 4 million years ago. This theory, proposed in 1979 by anthropologists Donald C. Johanson of the Institute of Human Origins and Timothy D. White of the University of California, both in Berkeley, suggested that *A. afarensis* was the ancestor of all later hominid species. According to this theory, between 2 million and 3 million years ago, the hominid line

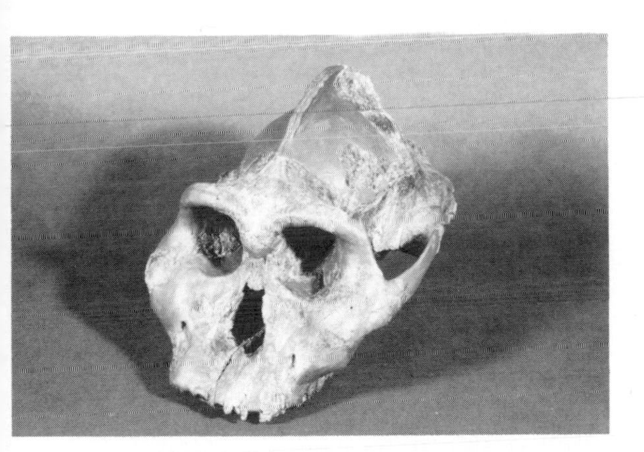

A New Family Tree?
The 2.5-million-year-old skull of a previously unknown type of *australopithecine* (apelike human ancestor), *above*, called WT 17000, has challenged accepted theories about the evolution of these prehuman creatures. The skull, whose discovery in Kenya was reported in August 1986, combines primitive characteristics with features thought to have developed much later. The new find suggests that the human family tree, *right*, has three branches rather than two as was previously thought.

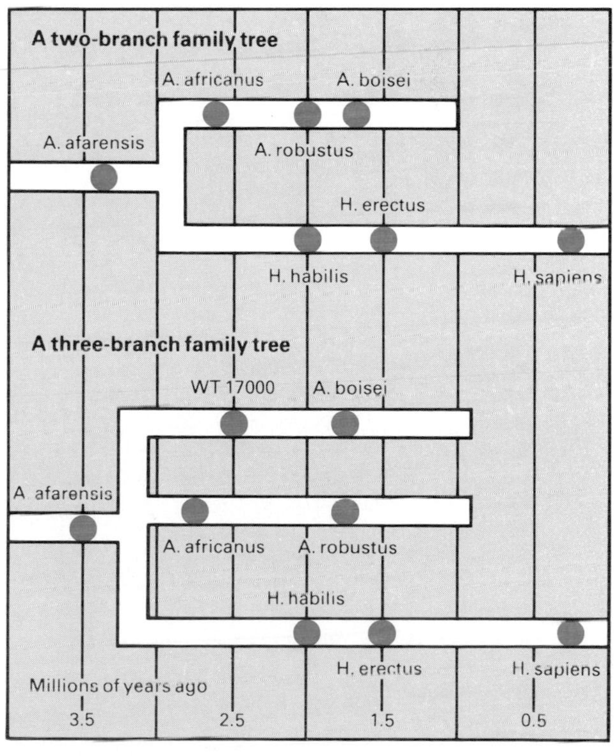

The Tasaday: Amazing Find or Clever Hoax?

It was called the anthropological find of the century: the discovery of the Tasaday, a tribe of primitive cave-dwelling people in the heavily forested mountains on the island of Mindanao in the Philippines in 1971. Described as the remnants of humanity's Stone Age past, the Tasaday lived a tranquil, simple life, with no weapons. Their language had no word for *warfare*. Unaware of agriculture, they lived on wild tubers and crabs, fish, frogs, and other small animals caught in nearby streams. But in 1986 and 1987, a strange international dispute erupted concerning the Tasaday's authenticity. Some scientists charged that the Tasaday are not a primitive, Stone Age-like tribe but a hoax devised for political reasons.

The Tasaday's first exposure to the modern world reportedly occurred in the early 1960's, when they came in contact with a trader from a nearby tribe named Dafal. Dafal supposedly introduced the Tasaday to spears, bows and arrows, traps, and cloth, and taught them to track and butcher large animals such as deer and pig. In June 1971, Dafal introduced several Tasaday to Manuel Elizalde, Jr., a friend of Ferdinand E. Marcos, then president of the Philippines. At the time, Elizalde was also the director of Panamin, a government agency that had been established to protect the rights of minority groups in the Philippines.

Elizalde broadcast the news of the Tasaday's existence, and the story made headlines around the world. Films made by a United States television network and the National Geographic Society presented the Tasaday as a people untouched by civilization. A book, *The Gentle Tasaday*, by journalist John Nance became an international best seller. To protect the Tasaday and their distinctive way of life, the Marcos government turned 19,000 hectares (46,300 acres) of virgin rain forest on the southern edge of Mindanao into a tribal preserve.

The story turned sour in spring 1986 when journalists from Switzerland and West Germany independently visited the Tasaday. In their articles about the tribe, the journalists reported that the portrayal of the Tasaday as a Stone Age tribe was false. Several anthropologists who studied the European journalists' evidence agreed.

Critics of the Tasaday's authenticity say that the tribe—which consisted of 25 people when first revealed to the outside world—was recruited from neighboring communities. The fake Tasaday, critics say, were told to wear primitive dress made from orchids and pretend ignorance of modern life. The critics also charged that the Tasaday's language was not unique but closely related to other languages of the region.

The critics accused Elizalde, who fled the Philippines before the fall of the Marcos regime in 1986, of orchestrating the hoax. They pointed out that he controlled all outside contact with the Tasaday. The only way to reach the tribe was by helicopter, which landed on a wooden platform built in the treetops. Critics say the people pretending to be Tasaday thus had plenty of time to prepare for any visit. The aim of the deception, according to the skeptics, was to divert attention from the imposition of martial law in the Philippines in 1972 by the Marcos regime and to depict Marcos as a benefactor, not a tyrant.

One of the chief critics is linguist and anthropologist Zeus Salazar of the University of the Philippines. He charged that the Tasaday had words for activities, such as planting and harvesting, they supposedly did not practice. In addition, he contended that the linguistic evidence was faulty because linguists who visited the Tasaday relied too heavily on interpreters who were not members of the tribe and did not speak the language well.

Another skeptic, anthropologist Gerald Berreman of the University of California in Berkeley, claimed that too few anthropologists visited the Tasaday for too short a time to judge

The cave-dwelling Tasaday, described as being like a Stone Age tribe with a simple, tranquil life when discovered in the Philippines in 1971, are accused of being a hoax orchestrated by the ousted Marcos government.

their authenticity. He called the "tools" depicted in films of the Tasaday a "joke" because, he said, they are totally unlike tools ever made anywhere else and seemed to have no use. Finally, the Tasaday were not very isolated, being within three hours' walk of agricultural communities. And the Tasaday area was frequented by traders. "The whole thing is implausible," Berreman said.

Supporters of the Tasaday's authenticity say that if the tribe is a hoax, it is one of the most elaborate and successful hoaxes in the history of science.

One of the tribe's chief defenders is linguist Carol Molony of Stanford University in Palo Alto, Calif., who visited the Tasaday for a total of 17 days on two occasions. Molony reported finding no evidence that the Tasaday's language is related to other languages of the region. She also said that during her visits the Tasaday never lapsed into other languages or used words from neighboring tribes. In addition, Molony found only one Tasaday word that may have been derived from another language.

"It's preposterous to think they could have been trained for something like this," Molony said. She concluded from an analysis of Tasaday vocabulary that the Tasaday had been isolated from outside influences for 600 years before their discovery. Molony also defended her research, contending that she used standard procedures in recording about 800 Tasaday words and that her interpreter, a young boy from a neighboring tribe, had lived with the Tasaday expressly to learn their language.

Some Tasaday supporters believe that attempts to discredit the tribe are due to resentment on the part of some Filipino anthropologists who were shut out of the research on the tribe. Tasaday supporters also point out that many of the tribe's critics have never visited the tribe personally and are basing their arguments on second-hand information provided chiefly by journalists, not scientists. There have also been charges that logging companies are involved in the controversy. If the Tasaday were declared a hoax, logging companies could gain access to the timber, particularly mahogany, on the Tasaday's preserve.

The issue may never be resolved satisfactorily. The Tasaday, Molony predicts, will soon intermarry with their neighbors to such an extent that they will lose their distinctive language and way of life. In fact, intermarriage has already raised the tribe's membership to 61, according to a recent report by Nance. But then, the critics say, the Tasaday were never distinctive in the first place. [Ian Anderson]

split. One branch eventually developed into modern human beings. The other branch led to *A. africanus*, *A. robustus*, and finally *A. boisei*. Scientists had thought that the second line became more robust over time.

But the discovery of the robust-appearing WT 17000 has called this theory into question. The most popular revised theory of hominid evolution suggests a three-branch tree with *A. afarensis* at the base. One branch gave rise to human beings. Another led to *A. africanus* and *A. robustus*. *A. boisei*, including WT 17000, occupied the third branch.

Apelike *Homo habilis*? The first discovery of arm and leg bones of *Homo habilis*, the first truly human species, suggests that these hominids had an apelike body rather than a more humanlike body. The discovery of the fossils in Tanzania by a team headed by Donald Johanson was reported in May 1987. The fossils were dated to about 1.8 million years ago. Previously, scientists had not found any limb bones that were confirmed as being from *H. habilis*.

Measurements of the arm and leg bones indicate that the creature—a female who stood about 1 meter (3½ feet) tall—had long, heavily built arms similar to those of modern apes. Scientists had thought that *H. habilis*, which had a relatively large brain and made stone tools, had a humanlike body. According to Johanson, the fossils suggest that the transition from an apelike to a humanlike body among hominids may have taken place later than scientists had believed.

New-found apes. The discovery of a collection of fossils of previously unknown ancient apes indicates that by 16 million to 18 million years ago, apes had evolved into a greater variety of forms than scientists had thought. The discovery was announced in November 1986 by Richard Leakey and Mary G. Leakey, also of the National Museums of Kenya. The fossils—skulls, lower jaws, teeth, and other bones—were found in 1985 and 1986 at Kalodirr, a fossil-rich site along the western side of Lake Turkana.

Ancient DNA. An analysis of 8,000-year-old brain tissue from skulls found in Florida produced the oldest known

Anthropology

Continued

samples of intact human deoxyribonucleic acid (DNA), the material of which genes are made. The analysis, reported in October 1986, was performed on skulls found at the Windover site, a small swampy pond about 24 kilometers (15 miles) west of Cape Canaveral. The skulls were found by archaeologist Glen H. Doran and his colleagues at Florida State University in Tallahassee. The DNA analysis was carried out by Philip J. Laipis and William W. Hauswirth of the University of Florida College of Medicine.

The ancient brain tissue produced only about 1 per cent of the DNA normally obtainable from fresh brain tissue, because the ancient tissue had been contaminated with plant materials and impurities from the soil in which it was buried. Although the recovered DNA strands are relatively short, they are long enough for future experiments to search for recognizable human genes or gene fragments. A study of the structure of the brains—representing four young men and one middle-aged woman—revealed that they are no different from modern human brains.

Pygmy stature. The pygmies of Africa, long studied by anthropologists, are small because they produce low levels of a hormone essential for growth, according to research reported in April 1987 by researchers at the University of Florida in Gainesville. The hormone is called insulin growth factor-I (IGF-I).

The scientists measured hormone levels in 64 pygmies living in the remote Ituri Forest of Zaire and found that pygmy children had normal levels of IGF-I until they reached puberty. During adolescence, however, the youngsters had only about a third as much of the hormone as did taller African Bantu and American teen-agers used as comparison groups. The researchers concluded that the pygmies' short stature apparently results from the absence of the growth spurt that normally occurs in adolescence in taller groups. [Charles F. Merbs]

In WORLD BOOK, see ANTHROPOLOGY; PREHISTORIC PEOPLE.

Archaeology, New World

The discovery of evidence of the earliest human presence in the New World was reported in June 1986 by French archaeologists. Their excavations at a rock shelter along a cliff in a remote area of northeastern Brazil indicate that people were living in this part of South America as early as 32,000 years ago. This is about 21,000 years before the time that most archaeologists believe people first arrived in the New World.

In the lowest—and oldest—layer of sediment in the cave, the archaeologists found several hearths—large circular cooking areas—with stone tools scattered around them. Wood charcoal from one of the hearths was radiocarbon-dated to from 31,700 to 32,160 years ago.

The evidence that such ancient people lived in the New World was made even more convincing because the archaeologists found tools and other artifacts in the upper—more recent—layers of sediment. At other sites in the New World believed to be older than 11,000 years, the archaeological evidence consists of only a few artifacts or artifacts in only one layer, representing only a single period of occupation. At the Brazilian cave, the archaeologists found evidence that seven groups of people had occupied the site after the earliest inhabitants.

The archaeologists also discovered that the cave's ceilings and walls are covered with prehistoric paintings. In one of the hearths, the archaeologists found a rock with two red lines—part of a painting—that had fallen from the ceiling or walls. Charcoal from this hearth was dated to about 17,000 years ago. So the scientists concluded that the rock is the oldest known evidence of cave art in the New World.

New light on Nazca lines. The results of recent field studies of the so-called Nazca lines in southwest Peru were reported in August 1986 by archaeoastronomer Anthony F. Aveni of Colgate University in Hamilton, N.Y. These lines, created in a rock-strewn desert valley, represent more than 300 geometric figures, mostly trapezoids but also figures of animals.

Archaeology, New World
Continued

Previous archaeological research revealed more than 150 sites where people once lived in hilly areas overlooking the valley. These sites date from between about 400 B.C. to about A.D. 1400. Aveni believes that the lines were made during this period.

Aveni's studies shed new light on the way in which the lines were laid out. To demonstrate a likely technique, 10 members of Aveni's team carried out an experiment in a remote part of a nearby desert valley. They used string to lay out a straight line 35 meters (115 feet) long and 1 meter (3 feet) wide. They then picked up the rocks, which are black, along the line and moved them to the side, exposing the pinkish-colored sand beneath. The finished line, which took only 1½ hours to construct, was similar to a line in one of the many rectangular shapes found among the Nazca lines. In fact, the scientists calculated that an average-sized Nazca trapezoid figure, covering up to 16,000 square meters (172,000 square feet), could be constructed in about a week without the need of elaborate surveying techniques or other special engineering skills.

One myth Aveni's team dispelled was the idea that the lines can be seen only from the air. The scientists found that most of the lines can be seen from ground level or from a range of low foothills ringing the desert valley.

Aveni, aided by archaeologist Thomas Zuidema of the University of Illinois in Urbana-Champaign, found that the Nazca lines resembled Inca *ceques*, a system of imaginary lines marked only by sacred places called *huacas*. Some of the ceques served as sight lines for observing the sky. Each huaca represented one day in the Inca agricultural calendar.

Aveni and Zuidema theorized that some Nazca lines, such as those that mark the sun's position on the horizon in spring when water starts to flow in the desert area's rivers and underground streams, may also have served as part of an agricultural calendar. But Aveni and Zuidema also theorized that many of the lines probably served as footpaths.

Prehistoric art decorates the wall of a cave in Brazil where archaeologists in June 1986 discovered evidence of the earliest known human presence in the New World. Charcoal from the hearth in the cave was dated to 32,000 years ago, and a decorated rock fragment that fell from the ceiling or a wall was dated to 17,000 years ago.

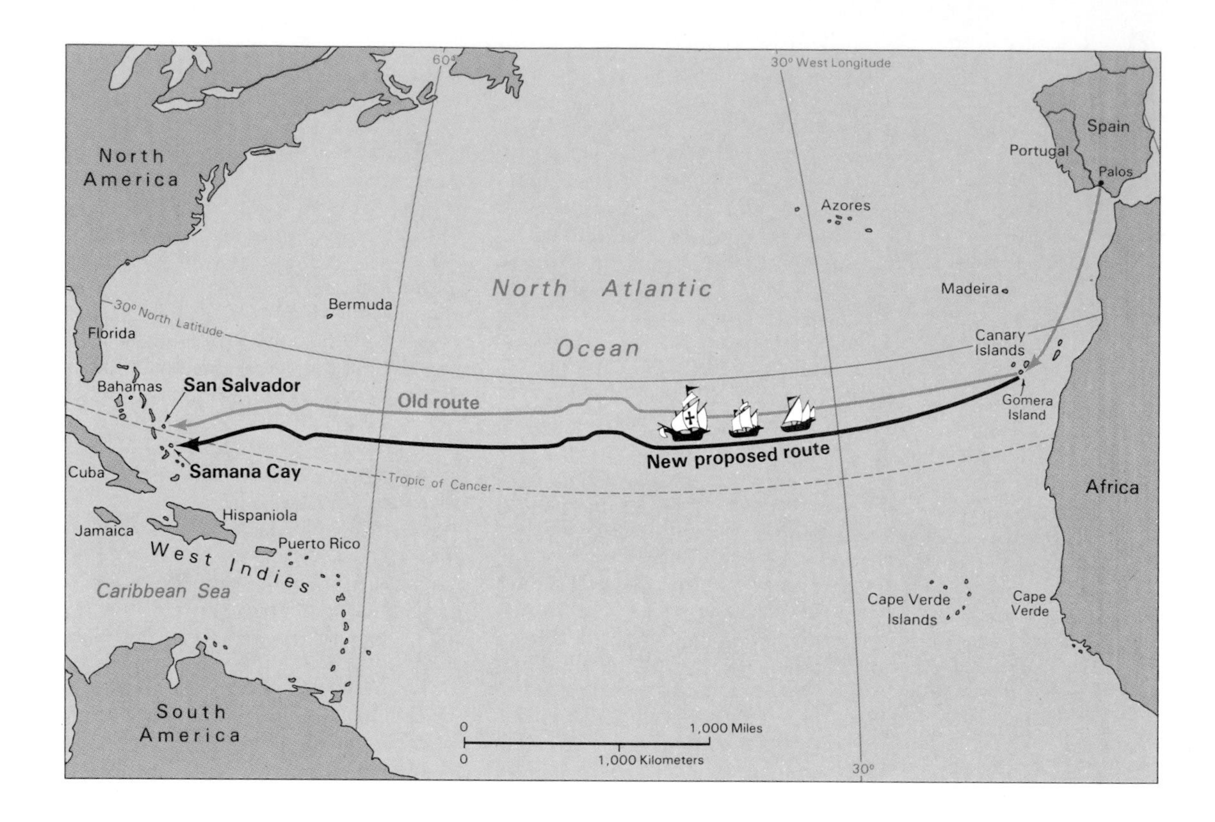

Archaeology, New World

Continued

A new translation of Christopher Columbus' log and a computer analysis of Atlantic Ocean currents and wind patterns suggest that Columbus may have made his first landfall in the New World on Samana Cay, an island in the Bahamas. The research, reported in October 1986, challenges the common belief that Columbus' first landing, on Oct. 12, 1492, occurred on San Salvador Island.

Chaco Canyon. An overview of archaeological discoveries made during a 15-year study of Chaco Canyon in a national park in northwestern New Mexico was reported in May 1987 by archaeologist Stephen H. Lekson of the University of New Mexico in Albuquerque. The study has revealed new information about the construction and purpose of Chaco Canyon's distinctive architecture, particularly its large apartment-style buildings called Great Houses. These structures are up to five stories tall, contain hundreds of rooms, and cover as much as 0.8 hectare (2 acres).

At first, scholars believed the Great Houses had been built by the ancient Aztecs of Mexico. Archaeological excavations that began in the early 1900's, however, revealed that the ruins had more in common with the dwellings of modern Pueblo Indians of New Mexico and Arizona. In the 1930's, tree-ring dating of the wooden beams used to construct the Great Houses confirmed that these buildings were put up between A.D. 900 and 1150, long before the Aztecs, who flourished in the 1400's.

In the 1950's and 1960's, most archaeologists came to believe that the Great Houses of Chaco Canyon were built by the Anasazi, the ancestors of the modern-day Pueblos. Some archaeologists, however, continued to argue for a "Mexican connection," linking the Great Houses to ideas introduced by ancient Toltec traders from central Mexico who had come north seeking markets for their goods. One goal of the study, which began in 1971, was to resolve this controversy.

Archaeological excavations revealed that in the A.D. 1100's—at the height of the Anasazi culture—Chaco Canyon was crowded with at least six Great Houses and hundreds of smaller dwellings. Crisscrossing the canyon was a system of broad, carefully leveled roads connecting the Great Houses. The roads also ran out from the canyon into the surrounding area.

Excavations at the Great Houses showed that they had not grown gradually larger by small additions over

Archaeology, New World

Continued

time, like other ancient and modern pueblos. Rather, complete blocks of as many as 50 rooms were built at one time. It also became clear that not all of the Great Houses were built at the same time. Three were constructed around A.D. 900; two others were built between 1020 and 1040. During the peak building period from 1075 to 1115, only one Great House was built, but massive additions expanded the five existing Great Houses.

The three earliest Great Houses were long, arc-shaped, multistoried structures with a plaza along the front of the arc. In these buildings, archaeologists found hearths and *middens* (trash mounds), indicating that people once lived there. The Great Houses and additions dating from the 1000's and 1100's, however, were largely uninhabited. The lack of evidence of human occupation and the design of these buildings—rectangular structures with tiny identical rooms—have led archaeologists to conclude that these Great Houses probably served as warehouses for the storage of agricultural products. On the basis of the roads, the design of the Great Houses, and the large quantities of imported pottery and other goods found in the canyon, archaeologists deduced that by 1100 Chaco Canyon was the center of a vast regional trading network.

By tracing the architectural developments in the canyon, archaeologists have also concluded that the design of all the Great Houses evolved from earlier forms of Anasazi architecture in the area. Later Great Houses, however, also show some Mexican influences, perhaps the result of communication brought about by trade.

The complex society of Chaco Canyon ended in about 1140. The record of rainfall in the later Great Houses' wooden beams indicates that the region suffered a severe 30-year drought. The drought may have been one reason the Anasazi abandoned the canyon. [Thomas R. Hester]

In the Special Reports section, see IMAGES FROM THE ICE AGE. In WORLD BOOK, see ARCHAEOLOGY; CLIFF DWELLERS; PUEBLO INDIANS.

Archaeology, Old World

The discovery in France of what may be the clearest evidence yet of ancient cannibalism was reported in July 1986 by anthropologist Paola Villa of the University of Colorado in Boulder. Villa excavated a pit in Fontbrégoua Cave in southeastern France that contained the 6,000-year-old remains of three adults, two children, and one individual whose age and sex could not be determined.

The scientists found the human bones in a shallow pit on the cave floor. Mixed with the human bones were animal bones. Both groups consisted chiefly of limb and shoulder bones. In addition, both groups had been discarded in the same way.

Villa and her associates determined that the human bones bore *cut marks*—microscopic scratches made by stone tools. The pattern of cut marks on the human bones was identical to that on the animal bones. Also, the limb bones of both groups had been split, apparently to obtain marrow.

Some scholars believe that ancient cannibalism occurred mainly as part of religious or warfare rituals. But Villa reported that there is no evidence of rituals at Fontbrégoua.

Stone Age abundance. Information challenging the belief that Middle Stone Age people had little and moved about continually in search of food was reported in March 1987 by archaeologists T. Douglas Price of the University of Wisconsin in Madison and Erik B. Petersen of the University of Copenhagen in Denmark. (People who lived by foraging—hunting, fishing, and gathering plants—after 8000 B.C., when farming became established, are known as *Mesolithic* [Middle Stone Age] people.) Price and Petersen found that foraging provided such an abundance of food for Mesolithic people living along the eastern coast of Denmark from 5200 B.C. to 3200 B.C. that they continued to follow this way of life for 500 years after groups elsewhere in Scandinavia began farming.

Between 1980 and 1983, Price and Petersen excavated Vaenget Nord, a small hill in a meadow about 1 kilometer (1.6 miles) from Øresund, a

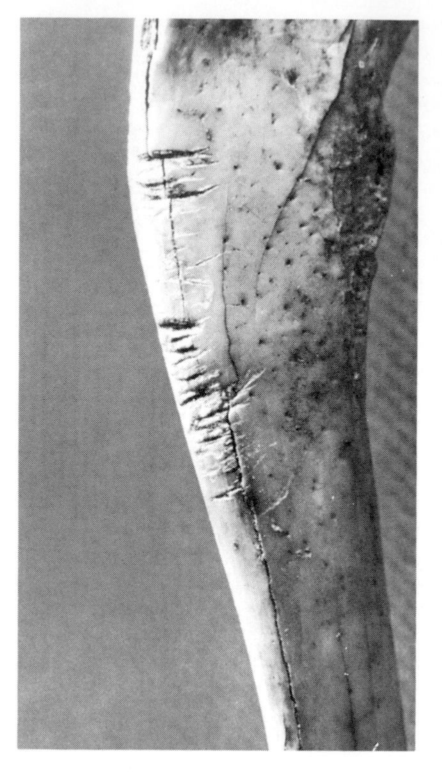

Microscopic scratches on a child's collarbone, *right,* and a child's skull, *far right,* made 6,000 years ago by a stone knife may be the strongest evidence ever found of ancient cannibalism. The bones were found in a cave in France among many human bones bearing similar marks of butchering, according to a July 1986 report.

Archaeology, Old World

Continued

sound (narrow stretch of water) that separates the Danish island of Sjaelland from Sweden. The site, about 40 kilometers (25 miles) north of Copenhagen, was occupied from about 5200 B.C. to 4800 B.C. During that time, however, sea levels were higher and Vaenget Nord was an island in an inlet of the sound, about 40 to 50 meters (130 to 165 feet) from shore.

Excavations since the 1920's have revealed evidence of at least 40 Mesolithic camp sites around Vaenget Nord. Stake holes and wooden posts found in these sites mark the location of dwellings, probably huts. Bones and other remains uncovered at the sites indicate that the people of the area ate fish as well as wild pigs, deer, and many other forest animals and plants.

Price and Petersen's excavation revealed that Vaenget Nord was probably a seasonal camp used during summer months. On the southwestern edge of the site, the scientists found a depression in the ground partly surrounded by stake holes, which they concluded represented the remains of a tent or cabin. Nearby, they found one area where flint tools and weapons were made and another where animals were butchered and skinned. Price and Petersen suggested that Vaenget Nord also served as a boat dock and a base for repairing hunting equipment.

The discovery of pottery and the remains of domesticated plants and animals at Vaenget Nord and nearby sites dated from about 3650 B.C. indicates that the people of the area had contact with farming groups. But they chose to remain foragers for nearly another 500 years, apparently because the area was so rich in foods.

The first numerals and writing. New theories about the origins of writing and numbers were presented in December 1986 by anthropologist Denise Schmandt-Besserat of the University of Texas in Austin. Scientists have long known that the earliest writing appeared about 3000 B.C. in Sumer, an ancient region of Mesopotamia (now southeastern Iraq). But the stages leading to the development of writing have been difficult to trace.

Schmandt-Besserat studied thousands of small pieces of baked clay called *tokens* found in the remains of ancient Mesopotamian villages and towns. There were two types of tokens. The first type, which appeared about 8000 B.C., had simple geometric shapes—spheres, disks, and cones—about the size of chess pieces. These simple tokens, which had unmarked surfaces, represented individual agricultural commodities, such as grains and animals. They were usually stored in clay "envelopes," after first being pressed against the outside of the envelope to indicate the contents.

The second type of token, which appeared about 3400 B.C., was larger and made in a greater variety of geometric shapes, as well as in the shape of tools and animals. While simple tokens continued to be used for raw products, complex tokens were used to record transactions involving manufactured goods, such as bread, oil, perfume, and furniture. The complex tokens were marked with a variety of signs representing the nature of the item counted. For example, a particular type of disk might represent textiles in general while markings on the disk indicated the type of fiber, cloth, or garment. According to Schmandt-Besserat, complex tokens were also used to record taxes, inventories, and other economic information. Complex tokens were perforated and strung together on a cord, which was then attached to an oblong piece of clay, called a *bulla*. Seals on the bulla identified the transaction or account.

Schmandt-Besserat traced the beginning of writing and numerals to the replacement of the bulky clay envelopes and bullae with more streamlined clay tablets. At first, the smaller simple tokens were pressed against the tablets to record transactions. The larger complex tokens, however, were too big and complicated to leave a clear impression on the tablets. As a result, Sumerian accountants began using the signs marked on the surfaces of the complex tokens to represent these tokens on clay tablets. Soon, the accountants began "writing" signs

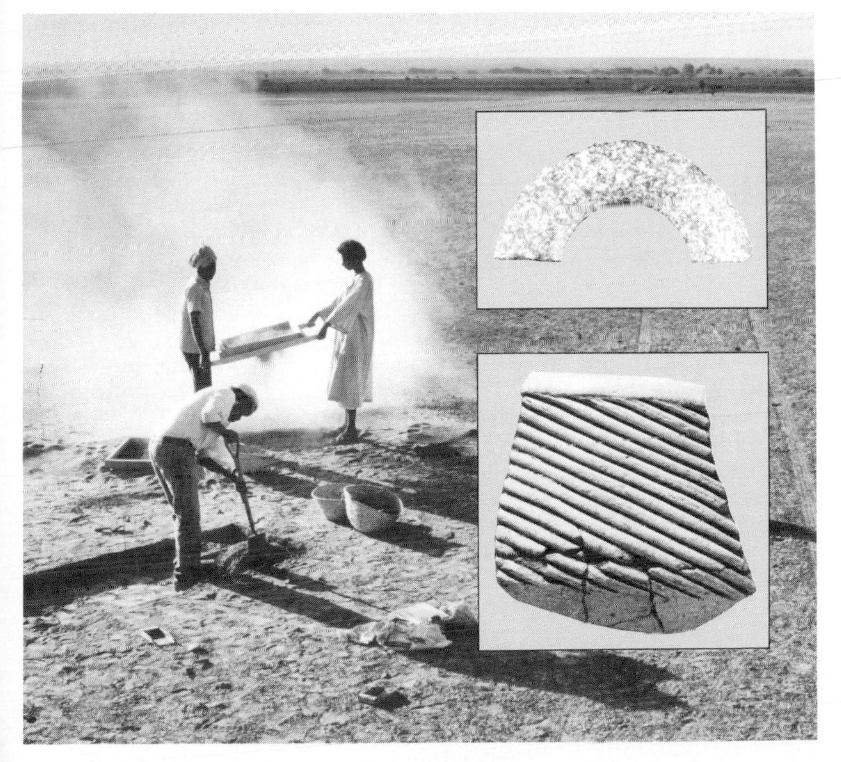

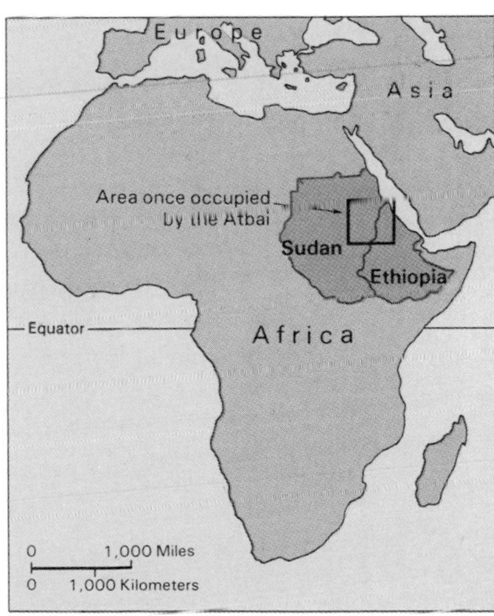

Workers dig and sift sand in a barren valley, *left*, looking for artifacts from the previously unknown Atbai Culture. These people inhabited what is now eastern Sudan and northwestern Ethiopia from about 5000 B.C. to 1000 B.C. Among the Atbai artifacts revealed in late 1986 were a section of a stone bracelet (inset, *top*) and a piece of a clay cup with a ribbed pattern (inset, *bottom*).

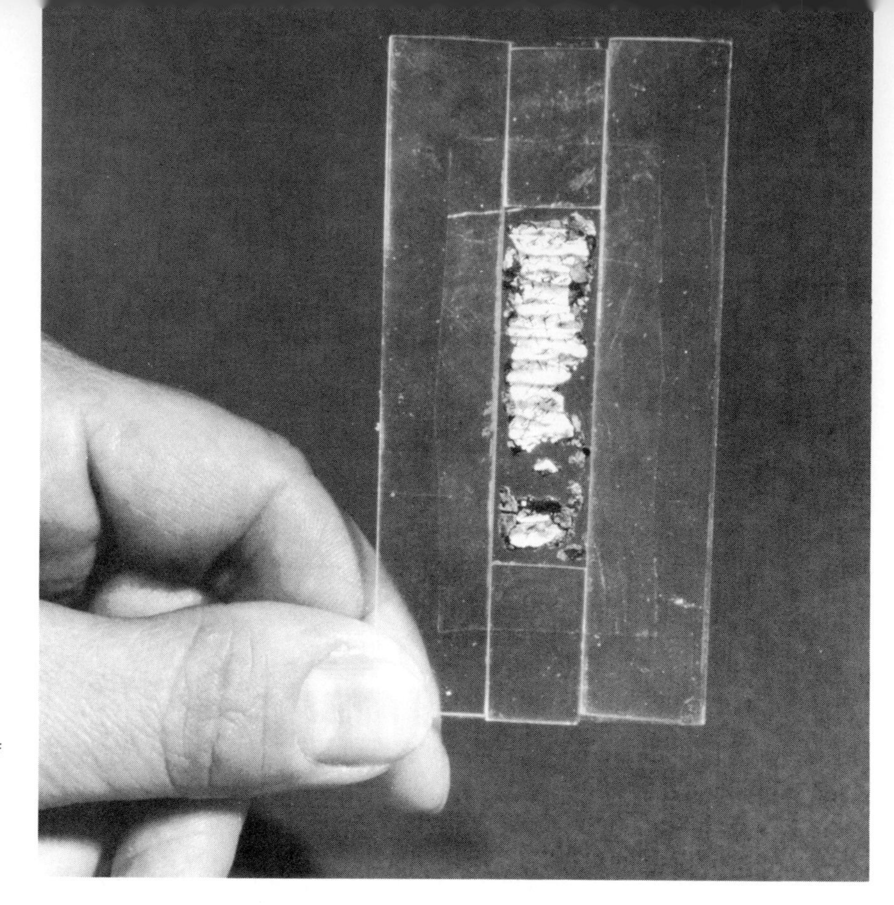

A tiny 2,600-year-old silver scroll, found in a burial cave in Jerusalem in 1979 and deciphered in 1986, contains the oldest known Biblical text, a blessing from the Book of Numbers.

to represent simple tokens as well. Schmandt-Besserat argued that the signs representing simple tokens eventually evolved into numbers, while the signs for complex tokens became signs used in writing words.

Light on the Dark Ages. Archaeological excavations in Italy have uncovered the ruins of an immense monastery and provided revelations about Europe's Dark Ages, according to a report published in January 1987 by archaeologist Richard Hodges of Sheffield University in England. The Dark Ages, a period lasting from about A.D. 400 to 900, occurred during the early part of the Middle Ages, which ended about 1500. During the Dark Ages, there were few towns, and community life centered on monasteries. Few monasteries from this period have ever been excavated, however.

Hodges and a team of Italian and English archaeologists spent five years excavating the monastery of San Vincenzo in central Italy. At its height in the early 800's, the monastery was occupied by about 1,000 people.

Among the many building ruins uncovered in the exacavation was a structure that had been about 30 meters (100 feet) long and 15 meters (50 feet) wide—immense by the standards of the time. Excavations revealed that the building had housed a waiting hall filled with large paintings of saints and prophets and a *refectory* (dining hall) that apparently had 30 colored windows. Near the large building was evidence of a workshop for making glass vessels, inlays, and windows.

Saracens—invaders from the Middle East—plundered San Vincenzo on Oct. 11, 881, killing about 500 monks. In the early 1000's, the monks dismantled the existing monastery buildings and built a new complex just to the south. But the monastery never regained its former prestige. By the late 1000's, San Vincenzo had been abandoned. [Robert J. Wenke]

In the Special Reports section, see EXCAVATING ROMAN BATH; IMAGES FROM THE ICE AGE. In WORLD BOOK, see CANNIBAL; MIDDLE AGES; PREHISTORIC PEOPLE; WRITING.

Astronomy, Extragalactic

A giant exploding star in a nearby galaxy burst into view in late February 1987, sending astronomers from all over the world to the telescopes of the Southern Hemisphere where the *supernova* (exploding star) was visible. This supernova is the first in modern times close enough to be observed with sophisticated astronomical instruments. See Close-Up.

Unusual arcs. The discovery of enormous glowing arcs in distant clusters of galaxies was reported at a January 1987 meeting of the American Astronomical Society (AAS) in Pasadena, Calif. These arcs represent a previously unknown physical structure in the universe. Astrophysicists Roger Lynds of the National Optical Astronomy Observatories (NOAO) in Tucson, Ariz., and Vahe Petrosian of Stanford University in California reported that they had found these huge arcs in the central regions of three galaxy clusters.

The arcs are more than 30,000 *light-years* wide and 300,000 light-years long. (A light-year is the distance light travels in one year—about 9.46 trillion kilometers [5.88 trillion miles].) Each arc appears to be a segment of a circle centered around bright *elliptical* (oval-shaped) galaxies.

Lynds first observed the arcs in 1977, but the images then were so faint that he was uncertain whether they were real, or false, images produced by interference from his instruments. In the fall of 1986, however, he observed the arcs with an extremely sensitive electronic camera that is less prone to interference. The camera was attached to a 4-meter (158-inch) telescope at Kitt Peak National Observatory near Tucson.

The astrophysicists believe the arcs are made up of young stars, because they produce blue images. The blue color indicates that a star is hot and massive, and such stars formed relatively recently. Understanding how the stars could have formed into arcs with such perfect and undisturbed shapes, however, is a puzzle. The arcs have distinct edges and end points. They are remarkably uniform in color and

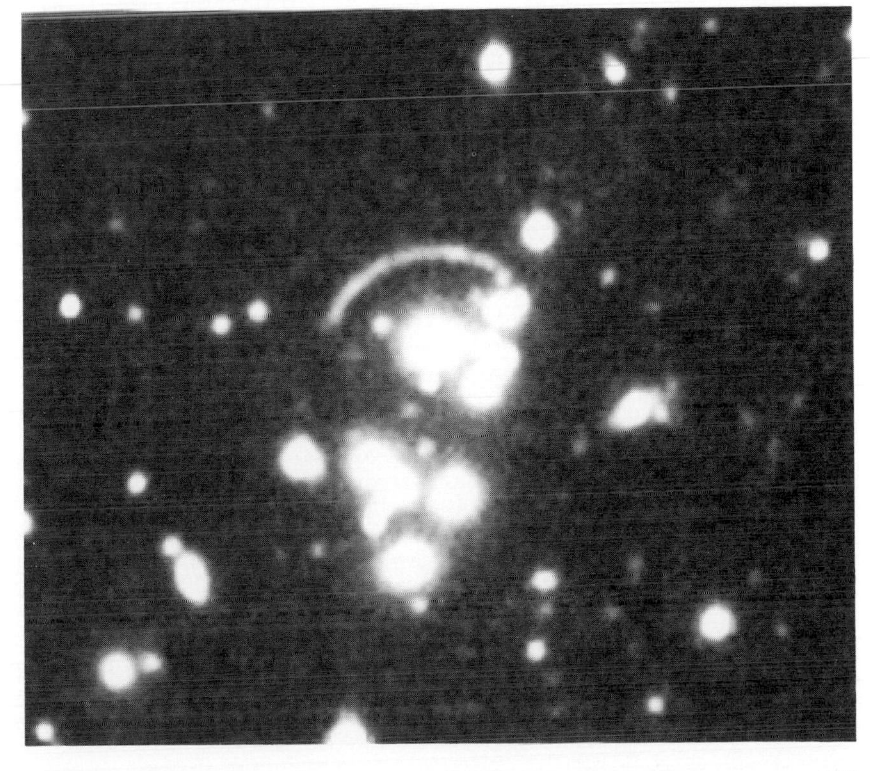

An arc, one of three found around distant clusters of galaxies, represents a physical structure never seen before in the universe, astronomers reported in January 1987. The arc is more than 30,000 light-years wide and 300,000 light-years long, more than three times the diameter of the Milky Way galaxy. (A light-year is the distance light travels in one year—about 9.46 trillion kilometers [5.88 trillion miles].)

brightness as well. This appears to rule out the possibility that the arcs formed when two or more galaxies collided, because such collisions create chaotic structures, rather than the uniform structures of the arcs.

The astrophysicists theorized that the arcs may have formed from *filaments* (threads) of gas left over from the formation of the galaxy clusters. These filaments were then compressed into an arc shape by some kind of shock wave traveling through space, and the compression of the gas began the process of star formation. If this is so, arcs may have been more common earlier in the history of the universe. So astronomers will look farther out in space—and therefore further back in time—for evidence that arcs formed early in the universe's history.

Infrared galaxies —galaxies that give off most of their energy in the form of *infrared radiation* (heat)—may provide clues to the origin of *quasars*, according to a January 1987 report. Quasars are extremely energetic compact objects often found at the centers of distant galaxies. They radiate much of their energy at infrared wavelengths. Many of the most luminous infrared galaxies are as bright as quasars, reported astrophysicist David B. Sanders of the California Institute of Technology in Pasadena, suggesting that there is a connection between these bright galaxies and quasars.

The infrared galaxies were first discovered in 1983 by an orbiting infrared telescope called the *Infrared Astronomical Satellite* (*IRAS*). Sanders followed up this *IRAS* discovery by studying 10 infrared galaxies with an optical telescope at Palomar Observatory near San Diego. Using a sensitive electronic camera attached to the telescope, Sanders and his colleagues obtained optical images showing that these galaxies have distorted shapes characteristic of colliding or merging galaxies.

Sanders also analyzed the infrared light given off by the galaxies to determine what chemical elements and compounds they contain. All elements in nature either give off or absorb radiation at characteristic wavelengths known as emission lines or absorption lines. By analyzing these lines, astronomers can determine the chemicals present in the objects.

The researchers found that carbon monoxide gas is present in molecular gas clouds in the galaxies. Carbon monoxide gas is regarded as a *tracer*—its presence indicates that there are even greater amounts of other gases that are more difficult to detect, particularly molecular hydrogen gas. Sanders calculated that the infrared galaxies have vast amounts of molecular hydrogen gas compared with typical galaxies. He concluded that this abundance of molecular hydrogen gas resulted from two galaxies merging. A single galaxy, he reasoned, could not contain that much molecular hydrogen gas.

Sanders also concluded that the colliding material in the two merging galaxies was triggering rapid star formation. Sanders believed the new stars accounted for the merging galaxies' strong infrared radiation. When new stars form, astronomers on Earth detect the energy they give off as mainly infrared radiation. The new stars are still surrounded by dust clouds left over from their creation, and the dust blocks the stars' visible light. But the heat the stars produce comes through as radiation. Eventually, the dust cloud blows away, and the star's visible light can be seen.

A similar situation might exist in terms of quasars and entire infrared galaxies, some astronomers theorize. Infrared galaxies may be the forerunners of at least some quasars. According to this theory, merging galaxies provide the fuel—in the form of abundant molecular hydrogen gas—that is needed to power an extremely energetic center, or nucleus. Like a quasar, an infrared galaxy's main source of energy is a compact spot at the nucleus of the galaxy. But this nucleus is not visible as light until a cloud of galactic dust surrounding the nucleus has blown away—just as the cloud around a star must blow away before the star can be seen shining. After the dust cloud has blown away, astronomers see a quasar. But before the galactic dust cloud blows away, astronomers see an infrared galaxy.

A galaxy forming. The first direct observation of the apparent formation of

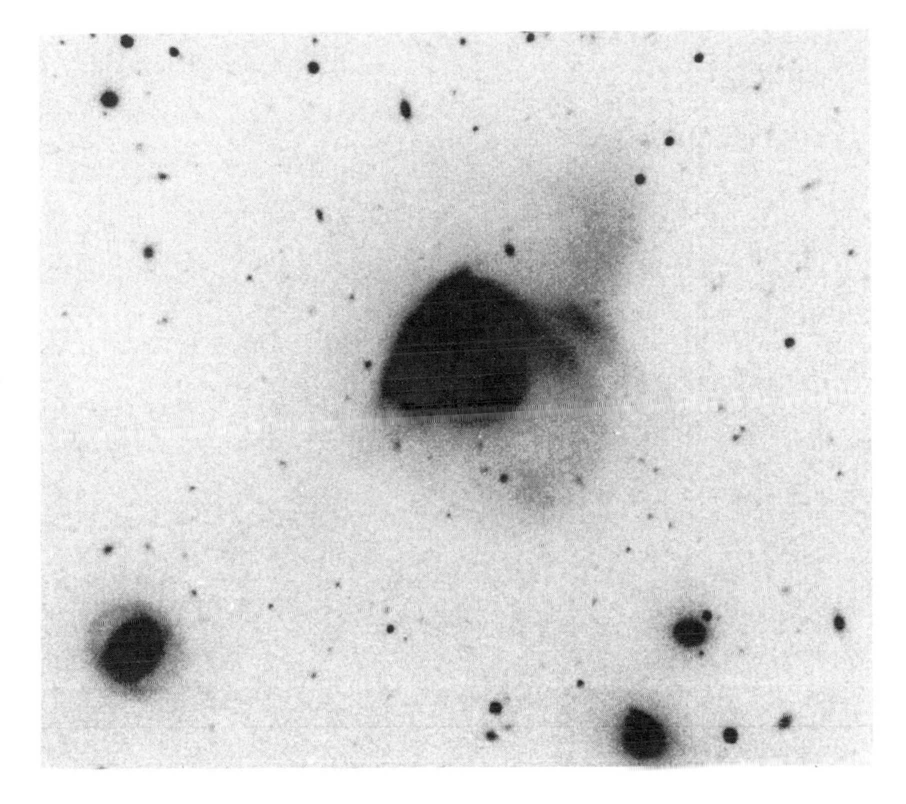

Two galaxies that have collided and merged may be the forerunner of a *quasar*, a compact object that gives off tremendous amounts of energy. Astronomers reported in January 1987 that this galactic merger, photographed with an optical telescope at Palomar Observatory in California, created an *infrared galaxy* that radiates more than 99 per cent of its energy at infrared wavelengths—as heat.

a galaxy was reported at the January 1987 meeting of the AAS by astronomer Patrick J. McCarthy of the University of California in Berkeley. McCarthy focused an optical telescope on an object that had been discovered earlier by radio telescopes. Using a special filter that is highly sensitive to *ionized* (electrically charged) hydrogen, McCarthy and his colleagues discovered a large cloud of ionized hydrogen gas surrounding the object. They calculated that the cloud is about 300,000 light-years in diameter.

Most of the cloud consists of ionized hydrogen. The astronomers, however, observed some very faint blue objects near the edge of the cloud and even fainter, widely scattered objects in the interior of the cloud. The blue objects could be hot, young stars.

The researchers concluded that the cloud is a young galaxy in the process of forming stars the size of our sun at a rate of 3,000 to 5,000 per year. By contrast, our fully formed Milky Way galaxy produces one star per year. Because most of the cloud still consists of hydrogen gas that has not yet condensed into stars, the bulk of the galaxy's stars have yet to be formed.

When galaxies formed. A technique that may help determine when galaxies first began forming was described in July 1986 by astronomers J. Anthony Tyson of Bell Laboratories in Murray Hill, N.J., and Patrick Seitzer of NOAO.

Using an electronic camera attached to a telescope, the two astronomers took a six-hour exposure of a small section of the southern sky. Tyson and Seitzer were able to photograph distant galaxies as faint as 27th *apparent magnitude*. (A star's apparent magnitude is the measure of its brightness as seen from Earth. The fainter a star is, the higher its magnitude.) The unaided, or naked, eye can barely see objects of the 6th magnitude. A 27th-magnitude object is about a billion times fainter than an object visible to the unaided eye.

The researchers found that there are so many galaxies in this part of the sky that almost a third of the area of

A Supernova Surprise

One of the most significant events in modern astronomy took place during the night of February 23 and the early morning hours of Feb. 24, 1987. During those hours, astronomers in the Southern Hemisphere discovered a bright *supernova*, an exploding star that increased in brightness a thousand times literally overnight—becoming bright enough to be seen with the naked eye. Canadian astronomer Ian Shelton, using the University of Toronto's 25-centimeter (10-inch) telescope at Las Campanas Observatory in northern Chile, was the first to recognize the supernova and report it.

The exploding star, which has been named Supernova 1987A, is located in the Large Magellanic Cloud, a small galaxy about 160,000 *light-years* from Earth. (A light-year is the distance that light travels in one year—about 9.5 trillion kilometers [5.9 trillion miles].) Supernova 1987A is the nearest and brightest supernova to Earth to be seen since 1604, when German astronomer Johannes Kepler observed a supernova within our Milky Way Galaxy. Astronomers often observe supernova explosions in distant galaxies, but they have been too far away and too faint to be studied in detail with the instruments of modern astronomy.

With the discovery of Supernova 1987A, astronomers have been able for the first time to test various theories about supernovae. In fact, within a month of its discovery, observations of Supernova 1987A had already yielded confirmation of a theory of how a massive star becomes a supernova.

Most astronomers believe that stars more massive than the sun will eventually end their lives as supernovae. According to the generally accepted theory, ongoing nuclear reactions involving gas in a star's core provide the energy that makes the star shine. This nuclear energy is produced when hydrogen changes into helium. Nuclear reactions also change hydrogen and helium into heavier elements, such as iron, until eventually the core of the star consists mainly of iron. At this point the star's core has "burned out," and it collapses under its own weight. This final collapse is a violent process that destroys the star's core in an enormous nuclear explosion. The explosion blasts the outer gaseous layers of the star off into space at speeds of thousands of kilometers per second. This expanding envelope of hot gas is what we see as a bright supernova.

Astronomers theorized that the explosive nuclear reactions that occur during the core's final collapse should send subatomic particles called *neutrinos* off into space. Neutrinos travel at or near the speed of light (299,792 kilometers [186,282 miles] per second) and have so little mass that they can pass through matter undisturbed. Supernova 1987A provided the first opportunity to test this theory by allowing scientists to look for the predicted neutrinos with detectors on Earth. Astrophysicists at the Institute for Advanced Study in Princeton, N.J., predicted the number of neutrinos from the supernova that would be detected at neutrino observatories located in mines in Japan and near Cleveland, Ohio. Physicists analyzed the number of neutrinos detected in late February and found the number agreed with the astrophysicists' prediction. This discovery from the new science of *neutrino astronomy* proved that the theory of how a massive star becomes a supernova is correct. But astronomers will be busy for months and perhaps years trying to find answers to other questions about the fabulous supernova. [John H. Black]

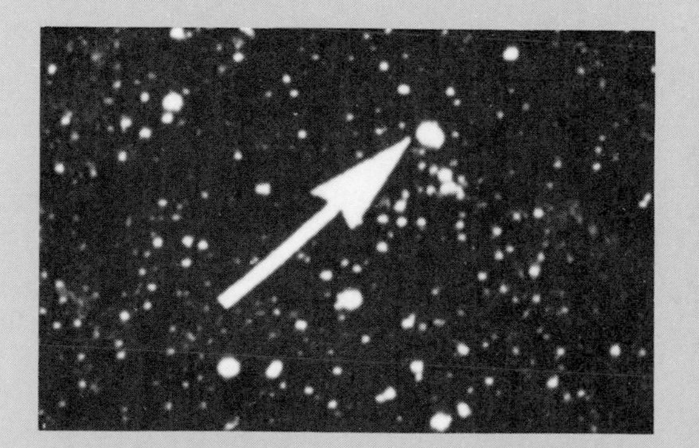

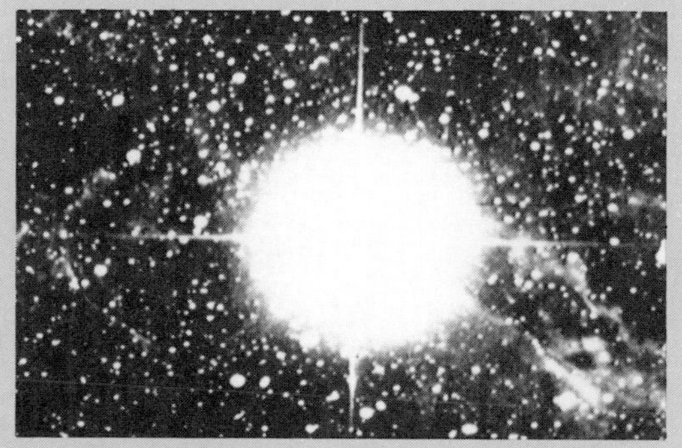

A star (arrow) in a nearby galaxy, *above left,* may be the one that was seen exploding as a supernova, *above right,* in February 1987.

Astronomy, Extragalactic

Continued

the image they obtained is occupied by galaxies. And there may be other galaxies so faint that the camera cannot detect them. But, Tyson pointed out, not much more sky area could be filled by these faint galaxies. There is a limit to how bright this and other areas of the night sky can be. If there were no limit, then the night sky would appear as bright as the daytime sky, because it would be filled with an infinite number of stars and galaxies giving off an infinite amount of light.

We know that this is not the case because when we look far out into the universe, we are also looking back in time, toward when the universe came into being and began to expand outward. This is because light travels at a finite speed. Light from an object a billion light-years away takes a billion years to reach us.

Sometime between 10 billion and 20 billion years ago there were no galaxies. So, as astronomers keep looking farther out into the universe—and therefore into the past—there must be a point where the number of galaxies no longer increases. Tyson and Seitzer believe that in detecting 27th-magnitude galaxies they may be close to seeing this point in time. They base their conclusion on the fact that between 70 and 80 per cent of the brightness of the area of the night sky that they examined can be accounted for by the galaxies in their photograph. In addition, the faint 27th-magnitude galaxies are bluer than the brighter galaxies, indicating that they contain mainly young stars. Since the light takes so long to reach Earth, astronomers are seeing these galaxies as they were long ago, when they were young.

The significance of this is that the approximate ages of the 27th-magnitude galaxies can be calculated. Some preliminary calculations by Tyson and Seitzer suggest that the 27th-magnitude galaxies formed between 9 billion and 19 billion years ago. So astronomers may be getting closer to answering the question of when galaxies first began to form. [Stephen S. Murray]

In WORLD BOOK, see ASTRONOMY.

Astronomy, Galactic

An unusual pair of stars that orbit each other every 11 minutes were discovered in our galaxy, the Milky Way, in 1987. These *binary* (double) stars are separated by a distance of only 130,000 kilometers (81,000 miles)—less than the distance between Earth and the moon. X-ray astronomers Luigi Stella and Nicholas E. White of the European Space Operations Center in Darmstadt, West Germany, and William C. Priedhorsky of Los Alamos National Laboratory in New Mexico reported in January that this is the smallest binary star system ever found.

The astronomers made the discovery while analyzing X-ray observations of the two stars made over a period of several years by the *European Space Agency's X-ray Observatory Satellite* (*EXOSAT*). Both stars are so hot that they give off their energy mainly in the form of X rays. These X rays can be observed only by telescopes orbiting above Earth's atmosphere, which blocks X rays from reaching Earth.

The scientists concluded that they had found a binary system after discovering that the brightness of the X rays observed by *EXOSAT* varied regularly every 685 seconds. Regular brightness variations such as this indicate that either two objects are orbiting each other every 685 seconds or an especially bright region on one star moves in and out of view as the star makes a complete rotation every 685 seconds. The astronomers noted, however, that the rotation of a single spinning star speeds up after a few years, and no change had occurred in the 685-second period, according to observations dating back to 1976. The researchers concluded that only two stars moving in and out of view as they orbit each other could account for the brightness variations.

A neutron star . . . The astronomers deduced that one of the stars in this binary system is a *neutron star*, a tiny star that has exhausted its nuclear fuel and consists only of subatomic particles known as neutrons. Neutron stars measure only about 20 kilometers (12½ miles) in diameter but are so dense and massive that a teaspoonful

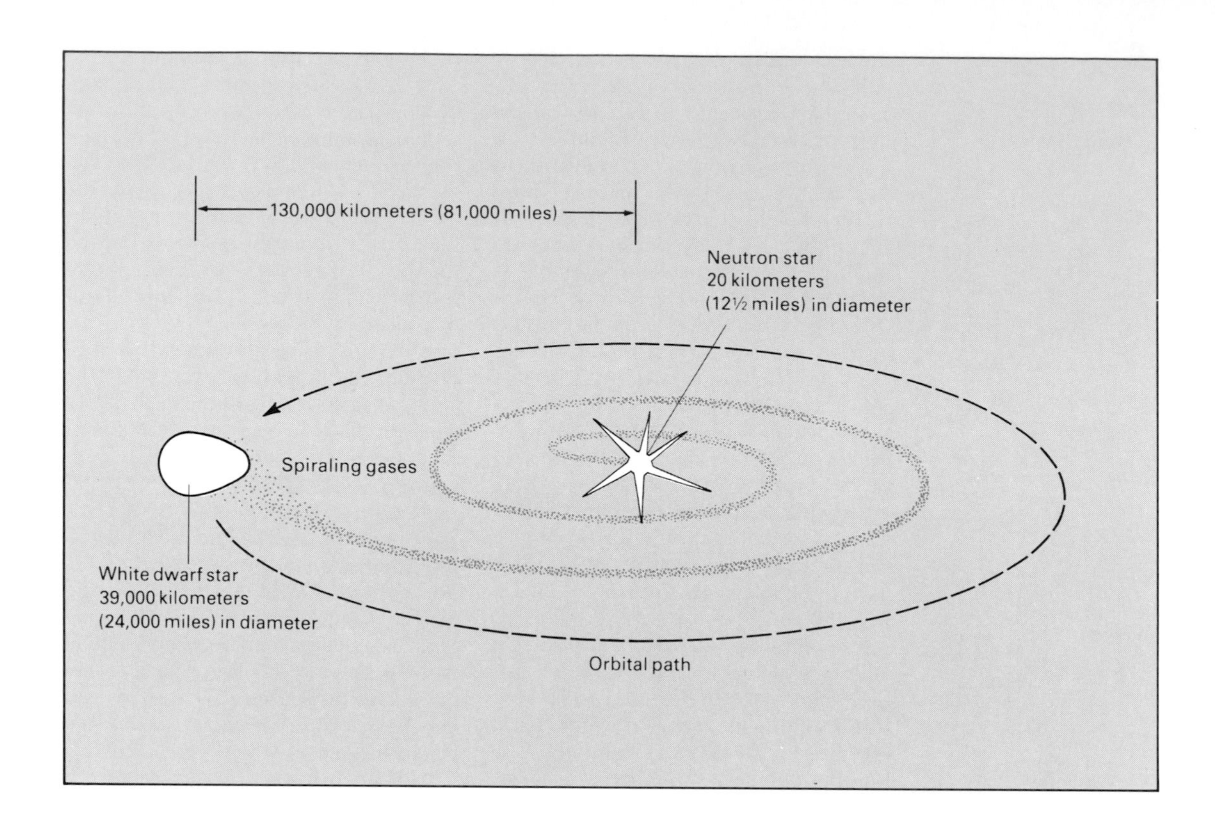

Astronomy, Galactic

Continued

Smallest binary system
X-ray astronomers in January 1987 reported the discovery of two stars that orbit each other every 11 minutes. The stars, separated by a distance of only 130,000 kilometers (81,000 miles), make up the smallest binary system ever found. One star is a neutron star, only 20 kilometers (12½ miles) in diameter. It pulls gases from its companion, a white dwarf star, 39,000 kilometers (24,000 miles) in diameter.

of their gases would weigh billions of tons on Earth.

The researchers based their deduction on observations of other binary star systems that release their energy in the form of X rays. In such systems, the gravitational attraction of a neutron star usually pulls gases from the outer layer of its companion star. The gases spiral around the neutron star and then fall onto its surface. The neutron star's gravitational force is so great—due to its compactness—that it heats the gas to temperatures of 10,000,000°C (50,000,000°F.). At these temperatures, the gas gives off X rays. Temperature measurements made by *EXOSAT* showed that gases of the binary system were in the range of 10,000,000°C, indicating the presence of a neutron star.

. . . and a white dwarf. The astronomers also deduced that the neutron star's companion was probably a *white dwarf*, the core of a large star that has shrunk to a smaller size. The researchers calculated that the companion star has a diameter of about 39,000 kilometers (24,000 miles) and has about one-tenth of the mass of the sun. Knowing the approximate mass of the two stars and how frequently they orbit one another enabled the researchers to calculate the distance between the stars.

Binaries in globular clusters. The research team noted that this binary system is located in a *globular star cluster*, a ball-shaped cluster of stars that may contain a million stars in a relatively small region of space. Stars are so closely packed together in the globular cluster where this binary system is located that several hundred stars can be found within 3 *light-years* of the binary system. (A light-year is the distance light travels in one year, about 9.5 trillion kilometers [5.9 trillion miles].) By comparison, there is only one other star within 4.3 light-years of our sun. Astronomers believe that collisions and close interactions between stars must be fairly common in a globular cluster because stars are so densely packed together in such clusters.

A theoretical explanation for the origin of a very small binary star system in a globular cluster—like that found by *EXOSAT*—was offered in January 1987 by astronomer Frank Verbunt of the Max Planck Institute for Extraterrestrial Physics in Munich, West Germany. Verbunt examined the motions and properties of different kinds of stars in globular clusters. He concluded that a very small binary system probably begins when a neutron star collides almost head-on with a *red giant star*, an extremely large star with a reddish glow. The mutual force of gravity between the two stars holds them in orbit. While the red giant star gradually loses its outer layers of gases, its orbit grows smaller, bringing the two stars closer together. Eventually, only the core of the red giant star remains, and this core becomes a tiny white dwarf.

Birth of a star. The first direct observation of a star in the process of formation was reported in October 1986 by a team of astronomers from the University of Arizona in Tucson and the University of Missouri in St. Louis. The astronomers described the object they observed as a *protostar*, the core of a collapsing gas cloud not yet hot enough or compact enough to be a star.

Astronomers believe that a star's formation begins when a cloud of interstellar gas and dust begins collapsing in on itself due to the force of gravity. Over millions of years the cloud collapses into a ball as gravity pulls it together. This causes the pressure of the gas to increase, heating the gas at the center of the ball. When the temperature at the center reaches about 1,100,000°C (2,000,000°F.), nuclear reactions begin, and a star is born.

The team of astronomers detected the protostar with the 11-meter (36-foot) radio telescope of the National Radio Astronomy Observatory located on Kitt Peak near Tucson. The object was first detected in 1983 by the *Infrared Astronomical Satellite* (*IRAS*)—an orbiting telescope that detects *infrared* (heat) radiation. But *IRAS* did not provide enough information to identify this object as a protostar. A protostar does not yet radiate light from nuclear reactions. Instead, the heat of the

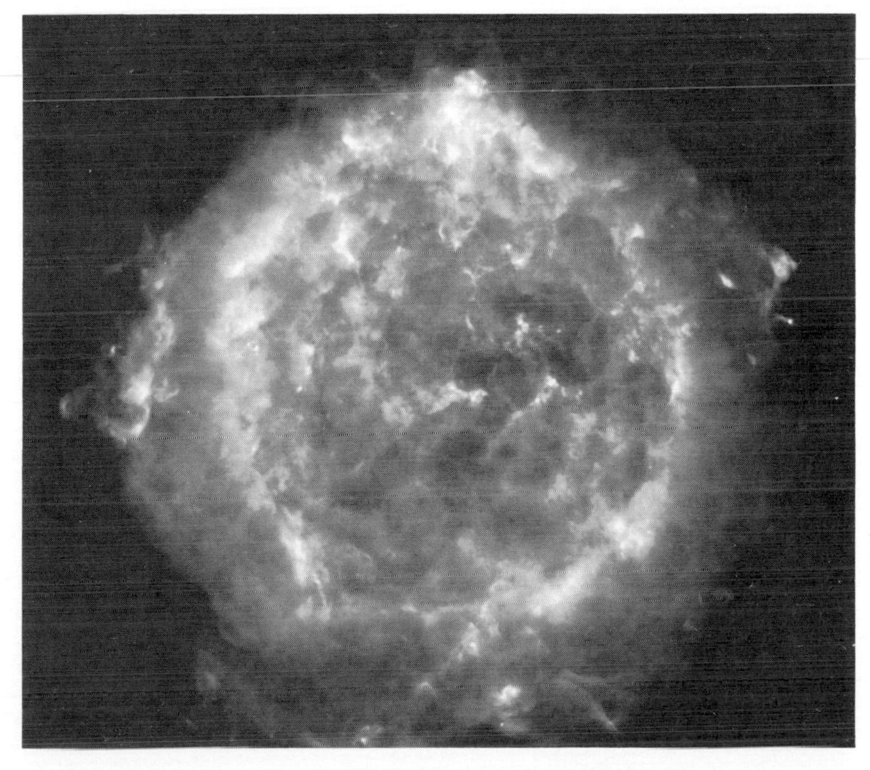

A detailed image of an expanding shell of gas left behind by a *supernova* (star explosion) was produced by radio astronomers in January 1987. The astronomers processed data from radio telescopes through a supercomputer to obtain an image of the shell more detailed than that provided by the most powerful optical telescope. The supernova that produced the remnant may have been witnessed by the British astronomer John Flamsteed in 1680.

Astronomy, Galactic

Continued

gases at its core gives off energy that can be detected by infrared and radio telescopes.

The astronomers found that gas in the cloud was collapsing inward by analyzing radio waves from a compound in the gas called carbon monosulfide (CS). Although the gas cloud is made up mainly of hydrogen, the astronomers chose to analyze CS molecules because they give off radiation only under conditions of high density, such as exist in the core of an interstellar gas cloud.

Each element gives off or absorbs radiation at its particular wavelength of the *electromagnetic spectrum*, the range of radiant energies that includes visible light and radio waves. The astronomers tuned their telescope so they would detect only CS molecules radiating at their particular wavelength in the radio band of the electromagnetic spectrum.

The researchers detected the speed and direction of these molecules by observing a phenomenon in the radio waves called the *Doppler effect*. The Doppler effect occurs where there is a change in frequency of electromagnetic wavelengths due to motion. For example, the wavelength of radiation from a molecule increases slightly when the molecule moves away from an observer and decreases slightly when the molecule moves toward an observer. Observing these Doppler shifts with their radio telescope, the astronomers noted that the CS molecules were moving inward toward the core of the interstellar cloud. The researchers calculated that the gas was collapsing inward at a speed of 0.8 kilometer (½ mile) per second, which is consistent with theories of how a star forms.

The astronomers also calculated that the mass of the protostar is about one-tenth that of the sun's mass. If gas keeps falling toward the center at its present rate, the mass of the protostar will equal the mass of the sun in 100,000 years, when nuclear reactions presumably will begin and a new star will be born. [John H. Black]

In WORLD BOOK, see ASTRONOMY; MILKY WAY.

Astronomy, Solar System

New findings that shed light on the nature of Uranus' magnetic field were reported in 1987. The magnetic fields that exist around the planets are among the most poorly understood natural phenomena in the solar system. Earth, Jupiter, Saturn, and Uranus have strong fields, but Mars, Venus, and Mercury have very weak fields, or none. Findings that showed why Uranus' magnetic field has eluded detection by astronomers using Earth-based radio telescopes were reported in March 1987 by space scientist D. D. Barbosa of the University of California, Los Angeles.

One of the ways astronomers study magnetic fields is by analyzing radio waves that are generated in a planet's *magnetosphere* (a region around the planet containing energetic electrons and *ionized* [electrically charged] atoms). Radio telescopes on Earth, however, have been unable to detect radio waves from Uranus' magnetosphere. Astronomers were puzzled when instruments on *Voyager 2* detected no radio waves until one week before the spacecraft's closest approach to the planet—and then, only occasionally.

By calculating the position of the *Voyager 2* spacecraft at the occasional times when it did detect radio waves, Barbosa concluded that the radio waves came from only the planet's north magnetic pole. Barbosa reported that the radio waves were detected only occasionally because they are beamed in a narrow, conelike pattern that scans across the sky like a searchlight as the planet rotates. The *Voyager 2* spacecraft did not detect these radio waves at first because it approached the planet over the south magnetic pole. As Uranus continues its path around the sun, however, its orientation toward Earth will change so that the radio beam from its north magnetic pole will be observable from Earth by the year 2000. Astronomers will then be able to study the Uranian magnetic field for the next 58 years. In the Special Reports section, see *VOYAGER*'S CLOSE LOOK AT URANUS.

Interstellar diamonds. A substantial amount of interstellar carbon is in the form of diamond grains, according to a March 1987 report by researchers at the Enrico Fermi Institute of the University of Chicago and the National Bureau of Standards in Washington, D.C. Carbon is one of the most interesting chemical constituents in the *interstellar medium* (the gas and dust between the stars) because carbon compounds are present in all plants and animals. Understanding the chemistry of carbon and its evolution in the interstellar medium may help tell us how life originated.

The scientists, led by Fermi Institute meteorite specialist Roy S. Lewis, discovered the diamond grains by studying the carbon-containing material found in certain types of meteorites. Most of the carbon in meteorites—which date from the formation of the solar system—is in the form of very complex molecules combined with nitrogen and hydrogen. The next most significant form of carbon in meteorites is graphite. According to Lewis and his colleagues, microscopic crystals of diamond embedded with atoms of other elements such as xenon and nitrogen represent a third type of carbon in meteorites. This was a surprising finding, because the formation of diamonds requires extremely high temperatures and pressures. The scientists found no sign of such conditions in these meteorites.

Lewis and his associates also discovered that the xenon and nitrogen embedded in the diamond grains were different *isotopes*, or forms, of xenon and nitrogen from those normally found in solar system material. The researchers theorized that these microscopic diamonds must have formed before the solar system did.

"Life" from cometary dust? An analysis of dust particles from Halley's Comet was reported in April 1987 by two West German researchers. The particles were collected by instruments on three spacecraft that flew by Halley's Comet in March 1986. Data about the dust particles were sent back to Earth by radio signals.

The West German researchers reported that the structure of many of

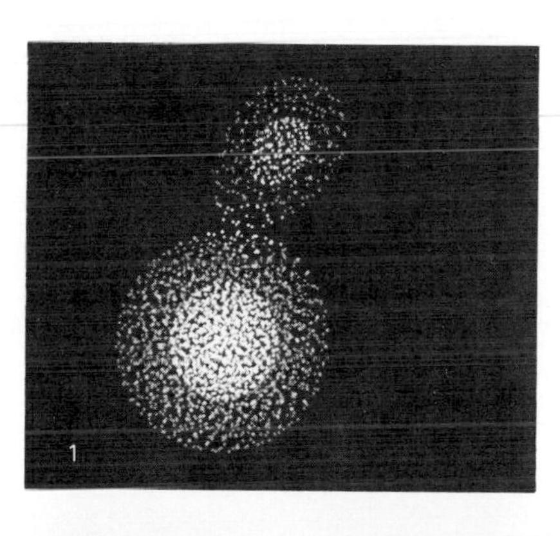

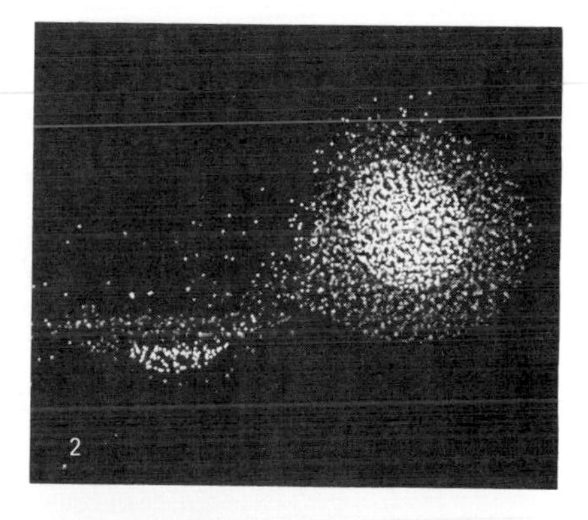

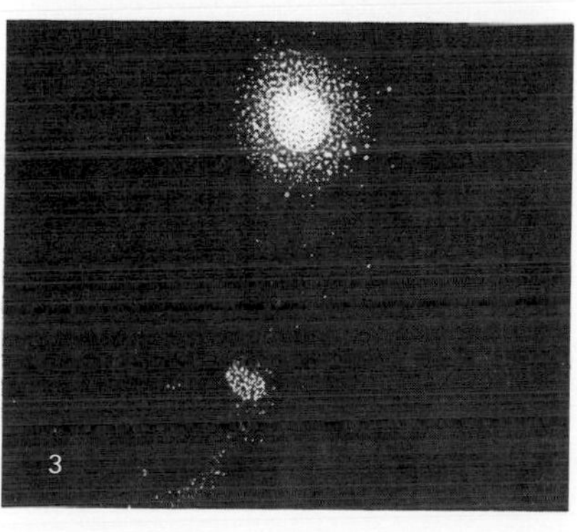

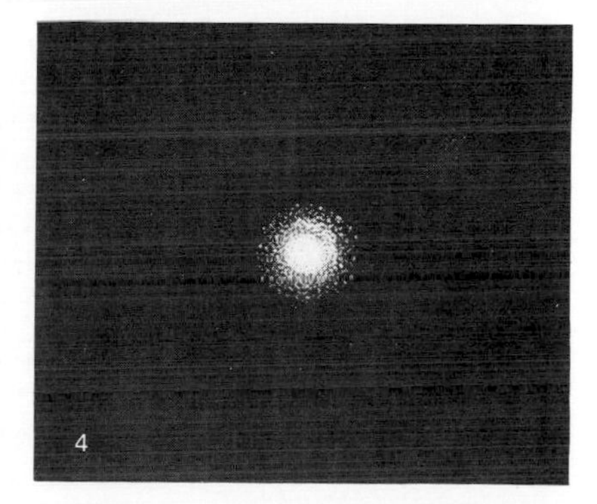

Making of the Moon? The moon may have formed after a collision 4½ billion years ago between a Mars-sized object and Earth, according to a theory reported in December 1986 and tested on a supercomputer. Computer images reveal that after the collision (1), parts of the object and Earth's surface were blown out into space (2). Earth's gravitational force pulled heavy iron in the blown-out material back to Earth (3). The lighter material—hot gas and particles—remained in space (4), gradually being pulled together to form the moon.

the dust particles resembles that of an onion. The particles consist of a core containing *silica* (a common mineral found in sand) surrounded by layers of material containing carbon.

These findings, the researchers claim, make the question of the origin of life in cometary matter "even more exciting." The scientists noted that if these carbon-containing materials come in contact with warm water, they can form sugars and the molecular prerequisites for *nucleic acids*—the building blocks of life. Moreover, the mineral cores of the particles could produce the correct chemical environment for such molecules to organize themselves into *ribonucleic acid* (RNA), a complex molecule that plays an important part in producing proteins important to all life forms.

Halley's Comet wobbles. A problem involving how long it takes the *nucleus* (core) of Halley's Comet to complete one rotation about its *axis* (an imaginary line around which the comet spins) captured the attention of astronomers in 1986 and 1987. While preliminary findings by many investigators failed to yield a satisfactory answer, the findings indicated that the axis of rotation may not be fixed. As a result, the nucleus may sometimes wobble, or nod up and down, rather than spin in a smooth circular motion.

Knowing the comet's rotation time is important in analyzing the thousands of observations of physical changes the comet underwent as it passed by Earth and the sun in 1985 and 1986. Astronomers need a timetable showing what parts of the nucleus were illuminated by the sun—and when—since sunlight is a major source of physical changes in comets.

Astronomers who observed the comet from Earth and studied images returned by the spacecraft in March 1986 at first thought the comet turned once on its axis every 2.2 days. But in December 1986, astronomers Robert Millis and Douglas Schleicher of the Lowell Observatory near Flagstaff, Ariz., reported a pattern of regular variations in the comet's brightness that indicated it rotated once every 7.4 days.

One possible explanation for the different observations of the rotational period is that the spinning motion of the nucleus may be very complex. As the nucleus loses gas and dust, the distribution of its mass changes, and this may in turn cause the axis of rotation to shift. As a result, the nucleus may gyrate wildly—as a spinning top does when it slows down. More observations of Halley's Comet will be necessary to calculate this complex motion and determine the true rotation period.

An inconstant sun. By examining observations of the sun made by astronomers in the late 1600's and early 1700's, a team of French astronomers from the Paris Observatory in March 1987 reported new evidence linking changes in the sun to Earth's weather.

The researchers examined records of the diameter of the sun and the position of *sunspots* (dark areas on the surface of the sun) between the years 1666 and 1719. This period coincides roughly with a period known as the *Maunder minimum*—from 1645 to 1715 —a time in which few sunspots were observed. Since sunspots are linked to magnetic activity, the reduction in sunspots indicated low magnetic activity. Some scientists have suggested that the reduced sunspot activity during the Maunder minimum was directly responsible for a period of extremely cold weather in Europe in the 1700's known as "the little ice age."

The French researchers found from historical records that the sun's diameter was 2,000 kilometers (1,200 miles) greater during the Maunder minimum than it is now. In addition, the movement of any existing sunspots across the sun's surface was slower. The researchers speculated that the low magnetic activity, large diameter, and slow movement of sunspots were connected. They concluded that for all these conditions to exist at the same time, the transport of heat from the interior of the sun to its surface must have been slowed for some reason during the Maunder minimum. If so, then the energy output of the sun must also have been reduced. A reduction in the amount of heat from the sun reaching Earth could explain the extremely cold weather of the "little ice age." [Michael J. S. Belton]

In World Book, see Astronomy; Halley's Comet; Sunspot; Uranus.

Books of Science

Here are 25 outstanding new science books suitable for the general reader. They have been selected from books published in 1986 and 1987.

Anthropology. *The Smithsonian Book of North American Indians: Before the Coming of the Europeans* by Philip Kopper summarizes what is known about the culture of New World peoples from the Arctic to Mexico. Kopper also discusses scientists' 200-year quest to understand them. Beautiful illustrations accompany the text. (Smithsonian Books, 1986. 288 pp. illus. $50)

Return to the High Valley: Coming Full Circle by Kenneth E. Read is a personal account of the author's return to the Asaro Valley in Papua New Guinea after a 29-year absence. Read describes the changes that have taken place for the Gahuku-Gama, a people he wrote about in his first book, *The High Valley* (Scribner's, 1965). (Univ. of California Press, 1986. 269 pp. $18.95)

Archaeology. *The Pyramid Builders of Ancient Egypt: A Modern Investigation of Pharaoh's Workforce* by A. R. David tells of the discovery and reexamination of artifacts excavated at Kahun, an ancient village occupied by workers who built the pyramid and temple of King Sesostris II in about 1895 B.C. (Routledge & Kegan Paul, 1986. 269 pp. illus. $24.95)

Astronomy. *The Cosmic Inquirers: Modern Telescopes and Their Makers* by Wallace Tucker and Karen Tucker identifies some of the world's foremost astronomers and examines their contributions to modern astronomy. The authors describe telescopes that are sensitive to radio waves, X rays, gamma rays, and infrared rays, as well as a proposed optical telescope in space. (Harvard Univ. Press, 1986. 221 pp. illus. $20)

Thunderstones and Shooting Stars: The Meaning of Meteorites by Robert T. Dodd shows how scientific analyses of the structure and composition of meteorites provide the keys to understanding the origins of the solar system and how meteorites have influenced life on Earth. (Harvard Univ. Press, 1986. 196 pp. illus. $24.95)

Biology. *The Blind Watchmaker* by Richard Dawkins discusses the various aspects of evolutionary theory, emphasizing the "sheer wonder of biological complexity." Dawkins argues that evolution is the only theory explaining the diversity of life on Earth. (Norton, 1986. 332 pp. illus. $18.95)

The Fabric of Mind by Richard Bergland argues that the functions of the brain are modulated by hormones, not electric impulses. Bergland believes that many kinds of illnesses could be more easily treated by understanding the hormone signals between the body and the brain. (Viking, 1986. 202 pp. illus. $16.95).

Memoir of a Thinking Radish: An Autobiography by Peter Medawar is a personal account of this Nobel Prize winner's scientific life and his important work on tissue rejection and organ transplantation. Medawar also comments on the impact of serious illness on his own life. (Oxford Univ. Press, 1986. 209 pp. illus. $17.95)

Origins: A Skeptic's Guide to the Creation of Life on Earth by Robert Shapiro explains what science does and does not understand about how life first began, using evidence found in fossil records. He also examines theories about how life began, including one that suggests an extraterrestrial origin. (Summit, 1986. 332 pp. $16.95)

Winston Churchill's Afternoon Nap: A Wide Awake Inquiry into the Human Nature of Time by Jeremy Campbell illustrates the way humans have adapted to the rhythms of the external world, and how various biological clocks give bodies and minds the ability to keep track of the passage of time. (Simon & Schuster, 1986. 432 pp. $18.95)

Earth Science. *Time's Arrow, Time's Cycle: Myth and Metaphor in the Discovery of Geological Time* by Stephen Jay Gould reveals how three writers in the last 300 years—Thomas Burnet, James Hutton, and Charles Lyell—contributed to the understanding of the immensity of geological time with their ideas about time's direction and time's cycles. (Harvard Univ. Press, 1987. 222 pp. illus. $17.50)

Ecology. *Ecological Imperialism: The Biological Expansion of Europe 900-1900* by Alfred W. Crosby argues that the successful spread of Europeans across the globe was due to biological and ecological factors as well as organizational and military achievements. Crosby discusses, for example, how the

BOOKS OF SCIENCE: A Science Year Close-Up

The 300th Anniversary of Newton's *Principia*

In the summer of 1687, the *Philosophiae Naturalis Principia Mathematica* (*Mathematical Principles of Natural Philosophy*), by English astronomer and mathematician Sir Isaac Newton, was published in England. Three hundred years later, it is still widely regarded as the greatest single contribution ever made to science.

In his *Principia,* Newton stated the laws of motion and the universal law of gravitation, all of which can be calculated mathematically. He showed how phenomena in space and on Earth—from the motions of comets to the behavior of ocean tides—are subject to these laws. Newton also introduced the concepts of *force* and *mass* that underlie almost all of modern astronomy and physics.

Before Newton, scientists relied on gathering evidence by making observations and doing experiments to arrive at physical laws. Newton, in the *Principia,* showed that mathematics could help reveal physical laws and that observations and experiments could then be used to confirm the truth of a mathematical formula.

The *Principia* is famous for introducing Newton's theory of universal gravitation. The origin of this theory lies in the famous story about Newton observing an apple fall from a tree while he was having tea in his garden. In a flash of intuition, Newton told a friend, he realized that the same force that makes an apple fall to the ground keeps the moon in its orbit around Earth. Newton called this force *gravity.* In the *Principia,* Newton showed mathematically that the force of gravity between two attracting bodies is proportional to the product of their *masses* (the amount of matter they each contain). He also showed that the force of gravity decreases in *inverse* (opposite) proportion to the distance *squared* (multiplied by itself) between the centers of the two bodies. So, if the distance between the centers of the objects doubles, then the force of gravity between them becomes four times weaker.

One of the great proofs of the *Principia* that helped establish its fame concerned Halley's Comet, named for English astronomer Edmond Halley. Halley observed a comet in 1682 and used Newton's concept of gravity to argue that the comet would return. He calculated that it would return in 1758. Halley's Comet returned as predicted, helping to confirm the validity of Newton's laws. There were many other tests, and nearly all proved the *Principia* to be correct.

During the next 100 years, scientists observed only one event—involving peculiarities in the orbit of the planet Mercury—that could not be explained by the principles in the *Principia.* Then, in 1915, the German-born physicist Albert Einstein announced his general theory of relativity, which provided correct calculations for Mercury's orbit. Einstein's general theory of relativity revolutionized scientific thinking about the nature of gravity and opened the door to the worlds of the very large universe and the very small atom, where the behavior of gravity sometimes differs from that predicted by Newton. But the principles expressed in the *Principia* are still valid in the everyday world and in much of the universe at large and are widely used by astronomers and engineers today.

After *Principia,* scientists realized that the universe works according to physical laws and that the test of a theory is its ability to predict physical effects by calculation. [Stephen P. Maran]

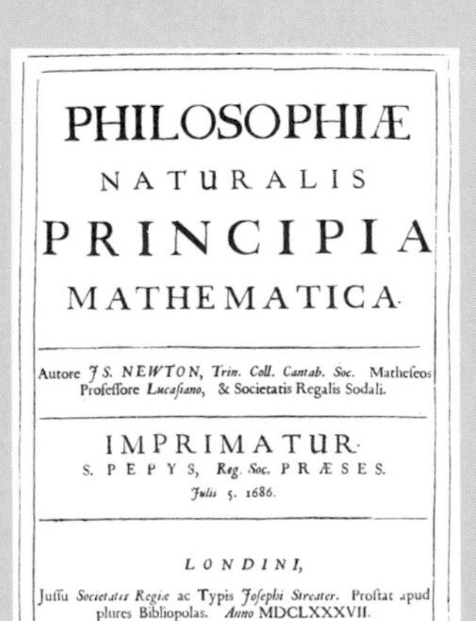

PHILOSOPHIÆ
NATURALIS
PRINCIPIA
MATHEMATICA.

Autore *J S. NEWTON, Trin. Coll. Cantab. Soc.* Mathefeos Profeffore *Lucafiano,* & Societatis Regalis Sodali.

IMPRIMATUR.
S. PEPYS, *Reg. Soc.* PRÆSES.
Julii 5. 1686.

LONDINI,

Juffu *Societatis Regiæ* ac Typis *Jofephi Streater.* Proftat apud plures Bibliopolas. *Anno* MDCLXXXVII.

English scientist Sir Isaac Newton 300 years ago published his masterwork, *Principia, far left,* still widely regarded as the single greatest contribution to science.

diseases carried by Europeans to new lands destroyed many native peoples. (Cambridge Univ. Press, 1986. 368 pp. illus. $22.95)

The Machinery of Nature by Paul R. Ehrlich explains the complex relationship among plants, microorganisms, animals, and humans. Ehrlich explores and explains the principles that govern the functioning of the entire ecosystem. (Simon and Schuster, 1986. 320 pp. $18.95)

General Science. *Dinosaurs in the Attic: An Excursion into the American Museum of Natural History* by Douglas J. Preston identifies the explorers, scientists, and collectors who helped to accumulate the museum's vast collections. Preston conducts the reader through vaults and storerooms and introduces the curators who are in charge of them. (St. Martin's, 1986. 244 pp. illus. $18.95)

Mind from Matter? An Essay in Evolutionary Epistemology by Max Delbrück examines Delbrück's theories on how speech, language, consciousness, and other human characteristics could arise from purely physical processes. Delbrück demonstrates that matter is a product of the mind and that mind is a property of the physical world. (Blackwell, 1986. 290 pp. illus. $29.95)

Secret House: 24 Hours in the Strange and Unexpected World in Which We Spend Our Nights and Days by David Bodanis chronicles the unseen world surrounding two people in their house. The unusual properties of aluminum, the action of a roof when it is heated by the sun, and the microscopic life that lives underfoot on the floor are among the topics covered. (Simon & Schuster, 1986. 223 pp. illus. $19.95)

Natural History. *The Chimpanzees of Gombe: Patterns of Behavior* by Jane Goodall is a detailed account of the author's 25-year study of all aspects of chimpanzee behavior. Goodall discusses communication, feeding, hunting, and social awareness in the chimpanzee community. (Harvard Univ. Press, 1986. 673 pp. illus. $30)

Coral Kingdoms by Carl Roessler beautifully illustrates and describes the unique and distinctive species of animals that inhabit the coral formations in such waters as the Caribbean Sea, the Red Sea, and the Indian Ocean. (Abrams, 1986. 216 pp. illus. $35)

Life Above the Jungle Floor by Donald Perry describes life in a tropical tree canopy, a rarely observed part of the jungle world, where two-thirds of its plants and animals spend their lives. Perry photographed this world from an observation platform anchored 34 meters (111 feet) above the jungle floor. (Simon & Schuster, 1986. 170 pp. illus. $16.95)

Mammal Evolution by R. J. G. Savage describes the life styles, diets, and anatomies of mammals that have inhabited Earth over the past 200 million years. (Facts On File, 1986. 259 pp. illus. $35)

Physics. *Einstein's Dream: The Search for a Unified Theory of the Universe* by Barry Parker describes the efforts of Albert Einstein and other scientists to develop a theory that unifies all the forces of nature. Parker explains what impact new ideas about such exotic concepts as supergravity, superstrings, and twistors have had on their efforts. (Plenum, 1986. 287 pp. illus. $18.95)

The Life It Brings by Jeremy Bernstein is a personal account of how the author became a physicist and of the influences of Albert Einstein and J. Robert Oppenheimer on his career. (Ticknor and Fields, 1987. 171 pp. illus. $16.95)

The Particle Explosion by Frank Close, Michael Marten, and Christine Sutton is a richly illustrated account of the methods used to create, discover, and investigate subatomic particles such as quarks, gluons, and the W and Z particles. (Oxford Univ. Press, 1987. 239 pp. illus. $35)

Technology. *Edison's Electric Light: Biography of an Invention* by Robert Friedel and Paul Israel reveals the steps leading up to Thomas Edison's creation of an electric lamp in October 1879. (Rutgers Univ. Press, 1986. 263 pp. illus. $27.95)

Made in Japan: Akio Morita and Sony by Akio Morita with Edwin M. Reingold and Mitsuko Shimomura chronicles the history of the Sony Corporation from its beginning 40 years ago. Morita, Sony's founder, discusses the development of products, such as the Walkman, and the application of his ideas to the management of the company. (Dutton, 1986. 309 pp. illus. $18.95) [William G. Jones]

Botany

A certain type of rust fungus identifies its victims by "touch," researchers at Cornell University in Ithaca, N.Y., reported in March 1987. Botanist Harvey C. Hotch and his colleagues said the fungus, one of many fungi that cause plant diseases called *rusts*, infects bean plants by detecting a specific shape on the plants' leaves.

Leaves have small pores on their surface. Each pore is surrounded by two special cells, called *guard cells*, that swell to open or contract to close the pore. When a spore from a fungus lands on a leaf, it sprouts a *hypha*—a minute filament—that grows along the leaf until it encounters a guard cell. The hypha then swells up to form a balloonlike mass known as an *infection structure*. The infection structure sends its own hypha through the pore.

The researchers found that the infection structure develops when the first hypha encounters a tiny ridge on a guard cell—but only if the ridge is very close to 0.5 micron high. (One micron equals 0.000001 meter or 0.000039 inch.)

The scientists discovered this in the laboratory using silicon wafers in place of leaves. Using an instrument that emitted a beam of electrons, they created microscopic ridges of different heights on the wafers. They then allowed spores of the rust fungus to sprout on the silicon wafers. When the hyphae from the spores encountered ridges of just the right height, they formed an infection structure. The investigators could not explain how hyphae, which lack a brain or a nervous system, can "feel" minute variations on a leaf or other surface.

Antifungus protein. Plants produce many different chemicals that protect against the growth of fungi and other pests. Most such compounds are small molecules of one kind or another, but in November 1986 botanists Angela Schlumbaum, Felix Mauch, Urs Vögeli, and Thomas Boller of the University of Basel in Switzerland reported that bean plants sometimes produce a complex molecule that inhibits the growth of a certain fungus.

The molecule is a protein called

Botanists at a U.S. Department of Agriculture research facility in North Carolina check the progress of snap bean plants exposed to varying amounts of ozone, a potentially harmful form of oxygen. The researchers reported in late 1986 that prolonged exposure to even low levels of ozone reduces the yield of snap beans and other crop plants.

chitinase (*KY tin ase*). Chitinase breaks down *chitin*, a stiff substance that forms the outer shell of lobsters, crabs, and many insects. Chitin is also a major building material of the cell walls of some fungi. The researchers showed that bean leaves increase their production of chitinase when they are infested with a fungus called *Trichoderma viride*.

The scientists added chitinase to culture dishes in which *T. viride* was growing. The protein stopped the growth of the fungus. If the researchers first damaged the chitinase by boiling it or causing it to react with other kinds of molecules, however, the fungus was again able to grow.

Another discovery relating to plants' resistance to fungi was reported in February 1987 by botanist Richard Karban and his colleagues at the University of California in Davis. They exposed cotton seedlings to the fungus *Verticillium dahliae* and found that the plants developed resistance not only to the fungus but also to tiny parasites called spider mites. Conversely, if the scientists first exposed the seedlings to spider mites, the plants became resistant to the fungus. The investigators said further research would be required to explain this dual resistance.

Roots that grow up. For well over a century, plant scientists have studied how most plant roots respond to gravity by growing downward while stems grow upward. But the roots of some tropical trees grow upward along the trunks of neighboring trees. In February 1987, botanist Robert L. Sanford, Jr., then at Stanford University in California, explained why this is so.

Sanford surveyed an area of Amazon rain forest in Venezuela. He found that at least 12 species of trees in the forest have roots that grow upward on nearby trees.

Sanford said the upward-growing roots are probably an evolutionary adaptation to the mineral-poor soil of tropical rain forests. In these forests, plants quickly absorb minerals from the soil through their dense root systems. Most of the minerals enter the forest in raindrops that contain dust particles and tiny salt crystals.

Upward-growing roots apparently allow some trees to "cut in line" in the competition for minerals, Sanford said. Rather than waiting for mineral-laden rain water to soak into the soil, those trees use their upward-growing roots to extract minerals from the water before it reaches the ground.

Photosynthesis inhibitor. The discovery of a sugarlike compound that *inhibits* (slows down) photosynthesis, the process by which green plants combine carbon dioxide and water to form complex compounds for energy and growth, was reported in November 1986 by investigators at several laboratories in Great Britain and the United States. The leader of the research team, botanist Steven Gutteridge of Rothamsted Experimental Station in Harpenden, England, said the substance may be the most widespread inhibitor molecule in all of nature.

Scientists had known for years that the production of carbohydrates and other compounds in photosynthesis is dependent on a protein called *rubisco*, probably the most abundant protein on Earth. They also knew that rubisco's activity drops off at night, but why that is so remained a mystery.

Gutteridge and his colleagues found the answer in chloroplasts, the green structures in plant cells where photosynthesis occurs. The researchers discovered that as the rubisco in chloroplasts becomes less active, another molecule—the long-sought inhibitor—increases in concentration.

The scientists purified the inhibitor from potato leaves that had been kept in darkness. Analysis of the compound revealed that it is a type of sugar. Its molecular structure is very similar to that of a sugar molecule that normally forms during photosynthesis when carbon dioxide and another kind of sugar combine with rubisco. Because of that similarity, the inhibitor attaches to rubisco just like the sugar molecule resulting from the chemical combination. In so doing, the inhibitor gets in the way so the normal steps in the photosynthesis reaction cannot be completed. The researchers said further experiments would be required to explain why the inhibitor loses its ability to interfere with rubisco in daylight. [Frank B. Salisbury]

In WORLD BOOK, see BOTANY; FUNGUS; PHOTOSYNTHESIS.

Chemistry

The dream of many dieters—a chemical that reduces the craving for food—may be on the horizon, according to an April 1987 announcement by chemist Michael J. DiNovi of the Monell Chemical Senses Center in Philadelphia. DiNovi revealed that a noncaloric carbohydrate called *2,5-AM* suppressed the appetites of laboratory rats. The carbohydrate is chemically related to fructose, a sugar found in corn syrup. Rats fed 2,5-AM before meals reduced their overall intake of food significantly.

Strangely, 2,5-AM can also stimulate the appetites of rats, depending upon when they consume it. Rats normally are active at night. In the morning, when they usually are resting, they get their energy from sugar stored in their blood. Rats fed 2,5-AM in the morning apparently could not use their stored energy, so they ate to obtain more energy, increasing their overall consumption of food.

DiNovi has not demonstrated that 2,5-AM works on human beings. But if it does, it could not only help control appetite in dieters but also stimulate the appetites of the elderly, of people suffering from liver disease, and of patients undergoing drug treatment for cancer.

The chemical appears to act upon the liver and thus may confirm a hypothesis that the liver is more important than the brain in controlling appetite. Appetite-decreasing chemicals that work on the brain, such as amphetamines, may be only weakly effective or have undesirable side effects such as psychological dependence and addiction. So far, 2,5-AM has caused no obvious ill effects on the rats.

DiNovi says the new chemical could be produced inexpensively from crab or lobster shells. Furthermore, 2,5-AM has a pleasant, sweet taste and could be directly added to a cookie or a candy bar without having to add other substances to mask the flavor. The food industry has already expressed interest in the chemical.

Bugs to the rescue. *Entomologist* (insect specialist) Jeffrey R. Aldrich announced in September 1986 that he and his colleagues at the United States Department of Agriculture (USDA) in Beltsville, Md., had developed the first *pheromone* (a communication chemical secreted by certain animal species) to attract a predator insect, the spined soldier bug. Spraying crop fields with the pheromone would attract the predators to the fields, where they would eat pests. Attracting the predators may reduce the need for chemical pest killers, reducing both the costs to farmers and the burden of poisons on the environment. The new pheromone is a sex lure made up of three simple chemicals that are produced naturally by the spined soldier bug but also are easily and economically produced commercially. The spined soldier bugs eat caterpillars and beetles.

Pheromones that attract pests have been known for years; they are used in so-called bug bags that trap Japanese beetles, for example. But pest attractants eventually lose effectiveness because the pests adapt themselves to ignore the attractants by means of natural selection. The pests that are the least attracted to the compounds are the most likely to survive and reproduce, and their offspring are likely to inherit a resistance to the compounds. This selection process is repeated generation after generation, and the compounds eventually become ineffective.

This adaptation would not occur in the case of predator pheromones because the attractant works in favor of the predator insects. The pheromones draw the predator insects to sites where food is readily available in the form of pests. Bugs that heed the call of the predator pheromones will thus multiply especially well, so predator pheromones should become more, not less, effective over time.

Dissolving gallstones. A new chemical compound that may dissolve gallstones was reported in April 1987 by medical chemist Ashok K. Batta and his colleagues at the University of Medicine and Dentistry of New Jersey in Newark. Approximately 20 million Americans suffer from gallstones, hard objects that form in the gall bladder or bile duct. Although gallstones are not in themselves life threatening, they can produce one of the worst abdominal pains of any illness. The standard treatment, removal of the gall bladder, requires major surgery.

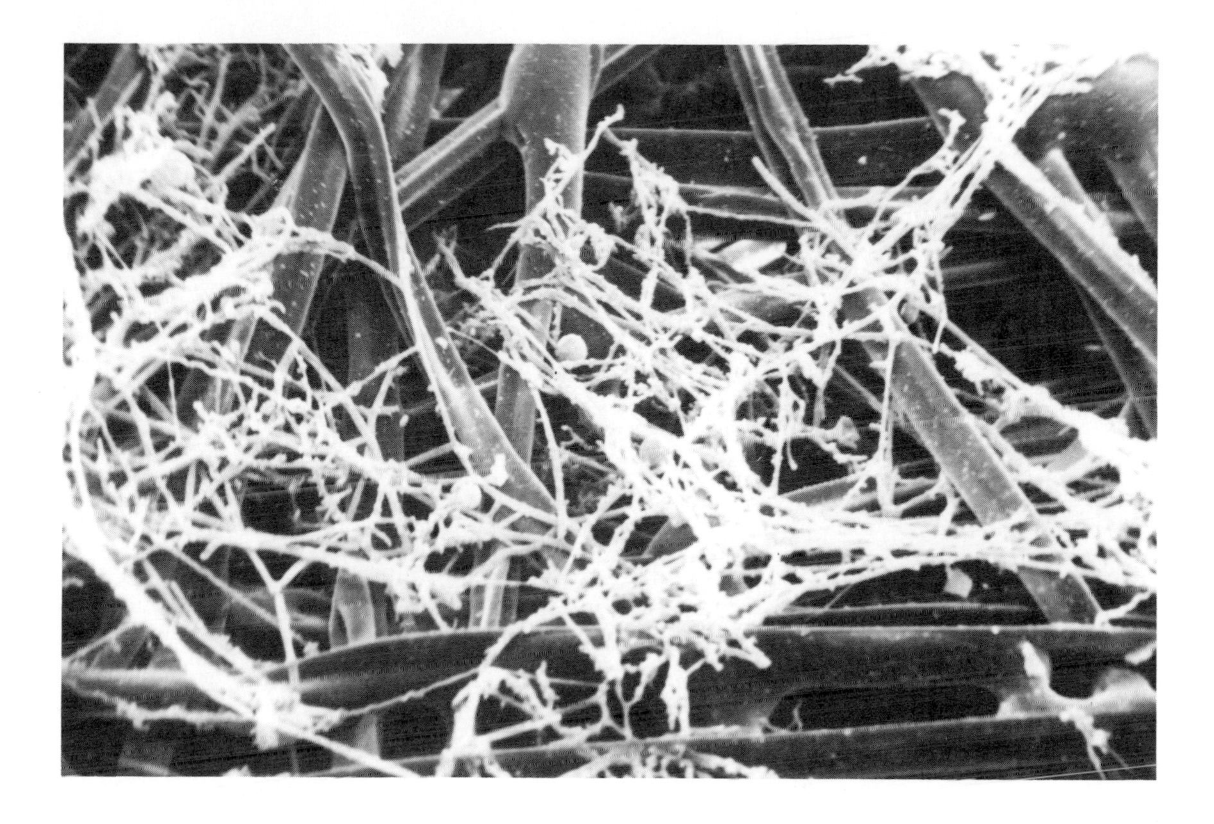

Chemistry
Continued

Scientists are experimenting with white rot fungus, *above* (greatly magnified), to see if it can be used to help clean up soil and water contaminated by the pesticide DDT, dioxins, and various cancer-causing chemicals. Researchers believe that the fungus, when injected into the ground, can destroy pollutants because it emits an enzyme that can break down their large molecules into simpler molecules that would then degrade into harmless carbon dioxide.

When the patient is not healthy enough to undergo surgery, drugs may be used to dissolve the gallstones.

Unfortunately, these drugs are expensive and typically take one to two years to work. And when the drug therapy ends, new gallstones tend to form.

The new compound, *ursocholic acid*, like currently used drugs, is based on bile acids that are naturally present in the gall bladder. In tests conducted in Italy, ursocholic acid completely dissolved gallstones in 4 of 10 people in just six months. The new compound is simpler chemically than today's drugs, so this compound might sell for less than 5 per cent of the cost of current drugs.

The new compound appears to work by forming liquid crystals, making cholesterol—a major component of most gallstones—more soluble in water so that it can be washed out in the urine. One problem with ursocholic acid, however, is that 30 per cent of it is washed out in the urine before reaching its target.

Batta and his co-workers focused their efforts on improving the compound. One improved form is made up of *amino acids* (the building blocks of proteins) chemically linked to the original compound. The amino acids guide the ursocholic acid toward the gallstones. By attaching amino acids, the researchers have already reduced the ursocholic acid waste to less than 10 per cent. A drug based on their work could be available commercially in 5 to 10 years.

Slow roaches. Chemist Ronald J. Nachman of the USDA's Research Center in Berkeley, Calif., and entomologists G. Mark Holman and Benjamin J. Cook of the USDA's Research Center in College Station, Tex., in September 1986 announced their attempt to turn the cockroach's brain chemistry against the roach.

Roaches are more than just pests. They carry at least 30 harmful bacteria and more than a dozen parasites. An estimated 7 per cent of the U.S. population has roach-associated allergies. Roaches have shown a remarkable ability to develop resistance to insecticides and are rapid breeders.

The scientists announced that they had found brain chemicals that stimulate a certain part of the insect's digestive tract. New compounds modeled on these brain chemicals might be used to either speed up or slow down the insect's digestion, making the roach sick and less able to survive.

Holman and Cook believe the roaches are less likely to become resistant to compounds based on their brain chemicals because these chemicals are essential to the basic activity of the roach brain. Also, the aim of the attack is to control rather than annihilate roaches. A compound intended to eliminate roaches might kill 80 per cent of a given population of the insects, for example, leaving the available food supply to the remaining 20 per cent. These would thrive, and their offspring would inherit some resistance to the compound. Succeeding generations would become even more resistant, eventually defeating the purpose of the compound.

Substances modeled on the brain chemicals of the roach, on the other hand, might kill only 40 per cent of the roaches, leaving a sizable number of sick insects to compete for food with resistant strains. This would prevent the resistant insects from increasing their numbers quite so rapidly.

The chemists must now find a way to administer the chemicals. Feeding roaches the chemicals in their present form will not work because they pass harmlessly through the insect's digestive tract. The scientists believe that they can modify the chemicals slightly to prevent this, resulting in an inexpensive, effective roach-control product in 5 to 10 years.

Seeing chemical bonds. For the first time, scientists in August 1986 "saw" the bonds that hold atoms together to form molecules. Bonds that exist at the surface of a silicon crystal appeared in images produced by *surface scientists* (researchers who study chemical reactions occurring on surfaces) at the International Business Machines Corporation's (IBM) Thomas J. Watson Research Center in Yorktown Heights, N.Y. The scientists used a scanning tunneling microscope (STM), the de-

A plastic sheet that contains up to 50 per cent cornstarch may help solve problems of how to dispose of plastic bags, mulches, and other products. The starch-based plastic can disintegrate in water or be devoured by microorganisms in soil.

CHEMISTRY: A Science Year Close-Up

Metals with a Memory

The development of metals that "remember" their shape has been one of the most exciting areas of materials science research in the 1980's. At room temperature, these memory metals are soft and pliable, and they can be formed into almost any shape. A memory-metal spring, for example, can be stretched to more than 10 times its original length. When heated, however, the metals become as strong as steel and snap back to the shape in which they were originally molded.

This phenomenon, called the *shape memory effect,* was first discovered in 1932 in an alloy of gold and cadmium. In the 1960's, scientists also found the effect in other alloys. Research showed that the effect is caused by an abrupt change in the crystal structure of the metal. Below a certain transition temperature—which can vary widely from one alloy to another—a memory metal can be easily reshaped because its atoms are easily shifted into new positions. But when the alloy is heated, the atoms suddenly shift back to their previous positions, restoring the rigid crystal structure and returning the piece of metal to its original shape.

For many years, shape memory was described as "a solution looking for a problem." Because the alloys were so difficult to produce, they were too expensive—approximately $2,000 a kilogram (2.2 pounds)—to be used in most devices and products. In the 1980's, though, more advanced processing methods brought the price of the alloys down to about $500 a kilogram.

As prices lowered and memory metals became widely available, the search for practical applications began in earnest. Between 1985 and 1987, patent applications for nearly 1,000 new uses for the remarkable alloys were registered in the United States, Japan, and other countries.

In Japan, a lingerie manufacturer in 1986 began to use a nickel-titanium alloy wire as the support in a brassiere. The wire gets twisted out of shape in the washing machine but recovers its original form when the bra is placed in a hot dryer.

Many dental and medical products are also being made with memory metals. A new type of orthodontic appliance, for example, uses memory wire to straighten teeth much more rapidly than is possible with conventional stainless-steel braces. The wire is engineered so that just the warmth of the mouth causes it to stiffen.

Other applications make use of the stresses that are generated when memory metal is prevented from fully recovering its original shape. Such stresses can be used to join pipes together, for instance. The sleeves into which the ends of pipes are inserted can be expanded while the alloy is in its soft state. Once the pipes are in place, applying heat causes the sleeves to tighten around the pipes with great force. Several recently developed alloys will maintain this tight seal after the external heat is removed.

Some new devices, called *actuators,* use memory metals to do work, such as lifting a weight or pumping water. One device opens and closes the louvers of an automobile fog lamp. When an electric current flows through a memory wire in the apparatus, the wire becomes hot and contracts, opening the louvers. When the current is shut off, the wire relaxes and the louvers close again.

Other actuators respond to the temperature of air or water. One device, for example, opens the windows in a greenhouse when the air inside the greenhouse rises to the alloy's transition temperature. Another actuator, a memory-alloy shower head, shuts off the flow of scalding water.

Looking to the future, researchers see potential uses for memory metals in robots, particularly for parts such as artificial muscles. And some researchers have already designed and built experimental engines that use memory metals to convert heat energy, particularly solar energy, to mechanical energy—to pump water, for example.

Unfortunately, these engines operate inefficiently, leading many scientists to predict that heat engines and other potential applications of memory metals will never be practical. But other researchers are more optimistic. They think the performance of heat engines and other devices will be improved and that memory metals will prove useful for many specialized tasks—including some that no one has yet envisioned.

[Tom Duerig]

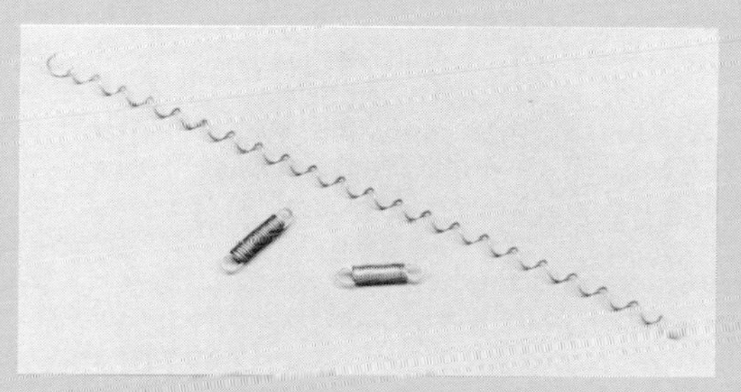

While cool, a memory-alloy spring stretches easily to 10 times its coiled length. But heating makes it snap back to its original shape.

vice that provided the first pictures of individual atoms in 1981.

To scan surface atoms, the STM runs a needlelike probe back and forth less than one-millionth of a millimeter above a surface. When a voltage is applied between the probe and the surface, electrons tunnel, or jump, from the surface to the probe, setting up an electric current that raises and lowers the probe depending on whether an atom is present. For example, the probe detects a surface atom as a bump and moves upward. Moving past the atom, the probe moves down.

By moving up and down in response to the current, the probe "feels" the surface without actually touching it. A computer translates the probe's vertical movements into an image that is displayed on a televisionlike screen or traced on paper.

By adjusting the voltage, the IBM scientists found they could control the tunneling process well enough to enable the probe to "feel" bonds as well as atoms at the surface. Images of bonds may help chemists understand phenomena such as how catalysts are able to speed up chemical reactions while themselves remaining almost unchanged.

The IBM researchers hope that the images of silicon crystals will lead to better silicon products such as microchips into which are built complete electronic circuits. STM images may help manufacturers control surface contamination of chips that can lead to the disruption of electronic signals. The STM also can modify a silicon surface with great precision by depositing atoms of metal onto the surface or etching thin tracks in crystals. These techniques would give manufacturers undreamed-of control over the surface of microchips. Manufacturers would be able to build electronic components on the surface literally atoms at a time. In the Special Reports section, see THE MICROCHIP—A MINIATURE MARVEL.

As STM's become more available, they should have a major impact on all areas of chemistry. [Peter J. Andrews]

In WORLD BOOK, see CHEMISTRY.

Computer Hardware

In a surprise announcement in September 1986, Compaq Computer Corporation of Houston introduced the DeskPro 386, which immediately became the world's most powerful personal computer. The machine's instant leadership was due to the fact that it was the first to use a *32-bit microprocessor chip*, called model 80386 and made by Intel Corporation of Santa Barbara, Calif. Industry experts had expected that International Business Machines Corporation (IBM) would be first on the market with a personal computer based on the 80386 chip.

Power in the chip. A chip is a tiny piece of material—usually silicon—onto which are built electronic circuits. The circuits on a computer's microprocessor chip perform essential computer functions such as arithmetic and logic operations.

A computer's power to process data and perform calculations depends heavily upon how many *bits*—0's and 1's in computer language—its microprocessor can handle at one time. Before the introduction of the DeskPro 386, the most advanced personal computers used 16-bit chips, which process data 16 bits at a time.

The DeskPro 386 with its 32-bit chip can run popular, standard *software* (programs) up to three times faster than the next most advanced computers. The DeskPro was offered in two versions. The less expensive machine, Model 40, at $6,499 can store 40 million *bytes*—40 megabytes—of data in its fixed disk drive. (One byte is one letter, numeral, or other single symbol.) Model 130 at $8,799 has a 130-megabyte disk drive. Both computer models are equipped with a 1-megabyte *random access*, or temporary, memory (RAM).

IBM-compatible Apples. After years of pleas from computer users and hardware and software firms, Apple Computer Incorporated of Cupertino, Calif., announced in March 1987 a family of Macintosh computers—Macintosh II and SE—that can be adapted to run software designed for IBM personal computers. According to market analysts, the inability of Apple com-

Computer Hardware

Continued

A new batch of more powerful personal computers was introduced in 1987. The Macintosh SE, *above left,* introduced in March, not only comes in models with a built-in 20-megabyte hard disk drive, but it also can be adapted to run software developed for International Business Machines Corporation (IBM) computers. IBM itself unveiled its powerful Personal System/2 computer, *above right,* which reportedly is difficult to imitate by makers of *PC clones*—machines that operate much like earlier IBM personal computers.

puters to run such software had limited Apple's growth in sales to business customers.

The $3,769 Macintosh II is designed around a 32-bit microprocessor made by Motorola Corporation of Schaumburg, Ill. The Macintosh SE, priced from $2,769, uses a Motorola 16-bit microprocessor. Both machines provide 1 megabyte of RAM and either two 800-*kilobyte* floppy disk drives or one 800-kilobyte floppy disk drive plus an internal 20-megabyte hard disk. (One kilobyte equals approximately 1,000 bytes.)

Enter IBM. In April 1987, IBM unveiled a family of eight personal computers, named Personal System/2 (PS/2), to revive sales lost to manufacturers of *PC clones*—machines built to operate like IBM's personal computers and to run software designed for these computers. Prices of PS/2 computers range from $1,695 to more than $10,000.

Five of the machines are built around 16-bit microprocessors, while the other three use Intel's 32-bit 80386 chip. The more expensive models include features IBM claims will be difficult for clone manufacturers to copy. Internal software necessary to make full use of the new computers was not expected to be ready until 1988. See Computer Software.

PC clones. Three firms introduced IBM-PC clones in early 1987. Atari Corporation of Sunnyvale, Calif., unveiled the Atari PC; Bondwell Computer of Hong Kong offered the X'Press 16; and Hyundai of South Korea announced the Blue Chip.

Commodore Business Machines Incorporated of West Chester, Pa., introduced their clones, called PC-10, in the United States in December 1986. These machines had been sold in Canada and Europe. In March 1987, the company announced an upgraded version of its model 128 computer and added more features to its Amiga computer. [Howard Bierman]

In the Special Reports section, see The Microchip—A Miniature Marvel. In World Book, see Computer; Computer, Personal.

Computer Software

A new generation of more powerful personal computers came on the market in 1986 and 1987. But for many of these new machines there was a lack of software. Compaq Computer Corporation of Houston, for example, introduced the DeskPro 386 in late September 1986. It was the first commercial computer to make use of Intel Corporation's powerful 80386 microprocessor chip. But as late as mid-1987, no software existed to take full advantage of the machine's outstanding features, such as the ability to handle several programs at once or to use its massive memory to do complex financial analysis and generate dazzling graphics.

The lack of software arose from the fact that MS-DOS, the *disk operating system* (DOS) commonly used with Compaq computers, cannot handle as much memory as the new machine can. And the producer of MS-DOS, Microsoft Corporation of Redmond, Wash., had not developed a more powerful DOS successor.

A DOS is software that acts as a "traffic cop" inside a computer, coordinating the activities of the disk drive, keyboard, monitor, and microprocessor. The DOS also acts as an "interpreter" between the computer, which uses machine language composed of 0's and 1's, and the human operator, who uses letters of the alphabet, numerals, and other symbols.

MS-DOS can handle only about 640,000 bytes (640K) of internal memory, but the DeskPro 386 can deal with 4 billion bytes. (One byte equals one letter, numeral, or other symbol.)

Software Link, Incorporated, of Atlanta, Ga., in late 1986 announced its PC-Modular Operating System/386, which the company claims can run several programs at once and handle tasks requiring large amounts of memory. Other software companies, such as Softguard Systems, Incorporated, of Santa Clara, Calif., introduced *extender* programs that supplement MS-DOS.

IBM's software problem. In April 1987, International Business Machines Corporation (IBM) announced Personal System/2 personal computers, the most powerful of which are also

Computer Software

Continued

built around 80386 microprocessors. But history repeated itself. Software was not ready for the new machines.

IBM planned to produce a new DOS, called OS/2, that would not only operate the new machines but also provide links to large IBM computers. Microsoft planned to provide application software developers with the tools to build programs based on OS/2 by August 1987.

Most major software publishers, including Lotus Development Corporation of Cambridge, Mass., and Ashton-Tate of Torrance, Calif., said they would create more powerful spreadsheet, database, and accounting programs to match the capabilities of IBM's new machines. But, they added, these products probably would appear six months after the release of OS/2, expected sometime in 1988.

Desk-top publishing systems continued to be a top priority for software makers. Apple Computer's Macintosh took an early desk-top publishing lead in the mid-1980's because several software companies created programs that would allow the Macintosh to merge text and graphics and create page layouts that could then be printed on a laser printer or even on professional typesetting equipment. In 1986 and 1987, hundreds of products, from page-layout programs to type-face and style selectors to laser printer programs, were introduced for the IBM PC and PC "clones" as well as for the Macintosh. Software Publishing Incorporated of Mountain View, Calif., announced its Click Art Personal Publisher in August 1986. This program enabled PC owners to design page layouts, create multiple text columns, and wrap text around graphics—capabilities that were previously available only to Macintosh owners.

Coping with copies. In 1986, software makers abandoned schemes to prevent their programs from being copied. They were convinced that the loss of good will among customers outweighed the loss in revenues from copying. [Howard Bierman]

In World Book, see Computer; Computer, Personal.

An "artificial reality" computer program tests police-officer trainees' judgment in a crisis situation. The program presents several situations, each with a variety of endings, on a TV screen. Some endings, such as the one shown *below*, call for the trainee to shoot a laser beam at the TV image. Other endings seem threatening at first but turn out to be safe (the woman takes a driver's license out of her purse), and the trainee should not shoot.

Deaths of Scientists

Harrison S. Brown

Fritz A. Lipmann

Robert S. Mulliken

Notable scientists and engineers who died between June 1, 1986, and June 1, 1987, are listed below. Those listed were Americans unless otherwise indicated. An asterisk (*) indicates that a biography appears in THE WORLD BOOK ENCYCLOPEDIA.

Berry, George P. (1898-Oct. 5, 1986), bacteriologist who was best known for his influence on medical education in the United States as dean of Harvard Medical School in Cambridge, Mass., from 1949 to 1966. Berry also contributed to the understanding of viral infections and was an early investigator of the theory that viruses may cause cancer.

Brown, Harrison S. (1917-Dec. 8, 1986), geochemist who played a major role in developing plutonium for the production of the first atomic bombs and who later became a leading opponent of the use of nuclear weapons. Brown was noted for pathbreaking research into the chemical makeup of meteorites and their links to the origin of the solar system. He was editor in chief of the *Bulletin of the Atomic Scientists* and a long-time member of the editorial advisory boards of THE WORLD BOOK YEAR BOOK and SCIENCE YEAR.

Carr, Archie F., Jr. (1909-May 21, 1987), zoologist who documented the exceptional navigational abilities of giant sea turtles, which migrate over thousands of kilometers. Carr's conservation efforts also helped save the turtles from extinction.

Cox, Allan V. (1926-Jan. 27, 1987), geophysicist who determined the history of Earth's magnetic field reversals through the study of rocks and thereby helped confirm the theory of plate tectonics. Cox worked with the U.S. Geological Survey from 1957 to 1967, when he became a professor at Stanford University in California. Cox served on the Editorial Advisory Board of SCIENCE YEAR in the 1970's.

***De Broglie, Louis Victor** (1892-March 19, 1987), French physicist and winner of the 1929 Nobel Prize for physics for his theory about the nature of electrons. He also helped lay the basis for the theory of quantum mechanics. In 1924, De Broglie proposed that electrons have characteristics of both particles and waves, a theory that was confirmed by experiments in 1927.

Doisy, Edward A., Sr. (1893-Oct. 23, 1986), biochemist and co-winner of the 1943 Nobel Prize for physiology or medicine for synthesizing vitamin K. Doisy also isolated two female sex hormones—estrone and estradiol. He was professor and chairman of the Biochemistry Department at St. Louis University's School of Medicine in Missouri from 1924 to 1965.

Ewing, Gifford (1904-Dec. 10, 1986), oceanographer who explored the use of remote-sensing satellites for the study of the oceans. He was a researcher at Scripps Institution of Oceanography in La Jolla, Calif., from 1948 to 1966 and at Woods Hole Oceanographic Institution in Massachusetts from 1966 to 1974.

Hammond, E. Cuyler (1912-Nov. 3, 1986), biologist and epidemiologist whose research in the early 1950's first established a link between cigarette smoking and lung cancer. Hammond found that cigarette smokers also face a high risk of death due to heart disease and other causes. In the 1960's and 1970's, along with other researchers, he also showed that asbestos and vinyl chloride can cause cancer.

Johnston, Richard B. (1930-Jan. 8, 1987), archaeologist noted for his excavation of prehistoric Indian burial grounds in Ontario, Canada, from 1956 to 1961 while he was a field director for the Royal Ontario Museum in Toronto.

Libby, Leona Marshall (1919-Nov. 10, 1986), physicist who worked on the first nuclear reactor, the only woman researcher on the Manhattan Project, which developed the first atomic bomb during World War II. Libby was later involved in the design of other nuclear reactors.

Lipmann, Fritz A. (1899-July 24, 1986), German-born biochemist who shared the 1953 Nobel Prize for physiology or medicine for his contribution to understanding *metabolism*, the conversion of food into energy. With co-winner Hans A. Krebs, a British biochemist, Lipmann discovered coenzyme A, one of the most important substances in the process of metabolism. In 1966, he was presented with the U.S. National Medal of Science, America's highest award for scientific achievement.

Deaths of Scientists
Continued

Marlin Perkins

Nikolai N. Semenov

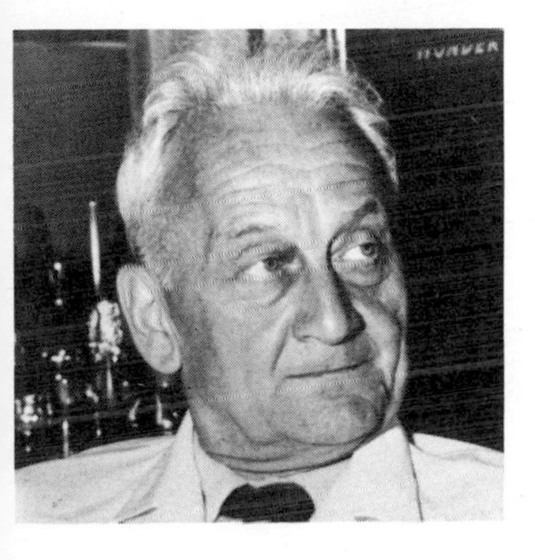

Albert Szent-Györgyi

Livingood, John J. (1903-July 21, 1986), physicist who helped design, build, and operate the earliest *cyclotrons* (atom smashers). Livingood also produced the first synthetic radioactive substance—radium E—and was instrumental in developing several radioactive isotopes, including radioactive iodine, which is used in research in physics, biology, and medicine.

Livingston, M. Stanley (1905-Aug. 25, 1986), physicist who built the first cyclotron in the early 1930's at the University of California in Berkeley in association with the cyclotron's inventor, Ernest O. Lawrence. Livingston was a co-author of the first nuclear physics textbook and taught physics at Massachusetts Institute of Technology in Cambridge from 1938 to 1970.

***Mulliken, Robert S.** (1896-Oct. 31, 1986), chemist who won the 1966 Nobel Prize for chemistry for his *molecular-orbital* theory of chemical structure, which explains how atoms combine to form molecules. Mulliken was known as the "father" of modern structural chemistry. His theory is used to study the structure of proteins, plastics, and other complex compounds. Mulliken taught chemistry at the University of Chicago from 1928 to 1961.

Nolen, William A. (1928-Dec. 20, 1986), surgeon who wrote several books on his medical experiences, including *The Making of a Surgeon* (1970). Nolen also served on the Editorial Advisory Board of THE WORLD BOOK HEALTH & MEDICAL ANNUAL.

Perkins, R. Marlin (1905-June 14, 1986), zoologist who hosted the award-winning television series "Wild Kingdom" from 1963 to 1985. Perkins also served as director of Lincoln Park Zoo in Chicago in the 1940's and St. Louis Zoo in the 1960's.

Rogers, Carl R. (1902-Feb. 4, 1987), psychologist who developed an approach to psychotherapy that allows patients to take the lead in their own treatment, rather than merely relying on the direction of the therapist. Rogers also helped pioneer "encounter group" therapy and was the author or coauthor of several books, including *On Becoming a Person* (1961).

Roth, Lloyd J. (1911-July 25, 1986), chemist who worked on the atomic bomb project at Los Alamos, N. Mex., during World War II. Roth later pioneered in the development of radioactive isotopes used to determine how drugs spread through the body. His work led to the development of *isoniazid*, which is considered the most effective drug against tuberculosis.

***Semenov, Nikolai N.** (1896-Sept. 25, 1986), Soviet chemist who shared the 1956 Nobel Prize for chemistry with British chemist Sir Cyril Hinshelwood for work that explained many complex chemical chain reactions, including those connected with explosions. Semenov was also credited with laying the foundations for a new branch of science known as *chemical physics*.

Severny, Andrei B. (1913-April 4, 1987), Soviet astronomer who pioneered in building telescopes on satellites that orbit in space. Severny had been director of the Crimean Astronomical Observatory since 1952, when he won the Stalin Prize for his research on solar flares.

Smyth, Henry D. (1898-Sept. 11, 1986), physicist who played a major role in the development of the atomic bomb and who later served on the U.S. Atomic Energy Commission. Smyth was the author of the U.S. government's official report on the atomic bomb, which became known as the "Smyth Report." It was released after the bombing of Hiroshima and Nagasaki, Japan, in 1945.

Szent-Györgyi, Albert (1893-Oct. 22, 1986), Hungarian-born biochemist who won the 1937 Nobel Prize for physiology or medicine for his discovery of vitamin C and for research on oxidation in tissues and fumaric acid. In 1954, Szent-Györgyi received an Albert Lasker Award for his research on heart muscle contraction, including identifying the biochemical *actomyosin* and discovering the role that it plays in muscle contraction.

Zacharias, Jerrold R. (1905-July 16, 1986), physicist who developed a new curriculum for teaching physics in high school and revolutionized science education in the United States in the 1960's. During World War II, Zacharias headed the engineering division of the Los Alamos atomic bomb project. He also designed the first atomic clock, the most precise tool for measuring time. [Rod Such]

Dentistry

Dental scientists at the University of Toronto in Canada reported in January 1987 that they had developed a method to test in human mouths the anticavity effects of substances in cheese. Dental researchers have known for some time that cheese protects against the development of cavities, and they believed that a number of substances found in cheese, such as calcium, calcium salts, phosphates, and protein might be responsible.

The researchers could not deliberately cause cavities in a person's teeth, so they used volunteers who wore *dentures* (false teeth). The researchers prepared small blocks of enamel from cows' teeth, which were sterilized and then attached to each side of a volunteer's lower dentures. Six times a day, for one week, the volunteers soaked one side of their dentures in water and the other side in a cheese extract for five minutes. With the dentures in their mouths, the volunteers rinsed twice a day with a sugar solution. The results, according to the researchers, showed that exposing the teeth to cheese extracts before rinsing them in the sugar solution significantly decreased demineralization of the *enamel* (outer layer) of the teeth. Demineralization, the dissolving of the minerals phosphate and calcium in a tooth's enamel, is caused by acid produced by bacteria in *plaque*, a film of bacteria that sticks to the teeth.

The researchers found that the cheese extract did not exert its protective effect by interfering with bacteria in plaque. Instead, the researchers suggested that cheese helps prevent cavities because the calcium it contains has a protective effect on the enamel. They noted that there was significantly more calcium in the plaque on the tooth enamel exposed to the cheese extract than on the tooth enamel exposed to the water.

Sugar and cavities. Eating sugar and sugary desserts or snacks three or more times a day is more likely to cause cavities than consuming these foods only once or twice per day, according to a study reported in December 1986.

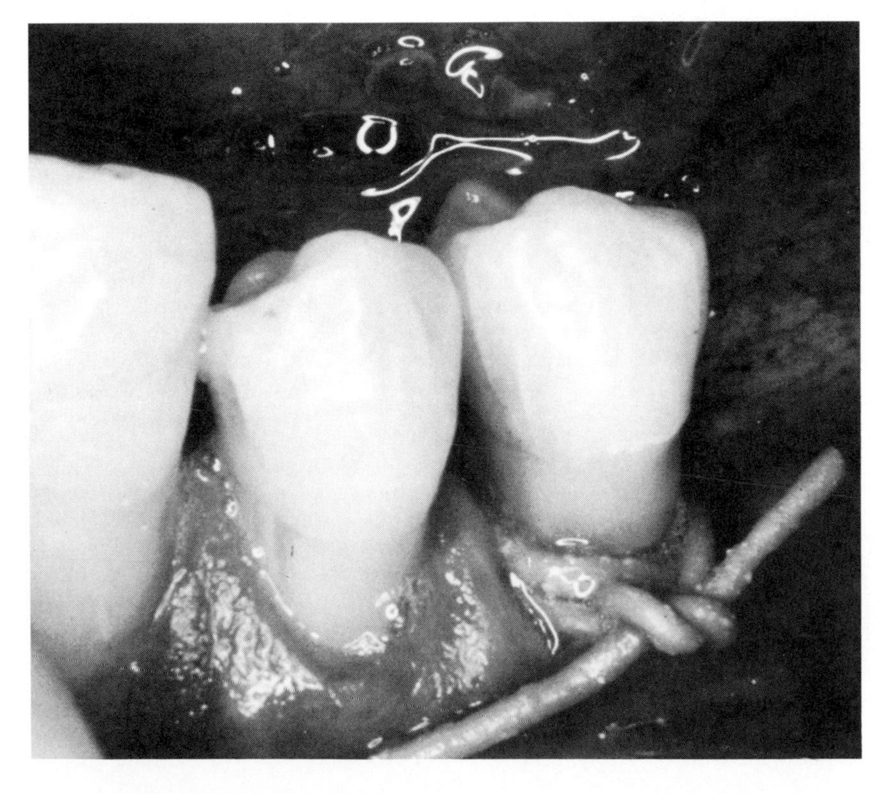

An antibiotic contained in a hollow thread is tied around a tooth in an experimental treatment for *periodontal disease* (gum disease), which can cause teeth to loosen and fall out. The thread, developed by the Alza Corporation of Palo Alto, Calif., and the Forsyth Dental Center in Boston, releases the antibiotic directly onto the bacterial infection that causes gum disease.

In the study, dental researcher A. I. Ismail of the University of Michigan School of Public Health in Ann Arbor analyzed the medical and dental information of more than 20,000 people in the United States who had participated in a health and nutrition survey between 1971 and 1974. The survey data also included a description of the participants' eating habits.

Ismail's analysis found a significant relationship between the number of cavities reported by the participants and the amount of sugar they consumed between meals. When table sugar or syrup was consumed at one or two meals a day, it did not seem to cause an increase in the number of cavities. Eating table sugar or syrup at three or more meals a day, however, caused a significant increase in the number of cavities. Ismail drew similar conclusions about the relationship between cavities and sugary desserts or snacks.

He concluded from these data that reducing the consumption of table sugar, syrups, and sugary desserts and snacks can help prevent and control cavities.

Test for cavity-causing bacteria. Research has shown that there is a strong correlation between the presence of *Streptococcus mutans*, the bacterium responsible for cavities, in people's saliva and the development of cavities in their teeth. In January 1987, scientists at the Forsyth Dental Center in Boston and the APO Diagnostic Company in Toronto, Canada, reported that they had developed a bacteria test kit that allows dentists to determine which patients are at high risk for developing cavities. The test detects and calculates the concentration of *S. mutans* in the patient's saliva. According to the scientists, the test can be used and analyzed in the dentist's office. The dentist can identify those patients most likely to develop cavities and take appropriate preventive action, such as applying sealants. [Paul Goldhaber]

In the Science You Can Use section, see SORTING OUT TRENDS IN ORAL HEALTH CARE. In WORLD BOOK, see DENTISTRY.

Drugs

After a review process of only two years, the United States Food and Drug Administration (FDA) in Washington, D.C., on March 20, 1987, approved the drug *azidothymidine* (AZT) as a treatment for patients with AIDS (*a*cquired *i*mmune *d*eficiency *s*yndrome)—a fatal virus infection that attacks the body's disease-fighting immune system. Developed by the Burroughs Wellcome Company of Triangle Research Park, N.J., and given the trade name Retrovir, AZT moved from laboratory tests on human beings to FDA approval in only two years. Normally, it takes new drugs almost six years to be reviewed and approved by the FDA.

The rapidly growing number of AIDS cases in the United States has tened the FDA's review of this drug. AZT appears to stop the reproduction of the *human immunodeficiency virus* (HIV), which causes AIDS. While AZT does not cure AIDS, it can prolong the lives of some AIDS patients.

How AZT works. HIV belongs to a family of viruses called *retroviruses*. Retroviruses differ from other viruses because they do not contain *deoxyribonucleic acid* (DNA), the carrier of the genetic code for all animals and bacteria and most viruses. Instead, HIV and other retroviruses contain another substance called *ribonucleic acid* (RNA). These viruses reproduce by making DNA from RNA, using an enzyme called *reverse transcriptase*.

AZT appears to stop HIV from reproducing by fooling the virus's reverse transcriptase enzyme. AZT's chemical structure is similar to that of *thymine*, a biochemical that is an essential part of DNA. When an HIV virus tries to make DNA, its reverse transcriptase uses AZT instead of thymine. AZT does not fit into the DNA molecule, however. It therefore disrupts the production of DNA, and prevents the virus from reproducing.

The quick approval of AZT was also due to the encouraging results of a study reported in September 1986 by scientists at the National Cancer Institute in Bethesda, Md., and the FDA. The study involved 282 AIDS patients.

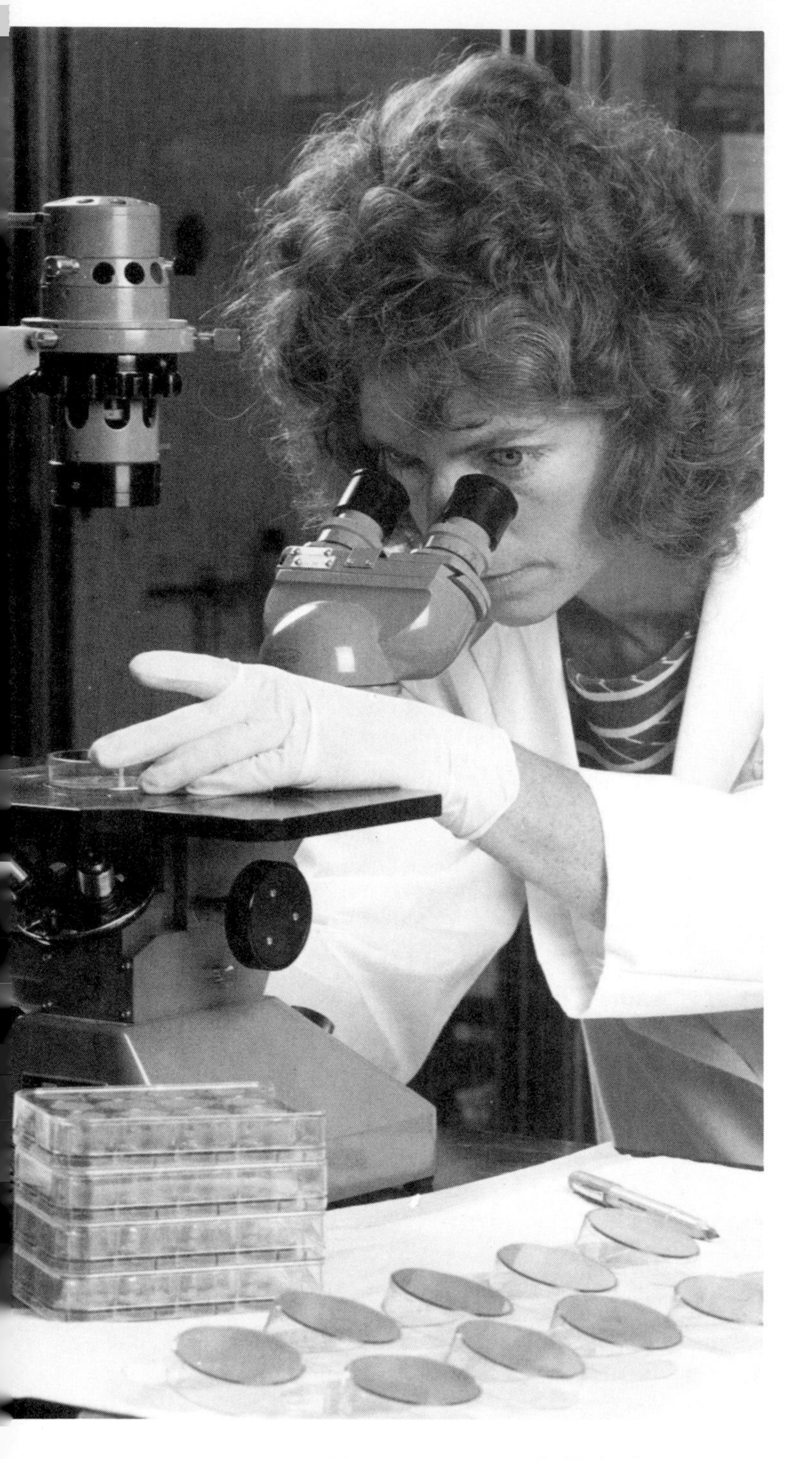

A researcher examines AIDS virus cultures treated with the drug azidothymidine (AZT). The drug, the first one proved to prolong the lives of AIDS patients—though it is not a cure and it has some potentially severe side effects—in March 1987 was approved for use by the U.S. Food and Drug Administration. It is manufactured by the Burroughs Wellcome Company.

AZT was given to 145 patients, while the rest were given a *placebo* (an inactive compound). During the study, only 1 patient who had received AZT died, compared with 16 deaths among those who had received the placebo.

But even with these positive results and FDA approval, AZT treatment is still considered dangerous. It can damage the bone marrow, slowing down the production of red blood cells and causing anemia and other complications. Also, the long-term consequences of AZT treatment are unknown. Another drawback is the high cost of AZT, estimated at $8,000 to $10,000 a year per patient.

AIDS and Peptide-T. Other scientists were looking for ways in 1986 and 1987 to prevent the AIDS virus from infecting cells in the body that are particularly vulnerable to the virus. These are white blood cells called *T-lymphocytes* as well as certain brain cells. The virus sticks or binds to places called *receptors* on the surface of these cells. The virus uses a particular receptor, the *T-4 receptor*, as a door through which it enters the cell. This T-4 receptor was first identified in the brain by neuropharmacologist Candace Pert and her colleagues at the National Institute of Mental Health (NIMH) in Rockville, Md., in 1986.

In early 1987, researchers at the Caroline Institute in Stockholm, Sweden, reported in *Lancet*, a British medical journal, that they had treated four AIDS patients with a substance called Peptide-T. Peptide-T seems to alter the ability of HIV to bind to the T-4 receptor on brain cells and lymphocytes. As a result, the virus is prevented from entering the cell.

Although the Swedish researchers were hopeful about the potential of Peptide-T as an AIDS treatment, they cautioned that their results were preliminary. The researchers planned to conduct more-complete studies of this drug later in 1987.

New antibiotics. A new antibiotic called *norfloxacin* became available in the United States in January 1987. This drug—the first of a new group of antibiotics called *quinolones*—may dramatically change the treatment of infections caused by bacteria called *pseudomonas*. Norfloxacin is marketed by

Drugs
Continued

Merck Sharp and Dohme of West Point, Pa.

Pseudomonas infections can affect the urinary tract, lungs, bones, and other areas of the body. The antibiotics used to treat these infections in the past were given to patients *intravenously* (into a vein) and were potentially poisonous. Often, patients were required to stay in the hospital for long periods of time.

Norfloxacin appears to be relatively safe and can be taken in pill form. It also may allow some patients with pseudomonas infections to be treated at home.

Another new antibiotic—aztreonam—became available in the United States in January 1987. This drug, marketed by E. R. Squibb & Sons, Incorporated, of Princeton, N.J., under the trade name Azactam, is used to treat a group of infections called *gram-negative infections*, which can affect the lungs, bowels, and other areas of the body.

Treating anxiety. In October 1986, the FDA approved the use of a tranquilizer called buspirone for anxiety. Unlike Valium, Librium, and other tranquilizers, buspirone relieves anxiety without causing drowsiness. This drug makes it possible for patients to safely perform many daily tasks, such as driving a car or working. Buspirone is marketed by the Bristol-Myers Company of New York City under the trade name BuSpar.

Drug to stop vomiting. In the summer of 1986, Roxane Laboratories of Columbus, Ohio, began marketing tetrahydrocannabinol (THC), the active ingredient in marijuana, as a treatment for vomiting caused by chemotherapy. The drug is sold under the brand name Marinol. This marked the first legal use of THC in the United States.

Scientists have been investigating THC since the 1970's, when physicians first observed that some marijuana smokers who were receiving chemotherapy seemed to vomit less than other patients. [B. Robert Meyer]

In WORLD BOOK, see AIDS; ANTIBIOTIC; DRUG.

Ecology

Four major studies in 1986 and 1987 addressed ecological concerns about the number of species on Earth and how rapidly those species are being exterminated by human activities.

Total species. Earth may contain about 30 million plant and animal species—10 times as many as previously estimated—according to an October 1986 announcement by entomologist Terry L. Erwin of the Smithsonian Institution in Washington, D.C. Erwin based his estimate on studies of the number of insect species found in the tropics.

Taxonomists (experts who classify species of plants and animals) have classified only about 1 million species. This is partly because most of Earth's species are insects, which few taxonomists study, and most of those insects live in the tropics, where few taxonomists live.

Erwin studied insects where their numbers were greatest—in the tropical rain forest *canopy*, the leafy treetops about 60 meters (200 feet) aboveground. He carried out his studies in Central and South America. Few other taxonomists had studied canopy insect species because of the obvious difficulties in collecting insects there. Researchers simply could not run around on slippery, vine-tangled branches high above the ground, trying to capture bugs with butterfly nets. Erwin solved this problem by using a special gun to shoot a rope over a high branch in the canopy. He then hoisted a bomb of insecticide spray. After setting off the bomb with a mechanism operated from the ground, Erwin collected the dead insects that dropped to the ground.

On the basis of his insect counts, Erwin calculated his rough estimate of 30 million total Earth species. The true number could be twice as large or only half as large. But his work indicates that the world is teeming with undescribed species and that most of them may be living in the tropical rain forest canopy.

Ecosystem size and extinction. Ecological changes created when human beings restrict the size of a wilderness

How Extinctions Are Related to Ecosystem Size
Patches of trees left standing in a harvested area in Brazil's rain forest, *below,* are part of an experiment to help ecologists learn how large an ecosystem must be to retain its species. All but the largest North American national parks, *bottom,* have lost some mammal species since they were established.

Loss of Mammal Species in North American Parks

Park	Size		Date established	Number of mammal species lost since park was established
	Acres	Hectares		
Bryce Canyon	35,835	14,502	1928	4
Lassen Volcanic	106,372	43,047	1916	6
Mount Rainier	235,404	95,265	1899	7
Rocky Mountain	265,200	107,323	1915	2
Yosemite	761,170	308,035	1890	4
Grand Teton-Yellowstone	2,530,306	1,023,978	1872	1
Banff-Jasper	4,328,960	1,751,873	1885	0

area cause the disappearance of many plant and animal species. This finding was announced in October 1986 by conservation biologist Thomas E. Lovejoy III of the World Wildlife Fund in the United States, a leading conservation association.

Forests throughout the world are being chopped down, and what happens to species in the remaining patches of forest is being examined by one of the most ambitious ongoing biological experiments in the world—the Minimum Critical Size of Ecosystems Project (MCS). The World Wildlife Fund is conducting this study with Brazil's National Institute for Amazon Research.

The project started because much of the Amazon rain forest, the most species-rich habitat on Earth, is being cut down to make room for development. Near the city of Manaus, Brazilian officials and local ranchers agreed to leave untouched areas of the forest ranging in size from 1 to 1,000 hectares (2½ to 2,500 acres). Each forest patch became as ecologically isolated as an island. In each area, MCS scientists cataloged species of birds, mammals, trees, butterflies, and other plants and animals before and after the surrounding rain forest was cut down.

The scientists found that the destruction not only eliminated the cleared areas as a habitat for forest species but also changed the ecology of the areas left untouched. Some forest bird and mammal species disappeared from the forest patches. Rare animal species were at high risk because an area might contain only a single female or male member of a rare species. If that animal died, the entire species would be unable to reproduce and would be doomed to die out unless members of the species came to the forest area from elsewhere. The MCS scientists also found that some species disappeared from the experimental areas because other species on which they depended disappeared.

The MCS study experimentally showed that it is not sufficient just to set aside *some* habitat. The habitat must be big enough to allow its species to survive.

Extinctions in national parks. All but one of the national parks in western

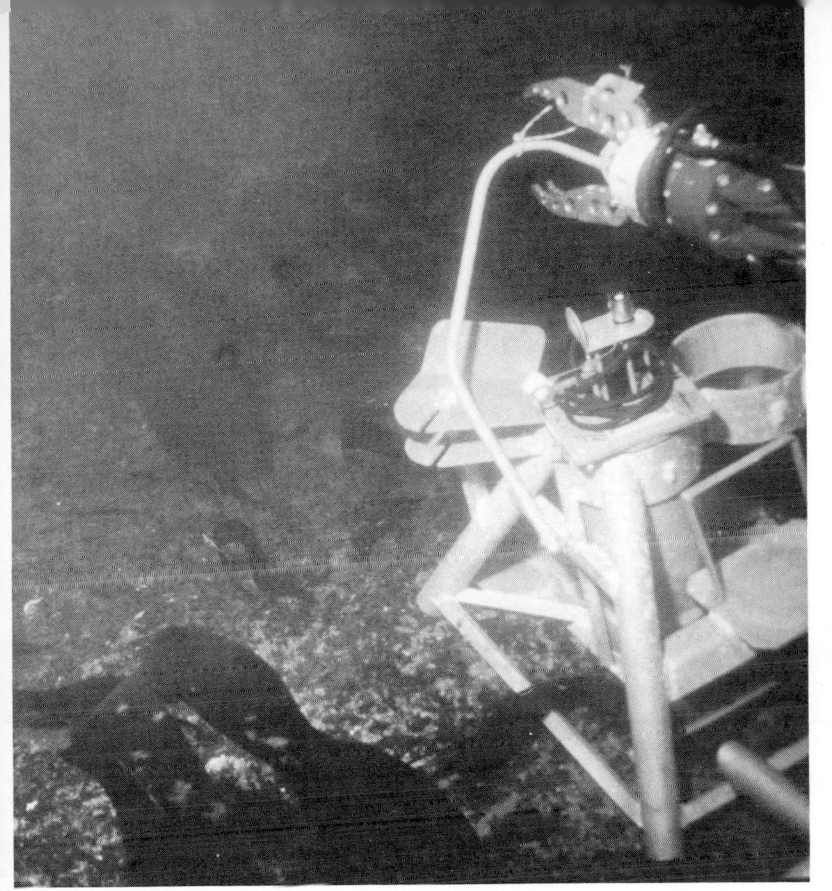

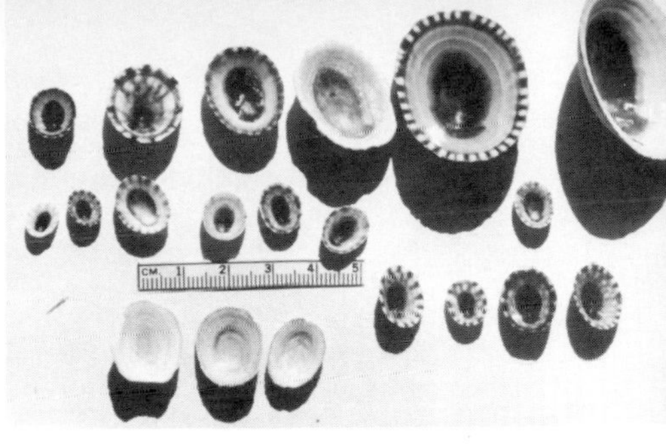

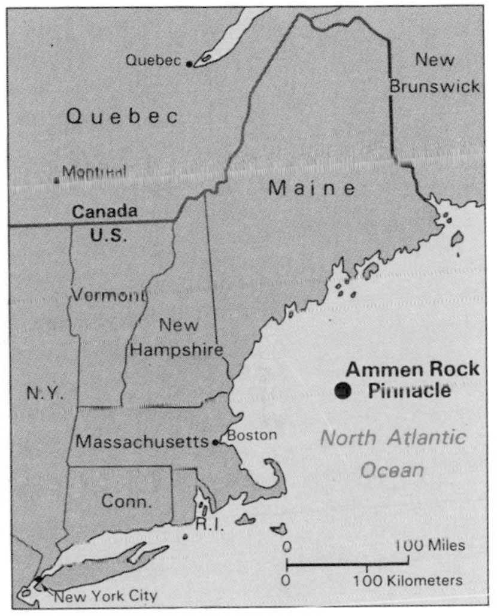

Ecology

Continued

A research submarine prepares to examine a unique type of kelp, *above,* at Ammen Rock Pinnacle, a 150-meter (500-foot) peak that lies 30 meters (100 feet) below the ocean surface off the coast of Maine, *above right.* The pinnacle ecosystem, first explored in 1986, is so isolated that many of its species—forced to inbreed—may have evolved into forms seen nowhere else. In addition to the kelp, scientists have found apparently unique types of limpets, *top right.*

North America are too small to support the number of species that originally lived in the area, reported biogeographer William D. Newmark of the University of Michigan in Ann Arbor in January 1987.

National parks in western North America range from giants such as Banff and Jasper in Canada, with a combined area of about 1,750,000 hectares (4,330,000 acres), to midgets such as Utah's Bryce Canyon, with about 14,500 hectares (35,800 acres). To determine how well the parks have fulfilled their conservation function, Newmark analyzed old and recent records of the numbers of large mammals at several parks. He concentrated his investigation on the disappearance of various species of rabbits, *carnivores* (meat-eaters), and hoofed animals, which, unlike birds and insects, cannot simply fly to new regions if their present habitat becomes inhospitable. In addition, the presence of these animals in a park is easy to detect—and so is their absence.

In all the parks Newmark studied—except the largest, Banff-Jasper—some mammal species that had been recorded as living in the area before it became a national park were now absent. The number of extinctions was highest in the smallest parks. Newmark found that while Banff-Jasper lost no species, the second largest park area, Yellowstone and Grand Teton national parks with a combined area of about 1,024,000 hectares (2,530,000 acres) lost only one—the wolf. But at the opposite extreme, Bryce Canyon lost four species, representing 36 per cent of its total mammal species. Lassen Volcanic National Park, occupying about 43,000 hectares (106,000 acres) in California, lost six mammal species, or 43 per cent of its original total.

Newmark pointed out that more mammal species, especially rare species, are likely to disappear from North American national parks in the future because the parks are too small. His study confirms the conclusions of the MCS study in Brazil.

Bioenergy overconsumption. Human beings use or waste 25 per cent of the biological energy Earth annually produces. That statistic comes from a February 1987 report by ecologists Peter M. Vitousek, Paul R. Ehrlich, and Anne H. Ehrlich of Stanford University in California and ecologist Pamela A. Matson of the National Aeronautics and Space Administration's Ames Research Center, located at Moffett Field, California.

Vitousek and his colleagues made their calculation by estimating the amount of biological energy produced each year by all species on Earth—that is, how much physical energy (mainly sunlight) is converted into biological energy, such as edible plants. The ecologists then estimated how much of this total human beings use directly, either as food for ourselves and our domestic animals, or as lumber, paper, and firewood. Direct consumption by human beings turned out to be 3 per cent of Earth's total biological energy—much more than our proportional share, but still a small fraction of the total.

Vitousek also estimated the percentage of energy wasted as a result of human activity. For example, more than half of a tree chopped down for lumber is wasted. He calculated that the total amount of energy wasted by human beings equals 16 per cent of Earth's total.

Finally, the ecologists calculated the amount of energy human beings irreversibly destroy. For example, human activity has turned some of Earth's productive forests and grasslands to deserts, which produce little energy, or to roads paved with asphalt, which produce no energy. About 9 per cent of Earth's energy has been destroyed in this way. Altogether, the scientists estimated that human beings use, waste, or destroy 25 per cent of Earth's total biological energy.

Vitousek's study suggests that since Earth's human population is increasing at a rate that will double the world population by about 2028, Earth will soon reach the biological limits of human population. [Jared M. Diamond]

In WORLD BOOK, see CLASSIFICATION, SCIENTIFIC; ECOLOGY.

Electronics

Manufacturers of semiconductor chips increased the capacity of their most powerful memory chips in 1987. (A semiconductor chip is a piece of material—usually silicon—into which electronic circuits are built. A typical chip is about the size of a fingernail.) Chips capable of storing 1 million *bits*, or 1 *megabit*, of memory had become available in 1985, and International Business Machines Corporation (IBM) began installing them in some of its computers in April 1986.

In February 1987, IBM and Nippon Telephone and Telegraph Limited (NTT) of Japan introduced experimental chips holding 4 megabits and 16 megabits, respectively. (Bits are the 0's and 1's that make up digital computer language.)

For high-capacity memory chips such as these to function in a computer, however, a way must be found to remove heat generated by the electric currents that pulse rapidly through the chips. Computer manufacturers usually connect tiny metal plates to chips to absorb heat and radiate it into the air. Some machines have fans built in to blow heat away from the chips.

Plates and fans take up valuable space inside a computer, so IBM and NTT used a more radical approach in the 4-megabit and 16-megabit chips. They lowered the voltage of the chip's electric power supply from the conventional 5 volts to 3.3 volts. Lowering the voltage not only cuts the generation of heat significantly but also minimizes stray currents that sometimes flow in—and cause malfunction of—such high-capacity chips.

Some computers, however, might need 3.3-volt memory chips and 5-volt chips. These computers would need two levels of power supply, which would be more complex to build and thus more expensive than single-level voltage supplies. In the Special Reports section, see THE MICROCHIP—A MINIATURE MARVEL.

Compact disk (CD) players "hit the road" in 1986 and 1987, with more than a dozen manufacturers offering players for automobiles. These players

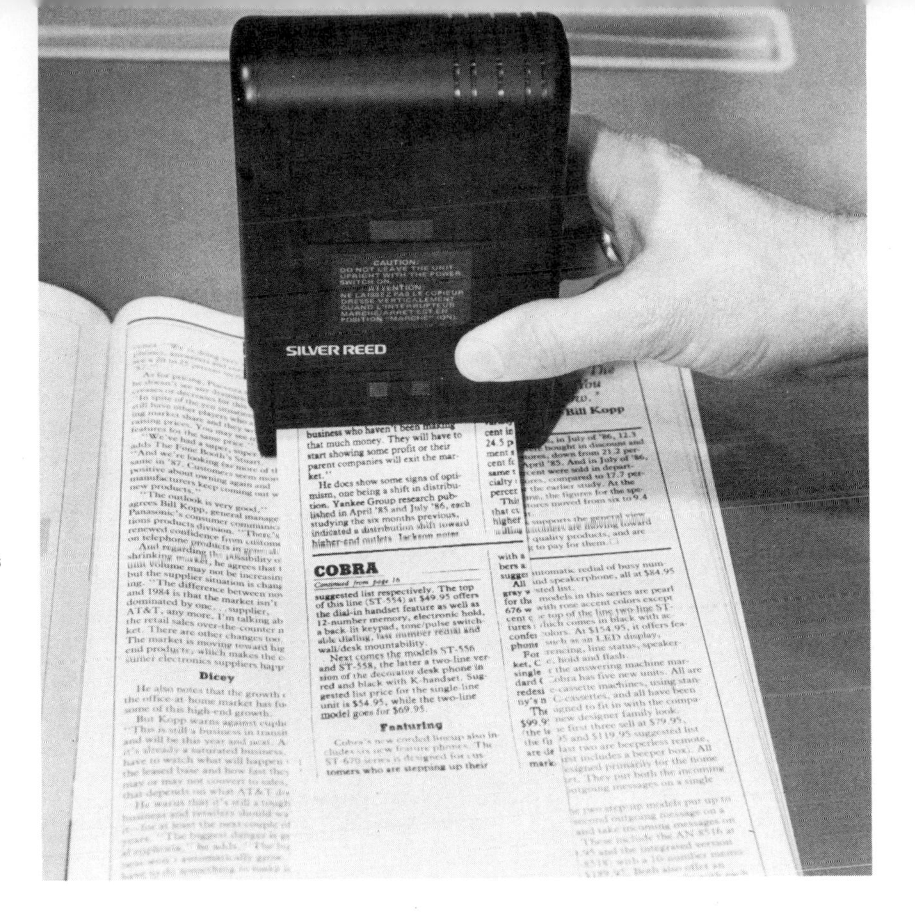

Porta Copy, a handheld copying machine, weighs only 1.7 kilograms (3 4/5 pounds) and produces copies 8.3 centimeters (3¼ inches) wide. Silver Reed of Torrance, Calif., introduced this battery-operated device in January 1987 at a price of $349.

are about the same size as a cassette deck but cost at least $500 and often more. They can cope with such road hazards as bouncing over potholes and turning on sharp curves and can withstand temperatures ranging from below freezing to sweltering tropical heat.

Auto CD players, like their counterparts in the home, use a laser beam to "read" a digitally encoded spiral track of microscopic pits etched in the bottom surface of a plastic disk. A typical auto CD player, however, has an extra track on each side of the main track to help keep the laser light focused when the car hits a bump. Normally, each of these three tracks is read by its own laser beam. When a bump shakes the car so that all three beams start to wander off track, the two lasers on either side send electric signals to a tiny motor that adjusts the main laser.

Digital audiotape (DAT) decks were expected to hit the U.S. market in the summer of 1987. Industry observers predicted that play-only decks would sell for about $600, while play-and-record models would cost more than $1,000. DAT offers sound quality equivalent to CD's and can play or record for two hours. By contrast, a CD plays for about one hour and cannot record.

Several Japanese and European firms were ready to ship DAT decks into the United States. But how soon the products would become available, and how successful they would be, hinged on a battle over copyright issues. The International Federation of Phonogram and Videogram Producers, a recording industry association, demanded that a device be installed on all DAT equipment to prevent DAT users from copying CD's, ordinary phonograph records, cassette tapes, and other DAT tapes.

At a December 1986 meeting with the federation, however, the Electronic Industries Association of Japan refused to comply with the music industry's plea for copyright protection. Failure to settle this impasse threatened to result in a boycott by producers of major record labels.

Credit Cards Are Getting Smarter

The traditional credit card may be destined for obsolescence. It soon may be replaced by the *smart card*, a piece of plastic no larger than a regular credit card, which can function as a miniature computer. Embedded in each smart card are *electronic chips*—tiny pieces of silicon or other material containing built-in electronic circuits. These chips can handle many of the mathematical functions of a large computer as well as store information.

Smart cards offer many advantages to credit-card companies, retailers, and the spending public. Thieves cannot use lost or stolen smart cards to make fraudulent purchases because the new cards, like the automatic teller cards issued by banks, are useless without the owner's personal identification number. This is a four- to six-digit number known only to the cardholder that must be punched into a keyboard before a transaction.

The smart card also keeps a running account of purchases and rejects any that exceed the cardholder's credit limit. Such limits help curb overspending and reduce the debt-collection problems of credit-card companies, who were burdened with more than $3.5 billion in uncollectable charge-card debts in 1986.

Stores and restaurants must install special card-reading devices called *terminals* to handle the new cards. Despite the expense of new terminals, retailers stand to gain from the new service because the smart card can authenticate transactions on the spot. There is no need for a clerk to make a long-distance telephone call to check a customer's credit rating.

An experimental "smart" credit card has a keyboard, two-line display, and built-in electronic chips to process and store data.

Other nations, especially France and Japan, are ahead of the United States in smart-card usage. The government of France has ordered all French retailers to install smart-card readers by 1990, and about 1 million smart credit cards are already in circulation in France. In Japan, many major banks, retailers, and government agencies are conducting field tests of smart cards.

MasterCard International Incorporated and Visa International Corporation are currently testing the smart card in the United States. Both the MasterCard and the Visa version contain an electronic chip called a *microprocessor* similar to those found in desk-top computers, together with a memory chip that can store data for about 200 purchases. These smart cards also contain circuitry to protect the chips against the static electricity generated by sliding the card from a wallet or pocket, which could wipe out the card's memory.

The smart card under development for MasterCard does not require a battery or other power source to keep data stored in its memory. The card communicates with outside terminals through eight metal contacts on its surface. When the card is inserted into a card-reader terminal, the terminal supplies electric power to the card through those contacts. Visa's smart card uses a tiny battery capable of powering the card for two years.

While experiments continue on smart cards, Visa is testing an even smarter version called UltiCard (*ulti*mate-transaction *card*). The same size as a credit card, the UltiCard houses a small battery; a liquid crystal display like the one on an electronic watch, which shows two lines of 24 characters each; microprocessor and memory chips; and a 10-key keyboard that enables the cardholder to enter information in the card's memory. The tiny plastic package not only serves as a credit card, it also stores information about at least two charge accounts and a checking account, addresses, telephone numbers, and health information such as allergies and blood type.

Skeptics, who doubt that the smart card will catch on, believe consumers will have trouble remembering their personal identification numbers and retailers will object to installing new card-reading terminals. But smart-card advocates believe retailers and consumers will adapt to the smart card once they experience its speed and convenience. Soon, they won't leave home without it. [Howard Bierman]

Chip wars. The marketing of Japanese-made memory chips caused a trade dispute between Japan and the United States in 1986 and 1987. American semiconductor firms accused Japanese companies of *dumping* chips (selling them below their cost of production) in world markets.

In mid-1986, U.S. and Japanese trade officials signed an agreement intended to prevent further dumping and force Japanese producers of business and consumer products to increase their purchases of chips from non-Japanese sources. But in March 1987, the United States concluded that Japan had failed to live up to the agreement, so it imposed tariffs on certain Japanese electronic imports, including television sets, personal computers, computer disk drives, and laser printers. In early June, the United States lifted about 17 per cent of the tariffs.

Aircraft anticollision system. In March 1987, during a flight from Greensboro, N.C., to Washington, D.C., Piedmont Airlines became the first commercial airline to use an electronic anticollision system in regular passenger service. The system warns pilots of a possible collision and flashes a recommended escape maneuver on the instrument panel. The U.S. Federal Aviation Administration (FAA) promised that, by the fall of 1987, it would propose a rule to make such systems mandatory on all commercial jet aircraft.

Earlier systems were hampered by the need for both planes in danger of colliding to be equipped with identical systems. Most owners of small planes cannot afford anticollision equipment.

Only one of the planes needs to have a system like that in the Piedmont craft, however, provided the other plane carries an *air-traffic transponder*, a relatively inexpensive device that receives radar signals from the anticollision device and immediately retransmits them. The device analyzes the retransmitted signals with a built-in computer and flashes the necessary alert. [Howard Bierman]

In WORLD BOOK, see ELECTRONICS.

A telephone with a built-in camera lens and a viewing screen sends and receives black-and-white still pictures over conventional telephone lines. The picture phone could be used for showing reports, electronic parts, or other objects to the person at the other end of the line. The device sells for $1,500, about 75 per cent less than other picture phones, according to the manufacturer, Luma Telecom of Santa Clara, Calif.

Energy

Electric power may become much less expensive due to a major breakthrough in energy research, physicist Paul C. W. Chu of the University of Houston said in February 1987. Chu and his colleagues at the University of Houston along with other physicists at the University of Alabama in Huntsville discovered that a compound made of the elements yttrium, barium, and oxygen could be used to conduct electricity more efficiently than any other known material.

The new conducting material is one of several *superconductors*. Electric currents pass through these substances without resistance, producing very little heat and thus wasting almost no energy. To be this efficient, however, each superconductor must be kept at its *critical transition temperature*, a temperature just low enough to cause the material's resistance to electricity to diminish. The most commonly used superconductor, the metal alloy niobium titanium, has a critical transition temperature of −264°C (−444°F.). To keep niobium titanium this cold, liquid helium, whose temperature never rises above −269°C (−453°F.), must be circulated around it. Because helium is a rare element, it is very expensive. It is so costly that superconductors are used only in exotic equipment, such as powerful *particle accelerators* (atom smashers).

Chu's discovery was an important breakthrough because the new material becomes superconducting at −180°C (−292°F.). Although still very low, this temperature can be maintained by using liquid nitrogen at a temperature of −196°C (−321°F.). Unlike helium, nitrogen is very abundant, making up 80 per cent of Earth's atmosphere. Compared with liquid helium, liquid nitrogen is about 20 times more efficient as a coolant and much less expensive.

The discovery of new superconductors means many new uses for electricity will be economically possible. If the new materials can be made into strong, pliable wire, they could be used to send electricity over high-voltage transmission cables—with no power loss—saving millions of dollars. High-speed trains held above the ground by powerful magnets would also be possible. See also PHYSICS, FLUIDS AND SOLIDS. In the Special Reports section, see EXPLORING THE STATES OF MATTER.

Fusion research milestone. An important development in nuclear fusion research occurred in July 1986 when scientists at the Princeton University Plasma Physics Laboratory in New Jersey obtained a temperature of 200,000,000°C (360,000,000°F.) in the Tokamak Fusion Test Reactor (TFTR). The temperature, the highest ever reached in a reactor, is 10 times higher than that of the sun's core.

Scientists obtained this record temperature while studying nuclear *fusion*, a way of producing energy by fusing together the *nuclei* (centers) of atoms. One way to cause this reaction is to heat the atoms to extremely high temperatures. For their fusion experiment, the Princeton scientists used an electric current to heat deuterium, a form of hydrogen, to approximately 10,000,000°C (18,000,000°F.). The deuterium became a gaslike plasma so hot that it had to be kept away from the walls of the reactor that contained it, or the reactor would melt. The TFTR scientists prevented the contact by generating a magnetic field that formed a barrier between the hot plasma and the reactor walls—creating a sort of "magnetic bottle." The plasma temperature was raised to the record high when the researchers beamed more high-energy deuterium atoms into the reactor.

Aluminum-air batteries. The usefulness of a new type of battery developed by Alcan International Limited of Montreal, Canada, with support from the Canadian government was demonstrated on Sept. 4, 1986, in Kingston, Canada. The new device, called an aluminum-air battery, powered an electric golf cart for eight hours, producing about 1,680 watts for each hour of the demonstration.

The battery generates electricity as aluminum slowly dissolves in a *saline* (saltwater) solution, producing aluminum hydroxide and *free electrons*, which are subatomic charged particles that have broken away from their normal orbit around an atomic nucleus. The free electrons move as an electric current through the saline solution, producing power that can operate ma-

Energy

Continued

Wind Star, the world's longest wind-powered vessel, made its maiden voyage in December 1986. The ship, 134 meters (440 feet) long, has masts that tower 62 meters (204 feet) above deck. Much of the work aboard ship is automated and the sails are computer-controlled, so only a few crew members are needed to sail the huge craft.

chines such as the experimental cart.

The battery is more efficient than the combustion of gasoline if fuel volume is the basis for comparison. An aluminum-air battery releases four times as much energy as the same volume of gasoline in an internal-combustion engine. The battery may prove useful for powering electric cars or providing power sources in remote wilderness areas.

Chernobyl update. Human error was responsible for the April 1986 nuclear power plant explosion in Chernobyl, near Kiev in the Soviet Union, according to a Soviet report. The report, made to the International Atomic Energy Agency in Vienna, Austria, in August 1986, stated that operators at the plant caused the accident while conducting an unauthorized test of the reactor's emergency systems.

Nuclear reactors generate heat by a controlled use of *nuclear fission* (the splitting of uranium or plutonium atoms). This process creates so much heat that for safety it must be carefully controlled by cooling systems that circulate water to remove the heat and control rods that slow the fission to a manageable level.

The Chernobyl operators shut down two of the reactor's control systems in order to make their test. First, they shut off the plant's emergency cooling system, which was intended to prevent the reactor from dangerously overheating if the main cooling system stopped working. They also turned off an automatic shut-down system, which would in effect turn off the reactor if it detected a too-rapid fission rate. When the reactor's power dropped very low—to 1 per cent of its capacity—the operators removed the control rods and decreased the flow of coolant water.

With the automatic shut-off system not operating, there was nothing to prevent a sudden power surge. Within three seconds, the reactor had heated up to 50 per cent of its heat-generating capacity. The operators had no time to reinsert control rods, and the reactor became so hot that the coolant water inside flashed to steam, blowing

Energy

Continued

off the reactor's 900-metric-ton (1,000-short-ton) "lid."

Energy experts believe the same type of accident would be unlikely in the United States. In most U.S. reactors, water is used as both coolant and *moderator* (a material that must be present for fission to occur). These reactors are safer because if too little water reaches the reactor as a coolant, this will cause fission to stop for lack of sufficient moderator. By contrast, the Chernobyl reactor used water as the coolant and graphite as the moderator.

Nuclear accident prediction. Experiments designed to show what might happen inside nuclear power plants during certain types of severe accidents were begun in fall 1986 under the direction of the U.S. Department of Energy's Sandia National Laboratories in Albuquerque, N.M. Scientists at the Surtsey test facility in Albuquerque conducted their experiments on a model one-tenth the size of an actual reactor. The researchers applied heat and pressure to the model reactor's steel shell, or pressure vessel. These stresses simulated accidents in which up to 81 kilograms (180 pounds) of molten fuel and other debris pooled in the bottom of the vessel. The scientists then created simulated explosions in which the molten fuel, gases, and steam burst through the bottom of a reactor's pressure vessel and poured into a concrete and steel containment building.

The experiments were designed to tell scientists whether such ejected material—whose temperature would be 1900° to 2200°C (3500° to 4000°F.)—would damage a containment building. If a containment building were damaged in a real accident, harmful radiation would enter the environment, as happened at Chernobyl.

Results of the Surtsey experiments, expected to be completed in 1988, will also help scientists predict the sequence of events that could occur during severe nuclear power plant accidents. [Marian Visich, Jr.]

In WORLD BOOK, see BATTERY; ELECTRIC POWER; ENERGY; NUCLEAR ENERGY.

Environment

A major environmental disaster occurred on Nov. 1, 1986, when about 27 metric tons (30 short tons) of hazardous chemicals spilled into the Rhine River near Basel, Switzerland. The chemicals killed almost all plant and animal life in a 125-kilometer (78-mile) stretch of the river between Basel and Karlsruhe, West Germany.

The spill occurred after a fire in a warehouse of Sandoz AG, a chemical company. More than 900 metric tons (1,000 short tons) of herbicides, pesticides, dyes, and fungicides were stored in the warehouse. Some of the water pumped into the burning structure by fire fighters mixed with the chemicals and washed into the nearby Rhine. Six days later, a temporary containment system in the warehouse burst, causing a second spill that sent the highly poisonous element mercury and other chemicals into the river.

Officials in West Germany and the Netherlands closed drinking-water plants along the Rhine and suspended fishing operations. To minimize the damage, divers used hoses to suction up some of the heavy chemical residue that had settled to the bottom of the river. Nonetheless, about 125 kilometers of the river's length was so severely polluted that environmentalists feared no life could survive.

Immediately after the accident, environmentalists predicted that the river's ecology would not recover until at least 1997. In January 1987, however, Swiss technical experts discovered that some bacteria and other microorganisms had survived the spill. The scientists predicted that these life forms may begin a new food chain in the river and become the basis for a more rapid reestablishment of the Rhine's higher plant and animal life.

Chernobyl aftermath. The environmental effects of an April 1986 explosion at the Chernobyl nuclear power station near Kiev in the Soviet Union were more serious than had been expected, according to a Soviet inquiry commission. In August 1986, Russia's State Committee for the Peaceful Uses of Atomic Energy reported that topsoil from a 2,600-square-kilometer (1,000-

Disaster on the Rhine

One of Western Europe's worst environmental accidents occurred in November 1986 when about 27 metric tons (30 short tons) of poisonous chemicals accidentally spilled into the Rhine River at Basel, Switzerland.

Fire at a Basel chemical warehouse, *above,* starts the tragedy on November 1. Fire fighters flood the building with water, washing tons of chemicals into the nearby Rhine. Six days later, a temporary containment wall gives way, spilling mercury into the river.

The chemical spill has a devastating effect on river life, with the most severe damage being done from Basel to Karlsruhe, West Germany, *above.*

The chemicals kill hundreds of thousands of eels, *left,* fish, and other forms of life in the Rhine. To remove contaminated sediment from the site of the spill, technicians, *above,* suction mud from the river bottom. It may take up to 10 years for the Rhine to recover from the disaster.

square-mile) area around the Chernobyl reactor was so contaminated by radiation that it had to be dug up and stored in a nuclear waste dump. The Soviets also predicted that radiation produced by the accident would cause 6,500 to 40,000 Soviet citizens to die of cancer by 2016.

Polluted U.S. wildlife refuge. Millions of chub fish in Nevada's Stillwater National Wildlife Management Area were killed by pollution, according to a report by U.S. government scientists in spring 1987. Tests showed that the fish died because their habitat in the area's marshy Carson Sink had become contaminated with high levels of salt due to improper irrigation practices by farmers in the region. Thousands of birds of many species were also found dead at the refuge. Preliminary reports suggested that the birds died of avian cholera, a disease that can occur in birds weakened by exposure to poisons. Autopsies showed that the birds had eaten or absorbed the poisonous chemicals boron, arsenic, and selenium, indicating that these pollutants contaminated parts of the reserve.

Acid rain may be the cause of a 65 per cent decrease in the population of the North American black duck, according to a February 1987 report by the Izaak Walton League. The league, an environmental organization based in Arlington, Va., kept track of the black duck population from 1955 to 1987. Acid rain can cause a decrease in wildlife populations by causing a lake's overall acid level to rise. Few aquatic plants can survive in highly acidic water, so animals such as ducks then have less food, and many starve.

Unless acid rain is curtailed within the next 50 years, approximately 300 lakes in the Northeastern United States will become acidic, the U.S. Environmental Protection Agency (EPA) reported in March 1987. The EPA scientists analyzed the acidity of water samples from various lakes. Then they based their prediction on the assumption that there would be a gradual increase in acid rain and that certain types of soil would neutralize acid-water runoff in some areas.

President Ronald Reagan in March 1987 announced plans to ask Congress for $2.5 billion over the next five years to combat acid rain. Reagan Administration officials said the announcement was made to assure Canadians of the President's concern about Canada's acid rain—50 per cent of which is said to result from sulfur dioxide and nitrogen oxides emitted by industries in the United States that burn fossil fuels. Environmental groups in both nations said, however, that $2.5 billion was not enough to solve the U.S.-Canadian acid rain problem.

A promising method for reducing acid rain and smog, reported in early 1987, was developed by chemist Robert A. Perry of the U.S. Department of Energy's Sandia National Laboratories in Livermore, Calif. The new method someday may be used in diesel trucks and coal-burning factories to reduce nitrogen oxide pollution. Nitrogen oxides combine to form ozone, a major component of smog, and nitric acid, a major component of acid rain.

In Perry's method, granular cyanuric acid—an inexpensive and nonpoisonous substance—is added to hot exhaust fumes. The cyanuric acid then turns into a gas. Nitrogen oxides in the exhaust fumes combine with the cyanuric acid gas and form less harmful compounds such as water and carbon dioxide.

Ozone disappearance. A mysterious decrease in Earth's *atmospheric ozone*—a form of oxygen about 24 kilometers (15 miles) above Earth—in the Antarctic region puzzled scientists in 1986 and 1987. The ozone layer is an important part of the atmosphere because it shields Earth from the sun's ultraviolet radiation, which can cause skin cancer and damage plants and animals. See Close-Up.

Radon threat. In 1 of every 70 U.S. houses, indoor radiation levels are equal to or higher than radiation levels in the average uranium mine, according to a November 1986 report by physicist Anthony V. Nero and other researchers in Lawrence Berkeley Laboratory's Indoor Environment Program at Berkeley, Calif. The source of the radiation—which can cause lung cancer—is *radon*, a naturally occurring gas that can seep into houses and other buildings. Nero based his estimate of the number of dangerously contaminated U.S. houses on the re-

ENVIRONMENT: A Science Year Close-Up

The Mysterious Hole in Earth's Ozone Layer

A mysterious hole has appeared in the atmosphere over Antarctica, and a team of scientists traveled there in late 1986 to learn more about it. The hole is in Earth's protective layer of *ozone*, a form of oxygen. The ozone layer, which is found in the atmosphere about 24 kilometers (15 miles) above Earth's surface, shields the planet from much of sunlight's harmful ultraviolet radiation. The hole is approximately as large as North America.

The first clue that there was trouble in the ozone layer came from British researchers who have continuously measured ozone levels in the atmosphere above Halley Bay, Antarctica, since 1958. In 1985, they announced that ozone levels apparently had decreased each year since the mid-1970's, most noticeably during September and October. The discovery caught atmospheric scientists by surprise, but the findings were soon confirmed by United States National Aeronautics and Space Administration (NASA) scientists at Goddard Space Flight Center in Greenbelt, Md. An examination of NASA's satellite observations clearly showed a thinning of the ozone layer over Antarctica.

Within months of the first reporting of the ozone "hole," scientists proposed theories to explain the phenomenon. Some believed that it was a natural occurrence caused by a change in the circulation patterns of the atmosphere. Jerry D. Mahlman of the Geophysics Fluid Dynamics Laboratory in Princeton, N.J., who specializes in the study of movement within the atmosphere, suggested that the hole was caused by changes in weather patterns near the planet's surface.

Mathematician Ka K. Tung of Clarkson University in Potsdam, N.Y., theorized that sunlight heated ozone-poor air near the ground, causing it to rise into the ozone layer, thus decreasing the concentration of atmospheric ozone. According to Tung's theory, recent volcanic eruptions and a general cooling of the *stratosphere* (the region of the atmosphere that contains the ozone layer) might have intensified the circulation of ozone-poor air and resulted in markedly lower atmospheric ozone levels.

Another theory suggested that the hole was caused by a series of chemical reactions. Some scientists, such as research chemist Susan Solomon of the National Oceanic and Atmospheric Administration in Boulder, Colo., and atmospheric physicist Michael B. McElroy of Harvard University in Cambridge, Mass., linked the appearance of the hole to the increased production of a group of chemicals called *chlorofluorocarbons* (CFC's).

Scientists have known since the early 1970's that CFC's could destroy some of the ozone layer. In 1978, the United States banned the use of CFC's in aerosol sprays, but they are still used in many industries. In early 1985—before the hole was reported—approximately 30 nations signed the Vienna Convention for the Protection of the Ozone Layer, and in 1987 governments throughout the world continued to discuss regulation of the amount of CFC's released into the atmosphere.

Unfortunately, due to the complete lack of data on the chemical composition of the Antarctic stratosphere, one theory could not be favored over another. Therefore, in August 1986, a team of scientists—including Solomon—quickly assembled for an expedition to McMurdo Sound, Antarctica. For three months the scientists studied the chemical composition of the Antarctic stratosphere, using equipment on the ground or attached to weather balloons. The scientists found that the chemical composition of the Antarctic stratosphere in the region of the ozone layer was highly unusual compared with the composition of the atmosphere at all other latitudes. They found great amounts of chlorine compounds that are capable of destroying ozone.

This lent strong support to the theory that CFC's caused the hole. The data are not complete enough, however, for scientists to reach a definite conclusion. Several U.S. agencies and the Chemical Manufacturers Association planned to conduct a more comprehensive investigation in late 1987. Data collected for this study by aircraft, satellites, weather balloons, and ground-based instruments should significantly improve our understanding of the Antarctic ozone hole. [Robert Watson]

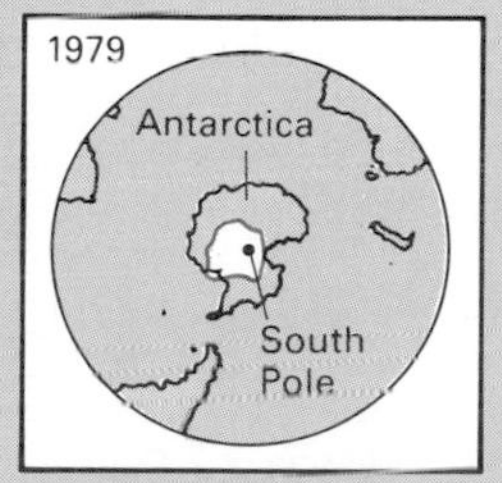

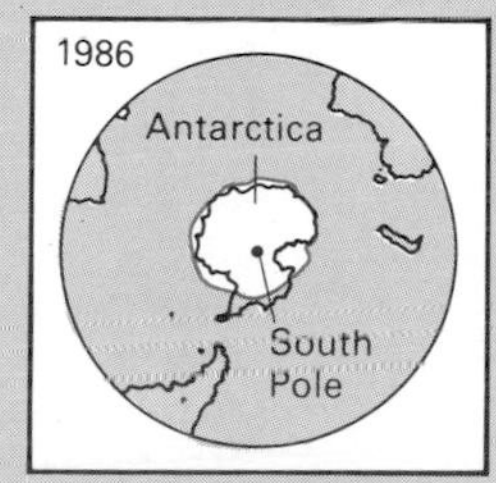

The puzzling hole (white) in Earth's atmospheric ozone layer above Antarctica grew dramatically from 1979 to 1986.

sults of radon surveys in 17 states. In the Special Reports section, see YOUR HOUSE CAN MAKE YOU SICK.

Chlordane. Environmental groups in 1986 and 1987 were unsuccessful in their calls for bans on chlordane, which is widely used to kill termites. The National Coalition Against the Misuse of Pesticides and environmentalist Samuel S. Epstein of the University of Illinois College of Medicine were among those calling for a complete ban of chlordane, citing evidence that the insecticide causes cancer and other serious diseases in humans. In January 1987, the EPA restricted the use of chlordane but did not ban it.

New organism released. The EPA and California officials also allowed the first outdoor testing of a genetically altered bacteria. The organism was designed to prevent frost from forming on fruits and vegetables. The test took place in Brentwood, Calif., in April 1987. See AGRICULTURE.

Some environmentalists, believing that genetically altered organisms could cause untold environmental damage, tried to obtain further court injunctions against the test but were unsuccessful. EPA experts said the bacteria did not threaten public health.

Condor capture. In April 1987, scientists from the Condor Research Center in Ventura, Calif., captured the last wild California condor in order to help save the species from extinction. The bird was caught at the Bitter Creek National Wildlife Refuge near Bakersfield and taken to the San Diego Wild Animal Park. Condors, which belong to the vulture family, were in danger of extinction because the animal carcasses they feed upon in the wild often contain lead bullets, which poison the birds.

All 27 remaining California condors are now held in captivity by the San Diego Wild Animal Park and the Los Angeles Zoo. Scientists hope to increase the condor population while the birds are in captivity and eventually return them to their natural habitat in the wild. [Jinger Griswold]

In WORLD BOOK, see ENVIRONMENTAL POLLUTION.

Genetic Science

The discovery of a gene that causes a severe mental disorder called *manic depression* was reported in February 1987 by researchers from the Massachusetts Institute of Technology (M.I.T.) in Cambridge; Yale University in New Haven, Conn.; and the University of Miami in Coral Gables, Fla. *Genes*, which contain the instructions for all the characteristics of cells, are located in structures in the cell called *chromosomes*. Human beings have 23 pairs of chromosomes, numbered for identification. The researchers traced the location of the gene for manic depression to an area on one particular chromosome.

Manic depression is among the most common of severe mental disorders. About 1 person in every 100 has the disease. Manic-depressive individuals have pronounced mood disturbances. Their moods alternate between periods of extreme optimism and activity and periods of deep depression.

Because manic depression is so common, the geneticists thought it likely that the disease might be associated with any one of a number of genes. In such cases, researchers often study a closely related group of people, including some who have the disorder. Among related individuals, it is likely that the same abnormal gene, passed down through the generations, is the cause of the disease in members of the group.

The researchers studied individuals in a religious community known as the Old Order Amish in Lancaster County, Pennsylvania. A team of psychiatrists headed by Janice A. Egeland of the University of Miami identified a number of large Amish families containing several people suffering from manic depression. Because many of them were related, it seemed likely that those affected would have the same abnormal gene.

After these families were identified, geneticists Daniela S. Gerhard and David E. Housman of M.I.T. and Kenneth K. Kidd and David L. Pauls of Yale began the genetic studies by extracting *deoxyribonucleic acid* (DNA) from blood samples taken from family

A tobacco plant glows in the dark after geneticists at the University of California at San Diego inserted into the plant's genetic material the firefly gene that makes fireflies glow. The glow is a telltale sign that the firefly gene is active—making the protein that allows the plant to glow—and thus allows scientists to monitor gene activity.

members. (DNA is the chemical of which genes are made.) Each patient's DNA was cut into many pieces by chemicals called enzymes. Each individual has a distinctive pattern of different-sized DNA pieces. Some of these fragments can be recognized as inherited from the individual's mother or father.

The patient's DNA pieces, called *restriction fragments*, were then placed on a *gel*—a jellylike slab—and subjected to an electric field. The field caused the DNA fragments to separate from each other, with the largest pieces at one end of the gel and the smallest at the other end. A small piece of DNA called a *probe*, tagged with a radioactive molecule, was added. Each probe attached to a matching piece of DNA in the sample. The radioactivity in the probe darkened a piece of photographic film placed over the gel, allowing scientists to determine the location of the matching DNA fragment in the gel. The scientists reasoned that a fragment that is always seen in a manic-depressive parent and child—but is rarely seen in that parent's unaffected children—probably contains a gene that plays a role in the disease.

The geneticists studied DNA obtained from the cells of 81 Amish family members, both normal and manic-depressive individuals. After examining several probes known to match areas on specific chromosomes, the geneticists concluded that the gene for manic depression in the Amish families is located near or within a segment of DNA on chromosome 11.

Only about 63 per cent of the individuals who inherit the gene for manic depression actually develop the disease. Scientists believe that additional elements—other genes or factors in the environment, such as stress—may determine which individuals with the gene will develop the disease and which ones will not.

Alzheimer's disease gene. The identification of a human gene that may be a key to the cause of Alzheimer's disease was reported in February 1987 by neurochemists at the National Institutes of Health (NIH) in Bethesda, Md., and geneticists from the University of Cologne in West Germany. Alzheimer's disease, also called senile

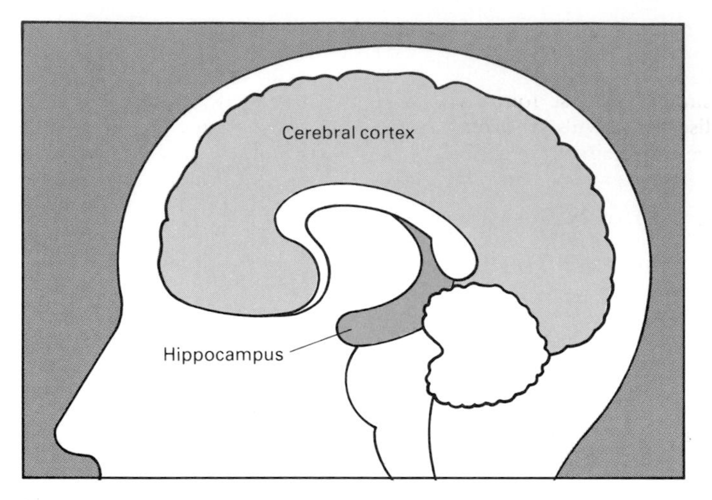

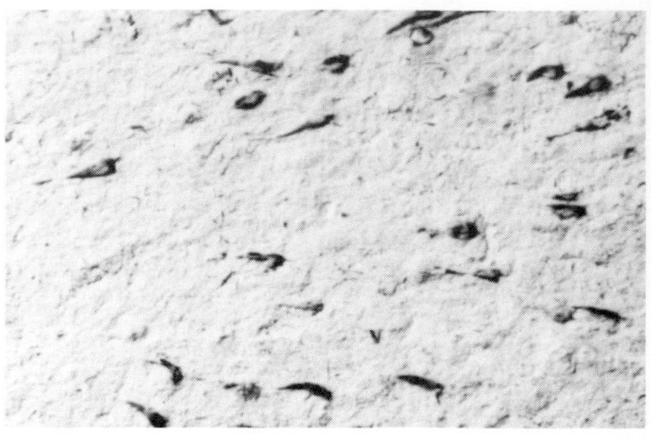

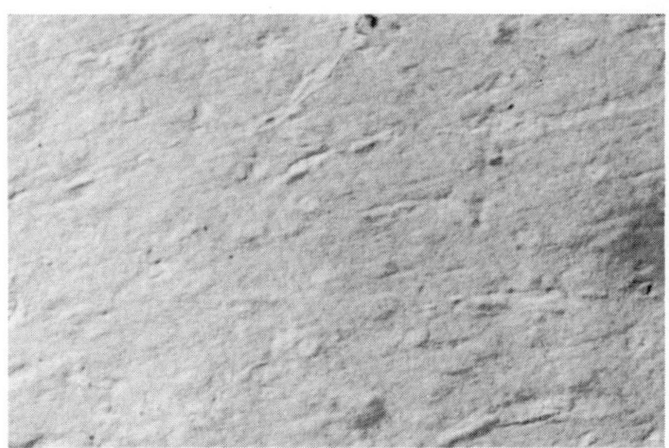

Alzheimer's Clue?
An abnormal protein, called A-68, found in the brain and spinal fluid of victims of Alzheimer's disease, may help to diagnose the illness, scientists reported in November 1986. The disease affects mainly the cerebral cortex and hippocampus of the brain, *above*. Brain tissue from an Alzheimer's patient, treated to detect A-68, reveals the protein (dark spots), *top right*. A sample from a normal elderly patient, *bottom right*, shows no dark spots.

dementia, is a degenerative brain disorder that causes memory loss and profound mental deterioration. It is believed to affect about 1 person in 20 over the age of 65. The brain of an individual suffering from Alzheimer's disease has certain abnormalities, including *plaques*—deposits—of a substance called *amyloid protein*. Amyloid protein is also found in smaller amounts in the brains of all elderly human beings.

Proteins are large, complex molecules made up of smaller units called *amino acids*. Each gene contains instructions for putting together a sequence of amino acids to make a particular protein. Scientists already knew the amino acid sequence of one end of the amyloid protein. So the two research groups chemically synthesized a DNA probe that contained the code for that end.

Each group used the DNA probe to identify fragments of chromosome containing the amyloid protein gene. They then determined the sequence of DNA units, called *bases*, that make up the gene. Knowing this genetic code, the researchers were able to deduce the sequence of amino acids in the entire amyloid protein.

The West German group suggested that the amyloid protein resembles a kind of cell-surface molecule called a receptor that allows the cell to take in various substances. What role the amyloid protein normally plays in the cell, and why it accumulates in Alzheimer's disease patients, is unknown.

The NIH group then determined the exact location of the gene for amyloid protein by using a technique that involves fusing mouse and human cells. This technique is used to isolate a particular chromosome, because when the hybrid mouse-human cells divide, they tend to lose all but 1 or 2 of the 46 human chromosomes usually present in a human cell. The researchers found that only hybrid cells containing human chromosome 21 had the amyloid protein gene.

Scientists were not surprised by the news that a gene associated with Alzheimer's disease had been found on

Genetic Science

Continued

chromosome 21. The brains of adults with a genetic disorder called Down's syndrome often contain amyloid protein plaques similar to those found in Alzheimer's disease. Down's syndrome results when cells contain three copies of chromosome 21 instead of the normal two copies.

Genetic engineering. A new genetically engineered vaccine against the virus that causes hepatitis B, a disease that can result in severe liver damage, was approved for use by the U.S. Food and Drug Administration in July 1986. About 200,000 new cases of hepatitis B are reported annually in the United States.

The new vaccine, called Recombivax HB, is manufactured by Merck, Sharp & Dohme of West Point, Pa. It is produced in genetically altered yeast cells that contain a gene from the hepatitis B virus. The gene codes for a protein, which the altered yeast cells produce. When this protein is injected into the body, it results in the formation of *antibodies*—immune-system molecules that help identify and destroy the virus. This produces immunity to hepatitis B infection.

In late 1986, tests of another product of genetic engineering—a human protein called *epidermal growth factor* (EGF)—showed promise in healing wounds. EGF, which occurs normally in the human body, but in tiny amounts, promotes the growth of certain cells, including skin cells. In 1982, geneticists at Chiron Corporation in Emeryville, Calif., transferred the gene that codes for EGF into yeast, making it possible to produce large quantities of the protein for study.

In the 1986 tests, supported by Johnson & Johnson Products of New Brunswick, N.J., EGF was used in eye drops to treat patients with eye injuries and corneal transplants, and after cataract surgery. Corneal transplants, which normally require several weeks to heal, may heal in only a week with EGF. Studies were also underway to determine EGF's effects in other kinds of injuries and burns. [Daniel L. Hartl]

In World Book, see Cell; Genetic Engineering; Genetics.

Geology

People living near Lake Nios (also spelled Nyos) in the central African country of Cameroon heard bubbling and rumbling sounds coming from the lake on Aug. 21, 1986. Soon afterward, a cloud of toxic gas rose from the lake and drifted over villages nearby. Within a few hours, about 1,700 people had been killed and another 300 had been injured. Scientists investigating the tragedy determined that it had been caused by a very large bubble of gas. See Close-Up.

Deep drill hole. Drilling began on a drill hole 5 kilometers (3 miles) deep in California on Dec. 7, 1986. Geologists plan to use the hole as an observatory to study the San Andreas Fault, which extends through much of California. If successful, the drill hole will be the deepest ever drilled in the continental United States for research purposes. The main goal of the project is to resolve questions about the source of friction in *faults* (deep cracks in Earth's crust). Scientists hope the answers to these questions will improve their ability to predict earthquakes.

The San Andreas Fault—1,200 kilometers (750 miles) long—marks the boundary between the North American Plate and the Pacific Plate, two of the *tectonic plates* that make up Earth's surface. Movement along this fault caused the great San Francisco earthquake of 1906 and many other lesser but also destructive quakes.

The hole is being drilled about 5 kilometers northeast of the fault in Cajon Pass, near San Bernardino. Drilling and scientific studies in the hole are expected to take more than one year and will cost about $8 million.

Bumpy core. Earth's core is not smooth and spherical as scientists have believed, but bumpy, with "mountains" and "valleys," according to research reported in December 1986 by several teams of scientists at California Institute of Technology (Caltech) in Pasadena. These features may be up to 1,000 kilometers (620 miles) long and hundreds of meters high or deep.

The findings were based on maps of the boundary between the liquid outer core and the semisolid mantle. The

Earth's core is not smooth and round but bumpy, according to a December 1986 report by geologists at California Institute of Technology in Pasadena. They analyzed waves generated by earthquakes to chart valleys in the core hundreds of meters deep and mountains hundreds of meters high. (In the artist's rendering, the core's features are exaggerated to make them more obvious.)

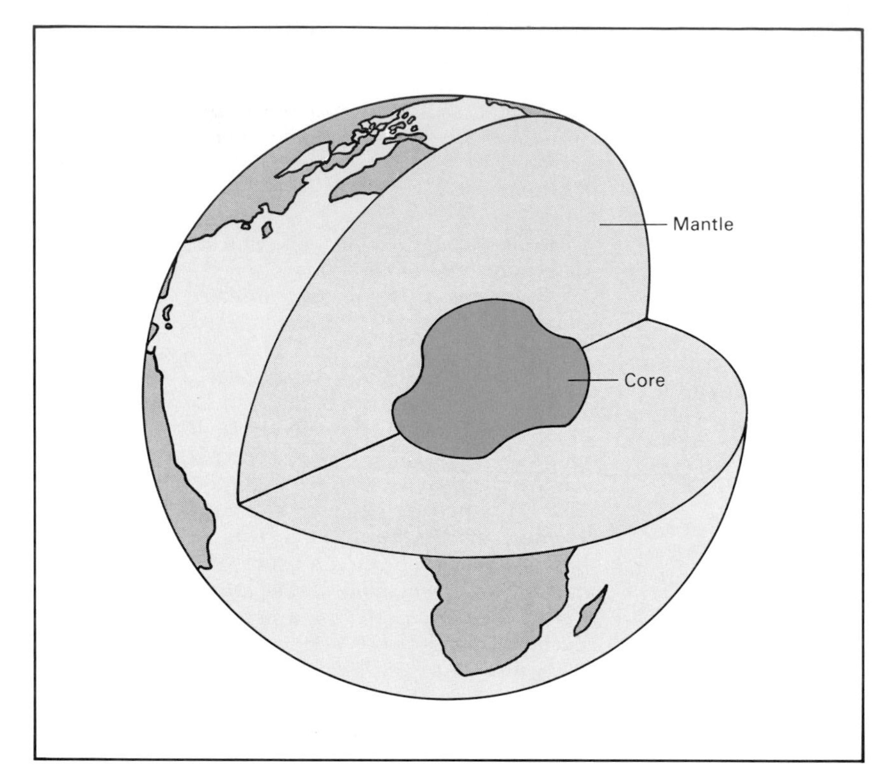

Geology

Continued

maps were made with a technique called *seismic tomography*. In seismic tomography, scientists compare the varying speeds of *seismic waves* (vibrations generated by earthquakes) passing through Earth to create computerized cross-sectional images of Earth's interior.

The Caltech scientists believe the bumps on the core may explain the "jerkiness" in Earth's rotation. The length of Earth's day varies by about five milliseconds each decade. The scientists theorized that the "sloshing" of the liquid core against the walls of the "mountains" and "valleys" could cause this variation.

But other seismic studies indicated that the bumpiness of the core cannot totally account for the variations in the speed of seismic waves passing through this region. In a December 1986 report, seismologists Kenneth Creager and Thomas Jordan of the Massachusetts Institute of Technology in Cambridge suggested that continent-sized masses are distributed irregularly along the boundary of the core. These masses are made up of material different from the liquid and semimolten rock that scientists believe make up Earth's core and mantle. These masses may have roots extending into the core. The scientists suggested that the irregular nature of the masses may disrupt the circulation of the liquid rock in the outer core, causing the changes in Earth's magnetic field that have occurred in the past.

Continental growth. Evidence that the continents are growing at the rate of about 1.35 cubic kilometers (⅓ cubic mile) per year was reported in July 1986 by geologists David G. Howell and Richard W. Murray of the United States Geological Survey (USGS) in Menlo Park, Calif. The scientists based their conclusion on measurements of cores of sediments brought up from the ocean floor.

Howell and Murray first used measurements of the amount of sediment in the cores to calculate how much sediment there is in the ocean basins. Nearly all of this sediment consists of rocky material eroded from the conti-

GEOLOGY: A Science Year Close-Up

A "Killer" Lake in Cameroon

A major disaster occurred in the West African nation of Cameroon on the night of Aug. 21, 1986. A huge, deadly cloud of gas billowed out of Lake Nios (also spelled *Nyos*) in Cameroon's northwestern mountains, killing almost 1,700 people and thousands of cattle in the area. During the next few months, a team of United States researchers determined that the gas cloud consisted of carbon dioxide released from the bottom of the lake, and the gas had asphyxiated the victims.

Lake Nios is a crater lake, lying in a basin or crater formed by past volcanic activity. The research team found that the lake water was saturated with almost pure carbon dioxide—containing so much dissolved gas that it could hold no more. Carbon dioxide concentrations above 10 per cent in the air can kill people. This is more than 300 times the concentration found in the air we normally breathe. Because they found no evidence that other toxic gases had been released from the lake, the researchers concluded that the victims had been smothered by carbon dioxide.

The researchers then faced the question of where the gas came from. Analyses showed that the dissolved carbon dioxide in Lake Nios was more than 35,000 years old—much older than the lake itself. This meant that the gas came from very old *magma*—molten rock. The researchers also found that water at the lake bottom was as cool as water at the surface. They concluded that the carbon dioxide had come from magma that was about 80 kilometers (50 miles) beneath Earth's surface rather than from a chamber of hot magma just below the bottom of the lake. But where was all that gas stored before it was released as a deadly cloud?

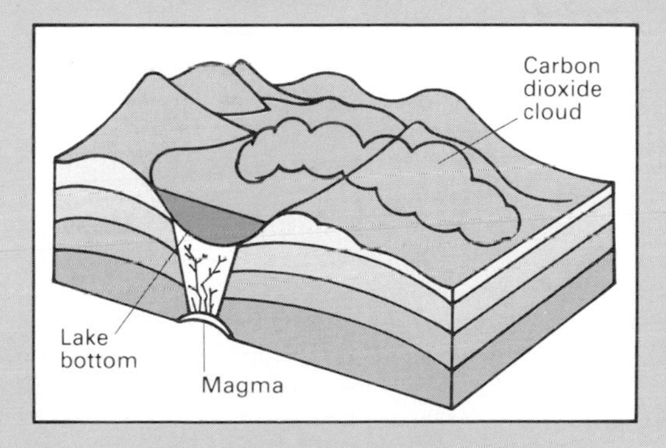

Experts believe that carbon dioxide filtered up from molten rock below Lake Nios, dissolved in ground water, then flowed into the lake bottom. The gas was held down by a "lid" of warm water until some disturbance triggered its violent release from the lake.

Researchers believe that the carbon dioxide gas gradually filtered up from the magma and dissolved in ground water, which then flowed into the lake bottom. Most lakes have an upper layer of warm water floating on top of a denser, lower layer of cold water. This *stratification,* or layering, keeps the upper and lower layers of water from mixing. In deep lakes like Nios, the stratification may be very pronounced, trapping the colder water at the bottom. In Lake Nios, the warm surface water acted like a giant lid, so that most of the gas entering into the bottom of the lake could not escape.

Scientists now agree that carbon dioxide accumulated at the bottom of the lake for hundreds of years, until the lake became completely saturated with gas, like a bottle of carbonated soda. If a soda bottle is opened very carefully, none of the carbon dioxide escapes. A slight disturbance, however, will cause the gas to form bubbles and escape. Once Lake Nios was fully saturated, almost any disturbance—such as a landslide, an earth tremor, or even a strong wind—could have triggered the violent release of the gas.

In 1984, two years before the disaster at Lake Nios, Cameroon was struck by a similar but smaller release of carbon dioxide from nearby Lake Monoun, killing 37 people. Both events occurred in August—a fact that scientists thought was significant. Once a year, when air temperatures are lowest, the surface waters in tropical lakes begin to cool. If the upper water cools to the same temperature as lower water there is no stratification; the lower water can mix with the surface water.

Limnologists (experts on lakes and rivers) from Duke University in Durham, N.C., found evidence that during August, the stratification is lessened in many Cameroon lakes, and the layers of water in some lakes may partially mix. The researchers proposed that such seasonal mixing could have taken the warm water "lid" off and triggered the explosive—and deadly—release of gas in the two lakes.

How can gas-charged lakes be prevented from causing harm? A group of U.S. scientists suggested placing large pipes in such lakes and gradually pumping saturated bottom waters to the surface, where the gas would safely escape into the atmosphere. [George W. Kling] nents and carried to the sea by rivers. They concluded there is 141.7 million cubic kilometers (34 million cubic miles) of sediment in the ocean basins. They then subtracted about one-third of this amount, which represented air space between the grains of sediment. Taking into consideration the fact that the average age of the ocean floor is 55.3 million years, they calculated that 1.65 cubic kilometers (⅖ cubic mile) of material eroded from the continents ends up on the ocean floor each year.

Howell and Murray then examined the area of oceanic *terrane* incorporated into the continents over the past 200 million years. Terranes are ancient volcanic islands, plateaus, and underwater mountains that ride on the ocean crust. When an ocean plate plunges under a continental plate, the terranes are scraped off the ocean plate and added to the edge of the continent.

Previous studies indicated that the area of oceanic terranes added to the continents in the past 200 million years is about 33 million square kilometers (12.7 million square miles). Howell and Murray assumed that the terranes have an average thickness of 20 kilometers (12 miles). Using this figure, they calculated that the terranes' total volume is 660 million cubic kilometers (145 million cubic miles). This means that about 3 cubic kilometers (7/10 cubic mile) of material is added to the continents each year. By subtracting the amount of material eroded from the continents each year—1.65 cubic kilometers—from the amount added—3 cubic kilometers—Howell and Murray calculated that the continents grow about 1.35 cubic kilometers per year or 43 cubic meters (56 cubic yards) per second.

Ancient Sahara rivers. Scientists confirmed that radar images of northern Africa taken from space revealed an immense ancient river system. The rivers crossed the Sahara, now one of the driest places on Earth, 10 to 30 million years ago. The results were reported in July 1986 by remote sensing specialists John F. McCauley, Carol S. Breed, and Gerald G. Schaber of the USGS in Flagstaff, Ariz.

The images were taken during shuttle flights in 1981 and 1984 by a special type of radar that can "see" several meters below the sandy surface of the desert. The radar data indicated that river valleys were buried below the sand.

McCauley and his team made three field trips to remote areas of the Sahara to find material deposited by streams that would confirm the presence of the river valleys. To locate landmarks in the desert, they made transparent copies of the radar images and laid these over satellite photos and standard maps. They also used a satellite navigation system, usually found on oceangoing vessels, to navigate through the desert.

The scientists succeeded in finding the stream deposits and also uncovered fossils of freshwater clams and snails. The scientists concluded that the Sahara river system had its headwaters in the hills on the Egyptian side of the Red Sea. The streams flowed 4,500 kilometers (2,800 miles) southwest through southern Egypt, northwestern Sudan, and Chad and finally reached the Atlantic Ocean at the site of the present-day Niger Delta. But about 6 million years ago, the headwaters of these rivers were diverted into the ancient Nile and their flow was diminished. When the climate of North Africa became very dry about 2 million years ago, the ancient river system disappeared completely.

Wandering poles. A review of evidence indicating that Earth's rotational poles have wandered over the planet's surface independently of the drift of the continents was published in April 1987. The review was performed by geophysicist Richard Gordon of Northwestern University in Evanston, Ill.

The rotational poles are the places where the *spin axis*—the imaginary line about which Earth rotates—intersects the surface. In fact, the spin axis—and the poles—remain fixed while Earth moves beneath them. For convenience, however, geophysicists describe the Earth as fixed and the poles and spin axis as wandering.

From studies of the magnetic fields in ancient rocks, scientists know that the position of the poles has changed over time. They attributed the movement of the poles to the movement of the tectonic plates that make up

Hubbard Glacier blocks the entrance to Alaska's Russell Fiord, background, in the summer of 1986. The huge glacier surged into Disenchantment Bay, foreground, at a speed of about 12 meters (40 feet) per day, creating a dam that trapped scores of seals, porpoises, and other marine animals in the fiord. The ice dam broke in October 1986, and water swept from the fiord into the bay, freeing the animals.

Earth's crust. More recent studies, however, revealed that not all polar wandering could be explained by plate movement.

In 1984, geophysicist Roy Livermore and his colleagues at the British Geological Survey in Keyworth, England, suggested that the movement of the entire *lithosphere*—the upper mantle and crust, including the plates—as a unit with respect to Earth's poles could explain some polar wandering.

To test their theory, Livermore and his colleagues compared the known positions of the poles in the past to the current positions of *hot spots*—plumes of magma that rise from deep within the mantle to form large volcanic areas on Earth's surface. The Hawaiian Islands sit on one hot spot. Because hot spots originate in the lower mantle, many scientists believe their positions remain fixed and are not affected by the movement of the plates. Instead, the plates move over them.

The scientists studied patterns of volcanic mountain ranges and other evidence of plates moving over hot spots in relation to the past positions of the poles and concluded that the lithosphere did not move very much during the last 65 million years. During this period, the lithosphere—and the poles—moved by only about 5 degrees (°). But between 70 million and 100 million years ago, the lithosphere and the poles shifted by about 14°.

The cause of this movement of the lithosphere is a mystery. Geophysicists agree that it must have something to do with redistribution of massive amounts of material on or within Earth. Earth is most stable when most of its mass is farthest away from the poles, at the equator. If a mass redistribution occurred elsewhere, Earth's lithosphere would shift and the poles would reorient themselves. Geologists do not agree, however, about the type of material being redistributed. Some have suggested it could have been the continents. [William W. Hay]

In the Special Reports section, see Drilling Under the Sea; Glaciers on the Go. In World Book, see Geology.

Immunology

The number of cases of AIDS (*ac*quired *i*mmune *d*eficiency *s*yndrome) continued to rise in the United States during 1987, reaching a total of 36,058 cases by June 1. This devastating disease is caused by the *human immunodeficiency virus* (HIV). HIV attacks a certain type of white blood cell called the *helper T cell*, which is important in fighting infection and tumors. There continued to be considerable efforts to develop an AIDS *vaccine*—a substance capable of preventing infection by stimulating the immune system to destroy the HIV virus.

AIDS vaccine research. A research team in Seattle reported in September 1986 that it had tested an experimental AIDS vaccine in animals. The researchers used genetic engineering techniques to insert key components of the AIDS virus into *vaccinia* viruses. The vaccinia virus is used to create smallpox vaccine.

The researchers found that the altered vaccinia virus produced two effects in the immune responses of the test animals. It aided the production of white blood cells, particularly T cells, and the production of *antibodies*. Antibodies help identify and destroy foreign invaders such as viruses. An important next step in the research would be to prove that the vaccine would actually prevent test animals from developing AIDS after being exposed to the HIV virus.

Because experts had predicted that an AIDS vaccine would not be ready for human testing for at least a year, they were astonished by reports in December 1986 that scientists in Zaire and France had started testing a vaccine on human beings. The scientists tested people who had been infected with the HIV virus but who had not experienced any of the characteristic symptoms of AIDS, such as *pneumocystis carinii* pneumonia or Kaposi's sarcoma, a rare type of skin cancer.

Experts predict that developing an AIDS vaccine could be difficult because many different variations of the HIV virus may exist. A single vaccine might not be effective against all of them. But the Zairian and French researchers hope that their experimental vaccine will stimulate the body to produce vast numbers of white blood cells called *killer T cells* that would be able to recognize and destroy any of the variant forms of HIV.

Although many details regarding the vaccine trials were not made public, scientists suspect that the trials were designed to test the safety and side effects of the vaccine. Experts predict that at least one year will be needed to evaluate the results of the trials.

Cancer-fighting technology. The results of new studies of a treatment for patients with advanced cancer were reported in April 1987 by researchers at the National Cancer Institute (NCI) in Bethesda, Md. The treatment, called *adoptive immunotherapy*, involves boosting the cancer-fighting ability of certain white blood cells by treating them with a natural tumor-fighting substance called *interleukin-2*. See MEDICAL RESEARCH.

A tumor-fighting substance made naturally by the body shows promise in tests against a variety of viruses, according to a report in June 1986 by researchers at Genentech, Incorporated, in South San Francisco, Calif. The substance, tumor necrosis factor (TNF), was discovered in the 1970's and identified as an antitumor substance made by a number of body cells. But it was difficult to study because only small amounts could be recovered from the body. In recent years, genetic engineering techniques have made it possible to produce large quantities of such rare natural substances. Scientists have inserted the gene that codes for TNF into bacteria that then produce TNF, making it possible to study this compound in greater detail.

The Genentech scientists reported that, in addition to its tumor-destroying ability, TNF is apparently able to protect cells from invasion and destruction by certain viruses, such as two strains of herpesvirus. They also found that when TNF was combined with interferon—another natural antiviral and antitumor substance that scientists can produce in large quantities in genetically altered bacteria—the combination was even more effective than either substance alone in preventing infection by a number of viruses. TNF also helped boost the tumor-fighting ability of interferon.

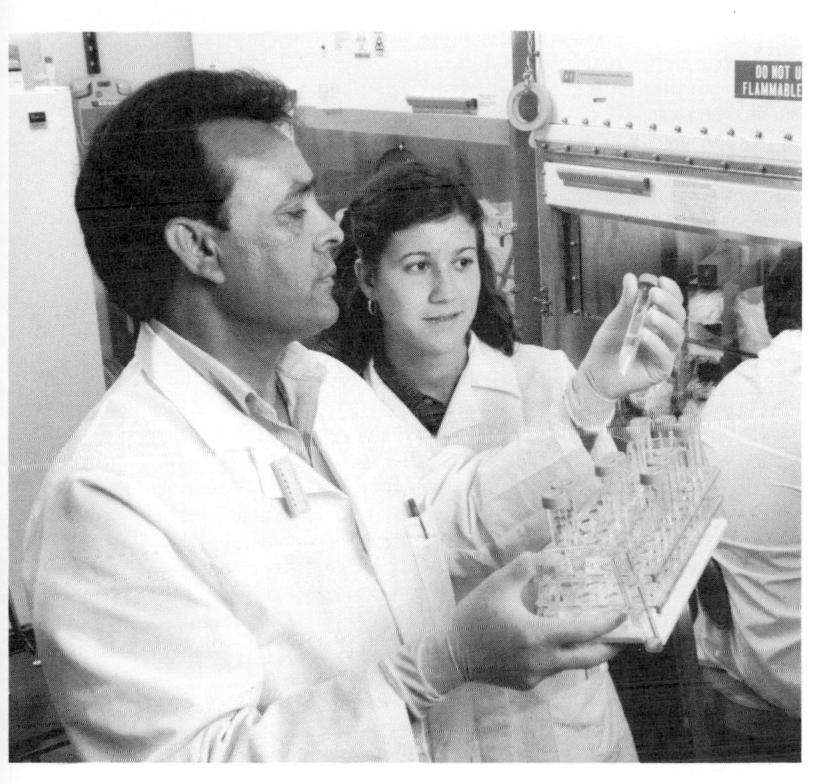

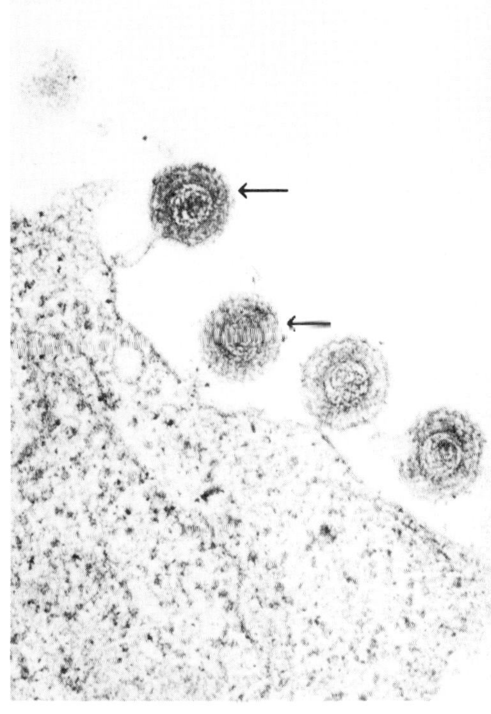

Immunology

Continued

Scientists at the National Cancer Institute in Bethesda, Md., *above*, study human B lymphotropic virus (HBLV), which, they reported in October 1986, may play a role in chronic mononucleosis syndrome, an illness that causes recurrent fatigue. The virus invades a white blood cell called a B-lymphocyte and uses the cell's reproductive equipment to make new virus particles (arrows), *above right*.

New virus discovered. The discovery of a new virus that infects white blood cells called *B-lymphocytes* was reported in October 1986 by NCI researchers. These B cells produce infection-fighting antibodies. Some scientists have proposed that the virus, called *human B lymphotropic virus* (HBLV), may be responsible for a condition called *chronic mononucleosis syndrome*, which causes chronic fatigue. They suggested that HBLV by itself may cause the illness or it may reactivate a herpesvirus responsible for infectious mononucleosis. This illness usually lasts three to six weeks and causes swollen lymph glands, fever, and fatigue. The HBLV virus has also been isolated from some patients with lymph-gland tumors called *lymphoma*. Experts said that further research is required to determine the precise role of HBLV in human disease.

Immunology of autism. Investigators at the Pitié-Salpétrière medical institution in Paris reported in June 1986 that children suffering from a severe developmental disorder called *autism* show some unexpected abnormalities in their immune systems. Autism is a disorder of unknown cause that begins before 4 years of age and is characterized by social detachment, unresponsiveness, and a lack of communication skills.

The French researchers measured levels of antibodies and white blood cell function, and found that these levels were abnormally high in autistic children. The researchers suggested two possible explanations for their findings. Either abnormalities in chemicals involved in the brain and nervous system might somehow stimulate the immune system, or abnormalities in the immune system may be a factor involved in the development of autistic behavior. A growing body of evidence suggesting a link between the brain and the immune system made the French team's finding particularly intriguing to both immunologists and neuroscientists. [Paul Katz]

In World Book, see AIDS; Autism; Cancer; Immunity; Interferon; Virus.

Medical Research

Medical researchers continued their quest for an effective vaccine or treatment for AIDS (acquired immune deficiency syndrome) in 1986 and 1987 (see DRUGS; IMMUNOLOGY; PUBLIC HEALTH). While these efforts often proved frustrating, research in other areas of medicine showed promise.

Brain grafts. In April 1987, surgeons at the National Autonomous University in Mexico City, Mexico, reported that they had reduced the symptoms of Parkinson's disease in two patients by transplanting cells from the patients' *adrenal glands*—small flat organs located on top of each kidney—into their brain. Parkinson's disease is a disorder of the nervous system that causes loss of muscle control. It occurs when certain brain cells die. These cells normally stimulate the release of *dopamine*, a substance that transmits nerve signals to muscles.

As the disease progresses, less dopamine is released. Parkinson's patients develop tremors in their arms and legs, lose facial expression, and have difficulty walking. Because drugs used to treat the illness are not always effective and may have very severe side effects, medical researchers have looked for other ways to treat the disease.

A surgical approach was first tried by Swedish surgeons in 1985. They removed dopamine-producing cells from the adrenal gland and implanted them deep in the brain of four patients with advanced Parkinson's disease. Although the implants did not harm the patients, the treatment did not reduce their symptoms.

The Mexican surgeons modified the Swedish technique. First, their patients were younger and had less severe symptoms than the patients treated by the Swedish group. Also, the Swedish surgeons had transplanted the adrenal cells into an arched structure called the *caudate nucleus*, which contains cells that control movements involved in walking and changing facial expression. The Mexican team decided to transplant the cells onto the top of the caudate nucleus, to ensure that the transplanted cells would be bathed in cerebrospinal fluid—a fluid that normally bathes part of the brain and the spinal cord. The Mexican team reasoned that this fluid would nourish the transplanted cells and increase the chances of a successful graft.

Before the operations, according to the surgeons, the two patients had severe tremors, fixed facial expressions, were unable to speak clearly, and were confined to wheelchairs. But a few months after surgery, both patients had improved markedly.

Since then, the Mexican surgeons as well as surgeons at Vanderbilt University Hospital in Nashville, Tenn., and at New York University School of Medicine in New York City have performed similar transplants on other Parkinson's disease patients.

Cancer detection test. In November 1986, researchers at Harvard Medical School and Beth Israel Hospital in Boston reported that they had developed a new blood test for detecting the presence of cancer. The test requires only a small blood sample, which is subjected to *nuclear magnetic resonance* (NMR) *spectroscopy*. This technique uses radio waves and powerful magnetic fields to analyze molecules.

Researchers place the blood sample in a powerful magnetic field. They then turn on a transmitter that emits radio waves, from which atoms in the blood absorb energy. When the researchers turn off the transmitter, the atoms release radio signals, which are picked up by a radio receiver. Since various kinds of molecules are made up of different atoms, they give off slightly differing signals. By analyzing these radio signals, researchers can determine what chemicals are present in the blood.

The Harvard researchers wanted to determine whether NMR could be used to detect some sign of cancer in blood samples. First the team developed a way to screen out the signal from water molecules, the major component of blood. This made it easier to identify signals from other substances. They were encouraged to find that the clearest signals were generated by *lipids* (fats). Cancer causes changes in the way the body stores and uses lipids. So the Harvard team reasoned that the lipids in blood taken from cancer patients might give off NMR signals that differ from those emitted by lipids in blood from cancer-free people.

To find out, the researchers took blood samples from 311 people and subjected each sample to NMR spectroscopy. Then they compared the *frequency* (cycles per second of a radio wave) expressed in units called hertz (Hz). The Harvard team decided that lipids giving off NMR radio signals of less than 33 Hz indicates the presence of cancer; lipid signals from all the patients with diagnosed cancer were in that category. The under-33 Hz category, however, also included two cancer-free groups—men with enlarged prostates and pregnant women. The researchers hypothesized that these patients—like cancer patients—have rapidly dividing cells that may provoke similar changes in lipids. As a result, the scientists do not recommend the test for diagnosing cancer in women who are pregnant or in men who have enlarged prostates.

The Harvard group reported in February 1987 that the NMR blood test had identified cancer patients among a second sample of 400 people. The researchers cautioned that more research is necessary before the test can be put into general use.

New heart muscle. Cardiac surgeons at the University of Pennsylvania in Philadelphia announced in December 1986 that they had created a heart-assist pump composed of back muscle and had tested it successfully in dogs.

The ultimate goal of surgeon Larry W. Stephenson and his colleagues was to design a pump made of muscle that will improve circulation in patients suffering from chronic congestive heart failure—a condition in which the heart's ability to pump blood is drastically reduced. The muscle pump would be used along with, or as an alternative to, drug therapy, heart transplantation, or mechanical pumps.

The surgeons reasoned that a pump made from a patient's own skeletal muscle would have several advantages. Unlike a transplanted heart, it would not present the risk of rejection, a process that occurs when the immune system recognizes and destroys "foreign" transplanted tissue. And, unlike an artificial heart, which requires an external power source, it could be completely contained within the body.

Thus, Stephenson designed and

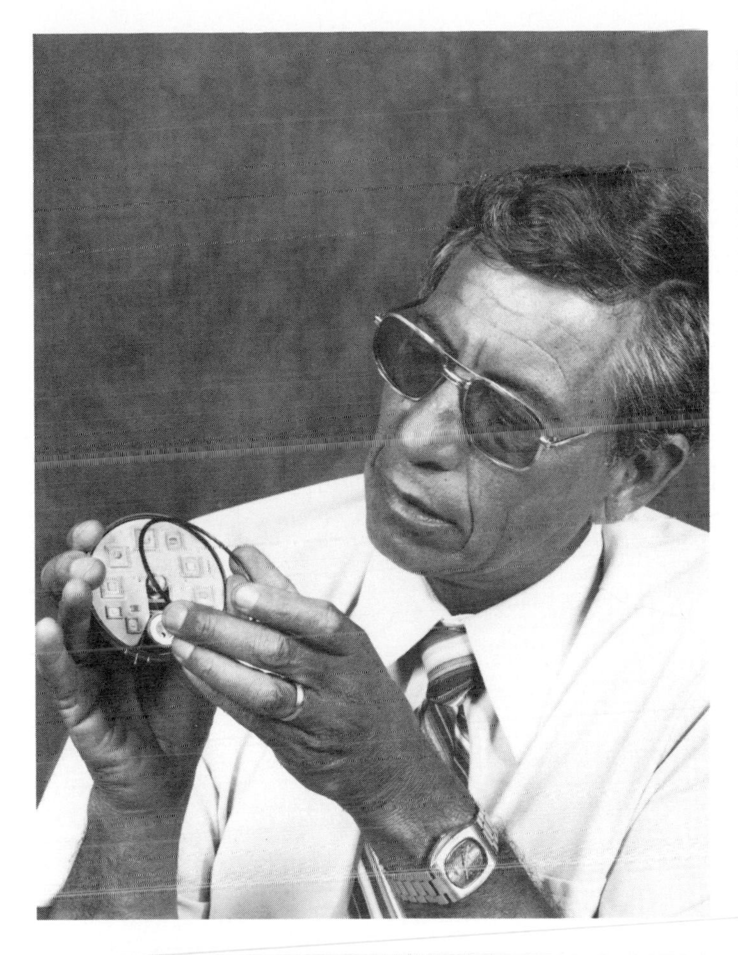

The first insulin pump designed to be implanted in the abdomen of a diabetic patient is held by its inventor, Robert E. Fischell of Johns Hopkins University, *top.* A handheld computer, *above* (left), sends radio signals that tell the pump, *above* (right), when to release preprogrammed doses of insulin. The pump is refilled four times a year with a hypodermic needle.

New Uses for Lasers in Medicine

The laser, *below*—a narrow, intense beam of light that can be focused with great precision—found an increasing number of uses in medicine in 1986 and 1987.

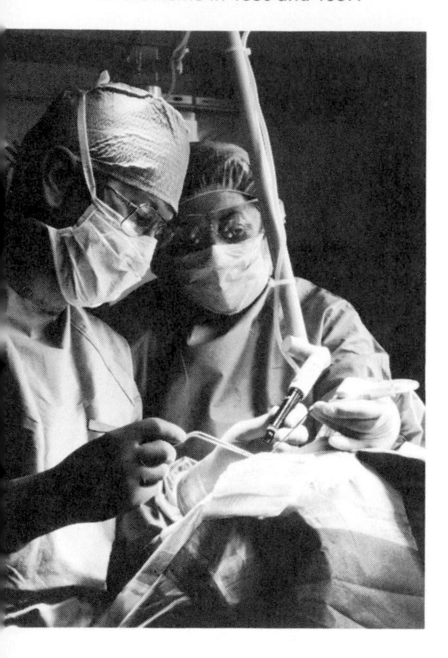

A new laser surgery technique may correct nearsightedness, *right,* which is usually caused by a cornea that is too long. Light rays entering the eye fail to focus on the retina to create a sharp image (1). A laser might be used to sculpt away a thin layer of corneal tissue to flatten the cornea (2). Light rays would then focus properly on the retina to create sharp images (3).

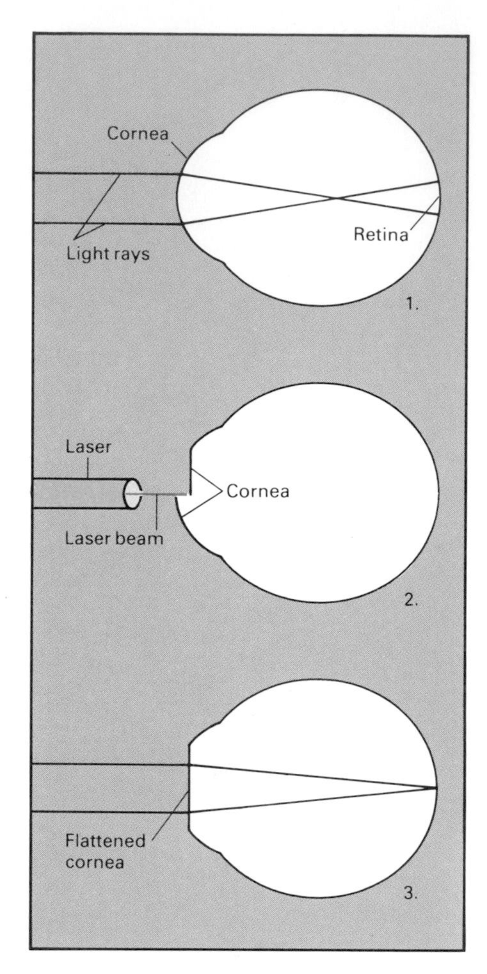

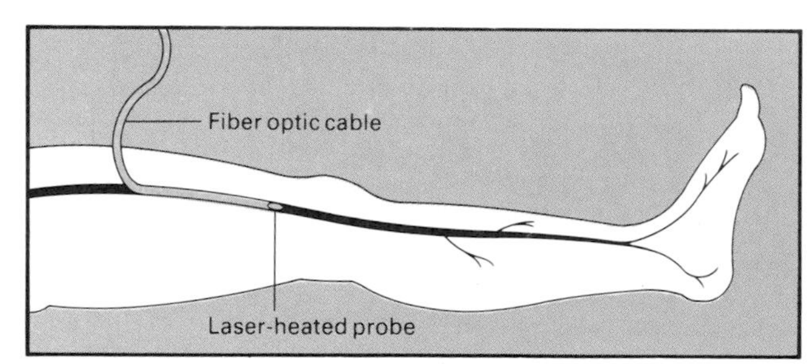

A metal probe heated by laser light traveling through a fiberoptic cable burns away blockages in leg arteries, *above*. Before treatment, *top right,* the blocked artery (arrow) appears very narrow in an X ray. Blood can flow through the widened artery, *below right,* and the blockage is gone (arrow) after the probe burned a tiny hole through the blockage and surgeons widened the opening by inserting and inflating a balloon.

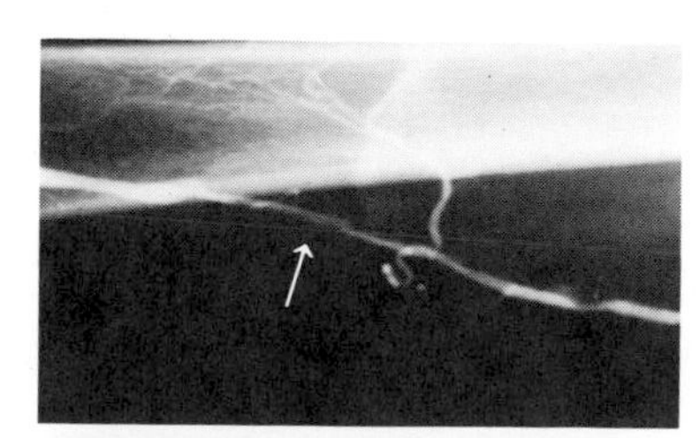

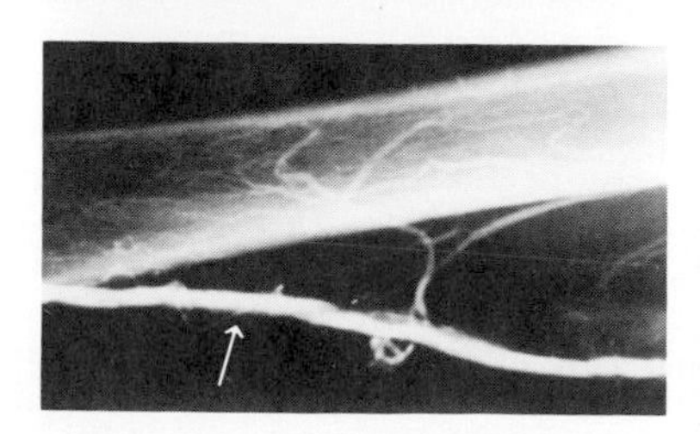

In a laboratory experiment, one pulse of a laser beam smashes into a kidney stone, *below,* and it begins to break into fragments (arrow). Medical researchers are studying whether the laser can replace stone-dissolving drugs or surgery for kidney stones that block the urinary tract.

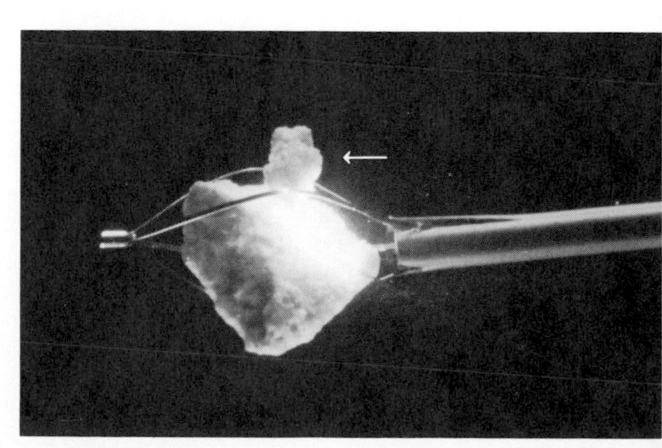

built a pump made of muscle for a dog. He took a strip of the dog's back muscle and shaped it into a pouch lined with Gore-Tex, a synthetic material used to make artificial blood vessels and outdoor clothing. He sewed the pouch to the dog's chest wall. He then connected the pouch to the dog's thoracic *aorta* (a major vessel leading from the heart) with two Gore-Tex tubes serving as artificial arteries. Thus, blood pumped by the heart would enter the aorta, pass through the muscle pouch, then return to the aorta for its journey through the body.

Skeletal muscle, however, tires more easily than does heart muscle. Stephenson knew that he would have to "train" the muscle to become more resistant to fatigue—like normal heart muscle—so that it could be used as a pump. He also knew from earlier experiments that constant electric stimulation could cause skeletal muscle to behave more like heart muscle.

To make the pouch beat like the heart so that it could help push blood through the arteries, Stephenson attached it to a *pacemaker*, a battery-operated device designed to deliver a series of small shocks to the heart. The shocks made the muscle pouch beat in time with the dog's heart, and also began the process of training the transplanted muscle to behave more like heart muscle.

Stephenson made heart-assist pouches for five other dogs. Although one dog survived for 11 weeks, the others died within a few weeks of surgery. All of the pouches were still functioning when the dogs died. But in at least three cases, many tiny blood clots had formed in folds of the Gore-Tex. These clots had become dislodged and had blocked blood vessels or passages in the kidney, causing death.

Stephenson speculated that treatment with anticlotting agents might prevent such blockages in future attempts. He also began experimenting with a muscular pouch made without the Gore-Tex lining, in an effort to eliminate the dangerous blood clots that formed in folds of the Gore-Tex.

Cholesterol study. In June 1987, researchers from the University of Southern California School of Medicine in Los Angeles reported the first clear evidence that heart disease can be reversed to a significant extent with a low-fat diet and cholesterol-lowering drugs. Cholesterol is a fatty substance found in animal food products. Excess cholesterol sets the stage for heart disease by clogging coronary arteries.

The scientists reported that about 16 per cent of patients who followed low-fat diets and took two cholesterol-lowering drugs had a significant reduction in existing fatty deposits in their arteries. The rate at which new deposits developed slowed as well. Federal health experts said that many of the 6 million people in the United States with coronary artery disease could benefit from the treatment.

Cancer treatment debate. Researchers in 1986 and 1987 continued to debate the merits of an experimental cancer treatment. The treatment involves the use of *interleukin-2* (IL-2), a substance called a *lymphokine*, which activates certain white blood cells of the immune system called natural killer cells. See IMMUNOLOGY.

The technique was severely criticized in December 1986 by cancer specialist Charles G. Moertel of the Mayo Clinic in Rochester, Minn., who charged that the treatment—a combination of IL-2 and lymphokine-activated killer (LAK) cells—should be abandoned because it had many toxic side effects. Some of the first patients to be treated suffered congestive heart failure and breathing difficulties, and two died as a result of the treatment.

In April 1987, however, two groups of researchers, including a group at the National Cancer Institute (NCI) in Bethesda, Md., published encouraging reports on studies using IL-2 with LAK cells to treat cancer patients. The NCI team reported that the treatment reduced the size of tumors by 50 per cent in 23 of 108 patients with various types of cancers. The second group—physicians at the Biological Therapy Institute in Memphis and at Biotherapeutics, Incorporated, in Franklin, Tenn.—reported that the treatment had reduced tumors by 50 per cent in 11 of 25 patients. Patients in both studies experienced no life-threatening side effects. [Beverly Merz]

In WORLD BOOK, see BLOOD; CANCER; PARKINSON'S DISEASE.

Medicine

Victims of radiation poisoning, resulting from an accident in April 1986 at the Chernobyl nuclear power plant near Kiev in the Soviet Union, receive transfusionlike bone marrow transplants. The radiation damaged their bone marrow—the source of red blood cells and many white blood cells. By August 1987, three patients had recovered enough for them to leave the hospital.

Two studies reported in 1986 and 1987 indicated that certain commonly accepted diagnostic techniques are widely employed when their use is unlikely to benefit the patient. These techniques include electronic fetal monitoring, in which the heartbeat of an unborn baby is checked with an electronic device, and skull X rays used in the diagnosis of possible head injuries.

Fetal monitoring. In most hospitals in the United States, all women in labor are fitted with electronic devices that monitor fetal heart rate. In September 1986, researchers at the Southwestern Medical School of the University of Texas Health Science Center at Dallas reported the results of a study indicating that electronic fetal monitoring is unnecessary in most cases.

Monitoring the fetal heart rate is one way to determine whether labor is progressing normally. An abnormal heart rate that lasts for several minutes is often a sign that the fetus is not getting enough oxygen. In such cases, doctors may decide to perform a Caesarean section—an operation in which the baby is removed through an incision in the mother's abdomen and uterus—rather than wait until the baby passes through the birth canal. Critics of the widespread use of fetal monitoring say that the technique sometimes falsely indicates that the fetus is in distress, prompting physicians to perform an unnecessary Caesarean section on the mother.

Until the mid-1970's, doctors monitored fetal heart rate periodically during labor, by listening through a stethoscope placed on the mother's abdomen, or by using an ultrasound probe—a device that uses sound waves to produce patterns of the baby's heartbeat. Since the late 1970's, however, continuous electronic monitoring has replaced periodic monitoring in most labor rooms.

The Dallas researchers wanted to find out whether this type of monitoring is necessary for all women in labor or only for those who have a high risk of complications. For a three-year period beginning in October 1982, two

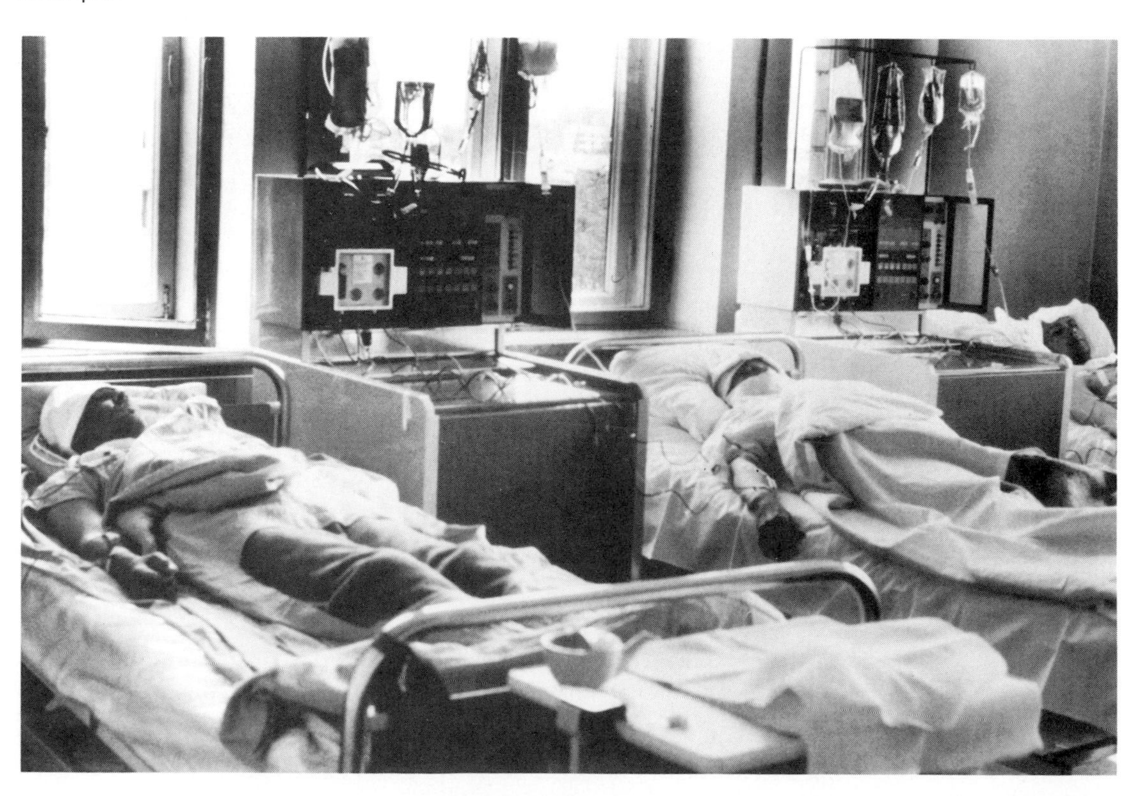

types of routine care were used for women in labor at a Dallas hospital. Every other month, all women entering the hospital in labor were attached to an electronic fetal monitoring device. On alternate months, only women whose labor or delivery was classified as "high risk" were electronically monitored during labor. The high-risk category included women who had abnormally long or short pregnancies, diabetes, or high blood pressure, or who had been given labor-inducing drugs.

The researchers compared 7,288 babies born during months when all fetuses were monitored with 7,330 babies born during months when only high-risk fetuses were monitored. They found that 551 fetuses in the first group—more than 7 per cent—were reported to have irregular heartbeats. Doctors had decided to deliver 64—about 1 per cent—of these babies by Caesarean section. Only 196 in the other group—less than 3 per cent—were reported to have irregular heartbeats, resulting in only 28 Caesarean deliveries—less than half of 1 per cent.

The research team found virtually no difference in how the two groups of babies fared. The researchers concluded that continuous electronic fetal monitoring is unnecessary unless there is a high risk of complications.

Skull X rays. In January 1987, a study coordinated by the U.S. Food and Drug Administration (FDA) concluded that there is no benefit in taking skull X rays of patients with head injuries when those patients have no symptoms of brain damage.

The study involved 7,035 patients with head injuries. The FDA researchers asked doctors at 10 medical centers to use certain guidelines to classify each head-injury patient as low-risk, moderate-risk, or high-risk for brain damage.

Low-risk patients were defined as those with scalp injuries and no symptoms other than headache and dizziness. Children less than 2 years old were defined as moderate-risk patients, as were all patients who had suffered severe facial injuries or skull damage or who had vomited, lost consciousness, suffered seizures, or developed amnesia after the injury. High-risk patients were defined as those with depressions or holes in the skull and who were losing consciousness and showing signs of decreased brain function.

The doctors participating in the study were asked to use their usual criteria to decide whether or not the patient should have a skull X ray or a *computerized tomography* (CT) scan, regardless of how they classified patients for the study. A CT scan is a kind of X ray that shows images of a cross section of the area examined and provides a more detailed image of soft tissue, thus indicating brain as well as skull injuries.

The researchers studied the records of all patients treated for head injury in the hospitals. Three months after the injury, they contacted the patients to determine whether they had sustained brain damage that had become apparent only after leaving the hospital. If the researchers could not reach the patient, they checked the National Death Index—a record of all registered deaths maintained by the U.S. Public Health Service—to determine whether the patient had died of a brain injury.

The researchers found that the doctors had ordered skull X rays for 2,795 of 5,254 low-risk patients. Of those X-rayed, 2,783 showed no skull injury and the remaining 12 had fractures that did not require treatment.

In the moderate-risk group, 1,129 of 1,610 patients were either X-rayed or had CT scans. Of these, 1,082 had no injury, 35 had fractures that did not require treatment, and 12 had the type of fractures that might produce brain injuries.

In the high-risk group, 144 of 171 patients underwent X ray or CT scanning. Of these, 113 had no fracture, 15 had fractures requiring no treatment, and 16 had the type of fractures that might produce brain injuries.

The researchers concluded that skull X rays are not necessary for low-risk patients. They recommended that these patients be given written information on the care of head injuries and sent home with someone who could watch them for signs of brain damage. They said that moderate-risk patients should be watched carefully—either in or out of the hospital—and undergo CT scanning if their condition deteriorated. CT scanning was recommended for all high-risk patients.

Self blood donors. A report published in November 1986 endorsed *autologous blood donation*—the practice of having surgical patients donate their own blood before surgery. The blood is stored for use during and after the operation.

In the report, the American Medical Association's (AMA) Council on Scientific Affairs recommended the use of autologous blood transfusions whenever possible, because the patient's own blood is safer than transfusions of blood from other donors. Self blood donations enable a patient to avoid certain infections that can be transmitted through exchange of blood—including malaria and a form of hepatitis. They also eliminate a potentially dangerous reaction that may occur when the patient's immune system recognizes and tries to destroy the various foreign cells and molecules in donor blood.

Transfusion practices. Researchers who surveyed blood transfusion practices being used at 18 major medical centers in the United States reported in February 1987 that autologous blood donation is not used nearly as much as it could be.

The researchers studied the records of 4,995 patients who had nonemergency surgery at the 18 medical centers during January and February 1986. They found that 1,287 had been diagnosed as likely to need transfusions. Although the researchers identified 590 of these patients as capable of making autologous blood donations, only 32 did so. And of these, only 4 required more blood than they had donated.

Among the remaining 558 patients, 199 required transfusions. The blood for all these transfusions was supplied by donations from other people. The researchers estimated that those 199 patients could have supplied 72 per cent of the blood needed for their transfusions, substantially reducing the demand on blood-bank supplies. [Beverly Merz]

In World Book, see Blood Transfusion; Childbirth; Medicine.

Meteorology

Thunderstorms can cause the spread of air pollution, according to a January 1987 report by a team of government and university meteorologists. The research was done by atmospheric scientists from the University of Maryland, College Park; Pacific Northwest Laboratory, Richland, Wash.; Brookhaven National Laboratory, Long Island, N.Y.; the National Center for Atmospheric Research (NCAR), Boulder, Colo.; and the University of Denver.

Their findings indicated that the strong upward drafts commonly produced by thunderstorms can suck surface-level air pollution to higher levels of the atmosphere. The lower temperatures at very high altitudes prevent harmful pollutants such as nitrous oxide and nitrogen dioxide from quickly breaking down into less dangerous compounds. The jet stream and other winds blow the pollutants to distant areas.

The meteorologists made these observations in a field experiment near the Oklahoma-Arkansas border. Flying in research aircraft, the scientists took several samples of atmospheric gases in and around a powerful thunderstorm. At the top of a large rain cloud, or *cumulonimbus*, about 10,000 meters (33,000 feet) above the ground, the researchers found levels of carbon monoxide, a common pollutant, that were twice as high as would be expected at that altitude if a thunderstorm were not present.

They also found higher than normal concentrations of other pollutants—nitrous oxide and ozone—in this region of the thunderstorm. Based on all the findings, they concluded that the *anvil*, or wide top of the cloud, contained polluted air that could have originated only near the ground. Thunderstorms may therefore be responsible for some of the global spread of pollutants and for acid rain falling many kilometers away from the source of pollution.

Ozone hole. The discovery of a thinning in Earth's ozone shield over Antarctica was reported in October 1986. (Ozone is a form of oxygen found

A microburst—a brief, violent downward rush of air near the ground—is revealed by rapidly evaporating rain falling beneath a cloud at Denver's Stapleton International Airport. Microbursts, or wind shears, which can cause airplanes to rapidly lose altitude or to crash on take-off or landing, were studied intensively in summer 1986 by U. S. government and university scientists.

about 24 kilometers [15 miles] above Earth's surface.) The thinning concerned many atmospheric scientists because of the ozone layer's importance in maintaining Earth's climate. See ENVIRONMENT (Close-Up).

Nuclear winter. The global cooling that would follow a nuclear war might not be as severe as previously predicted, according to a June 1986 report by climate researchers Starley L. Thompson and Stephen H. Schneider of the NCAR. This report challenged the findings of other atmospheric scientists who, since 1982, had studied the possibility of a nuclear winter.

In earlier reports, scientists such as astronomer Carl Sagan of Cornell University in Ithaca, N.Y., predicted that nuclear explosions would not only cause great destruction on the ground but also send thick clouds of smoke and dust into the atmosphere. Most scientists believed this would cause temperatures to fall by as much as 25°C (45°F.), because sunlight could not fully penetrate the clouds. Such a drastic temperature change—especially if it occurred in the summer—would cause crop failure and famine.

Thompson and Schneider's findings suggest that the temperature drop would be less severe—about 14°C (25°F.)—and would last only one or two weeks rather than the one or two months estimated earlier. According to Thompson and Schneider, temperatures would probably not drop below freezing after a summertime nuclear war.

Like earlier researchers, Thompson and Schneider used computer modeling as a basis for their predictions. Their model included information neglected in earlier studies, however, such as the facts that rain would wash dust and smoke out of the air and that wind could be expected to disrupt the cloud, so that a continuous "blanket" over large areas would be unlikely. Thompson and Schneider also included calculations for the heat stored in oceans, which warms the atmosphere even during long periods of overcast skies.

Other scientists working on the nuclear winter problem, including atmospheric scientist Michael C. MacCracken of Lawrence Livermore Na-

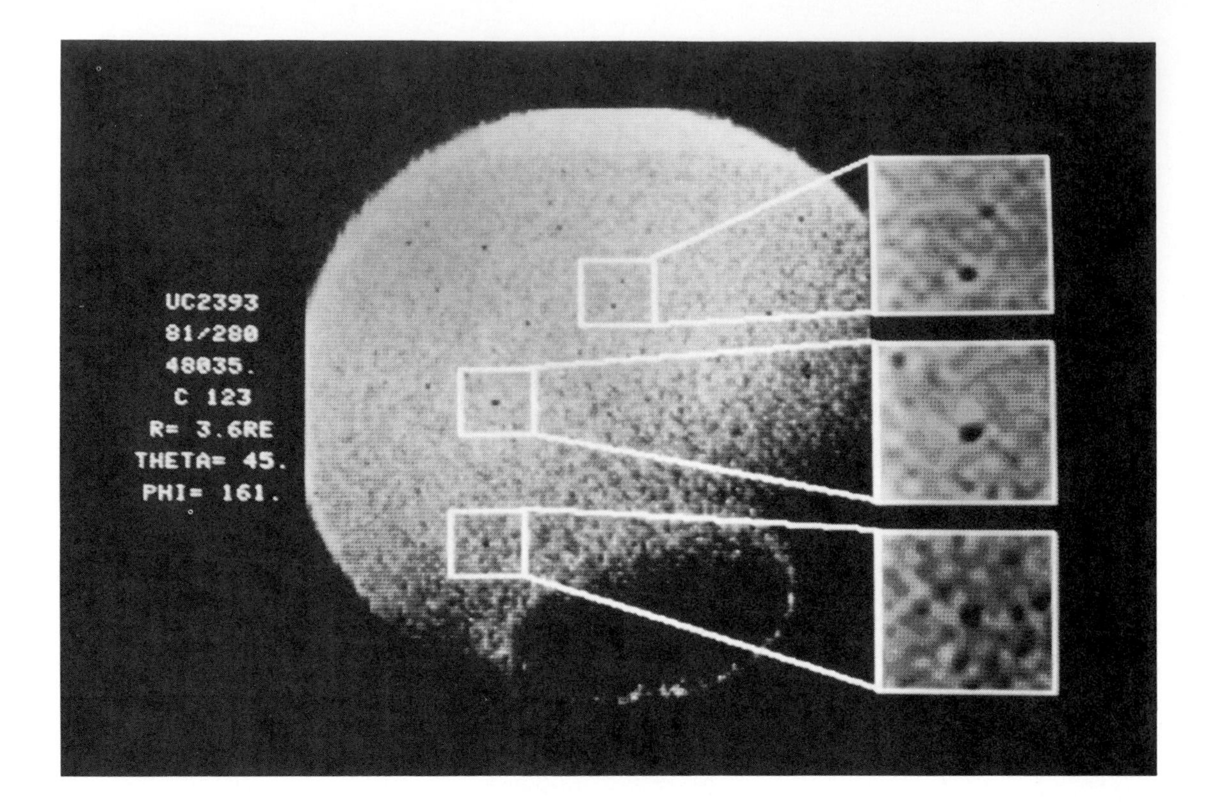

Holes in Earth's atmosphere, detected by an orbiting satellite, were reported in mid-1986 by geophysicists at the University of Iowa in Iowa City. The holes—which appear as dark spots in the image above—may be created by icy comets that pierce the atmosphere and then, heated by friction, turn into steam. The scientists suggested that vapor from such comets could, over billions of years, have been the source of Earth's water.

tional Laboratory in California, agreed with the new predictions. MacCracken's study also predicted a less severe atmospheric aftermath to nuclear war, and he suggested that *nuclear fall* might be a more appropriate name for the phenomenon.

El Niño predictions made in March 1986 turned out to be more accurate than expected. A moderate-to-weak El Niño in the central Pacific during the winter of 1986-1987 was apparently predicted by oceanographers Mark A. Cane and Stephen E. Zebiak of Columbia University's Lamont-Doherty Geological Observatory in Palisades, N.Y. In June 1986, a similar prediction was made by oceanographer James J. O'Brien of Florida State University in Tallahassee. El Niño is a warm stream of water in the Pacific Ocean that appears off the coast of Peru every two to seven years, disrupting normal weather patterns.

All El Niño predictions, including Cane and Zebiak's, were in some ways inaccurate, however. Cane and Zebiak, for example, predicted a warming of ocean waters near South America that did not occur, and they also believed the El Niño warming would be larger than it actually was.

The inaccuracies may be blamed on the limited data used in making the predictions. The scientists based their calculations on the strength of *trade winds* (strong winds from the east). These winds can influence the appearance of El Niño. Under normal conditions, they push the warm upper layer of the Pacific Ocean toward the west as far as Indonesia. During the early stages of an El Niño, scientists believe, the trade winds die down so that the warm water pushed to the west flows back toward the east. After about six months, this warm water reaches the South American coast.

Other factors need attention, however, if El Niño is to be more accurately predicted. In addition to examining how wind changes the surface layer of the ocean, scientists should include the ocean temperature's influence on wind in their calculations. This influence, along with the effect of ocean currents, is a critical factor in El Niño prediction.

Improved understanding of El Niño and similar occurrences in the Atlantic and Indian oceans may result from an international research program launched in 1986. In the Tropical Oceans and Global Atmosphere program (TOGA), scientists from Australia, China, France, India, Indonesia, Japan, the United Kingdom, and the United States are measuring the temperatures and currents in the upper layers of tropical oceans.The study is part of the ongoing World Climate Research Program.

The wind profiler—an instrument that measures wind speeds at different altitudes—may become an important part of weather-observing systems of the future, according to a January 1987 report by meteorologists John A. Augustine of the National Oceanic and Atmospheric Administration's Weather Research Program and Edward J. Zipser of the NCAR, both in Boulder, Colo. A wind profiler uses a special kind of *Doppler radar*, which sends out radio waves that strike moving air molecules and bounce back. As the speed of a windblown air molecule increases or decreases, characteristics of the returning radio waves also change. By measuring these changes, the wind profiler calculates the speed of windblown molecules—and thus continuously measures wind speed.

Augustine and Zipser conducted field tests during May and June 1985 in Kansas and Oklahoma. There, several wind profilers were added to the National Weather Service's usual observation equipment, such as weather balloons and storm-sensing radar. The profilers were able to provide detailed measurements of wind speeds at several altitudes as a *squall line* (a violent lifting of air caused by an approaching cold front) passed through the region. The profilers' data included new information about wind speeds inside a squall and gave the first measurements of wind speeds at the back of a squall line. [W. Lawrence Gates]

In WORLD BOOK, see METEOROLOGY; WEATHER.

Molecular Biology

How an enzyme recognizes a particular site on a molecule of *deoxyribonucleic acid* (DNA) and cuts the DNA at that site was reported by a team of molecular biologists from the University of Pittsburgh and the University of California at San Francisco in December 1986. DNA and proteins such as enzymes are *macromolecules*—very large and complex molecules that contain thousands of atoms. One of the major goals of molecular biology is to learn how a macromolecule's structure determines its biological function. Scientists are particularly interested in learning how the structures of such molecules are altered when they interact.

The researchers, led by molecular biologist John M. Rosenberg of the University of Pittsburgh, determined the complete three-dimensional structure of an enzyme called Eco RI, a member of a large group of enzymes called *restriction enzymes*, which act like chemical scissors. Restriction enzymes recognize specific places, *recognition sites*, in a length of DNA, and cut the DNA at those sites. Eco RI, discovered in 1971, was the first DNA-cutting enzyme to become a basic tool for genetic engineering.

A DNA molecule is made up of units called *bases*, which are connected to each other in a molecular chain. These chemically distinct bases are guanine, adenine, thymine, and cytosine (usually abbreviated G, A, T, and C). Each restriction enzyme recognizes a particular sequence containing these four bases. The Eco RI enzyme cuts a DNA chain whenever it encounters the sequence of bases GAATTC.

Even though the GAATTC sequence might occur only once in a length of DNA that includes thousands or tens of thousands of bases, the Eco RI enzyme will locate its GAATTC recognition site and cut the chain there and only there. From their general knowledge of the way that enzymes work, molecular biologists had theorized that in order for Eco RI to recognize this particular sequence of bases, there must be a precise fit between the three-dimensional shape of a segment of DNA—particularly the recognition site—and a site on the surface of the Eco RI protein.

The research team used a technique called *X-ray crystallography* to study what happens when Eco RI recognizes and cuts a piece of DNA. This method uses X rays to analyze the arrangement of atoms in a crystal. When exposed to X rays, the crystals scatter, or *diffract*, the radiation in a pattern reflecting the arrangement of the atoms in the crystal. A computer analysis of the pattern permits scientists to determine the positions of the thousands of atoms in the crystal.

By evaporating a solution containing Eco RI and DNA, the research team created crystals of Eco RI bound to DNA. When an enzyme binds to the substance on which it acts, the two molecules form a temporary structure called a *complex*. The X-ray analysis of the DNA-Eco RI complex revealed that the DNA molecule fits precisely into a "pocket" in the surface of the Eco RI enzyme—just as the researchers had predicted.

The scientists also discovered that one part of the enzyme molecule is responsible for recognizing the GAATTC sequence and another part for cutting the chain between the G and the A. Furthermore, they uncovered an explanation for how Eco RI specifically recognizes and cuts only this particular sequence of six bases. GAATTC is the only six-base sequence whose structure permits Eco RI to form 12 contacts or bonds with parts of the DNA molecule. These bonds appear to be crucial for Eco RI's ability to recognize and cut DNA in places where the six-base sequence occurs.

Biological milestone. Experts said that the analysis of how Eco RI acts on DNA represented a milestone of modern molecular biology, and they predicted that the study was merely the first in a long series of studies of complexes between DNA and restriction enzymes. Ultimately, such research may permit scientists to design enzymes that cut DNA wherever they wish.

Insights into how enzymes recognize special areas of DNA are also expected to help scientists understand many other processes in biology that depend on the interactions of proteins with

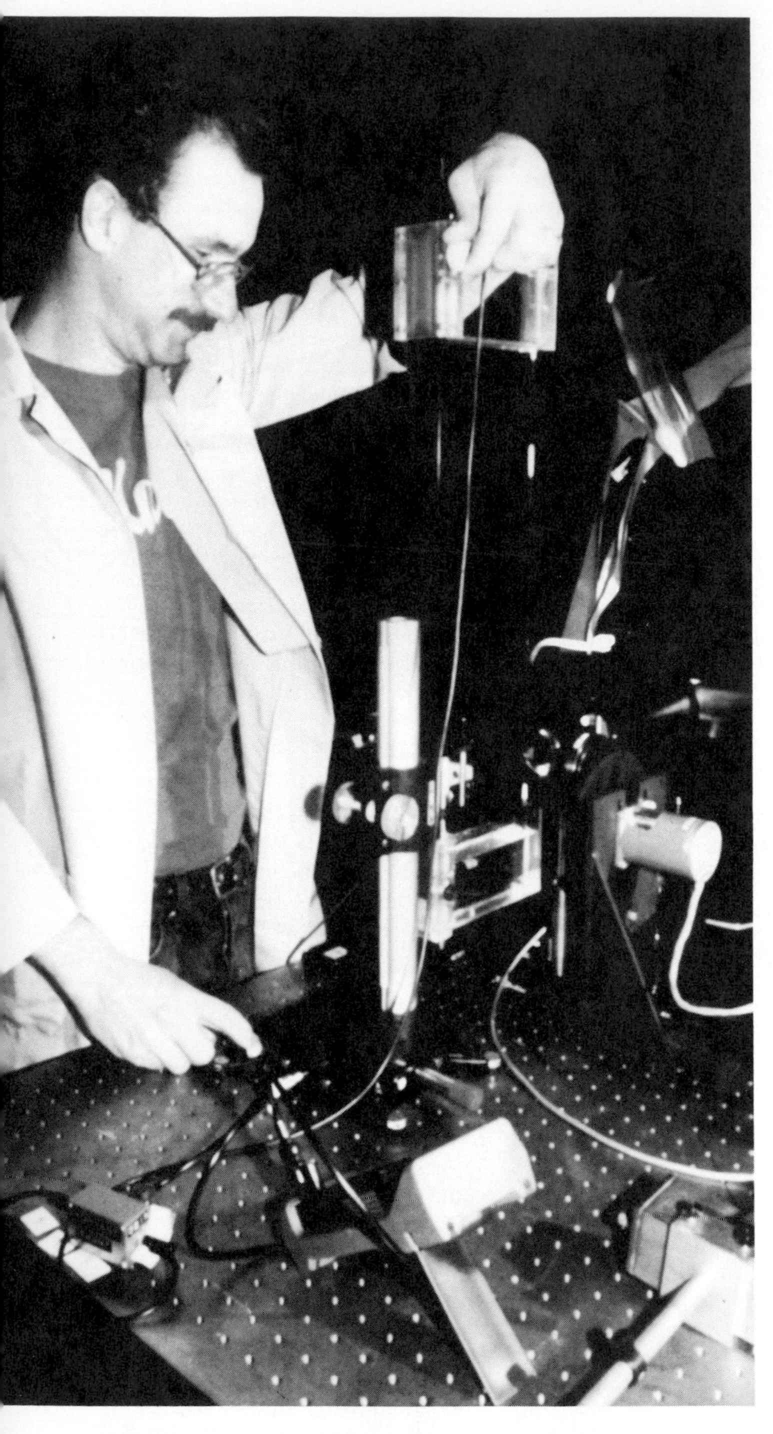

A scientist adjusts an automatic gene sequenator developed at California Institute of Technology in Pasadena, Calif., in June 1986. The machine uses a laser, colored dyes, and a microcomputer to automatically analyze the structure of a molecule of DNA, the substance of which genes are made.

specific DNA sequences. For example, such interactions are known to play a central role in ensuring that a gene will instruct the cell to make the protein for which the gene codes only at appropriate times.

"Genetic surgery" was used by a team of molecular biologists in February 1987 to cure an inherited neurological defect of mice that causes tremors, convulsions, and early death. They transplanted into fertilized mouse eggs a gene that normal mice possess and that the affected mice lack. The aim of this study was to understand a disease at the molecular level—to analyze the defective molecule or molecules responsible for the effects of the disease and to use that knowledge to design a treatment or cure.

The research team, which included scientists from California Institute of Technology in Pasadena, Calif., and Harvard Medical School in Boston, used a strain of mice with a genetic defect called "shiverer." The defect prevents shiverer mice from producing a molecule called *myelin basic protein* that is essential for the normal transmission of nerve impulses. Myelin basic protein is a component of the myelin sheath—a white, fatty substance that surrounds and insulates *axons*. The axon, also called the nerve fiber, is the part of a nerve cell that conducts nerve impulses.

Previous studies had shown that the inability of shiverer mice to produce myelin basic protein is due to a *genetic mutation* (change in a gene) that results in the loss of most of the DNA that codes for this protein in normal mice. DNA is the molecule of which genes are made. It contains the instructions for all the characteristics of cells. Each gene acts as a blueprint for making a specific protein, including myelin basic protein.

The researchers theorized that they could correct the shiverer defect by injecting a piece of DNA containing the normal myelin basic protein gene into fertilized eggs from shiverer mice. First, they obtained the gene to be injected from normal mice and, by means of genetic engineering tech-

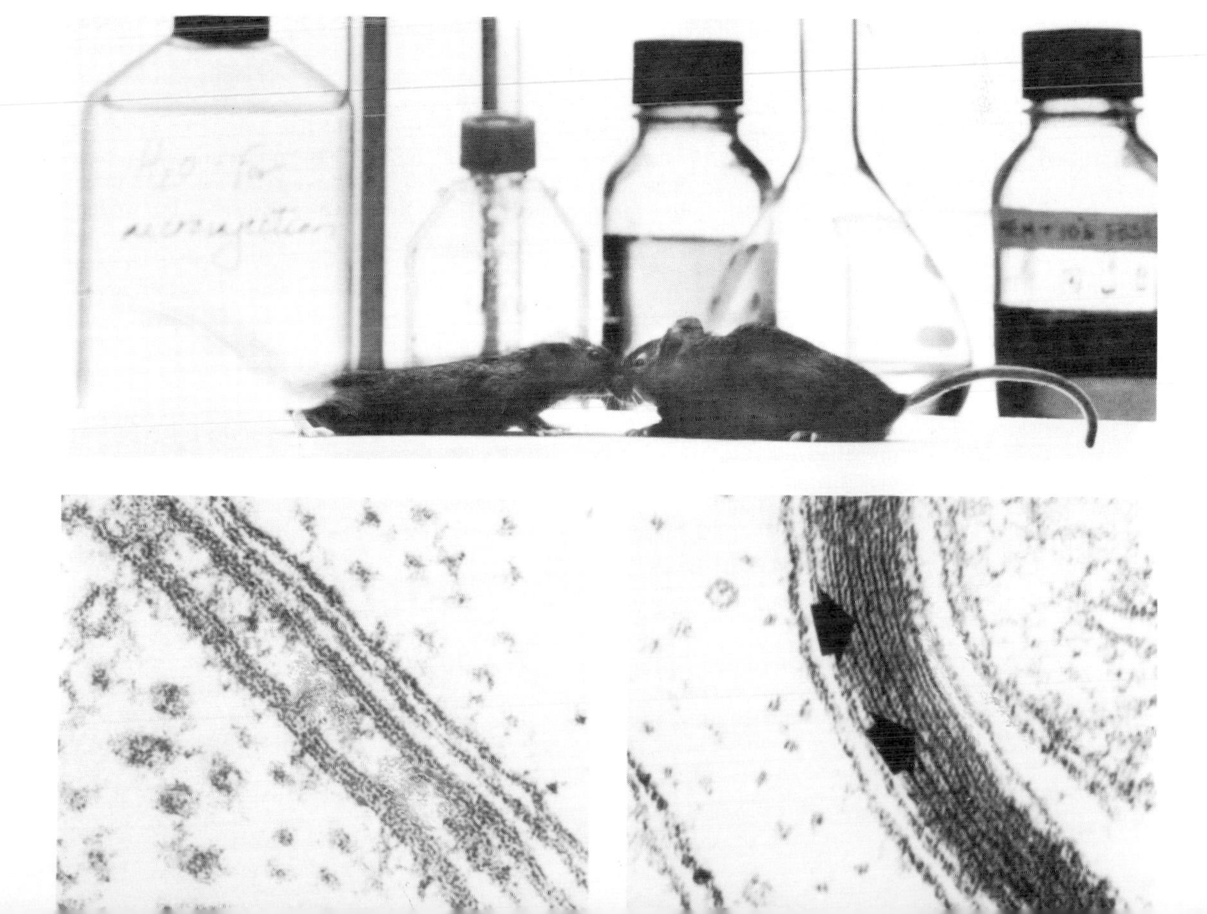

The trembling "shiverer" mouse, *below* (at left), has an inherited disease that can cause early death. The mouse lacks a gene required to help make the sheaths that insulate nerve cells. The cured shiverer mouse at the right developed from a fertilized mouse egg into which scientists inserted the sheath-making gene. A nerve cell from a cured mouse, *bottom right,* contains a well-formed sheath (arrows), while a nerve cell from an untreated mouse, *bottom left,* has a poorly formed sheath.

niques, grew it in large quantities in laboratory cultures of bacteria.

The scientists then injected tiny amounts of DNA into fertilized mouse eggs. In this procedure, they used very thin glass needles controlled by a mechanical device called a *micromanipulator*. Once injected with the gene, the eggs were inserted into the uterus of a female mouse.

Some of the mice that were born following the gene transplant had incorporated the gene for myelin basic protein into one of their *chromosomes* (the structures in the cell nucleus that contain the genes). The catastrophic effects of the defective gene largely disappeared in these animals. But the mice still displayed a slight tremor during certain types of movement—probably because the transplanted gene did not function quite as well as would a natural myelin basic protein gene.

The "cure" for "shiverer" can be passed along by breeding a mouse that contains the transplanted myelin basic protein gene. At least some of the offspring of that mouse will contain the transplanted gene and will be spared the severe symptoms of the shiverer defect.

Little is known about the normal role of myelin basic protein in the conducting of nerve impulses, or precisely why an absence of this substance causes severe problems. The next step in studying these questions would be to make slight changes in the injected DNA and determine the effects of these alterations on the DNA's ability to correct the defect in shiverer mice receiving the transplanted DNA.

For most proteins that play roles in such complex settings as the nervous system, scientists lack ways to test how a change in the protein's structure affects its function in a living animal. Experts noted that the techniques used in the shiverer mouse study point the way toward using a similar technique to study other proteins that play complex roles in the processes of health and disease. [Maynard V. Olson]

In WORLD BOOK, see BIOCHEMISTRY; CELL; MOLECULAR BIOLOGY.

Neuroscience

A new finding that helps explain why tissue transplanted from one brain to another seems to lose its effectiveness was announced in February 1987 by neuroanatomist Jeffrey M. Rosenstein of George Washington University Medical Center in Washington, D.C. Rosenstein and other investigators hope that research on brain-tissue transplants will lead to the development of treatments for many neurological disorders. Such treatments would depend on transplanted tissue producing needed brain chemicals.

Rosenstein transplanted brain tissue from rat fetuses into the brains of 34 anesthetized rats. He later found that certain proteins in the rats' blood, which are usually prevented from entering the brain by the *blood-brain barrier*, had penetrated into the transplant. The blood-brain barrier is a protective filtering system in the normal brain that prevents potentially harmful substances in the blood from entering brain cells.

Rosenstein's finding showed that implanting foreign tissue into a brain causes the blood-brain barrier to lose much of its effectiveness. As a result, material from circulating blood can enter the new tissue and disrupt its functioning. After spreading through the implant, the invading substances move into the surrounding brain tissue. As scientists gain a better understanding of this process they may be able to develop new ways of making brain-tissue transplants more effective.

Parkinson's disease. The first brain implants to relieve the symptoms of a neurological disease in human patients were announced by physicians in Mexico City, Mexico, in April 1987. The doctors reported that implants of adrenal-gland tissue relieved the symptoms of two patients suffering from Parkinson's disease, a progressive neurological disorder marked by tremors and loss of muscular control. See MEDICAL RESEARCH.

Schizophrenia and dopamine. A new insight into schizophrenia, the most common and disabling of the major mental illnesses, was reported in December 1986 by scientists at Johns

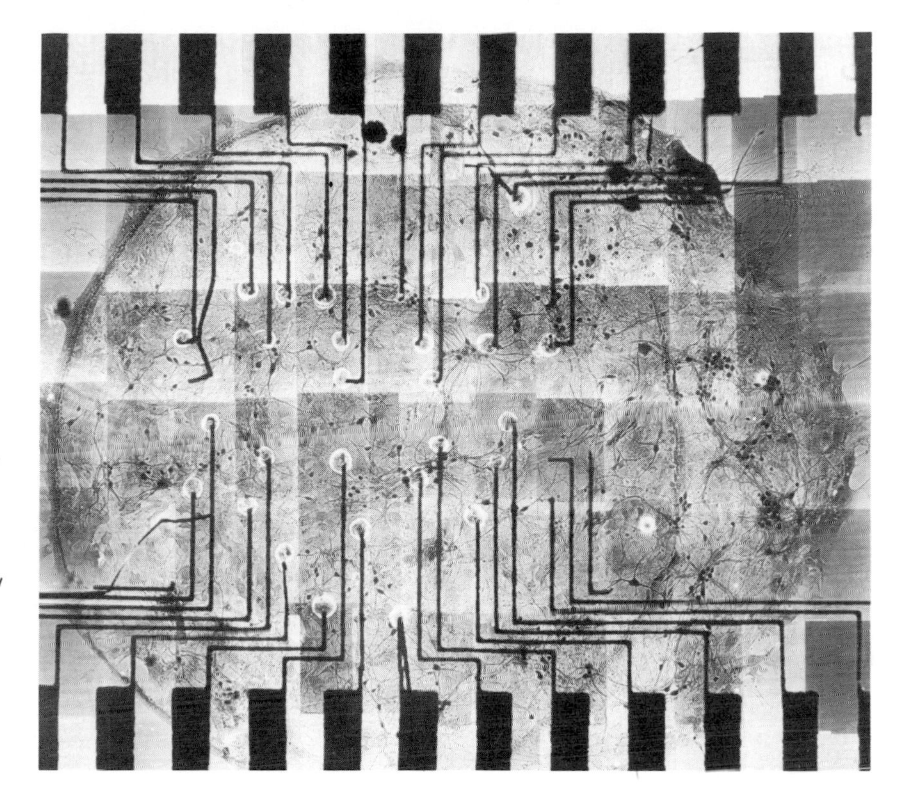

Mouse nerve cells grow on a computer chip 1.8 millimeters (0.07 inch) wide developed at North Texas State University in Denton. Electrical activity within the nerve cells is detected with microelectrodes (black lines). This type of study may help neuroscientists learn how networks of nerve cells process and store information.

Hopkins Medical Center in Baltimore. The researchers took brain scans of schizophrenia victims and found that the patients had a greater than normal number of *dopamine receptors*—molecules on the surface of brain cells that enable a brain chemical called dopamine to enter the cells. The researchers found large numbers of dopamine receptors in a part of the brain called the caudate nucleus whether or not a patient was taking drugs to control the schizophrenia. Such drugs work by blocking dopamine activity.

Earlier, researchers performing autopsies on the brains of schizophrenics had also found extra dopamine receptors but thought the excess must have been caused by drugs the patients had been taking. The new finding showed, however, that an excess of dopamine receptors is part of the disease itself. This knowledge may help researchers develop new medications for treating schizophrenia.

Genes and the brain. Researchers reported in 1987 that they had discovered human genes linked with two severe illnesses—manic-depressive disorder and Alzheimer's disease. Manic-depressive disorder is marked by extreme mood swings. In February 1987, researchers from three universities in the United States reported finding a gene that is directly related to the disease.

Also in February, investigators in the United States and West Germany announced that they had identified a gene that may play a part in the development of Alzheimer's disease, a disorder that destroys brain cells and causes severe mental deterioration. See GENETIC SCIENCE.

In a related finding, neuroscientists at Albert Einstein College of Medicine in New York City and Dartmouth-Hitchcock Medical Center in Hanover, N.H., in November 1986 announced that an abnormal protein found in the brains and spinal fluid of Alzheimer's disease victims may be a reliable *marker* (diagnostic indicator) for the illness. An accurate marker would be of great value to physicians because, without such an indicator, Alzheimer's disease

Neuroscience

Continued

cannot be diagnosed with absolute certainty until the patient's brain is examined after death for several specific abnormalities.

The researchers examined the fluid of nine patients diagnosed by their physicians as probably having Alzheimer's disease. The neuroscientists found a protein, which they called A68, in the fluid of eight of the nine patients. In 1985, the investigators had found high levels of this same protein in the brains of people who had died with Alzheimer's disease. The protein does not appear in the brains or spinal fluid of normal individuals. In addition to perhaps providing a sure diagnosis for Alzheimer's disease, further study of A68 may lead to new insights into this devastating and still incurable disorder.

Sonar for blind infants? Blindness in newborn infants is a devastating handicap and one that prevents certain parts of the brain from developing normally. But psychologist Bobby J. Sonnier of the University of California at Riverside reported in November 1986 that sonar may help substitute for a blind infant's eyes.

Sonnier and his colleagues fitted blind infant macaque monkeys with a sonar device that emitted sound signals and picked up the echoes of those signals bouncing off nearby objects. The device allowed the animals to sense their surroundings by hearing the echoes from the objects. The blind monkeys moved around their cages examining objects almost as actively as monkeys with normal vision. Blind monkeys without the sonar device moved about very little.

The scientists later compared the brains of the monkeys that had worn the sonar device with the brains of normal monkeys. All the monkeys' brains showed equal development, including the areas controlling spatial perception and movement.

Sonnier said the sonar "vision" research could lead to the development of sonar devices for blind human infants. [George Adelman]

In World Book, see Brain; Nervous System.

Nobel Prizes

The Royal Academy of Science and the Caroline Institute in Stockholm, Sweden, in October 1986 awarded Nobel Prizes in chemistry, physics, and physiology or medicine to eight scientists from Canada, Italy, Switzerland, the United States, and West Germany. Winners in each category shared prizes of $290,000.

Chemistry. The Nobel Prize for chemistry was shared by Dudley R. Herschbach of Harvard University in Cambridge, Mass., Yuan T. Lee of the University of California in Berkeley, and John C. Polanyi of the University of Toronto in Canada. The three were honored for developing techniques for studying chemical reactions at the molecular level.

Herschbach and Lee devised a method of colliding molecules in a vacuum and analyzing how the molecules link up. Polanyi invented a technique for studying the faint infrared light given off when molecules combine.

Herschbach was born in San Jose, Calif., in 1932. He earned his doctorate in 1955 at Harvard.

Lee, a naturalized U.S. citizen, was born in Taiwan in 1936. He received his doctorate at Berkeley in 1965.

Polanyi was born in Berlin, Germany, in 1929 but later moved with his family to England. He earned his doctorate at the University of Manchester in England in 1952 and has been at the University of Toronto since 1956.

Physics. The Nobel Prize for physics was shared by Ernst Ruska of the Fritz Haber Institute in West Berlin, West Germany, and Gerd Binnig and Heinrich Rohrer of International Business Machines Corporation's (IBM) research laboratory in Zurich, Switzerland. Ruska received half of the $290,000 award for designing the first electron microscope in the 1930's.

Binnig, a citizen of West Germany, and Rohrer, a Swiss citizen, shared the other half of the prize. They were cited for their 1981 design of another type of electron microscope, the scanning tunneling microscope.

The electron microscope illuminates specimens with a beam of electrons that is focused by magnetic "lenses."

Nobel Prizes

Continued

Biochemist Stanley Cohen, *above left,* of Vanderbilt University and neurobiologist Rita Levi-Montalcini, *above right,* of the Institute of Cell Biology in Rome, shared the 1986 Nobel Prize for physiology or medicine. Cohen and Levi-Montalcini were honored for research on substances that promote and help regulate the growth of cells.

The scanning tunneling microscope maps a surface with a needle that "floats" on a cushion of electrons.

Ruska was born in Heidelberg, Germany, in 1906. He earned his doctorate at Berlin Technical University in 1933 and joined the Haber Institute in 1955.

Born in Frankfurt, West Germany, in 1947, Binnig is the youngest of the 1986 Nobel laureates. He earned his doctorate at the University of Frankfurt before joining IBM in 1978.

Rohrer was born in Switzerland in 1933 and received his doctorate at the Swiss Federal Institute of Technology in 1960. He joined IBM in 1963.

Physiology or medicine. The Nobel Prize for physiology or medicine was awarded to biochemist Stanley Cohen of Vanderbilt University in Nashville, and neurobiologist Rita Levi-Montalcini of the Institute of Cell Biology in Rome. They were honored for their research on cell growth in the 1950's at Washington University in St. Louis, Mo.

Cohen and Levi-Montalcini discovered chemicals called *cellular growth factors* that promote and help regulate the growth of various kinds of cells. Levi-Montalcini was cited in particular for her discovery of a substance that stimulates the growth of nerve cells. Cohen, her collaborator in that research, later discovered a similar substance that plays a role in regulating the growth of a number of other kinds of cells. Their findings led to a better understanding of how cells operate and how they can malfunction.

Cohen was born in New York City in 1912. He earned his doctorate at the University of Michigan in 1948 and did postdoctoral work at Washington University. He joined the Vanderbilt faculty in 1959.

Levi-Montalcini, who is a citizen of both Italy and the United States, was born in 1909 in Turin, Italy. She earned an M.D. degree from the University of Turin in 1936 and moved to the United States in 1947. After retiring from the Washington University faculty in 1977, she returned to Italy to live and work. [David L. Dreier]

In WORLD BOOK, see NOBEL PRIZES.

Nutrition

The safety and effectiveness of dietary supplements continued to be topics of interest in nutrition during 1986 and 1987. Three supplements that aroused particular concern were fish-oil capsules, calcium supplements, and vitamin B_{12} nasal gels.

Fish-oil fad. Use of fish-oil supplements to lower the risk of heart disease gained greatly in popularity. Consumers in the United States spent about $30 million on fish-oil capsules in 1986, and sales were projected to increase by as much as $200 million in 1987. The popularity of fish oil stemmed from research in the 1970's indicating that fish-eating Eskimos have few heart problems. More solid evidence came out in 1985, when two studies published by *The New England Journal of Medicine* suggested that eating fish or fish oil reduces the risk of heart attack.

The substance in fish that apparently protects against heart disease is an element of the fish's body fat known as *omega-3 fatty acids*. These fatty acids, like corn oil, are *polyunsaturated fats*, a healthful kind of fat that lowers the blood's levels of cholesterol and triglycerides, fatty compounds that may contribute to heart disease.

Omega-3 fatty acids also decrease the stickiness of the blood cells involved in clotting and thus reduce the ability of the blood to form clots. This effect may benefit individuals threatened by *coronary thrombosis*, a blood clot that blocks an artery of the heart, which is a primary cause of heart attacks. The anticlotting effect is a disadvantage, however, to people with high blood pressure who are at risk for strokes. Many strokes result when a blood vessel in the brain bursts and causes bleeding into the brain. Such strokes will be more severe if blood clots, which stop the bleeding, fail to form quickly.

The effect on clotting is not the only reason scientists are concerned about the excessive use of omega-3 fatty acids. Endocrinologist Alan Failor of the University of Washington in Seattle reported in 1986 that omega-3 fatty acids might harm patients with a common inherited disorder characterized by high cholesterol levels. Use of omega-3 fatty acids actually increased the risk of heart disease in some of these patients by increasing their levels of *low-density lipoproteins*, the substances that deliver cholesterol to the cells.

Calcium supplements were another popular but controversial product in 1986 and 1987. Sales of calcium supplements in the United States jumped from $50 million in 1983 to nearly $140 million in 1985 and continued to grow. This interest in adding calcium to the diet stemmed largely from concerns about *osteoporosis*, a disorder that causes bones to become brittle and fracture easily. Osteoporosis is a major health problem for elderly women. Because of the high concentration of calcium in bones, many people have assumed that increasing their calcium intake will decrease their risk of osteoporosis or at least the rate at which osteoporosis may develop.

In 1986 and 1987, however, scientists expressed doubts about whether eating more calcium will prevent osteoporosis. Many other nutritional factors are involved in the development and maintenance of healthy bones.

Biochemist Paul D. Saltman of the University of California at San Diego presented evidence at the September 1986 meeting of the American Chemical Society in Anaheim, Calif., that the mineral manganese may have an important role. Saltman found that manganese affects the rate at which calcium is deposited in bone and reabsorbed from it. He also found that adequate manganese intake is essential to maintain the framework of protein fibers that surround the mineral matter in bone. These processes help prevent or control osteoporosis.

Another important nutritional factor in preventing osteoporosis is adequate intake of vitamin D. Biochemist Hector F. DeLuca of the University of Wisconsin-Madison reviewed the importance of this vitamin at the annual meeting of the Federation of American Societies for Experimental Biology in Washington, D.C., in April 1987. Scientists have long known that vitamin D is essential in order for calcium to be absorbed and used by the body.

The female sex hormone estrogen also plays a major role in preventing osteoporosis. Older women use calcium less efficiently than do young

A U.S. Department of Agriculture chemist compares a cake made partly with a new no-calorie, high-fiber filler, foreground, with a cake made with only ordinary flour. The new filler could lead to commercial baked goods containing fewer calories.

women because of a decrease in body estrogen following *menopause*—the time in a woman's life when menstruation stops. Probably because of this lack of estrogen, calcium supplements are not effective in preventing calcium loss from bone in older women, according to research by Danish scientists reported in *The New England Journal of Medicine* in January 1987. In a two-year study conducted at the University of Copenhagen, Denmark, postmenopausal women lost bone rapidly even when they took 2,000 milligrams of calcium daily—more than four times the average daily intake. Their bone loss was nearly equal to that of women taking *placebos*, look-alike tablets with no active ingredients. Women who received estrogen therapy but no supplemental calcium had no bone loss.

Vitamin B_{12} and energy. At the First International Congress of Vegetarian Nutrition in Loma Linda, Calif., in March 1987, physician and nutritionist Victor Herbert of Mount Sinai Medical Center and Bronx Veterans Administration Medical Center in New York City strongly criticized the sale and use of vitamin B_{12} nasal gels as an energy booster. Health food stores in California first marketed the gels, which became a nutritional fad that spread throughout the United States. Herbert acknowledged that vitamin B_{12} probably can be absorbed through the soft, moist mucous membrane lining the nasal passages and can thus reach the bloodstream. Whether vitamin B_{12} comes from nasal gels, food, or vitamin supplements, however, it does not increase an individual's energy level. As a result, Herbert warned, money spent for vitamin B_{12} energy boosters is money wasted.

Vitamin B_{12} occurs naturally only in foods of animal origin, but it is commonly added to ready-to-eat breakfast cereals and other enriched food products. As a result, even pure vegetarians who eat no meat, eggs, fish, or dairy products are unlikely to develop actual deficiencies of the vitamin. Microorganisms in the human intestinal tract also manufacture B_{12}, further reducing the risk of deficiency, according to Herbert. [Constance Virginia Kies]

In WORLD BOOK, see NUTRITION.

Oceanography

An unusually large undersea *plume*—a column of heated water, gases, and minerals that rises from an opening in the sea floor—was discovered in August 1986 over the Juan de Fuca Ridge. This undersea mountain range lies about 400 kilometers (260 miles) off the coast of Oregon. Oceanographers from the Pacific Marine Environmental Laboratory in Seattle found the huge plume while aboard the National Oceanic and Atmospheric Administration's research vessel *Discoverer.* Using instruments in a towed submersible vessel, the oceanographers determined the water temperature and mineral composition of the plume.

The plume extended to about 1,000 meters (3,300 feet) above the sea floor, much higher than most plumes. The plume measured about 18 kilometers (11 miles) in diameter and some 600 meters (2,000 feet) deep at its thickest point.

Water temperatures in most plumes are no more than a few hundredths of a degree higher than the temperature of the surrounding seawater. But water in the giant plume was as much as one-fourth of a degree higher than the surrounding water temperature. Although the temperature difference was small, in such a large body of water it represents a huge amount of heat energy. The scientists calculated that the heat contained by the water was equal to 10 billion kilowatt-hours of energy, or nearly twice the annual energy production of a large hydroelectric power plant.

The water mass was also unusual because it carried some 5,400 metric tons (6,000 short tons) of mineral particles, or 20 times more particles than are normally found in deep ocean water.

The scientists attempted to explain their finding by proposing that, in contrast to the usual slow and steady release of heated water from cracks in the ocean floor, the giant plume was the result of a single, very large undersea volcanic eruption. The plume might also have been created by the release of a large pulse of heat and particles from a field of sea-floor *vents* (volcanolike formations resembling chimneys).

The finding may cause oceanographers to change their theory that the formation of volcanoes is a gradual process. Undersea volcanoes may actually erupt for a few years, then lie dormant for many years. In addition, if giant plumes are common and release large amounts of gases, minerals, and heat into the oceans, then plumes may have played a significant role in determining Earth's climate.

Gas diets. Scientists have for the first time identified an organism that draws its food energy from natural gas, according to a September 1986 report by oceanographers at the University of California in Santa Barbara and Texas A&M University in College Station. The scientists found a type of mussel that seems to meet its needs for carbon, an element necessary to sustain life, by consuming methane, the major component of natural gas. The mussels were found at a depth of 700 meters (2,300 feet) near the Louisiana Slope of the Gulf of Mexico, where oil seeps out of cracks in the ocean floor. Survival there calls for extraordinary adaptations to the environment. The water contains high concentrations of chemicals such as sulfur and hydrocarbons, and it is too dark for *photosynthesis* (the process by which plants convert sunlight into food energy).

Laboratory study of the mussels showed that their gills were covered with bacteria. The oceanographers discovered that the bacteria consumed methane and changed it into organic carbon in amounts large enough to fill the carbon needs of both the bacteria and the mussels. The oceanographers' finding adds to the growing list of deep-sea animals that derive food energy from chemical conversion rather than from photosynthesis.

Stable icecaps. Sample cores of sea-floor sediment taken by oceanographers on the research drill ship *JOIDES Resolution* suggest that polar ice sheets may be more stable than previously thought, according to an April 1987 report. Oceanographers from Birmingham University in England and the University of Rhode Island in Kingston directed the recovery of cores from nine locations in the Weddell Sea off Antarctica. The composition of the cores indicated that the extensive ice sheets covering Antarc-

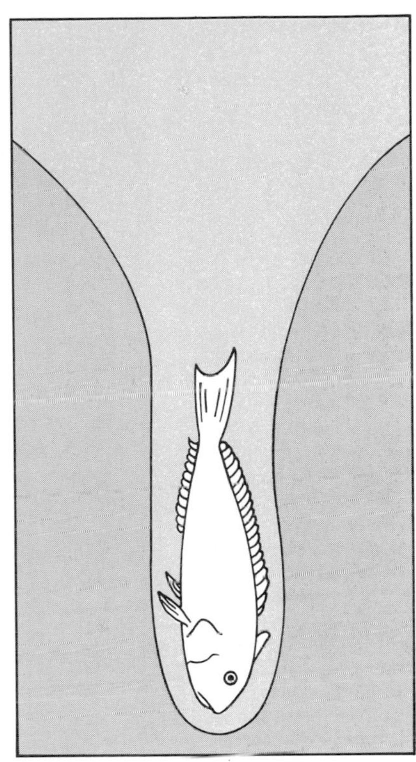

Oceanography

Continued

A tilefish, *above left,* prepares to enter its burrow, where it hides from sharks, as shown in the drawing *above right.* In 1986, oceanographers were surprised to find that tilefish—which probably use their mouths to scoop out burrows—had eroded and reshaped the continental shelf extending from Cape Cod to North Carolina. Far from being a smooth layer of sediment, as scientists previously thought, the shelf is in some areas pocked with more than 1,200 burrows per square kilometer (3,000 per square mile).

tica formed gradually over millions of years rather than suddenly. The ice also seemed to be more stable—that is, less likely to melt—than ice found in frigid locations on the North American continent.

The cores showed that the great ice sheets over the pole began to form about 15 million years ago. The icecap first covered the land mass on East Antarctica; several million years later, ice formed over West Antarctica, an *archipelago*, or chain of islands. About 7 million years later, the West Antarctic ice sheet melted and then froze again several times. It stabilized in its current form about 5 million years ago.

If this ice sheet is as stable as the core evidence suggests, climate scientists may need to revise their estimates about the impact of global warming. These scientists have predicted that a slight warming of Earth's atmosphere caused by air pollution may melt some of the polar ice sheet. This would cause sea levels to rise, flooding coastal plains. In fact, the stable polar ice sheet may melt more slowly than climate scientists have predicted—producing a less drastic outcome.

Lead pollution. The first accurate measurement of the lead content of sediment in deep parts of the North Atlantic Ocean was reported in March 1987 by French scientists who used the research vessel *Le Suroit* to collect a sediment core from the deep North Atlantic Ocean, about 400 kilometers (250 miles) from land at a depth of 3,650 meters (11,000 feet).

The oceanographers found increased levels of lead in the sediment core, suggesting that a large percentage of lead has built up in the North Atlantic. The findings may help scientists to chart the movement of lead more accurately in the ocean and determine the benefit to the ocean environment of the gradual elimination of the use of leaded gasoline in the United States. [Lauriston R. King]

In the Special Reports section, see DRILLING UNDER THE SEA. In WORLD BOOK, see ENVIRONMENTAL POLLUTION; OCEAN.

Paleontology

The Newest Oldest Bird

Fossils of a birdlike animal that lived 225 million years ago, called *Protoavis, below,* may represent the oldest known bird. *Protoavis'* thin birdlike hollow bones and breastbone for the attachment of flight muscles, combined with dinosaurlike hind legs and tail, *below right,* are strong evidence that birds evolved from dinosaurs. *Protoavis* may replace *Archaeopteryx, bottom right,* previously the oldest known bird, as the theoretical link between dinosaurs and birds.

One of the most exciting and significant paleontological findings of 1986 and 1987 was the discovery in Texas of what appears to be the oldest fossil bird. The discovery of two incomplete skeletons of a birdlike animal now called *Protoavis* was announced in August 1986 by paleontologist Sankar Chatterjee of Texas Tech University in Lubbock. Found in 225-million-year-old rocks from the Triassic Period, the crow-sized skeletons are about 75 million years older than *Archaeopteryx,* previously the oldest known fossil bird. According to Chatterjee, the Texas fossils are strong evidence for the theory that birds evolved from dinosaurs.

Many paleontologists have long considered *Archaeopteryx* the "missing link" between dinosaurs and birds. Chatterjee contended, however, that *Protoavis* had more birdlike characteristics than *Archaeopteryx.* These characteristics include hollow bones and a large breastbone for the attachment of flight muscles. Although Chatterjee found no fossil impressions of feathers, he did report finding knobs on some of the bird bones to which he believes the quills of feathers may have been attached. *Protoavis'* hind legs and tail resembled those of dinosaurs. Chatterjee also argued that *Archaeopteryx,* which seemed to have been less suited for flight than *Protoavis,* was a side branch on the tree of bird evolution.

Some paleontologists, however, said that *Protoavis* is as old as the first dinosaurs. They argued that *Protoavis* is actually evidence that birds and dinosaurs evolved at about the same time.

Oldest land animals. Evidence of what may have been the first land animals was reported in January 1987 by geologists Gregory J. Retallack and Carolyn R. Feakes of the University of Oregon in Eugene. The evidence consists of a network of cylindrical fossil burrows from about 2 to 21 millimeters (0.08 to 0.83 inch) in diameter found in central Pennsylvania.

The burrows, dated to 440 million years ago during the Ordovician Period, are from 20 million to 25 million years older than any previous evidence for animal life on land. Because the

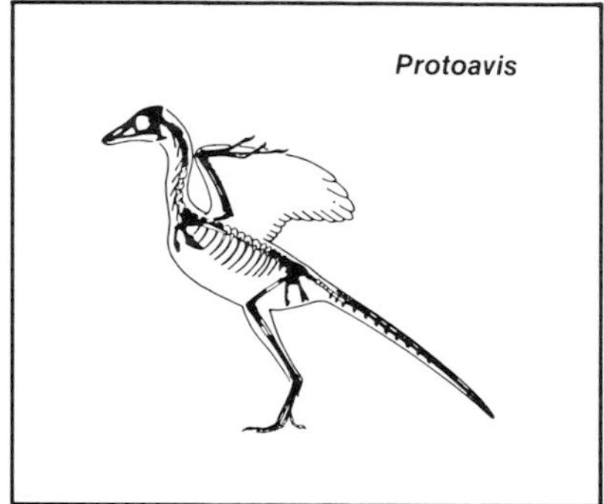

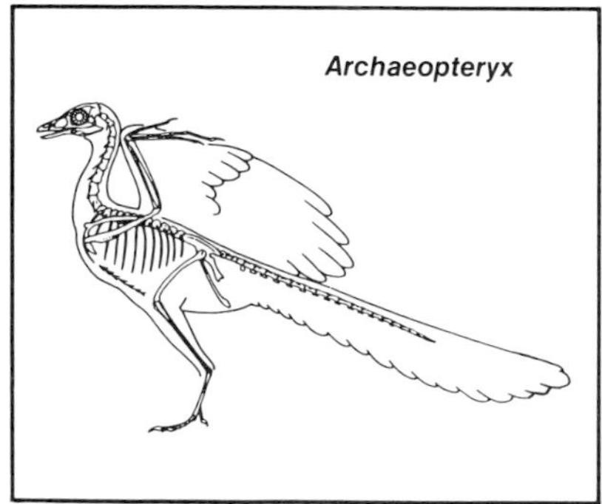

Paleontology

Continued

A debate about human-like footprints found in a Texas stream bed along with dinosaur tracks came to an end in mid-1986. A dinosaur expert who examined the footprints found faint impressions of dinosaur toes (arrow), proving the tracks were not human. Both sides then agreed that the tracks were not made by people but by an unknown type of dinosaur that walked on its heels and soles rather than only on its toes, like other dinosaurs.

fossil burrows resemble burrows made by modern millipedes, Retallack and Feakes suggested that they may have been made by an ancient millipede.

Dinosaur tracks revealed. Mysterious footprints mingled with dinosaur tracks in a Texas riverbed were not made by human beings as some scientific creationists have argued but by an unknown type of dinosaur with an unusual way of walking. That was the conclusion of an analysis of the footprints reported in June 1986 by Glen Kuban of North Royalton, Ohio, an amateur expert in dinosaur tracks.

The footprints, which somewhat resemble human footprints, were found in 1938 in the bed of Paluxy Creek near Glen Rose, Tex. Scientific creationists, who believe the world began through an act of creation perhaps only thousands of years ago, have cited the footprints as evidence that human beings and dinosaurs were created and existed at the same time. Nearly all scientists believe that dinosaurs became extinct long before the appearance of human beings or their remote ancestors. Kuban's discovery of faint impressions of dinosaur toes led both religious fundamentalists and scientists to agree that the footprints were not those of a human being. They apparently were made by a type of dinosaur that walked on its heels and soles, as do human beings, rather than on its toes, the way most dinosaurs did.

A new dinosaur. The discovery of a nearly complete skeleton of a very large, previously unknown, flesh-eating dinosaur was reported in November 1986 by paleontologists at the British Museum in London. Found in a clay pit in southern England, the skeleton was dated to the early Cretaceous Period, about 125 million years ago. Few other large dinosaur fossils from this period have been found.

The dinosaur—named *Baryonyx walkeri* after William Walker, the amateur fossil collector who found it—was about 9 meters (30 feet) long, stood 3.5 to 4.5 meters (12 to 15 feet) tall on its hind legs, and weighed nearly 1.8 metric tons (2 short tons). The dinosaur had enormous curved claws,

Paleontology
Continued

which may have been used to spear fish the way grizzly bears do today.

Mass extinctions. Research in 1986 continued to strengthen the link between mass extinctions at the end of the Cretaceous Period about 65 million years ago and the collision of a large meteorite or asteroid with Earth.

In November 1986, paleobotanists Jack A. Wolfe and Garland R. Upchurch of the United States Geological Survey in Denver reported evidence that the Western United States experienced a period of abrupt cooling at the end of the Cretaceous Period. Some scientists have suggested that the meteorite's impact would have thrown up a huge cloud of dust that blocked much of the sun's light and heat.

The scientists charted the extinction patterns of plants at the end of the Cretaceous by studying the types and shapes of fossil leaves found at the Cretaceous-Tertiary (K-T) boundary in the Western United States. (The K-T boundary is the dividing line between the Cretaceous Period and the Tertiary Period, which followed. It is usually marked by a layer of clay.) They found that the proportion of cool-climate plants, compared with warm-climate plants, increased during this period. They also found that more plants became extinct in the Southwest than in the Northwest. They concluded that this occurred because plants in the South were more susceptible to a sudden drop in temperature.

Major changes in plant life at the K-T boundary also occurred in Japan, according to research reported in September 1986 by a team of geologists at Yamagata University in Japan. In the thin clay layer just above the K-T boundary, the researchers found that the proportion of pollen from flowering plants and conifer trees temporarily decreased sharply while the spores of ferns, which prefer damp, shady areas, suddenly became common. Shortly afterward, however, species of flowering plants and conifers again became more abundant, according to the researchers. [Carlton E. Brett]

In WORLD BOOK, see DINOSAUR; PALEONTOLOGY.

Physics, Fluids and Solids

What may be one of the most important breakthroughs in the history of the physics of solids occurred in late 1986 and early 1987 when researchers developed materials that become *superconductors* at relatively high temperatures. A superconductor conducts or carries electric current (a flow of electrons) without resistance when chilled to a temperature far below 0° Celsius or Fahrenheit.

This breakthrough could bring about tremendous savings in electrical energy. It could also lead to the development of a wide range of sophisticated scientific and medical devices and commercial products. But first, researchers must make practical conductors such as wires and *films* out of the new materials. (A conducting film is a thin layer of conducting material that could be deposited on a computer chip.)

Overcoming resistance. A conventional conductor, such as a copper wire, resists current by absorbing energy from the flowing electrons. This happens because electrons collide with atoms in the crystal structure of the conductor. Each collision transfers energy from an electron to an atom, causing it to vibrate more, thus heating the conductor.

Whatever is supplying the electricity—usually an electric generator or a battery—must work harder to make up the lost energy. Furthermore, because the absorbed energy heats the conductors, many electrically driven devices would burn out if they did not have cooling systems using fans or blowers to circulate air, or pumps to circulate liquid coolants. These cooling systems also require energy.

Electrons flowing in a superconductor do not collide with the material's atoms. A superconductor therefore conducts electricity without resistance. With zero resistance there is no energy loss to make up and no need for devices to carry away heat.

There is, however, a "catch" to superconductivity. A superconducting wire will not actually superconduct unless it is cooled to almost *absolute zero* (−273.15°C or −459.67°F.). This re-

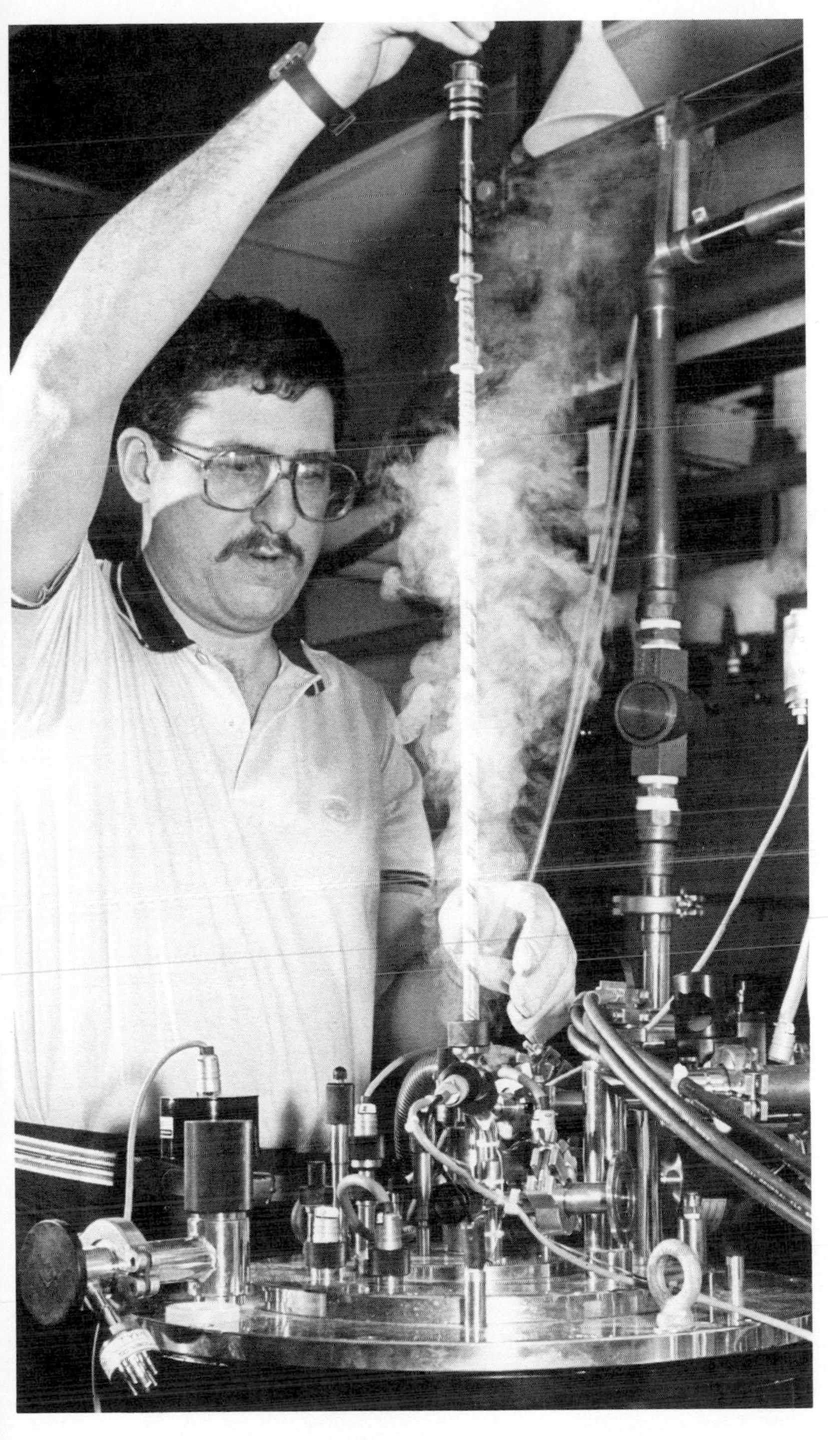

A physicist at Argonne National Laboratory pulls a ceramic superconducting wire from a vat of cold liquid nitrogen. A variety of new superconductors, which carry electricity without resistance, were introduced in late 1986 and early 1987. The new materials become superconducting at temperatures of at least −180°C (−292°F.), much higher than previously possible.

quirement calls for a cooling system far beyond the usual fans, blowers, and pumps. In fact, until the new superconductors were developed, only one substance—liquid helium—could cool materials efficiently to temperatures necessary for superconductivity.

Liquid helium, which boils at −269°C (−452°F.), is expensive, costing $3 per 1 liter (approximately 1 quart). In addition, liquid helium cooling systems are expensive because they must be designed to prevent the heat in the surrounding air from boiling the helium.

Because of the expense of cooling, conventional superconductors have found few applications other than in electromagnets for *particle accelerators* (atom smashers) and magnetic resonance imaging machines for producing pictures of tissue in the human body. In addition, superconducting electromagnets have been used to elevate experimental trains a few centimeters above the rails so that the trains run without friction.

By developing the new materials, scientists have jumped a technological hurdle. The new superconductors can be cooled with liquid nitrogen, which boils at −196°C (−321°F.). Liquid nitrogen sells for only 6 cents per liter and is much easier to maintain below its boiling temperature than is liquid helium.

Rising temperatures. The first announcement of a high-temperature superconductor was made in October 1986. Physicist K. Alex Müller and crystallographer J. Georg Bednorz of the International Business Machines Corporation (IBM) Research Center in Zurich, Switzerland, reported that a mixture of the chemical elements lanthanum and barium and the chemical compound copper oxide has a *critical temperature* (the temperature at which a material becomes superconducting) of −243°C (−405°F.). This announcement triggered a flood of experiments in laboratories throughout the world.

The second major announcement came in January 1987. Physicist Paul C. W. Chu and his colleagues at the University of Houston reported that they had raised the critical temperature of a similar mixture to −220.7°C (−365.3°F.).

A Flight into History

Aviation history was made on Dec. 23, 1986, when test pilots Richard G. Rutan and Jeana Yeager became the first to fly around the world nonstop without refueling. They made their record-setting flight in a custom-built aircraft called *Voyager*. *Voyager*'s odd-shaped body and wings contained no metal, making it light enough to carry sufficient fuel for its around-the-world journey.

Voyager was designed by Richard Rutan's brother, Burt, an aircraft designer. The structure of many of Burt Rutan's airplanes contain some *composites*—strong, lightweight plasticlike materials. But *Voyager* was the first aircraft ever built entirely of composites.

Voyager's outer skin and skeletal structure were formed from layers of two strong composite materials—a fiber and a stiff honeycomb, both made from *graphite*, a form of carbon. The aircraft has two long wings, with a total wingspan of 33.77 meters (110.8 feet). This is almost four times the wingspan of a typical single-engine private airplane.

Its cockpit, only 2.3 meters (7.5 feet) long and 1 meter (3.3 feet) wide, is located at the end of the central tube-shaped *fuselage* (body). *Voyager*'s two engines—one at the front and one at the rear—received fuel from tanks located in the wings and fuselage. The full tanks held more than 3,400 liters (900 gallons) of fuel, weighing approximately 3,200 kilograms (7,000 pounds). *Voyager*'s structure weighed only 425 kilograms (938 pounds), but radar and other instruments, the fuel, and the pilots increased *Voyager*'s weight to 4,427 kilograms (9,760 pounds) at take-off.

The plane was so heavy that it almost did not make it off the ground as it rolled along the desert runway at Edwards Air Force Base in California on December 14 just after 8 A.M. With Rutan at the controls and Yeager behind him, *Voyager* slowly gathered speed. Its two engines—a 130-horsepower (97-kilowatt) air-cooled engine in front for take-off and maneuvering, and a 110-horsepower (82-kilowatt) water-cooled engine in the rear for take-off and cruising—strained at full power.

When *Voyager* moved forward down the runway, its wing tips drooped, and as the plane gathered speed they began dragging on the runway. Pieces of the composite outer skin tore away from the wing tips, exposing their blue foam core. A navigation light at the tip of the right wing ripped off.

With only about 300 meters (1,000 feet) of runway left, Rutan pulled back on the controls, and *Voyager* finally lifted off. Although initially worried about the damage caused during take-off, the pilots reported that *Voyager* handled well enough to continue the flight. Rutan shut down the forward engine as planned. The engine in the rear was enough to keep the plane at an altitude of between 2,100 and 3,300 meters (7,000 and 11,000 feet).

For the next nine days, Rutan and Yeager fought wind, storms, and exhaustion. They also fought *Voyager*. Rutan later said that *Voyager* was a difficult airplane to fly. A major problem was keeping *Voyager* balanced by pumping fuel back and forth among its 17 tanks.

Voyager's route took them west across the Pacific Ocean, then over Malaysia. To avoid an encounter with Typhoon Marge, they were forced to turn north, across India. But the stormy weather had some benefit—it gave *Voyager* unexpectedly strong tail winds during most of its flight, thus saving on fuel.

On December 19, the plane reached Africa. Tropical storms forced Rutan and Yeager to zigzag across the continent. The turbulence continued as they crossed the Atlantic Ocean. *Voyager* turned northwest when it reached the coast of South America and crossed Central America before turning north to California for the final leg of the trip up the Pacific coast.

At 8:06 A.M. on December 23, *Voyager* landed back at Edwards Air Force Base. *Voyager* had circled the world in 9 days, 3 minutes, and 44 seconds. It flew 40,253 kilometers (25,012 miles) nonstop.

In May 1987, *Voyager* made its last flight—inside a United States Air Force cargo plane that delivered it to the Smithsonian Institution in Washington, D.C. There it joins the Wright brothers' plane, *The Spirit of St. Louis*, and other craft that made aviation history. [Jim Schefter]

Approaching Edwards Air Force Base, Calif., on Dec. 23, 1986, *Voyager* nears the end of its record-setting flight around the world.

In March 1987, both Chu's team and physicists led by Mau-Kuen Wu of the University of Alabama in Huntsville announced the discovery of superconductivity in a mixture in which lanthanum was replaced by the element yttrium. The Houston and Alabama scientists obtained a critical temperature of −180°C (−292°F.).

In experiments at other laboratories, similar mixtures showed signs—but not convincing evidence—of superconductivity at even higher temperatures. Labs reporting such indications included the Chinese Academy at Beijing (Peking); the National Research Unit for Metals in Tokyo; Tokyo University; The University of California in Berkeley; AT&T Bell Laboratories in Murray Hill, N.J.; Tata Institute of Fundamental Research in Bombay, India; Bhabha Atomic Research Centre in Trombay, India; and Clarendon Laboratory, located in Oxford, England.

A new type of solid called a *quasicrystal, above* (highly magnified), continued to puzzle physicists in 1986 and 1987. Groups of atoms in this material form crystallike shapes in which five flat surfaces meet at one point. According to the laws of geometry, however, it is impossible to create a crystal with this so-called *fivefold symmetry.*

The magnitude of the scientific interest in superconductors was reflected in the attendance at the Special Panel Discussion on Novel Materials and High-Temperature Superconductivity organized by the American Institute of Physics in New York City on March 18, 1987. About 2,000 physicists crowded into a room to hear a series of five-minute presentations on superconductivity while many others watched television screens outside. They heard Constantin Politis of the Karlsruhe Nuclear Research Center in West Germany report indications of superconductivity at a temperature of −148°C (−234°F.).

In May 1987, Juei-Teng Chen and his colleagues at Wayne State University in Detroit reported some evidence of superconductivity in a complex substance chilled to −33°C (−27°F.). The effect apparently occurred in microscopic "sandwiches" made up of insulating material located between layers of a superconductor.

Superconducting ceramics. The new superconductors are *ceramics*—materials that are neither metals nor plastics. The superconductors are brittle and are difficult to make into films or wires suitable for technical applications.

In spite of the difficulty, researchers at IBM's Yorktown Heights Laboratory and Stanford University in Palo Alto, Calif., have produced superconducting films. The Stanford researchers have created superconducting *tunneling devices,* components that probably will be used as electronic switches in future computers. Scientists at Argonne National Laboratory near Chicago and Toshiba Corporation in Japan have made wires and tapes that are somewhat flexible, but still very brittle.

Super dilemma. As a result of the new superconductivity developments, the U.S. Department of Energy (DOE) faced a dilemma in the planning of a particle accelerator called the Superconducting Super Collider (SSC). This machine would accelerate two beams of subatomic particles in opposite directions around a ring about 84 kilometers (52 miles) in circumference, then force the beams to collide. Aided by gigantic detection devices and sophisticated computer systems, physicists would examine the debris of the collisions for new information on the fundamental particles and forces of nature.

Thousands of electromagnets would "steer" the accelerated particles around the circular path. Accelerators need extremely strong magnets, because the stronger the magnets, the smaller the required circumference of the path.

Present plans for the SSC call for wire made of a conventional superconductor in the electromagnet coils. Several scientists, however, fearing that the SSC magnets would be rendered obsolete by the development of the new ceramic superconductors, have proposed that construction of the SSC be delayed for a few years. They believe that the brittleness and other limitations of the ceramics will be overcome soon and the new materials will become usable in superconducting magnets.

Many other scientists oppose a delay. They favor going ahead with a proven technology rather than waiting for the development of an uncertain technology. [Alexander Hellemans]

In the Special Reports section, see EXPLORING THE STATES OF MATTER. In WORLD BOOK, see PHYSICS; SUPERCONDUCTIVITY.

Physics, Subatomic

Two huge underground particle detectors built to test the stability of matter gave subatomic researchers an unexpected and exciting bonus at 12:35 P.M. Eastern Standard Time on Feb. 23, 1987. They detected, for the first time in history, neutrinos that originated outside the solar system.

What made this event so exciting was the source of these neutrinos—supernova 1987A, the first stellar explosion visible to the naked eye since 1604. Supernova 1987A is located in the Large Magellanic Cloud, a small galaxy about 170,000 *light-years* from Earth. A light-year is the distance light travels in one year—about 9.46 trillion kilometers (5.88 trillion miles). Because it took the light and neutrinos 170,000 years to reach Earth, scientists know the explosion took place about 170,000 years ago. See ASTRONOMY, EXTRAGALACTIC (Close-Up).

A burst of neutrinos. Neutrinos are subatomic particles closely related to the electron. Unlike the electron, however, they carry no electric charge. Because of this, they interact only feebly with matter and so are extremely difficult to detect. A beam of neutrinos can pass through Earth without any significant loss in strength.

The two particle detectors that made history in February 1987 are large pools of water located in deep mines. One is in a salt mine near Cleveland; the other, in a lead mine near Kamioka, Japan. Covering the walls of each detector are *photomultiplier tubes*, electronic "eyes" sensitive to the faint flashes of light produced when electrically charged particles pass through water. Signals from the tubes are recorded by computers.

On February 23, millions of billions of neutrinos swept through each detector in a span of a few seconds. The neutrinos themselves produced no light. A few neutrinos, however, interacted with protons in the water, picking up the protons' positive charge to become *positrons*, the antimatter counterparts of electrons. (An antimatter particle is exactly like its matter counterpart but has an opposite electric charge.) The positrons then flashed through the water, radiating light.

These underground particle detectors were built in the early 1980's to search for the breakup of protons. A theory had predicted that each year, 1 proton in 10^{30} would disintegrate. (The number 10^{30} is the same as the numeral 1 followed by 30 zeros.) One metric ton of matter contains about one-third this number of protons, so the detectors had to contain many tons of water—1,500 metric tons at Kamioka, and 8,000 metric tons in the salt mine. (One metric ton equals 1.1 short tons.) The detectors were built in deep mines to avoid being overwhelmed by the constant hail of radiation that reaches Earth from space.

The neutrinos were released from the distant star at the precise instant that the supernova began. So the neutrinos' time of arrival on Earth provides a comparison point for researchers to use in measuring the duration of various processes involving neutrinos that occur in supernovas. For example, astronomers believe that an explosion such as the one that produced the neutrinos starts when the core of a massive star collapses under its own weight. This collapse releases a gigantic amount of energy, more than 90 per cent of which is carried by neutrinos. The details of the neutrino burst from supernova 1987A provide the first data ever from a collapsing core.

In the explosion that formed the supernova, the outer layers of the star blew off into space at 17,000 kilometers (10,500 miles) per second. Upon first reaching Earth, the visible light from the cosmic fireball took nearly 12 hours, however, to grow bright enough to catch the attention of astronomers. Having the starting time neatly marked by the neutrinos will make it easier to reconstruct exactly what happened and when.

The detection of the neutrinos also shed some light on the question of whether neutrinos have any mass. All the neutrinos arrived within a few seconds, indicating that they all moved at the same speed. Their speed was, at most, 10-billionths of 1 per cent slower than the speed of light. Objects that have mass travel slower than light. According to calculations, neutrinos must be at least 50,000 times lighter than electrons, if they have any mass.

Big year for accelerators. Research with colliding beams of high-energy

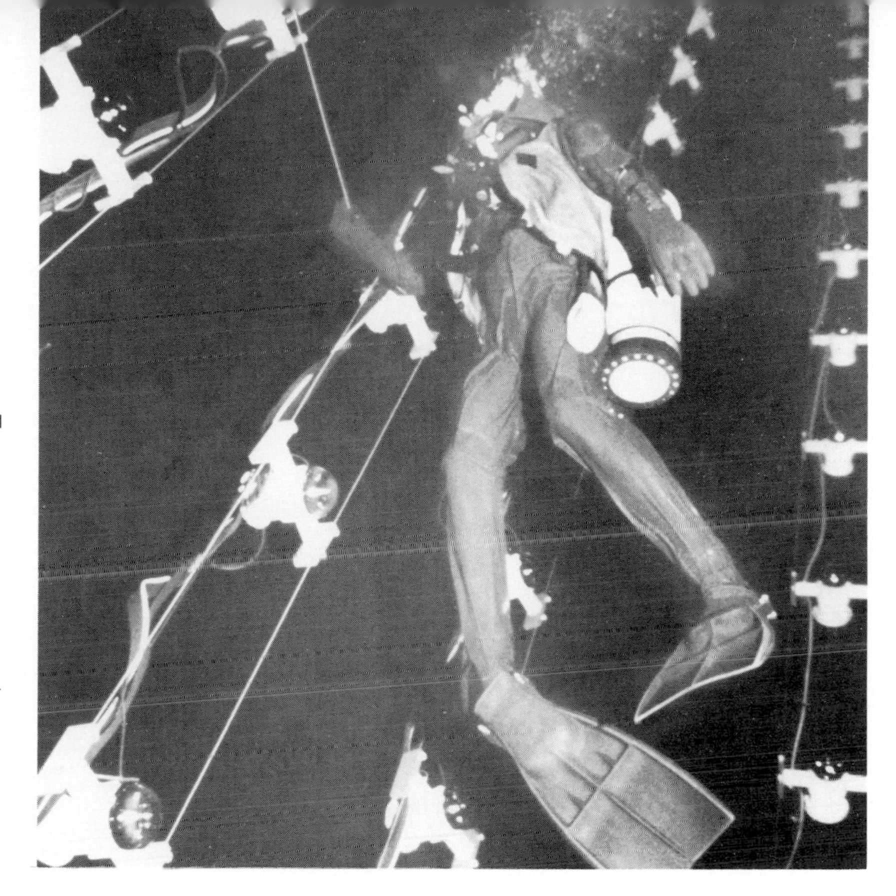

A diver examines neutrino sensors in a tank of water in a salt mine near Lake Erie. These subatomic particle sensors on Feb. 23, 1987, detected a burst of neutrinos from a spectacular *supernova* (exploding star) in a nearby galaxy. Neutrinos are difficult to detect because they have no electric charge, travel almost as fast as light, and can pass right through Earth. Finding this burst of neutrinos will help astrophysicists determine what happens when a star explodes and dies.

particles, the main source of information about the subatomic world, had a period of unprecedented growth in 1986 and 1987. Three new colliding-beam particle accelerators began operating, and three more were under construction in Europe and scheduled to begin operating by 1990. The United States announced plans for an even more ambitious machine.

These "atom smashers" accelerate two beams of particles to a speed approaching the speed of light, then force the beams to collide head-on. Most such machines collide a beam of matter particles with a beam of their antimatter counterparts. When a matter particle collides with an antimatter particle, the two particles annihilate each other in a flash of pure energy. Much of this energy immediately forms a vast swarm of new particles.

At Fermi National Accelerator Laboratory (Fermilab) in Batavia, Ill., the world's most powerful collider began operating in January 1987. This machine, called the Tevatron, circulates a beam of protons and a beam of their antimatter counterparts, *antiprotons*, in opposite directions around a circular track 6.3 kilometers (3⁹⁄₁₀ miles) long.

The beams circulate for hours, held on course by powerful magnets driven by electric currents in wires made of *superconductors*, materials that conduct electricity without resistance. Each time around the loop, a few particles collide head-on, with a combined energy of 1.8 trillion electron volts (TeV). (One electron volt is the amount of energy an electron gains when it moves across an electric potential of 1 volt.) The previous record had been 0.63 TeV in the Proton-Antiproton Collider at the European Laboratory for Particle Physics (CERN) near Geneva, Switzerland.

A proton is made up of three particles called quarks, while an antiproton is composed of three antiquarks. A proton-antiproton collision involves a complex interaction of many particles. Experimenters observe the debris from the collisions with the aid of huge assemblies that include magnets weighing hundreds of tons,

Physicists at the Bevatron, an "atom smasher" at Lawrence Berkeley Laboratory in Berkeley, Calif., examine a new kind of subatomic-particle detector: glass and quartz rods designed to identify massive atoms created when two other atoms collide and break apart in a shower of particles.

thousands of particle detectors, and dozens of computers. All of this sophisticated hardware and software must be tuned and debugged, and machine operators and experimenters must polish their skills. This can take many months.

Nonetheless, by March 1987, the Fermilab experimenters had seen their first example of a W particle, an unstable particle 87 times heavier than a proton. The W is involved in nuclear *beta decay* (a form of radioactivity) and certain reactions that transform one kind of subatomic particle into another. The production of the W at Fermilab was an important milestone, because the Tevatron Collider was designed in large part to study the W and a similar particle called the Z particle, by cataloging the lighter particles that emerge when W's and Z's break up.

Japan sets a record. In mid-November 1986, the TRISTAN accelerator at Japan's National Laboratory for High Energy Physics (KEK) in Tsukuba set a record for electron-positron colliders. TRISTAN accelerated a beam of electrons and a beam of positrons to a combined energy of 50 billion electron volts (GeV), surpassing the old record of 46.7 GeV held by the PETRA collider at the German Electron Synchrotron Laboratory (DESY) near Hamburg, West Germany. DESY had taken PETRA out of service about a week earlier, after eight years of operation.

TRISTAN is a ring of magnets about half as big as the Tevatron. But TRISTAN reaches only one-twentieth of the Tevatron's energy, because beams of electrons and positrons continually lose energy as they coast around a ring. Even so, electron-positron collisions are useful, because they are easier to interpret than the Tevatron's proton-antiproton collisions.

By April 1987, experimenters at KEK announced that they had observed the first head-on collisions in TRISTAN's new energy range. TRISTAN should eventually reach even higher energies, but it will probably not enjoy a long reign as electron-positron energy leader. Its successor almost certainly will be the SLAC Linear Collider (SLC) at the Stanford Linear Accelerator Center in Palo Alto, Calif. The SLC, which began tune-up operations in March 1987, is designed to reach 100 GeV.

SLC is the first of a new style of machine that does not store beams on a magnetic race track. Instead, the particles make a single pass along a straight track, then divide into two beams of opposite charge. The beams enter a small loop at the end of the track and collide halfway around the loop. This design avoids the energy losses of stored beams, but it requires very precise steering and focusing of the beams to ensure an adequate rate of head-on collisions.

Atom smashers under construction. The SLC should in turn be surpassed by the LEP collider under construction during 1987 at CERN. This machine resembles TRISTAN but is nine times larger. It is designed to reach about 150 GeV. The 27-kilometer (17-mile) tunnel that houses this machine was completed in early 1987.

Under construction at DESY was a new type of accelerator called HERA. It will collide electrons with protons. The electrons will be used to probe the inner workings of protons.

Fermilab's Tevatron will eventually be eclipsed by the Soviet Union's UNK, a 21-kilometer (13-mile) proton-antiproton collider under construction near Serpukhov, about 100 kilometers (60 miles) south of Moscow. UNK's beams should reach a combined energy of 6 TeV.

Super collider. The United States was the pioneer nation in subatomic particle research, and the U.S. government places a high priority on regaining world leadership. President Ronald Reagan in February 1987 announced his support for building a machine called the Superconducting Super Collider (SSC). The SSC is to be an 84-kilometer (52-mile) oval track for two proton beams with a combined energy of 40 TeV. [Robert H. March]

In World Book, see Atom; Particle Accelerator; Particle Physics; Supernova.

Psychology

Evidence that human beings can learn to be helpless and that this kind of helplessness may be linked to depression was reported by psychologist Martin E. Seligman of the University of Pennsylvania in Philadelphia in August 1986. Seligman, after 20 years of research, found that people have characteristic ways of explaining unpleasant events in their lives and that this explanatory style can affect their future behavior and their health.

Seligman first developed a theory that animals could learn to be helpless in 1966, after observing an experiment in which a group of dogs were being taught to escape pain. The pain was caused by an electric shock that came through the floor of their cage. Most of the dogs learned rapidly to avoid the pain by jumping over a barrier. But one group of dogs did not even try to escape the shock. Seligman found out that these animals had previously been exposed to a shock from which they could not escape. He suggested that the experience had taught them that it was useless to try to avoid pain. It was not the shock that interfered with the dogs' response but the expectation that they would have no control over it. In other words, these dogs had learned to be helpless.

Seligman envisaged a similar type of link between learned helplessness in humans and how they explain events. Seligman theorized that all people explain unpleasant events as being caused by something either stable or unstable (referring to the future), global or specific (in thinking about the world), and internal or external (referring to self). If a relationship breaks up, for instance, a person explaining it as something *stable* (unchanging in their life) would say, "I *always* mess up my personal relationships." This person will expect the situation to happen again in the future and will show signs of helplessness in dealing with future relationships. Someone viewing the event as having an *unstable* cause might say, "Well, I guess *this* relationship was not a very good one." They will not expect it to happen again in the future.

A person explaining the breakup as global might say, "I'm incapable of doing *anything* right." This person may expect bad things to happen in every area of his or her life—not just in relationships. A person with a specific explanation might say, "I guess I made a mistake *this time*."

A person who explains the breakup as internal may say, "It was all *my* fault, not my partner's." This person is likely to show signs of low self-esteem. A person with an external explanation might say, "My relationship broke up because my partner got a new job."

Depressing explanations. According to Seligman, people who constantly explain unpleasant events in stable, global, or internal terms are more likely than other people to become depressed when unpleasant events occur. He and his colleagues developed the Attributional Style Questionnaire (ASQ) to test this theory. They had students fill out the ASQ and then take a test for depression. As expected, the students who gave stable, global, or internal explanations for events listed on the ASQ were more depressed than those who had other explanations for misfortune.

Seligman also found that explanatory style may predict a person's future achievement, illness, and even death. People with a negative explanatory style, he said, are more likely to give up after a failure. Working with graduate student Leslie Kamen, he found that explanatory style predicts achievement in school. College students with the more positive unstable, specific, and external explanatory styles got better grades than were predicted by their Scholastic Aptitude Test scores and high school grades.

Explanations and health. Seligman thinks the way people explain unpleasant events in their lives can even affect their physical health. Feelings of helplessness, he believes, appear to impair the body's ability to combat disease because such feelings increase stress levels. Other researchers have found evidence that stress can weaken the body's immune system, increasing a person's vulnerability to disease.

Seligman conducted several studies on learned helplessness and health, including one in which he and his colleagues examined the medical histories of 18 men who had been students at Harvard University in Cambridge, Mass., between 1939 and 1944. They determined the explanatory styles of these men by analyzing questionnaires the men had answered in 1946 about their experiences in World War II. In their research 40 years later, the researchers contacted the men to find out the state of their physical health. Although the results are preliminary, Seligman said that the men who had negative explanatory styles when they were younger seemed to have more health problems than those who had more positive explanatory styles. It appears that explanatory style may actually predict a person's physical health in the future, he said.

Seligman warned, however, that psychology plays only a minor role in physical illness. "If a crane falls on you," he says, "it doesn't matter what you think. If the magnitude of your cancer is overwhelming, your psychological outlook counts for zero. On the other hand, if an illness is just beginning, your psychological state may be critical."

Although people tend to have a characteristic explanatory style, Seligman said, they can learn to change it. He believes that *cognitive therapy*, a common treatment for depression, is the best way to accomplish this since it considers depression a result of distorted thinking about the world, the future, and oneself. Cognitive therapy works to change the explanations people use when they are depressed, and encourages them to think about different kinds of causes that may be responsible for their problems.

Confusing numbers. Why do Asian students consistently do better on tests of mathematical ability than their peers in the United States? Psychologist Irene T. Miura of San Jose State University in California suggested a partial explanation at an American Psychological Association meeting in Washington, D.C., in August 1986. Miura believes the numbers 11, 12, and 13 are confusing to children learning math. In comparing 21 Japanese-speaking children with 20 English-speaking children, she found that the children's primary language

"I'd say you're coming along fine—already you've progressed from 'everybody's out to get me' to 'nobody cares about me.' "

seemed to make a difference in their understanding of numbers.

According to Miura, when American children look at these figures they know what they mean, but when they hear them spoken aloud or see them spelled out, the numbers do not make sense. For example, the word *sixteen* looks and sounds like the words *six* and *ten,* which is what it actually means. But there is no such clear association with *eleven* or *twelve*. Unlike English, Asian languages, such as Japanese, have words for these numbers that precisely define their meaning: for example, *ten-one* means 11 and *ten-two* means 12.

Love theory. Psychologist Robert J. Sternberg of Yale University in New Haven, Conn., well known for his research on intelligence, proposed a theory that might help clarify how and why people love. His three-sided, or triangular, theory of love was reported in the September 1986 issue of *Psychology Today*. It was based on the results of questionnaires completed by 35 men and 50 women between the ages of 18 and 70 and on Sternberg's analyses of various relationships.

Sternberg concluded that love is a three-sided combination of differing amounts of intimacy, passion, and commitment. Intimacy is the emotional side; passion is the physically motivating side; and commitment is the rational, thinking side.

Alone and in combination, these three elements compose eight different kinds of love relationships, according to Sternberg, even one he calls *nonlove*. This is the absence of all intimacy, passion, and commitment, and it accounts for most casual relationships. *Liking* involves only closeness, as in friendship; *infatuation* involves only passion; *empty love* has commitment but lacks both intimacy and passion.

Romantic love, according to Sternberg is a combination of intimacy and passion; *fatuous love* involves passion and commitment without intimacy; and *consummate* or *complete love* is a combination of intimacy, commitment, and passion. [Robert J. Trotter]

In WORLD BOOK, see PSYCHOLOGY.

Public Health

Outbreaks of salmonella, a type of food poisoning caused by eating food contaminated with salmonella bacteria, have increased in the United States from 649 reported cases in 1945 to more than 65,000 cases in 1985. Although the dramatic jump in 1985 was due to one large outbreak in the Midwest, the steady rise in the number of cases raised public concern in 1986 and 1987 about how meat and poultry, dairy products, and other foods are processed.

The frightening new disease AIDS (*ac*quired *i*mmune *d*eficiency *s*yndrome) thrust a much older disease, tuberculosis, back into public health attention in 1986 and 1987. The number of cases of tuberculosis in the United States had dropped steadily from the early 1900's until 1986, when there was an unprecedented increase in the disease. This increase was partially due to AIDS, a viral disease that cripples the body's disease-fighting immune system.

Victims of AIDS are susceptible to infection by a variety of organisms, including a group of bacteria called *mycobacteria*. Many AIDS patients develop the common form of tuberculosis, which is caused by an organism called *Mycobacterium tuberculosis*, also known as the tubercle bacillus. Another infectious organism seen in AIDS patients is *Mycobacterium avium*, which produces tuberculosis in hogs and poultry but had previously been extremely rare in human beings.

The rise of these diseases brought a new surge of research. The federal Centers for Disease Control in Atlanta, Ga., and several local public health departments were investigating the prevalence of tuberculosis in patients who have AIDS or who have been infected with human immunodeficiency virus (HIV), the AIDS-producing virus. Researchers are also studying whether patients with HIV infection and tuberculosis are more likely—or less likely—to transmit tuberculosis to others and are testing various treatments.

New cases of tuberculosis were expected to reach about 23,000 per year in the United States—9.1 per 100,000 population—by 1988. The worldwide toll is more than 10 million cases annually, with 3 million deaths directly attributable to the disease.

AIDS prevention. Surgeon General of the United States C. Everett Koop issued a report on AIDS in October 1986. The report focused on prevention of AIDS through safe sexual practices and stressed the importance of education about safe sex as the only current means of disease prevention.

Koop acknowledged that some peo-

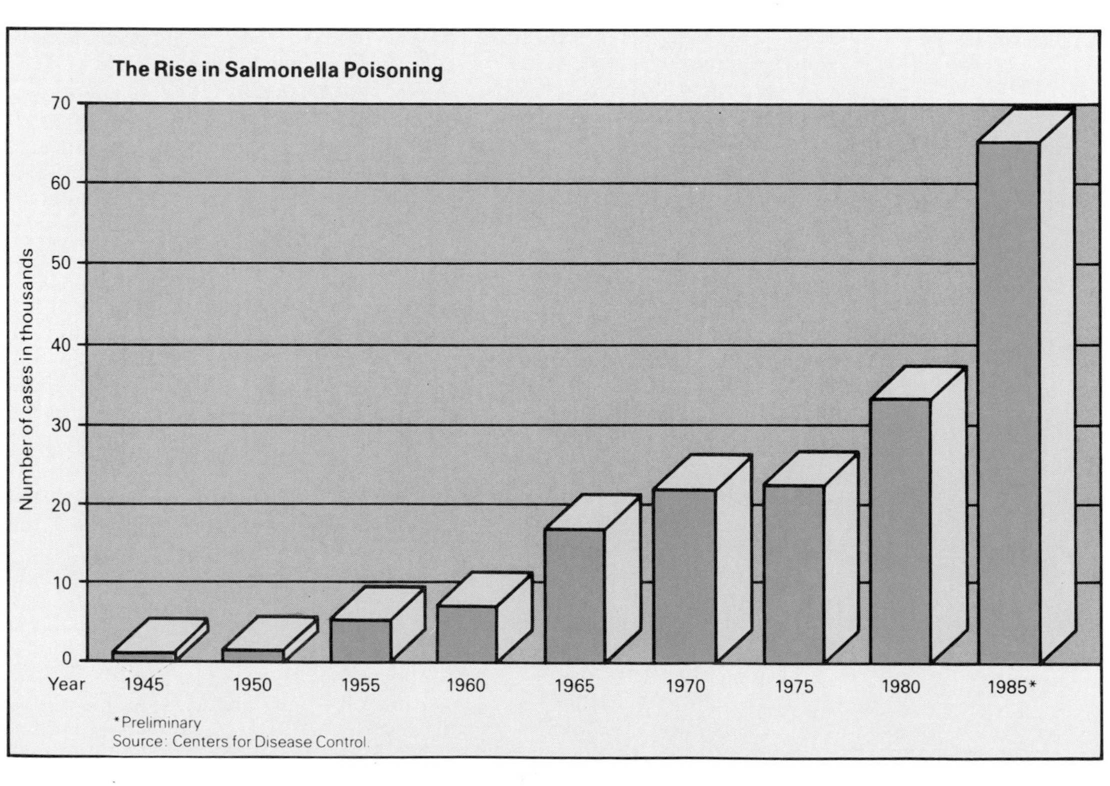

ple have difficulty dealing with the subject of sex, and particularly homosexual practices. But he insisted that all such issues must be discussed to explain the health consequences of sexual behavior.

Koop stressed that education about AIDS should start in early elementary school and at home so that young children grow up understanding how to protect themselves against the AIDS virus. He gave parents and educators the responsibility of creating an acceptable sex-education curriculum that would include information on heterosexual and homosexual relationships, with a heavy emphasis on prevention of AIDS and other sexually transmitted diseases.

Koop targeted teen-agers and preteens as a group needing special attention. Such young people are exploring their own sexuality and may even experiment with intravenous drugs, which makes them especially vulnerable to AIDS. AIDS can be spread among drug abusers sharing contaminated needles.

The surgeon general's report estimated that 54,000 people in the United States will die from AIDS in 1991. That estimate included people who are not yet infected with the AIDS virus, however. With proper information and education, 12,000 to 14,000 people could be saved.

Seat-belt laws. By the end of 1986, 24 states and the District of Columbia had mandatory seat-belt laws. The first law in the United States requiring the use of safety belts had passed in New York state in July 1984 and took effect on Jan. 1, 1985. State officials quickly pronounced the law a success. There were 2 per cent fewer traffic fatalities in New York during the first five months of 1985 than during the same period in 1984, and fewer than the average during the same period over the previous five years. As a result, many states passed similar laws.

To study whether seat belts really do save lives, economists Eric A. Latimer and Lester B. Lave of Carnegie-Mellon University in Pittsburgh, Pa., used a statistical model to estimate the effectiveness of the New York state law. Their study, reported in the *American Journal of Public Health* in February 1987, took into account such factors as weather, number of miles driven, and driver behavior, such as drunken driving.

In a study of 12 metropolitan areas in New York state, Latimer and Lave estimated that the seat-belt law prevented 220 deaths and 8,700 injuries during the first six months it was in effect. They calculated that if legislation increased seat-belt usage to 60 per cent or more in the United States, as many as 9,000 deaths and 300,000 injuries could be avoided each year.

Bans on public smoking. Efforts to curb cigarette smoking in public places increased during 1986 and 1987. More than 200 cities and counties had enacted antismoking ordinances, most of them since 1985. Twenty-seven states had established some restrictions.

This legislation has come from growing awareness that breathing second-hand smoke is harmful and from resentment on the part of *passive smokers*, nonsmokers who are forced to breathe smoke-filled air.

The U.S. surgeon general's annual report on smoking and health, released in December 1986, focused entirely on the ill effects of passive smoking. In releasing the report, Koop called for smoke-free workplaces for nonsmokers.

In July 1986, the U.S. Army prohibited recruits from smoking during basic training. The Army also banned smoking on Army facilities, in military vehicles, and in aircraft—outside of certain designated areas. The Army said the policy was necessary because smoking reduces combat readiness by impairing physical fitness and by increasing illness, absenteeism, premature death, and health care costs.

In February 1987, the United States government prohibited smoking by its employees except in designated areas of government buildings.

Private industry has also responded to demands for a smoke-free environment. A 1986 survey of 662 employers by the Bureau of National Affairs in Washington, D.C., which issues reports on economics, found that 36 per cent had limited smoking among workers.

[Michael H. Alderman]

In WORLD BOOK, see PUBLIC HEALTH.

Science Education

More than 85 per cent of the people in the United States are in favor of improving science education in American public schools, according to public-opinion polls in 1986. That percentage surpassed the previous peak in public support for science education, which occurred in the late 1950's after the launching of the Sputnik satellite by the Soviet Union.

Upgrading science teaching. By 1987, state governments throughout the United States were taking an active role in upgrading the teaching of science. Between 1982 and 1987, the states passed more than 700 laws establishing tougher certification standards for teachers and making other changes in state educational policies.

Many of the new laws set specific requirements for certification at the elementary, junior high school, and high school levels. A number of states, including California, Florida, New York, and Texas, also established minimum science-competency requirements for students. In those states, students must pass a standardized written test of science knowledge before they can graduate or, in some cases, advance from one grade to the next.

Critics protest science efforts. Other Americans, however, protested efforts to strengthen science education. A frequent objection was that the growing emphasis on science threatened to force all religious values out of school curriculums. Many textbooks, critics charged, were already ignoring the role of religion in human society.

In March 1987, U.S. District Judge W. Brevard Hand ruled in Mobile, Ala., that more than 40 textbooks in Alabama public schools not only disregarded traditional religion but also promoted a different sort of "religion"—*secular humanism.* Secular humanism is a label applied by fundamentalist Christians to the philosophy that humanity must take responsibility for its own destiny, not rely on the help of a Supreme Being.

Hand ordered the offending textbooks removed from Alabama schools. Although none of the books were science texts, some educators expressed concern that an increased emphasis on religion in public schools might harm science education. A *secular* (nonreligious) point of view, they said, is essential to the proper teaching of science.

Federal funding. After much debate on whether to increase its funding for science-education programs, the U.S. Department of Education in 1987 delayed undertaking any new initiatives. The department actually reduced so-called Title II funds for the support of science education in the states from $100 million to $55 million. Legislation introduced in Congress in 1987 called for spending increases totaling more than $400 million for several new science-education projects. Secretary of Education William J. Bennett, however, opposed the spending hikes for budgetary reasons.

Meanwhile, the National Science Foundation (NSF) increased its support for science education to $99 million in fiscal 1987, $45 million more than it spent in 1986. The NSF asked Congress for an additional $20 million for fiscal 1988 to support undergraduate science education in colleges.

During 1987, the NSF funded teacher-education programs at nine U.S. universities, including Boston University and the University of Georgia in Athens. The purpose of the programs was to develop guidelines that will help colleges and universities establish course requirements for students wishing to pursue careers as science teachers.

The NSF planned to spend a total of $18 million over a five-year period on the teacher-education project. Additional funding was being provided by private companies and educational foundations.

The private sector also funded a number of its own teacher-education programs. The Standard Oil Company of Ohio, for example, in 1986 and 1987 continued its support for a project called Science Resources for Schools, sponsored by the American Association for the Advancement of Science. A number of utility companies and nonprofit institutions increased their support for the development of educational materials and for teacher training. [Robert E. Yager]

In the People in Science section, see HIGH SCHOOLS OF SCIENCE. In WORLD BOOK, see EDUCATION; SCIENCE PROJECTS.

An Olympics of the Mind

International athletic competitions such as the Olympic Games and the Pan American Games have inspired countless young people to improve their athletic skills. But what about the young person whose greatest talent is academic? To provide such a person with recognition and with a chance to compete with gifted students from other nations is the goal of international academic competitions such as the International Math Olympiad, the International Chemistry Olympiad, and the International Physics Olympiad.

The International Physics Olympiad originated in Eastern Europe in 1967 and gradually expanded to include most nations of Western Europe. Canada began to participate in 1985, and, in 1986, China and the United States entered the competition.

Members of the American Association of Physics Teachers (AAPT), led by Jack M. Wilson, the association's executive officer, considered U.S. participation in the Physics Olympiad to be an important step toward improving physics education in the United States. A U.S. physics team would draw attention to high school physics, encourage students to excel, and give long-overdue recognition to the most accomplished students.

Wilson and two AAPT academic directors—Arthur Eisenkraft, a high-school teacher from New York state, and Ronald D. Edge, a physics professor from the University of South Carolina in Columbia—began raising funds. They gathered contributions from a variety of sources, including physics organizations, industry, and publishing companies.

The U.S. physics team and its advisers pose with a statue of Albert Einstein in Washington, D.C., during training for the Physics Olympiad in 1986.

Simply finding the brightest physics students in a nation of more than 240 million people without a centralized school system proved challenging. First, the organizers asked teachers throughout the United States to nominate their best physics students. The 185 students who were nominated then took two difficult qualifying examinations. The 20 students with top scores became the U.S. team.

On June 30, 1986, the 20 U.S. team members, ages 15 to 18, converged at the University of Maryland in College Park, near Washington, D.C., for two weeks of intensive training in all facets of physics. During the training, the team members still found time for sports, music, cookouts, and other activities. One member of the team broke his arm during a vigorous game of Frisbee.

At the end of the training camp, the students were given another series of tests to select five students who would represent the United States in the 1986 Physics Olympiad at Harrow School, near London, from July 13 to 20. The five team representatives were Howard Fukuda of Hawaii, Paul Graham of Colorado, Philip Mauskopf of North Carolina, Srinivasan Seshan of Virginia, and Joshua Zucker of California.

Within days, the team assembled in London, accompanied by Eisenkraft, Edge, and Wilson of the AAPT. The U.S. team visited tourist attractions, played sports, and, despite language barriers, discussed physics with their 100 counterparts from 20 other nations, including Cuba, East and West Germany, and the Soviet Union. Because nearly every Olympiad student will become a leading scientist—and science is now a truly international profession—team members were actually meeting their future colleagues.

The two days of Olympiad competition were grueling. On the first day, the students were asked theoretical questions about physics. On the second day, they were tested on their laboratory and computer abilities.

When the Olympiad scores were tallied, three Soviet students won gold medals, and the fourth gold medal went to a student from Romania. The results also revealed an astounding achievement for the U.S. team. They had compiled the best record for any team competing for the first time in the Physics Olympiad. Three of the five U.S. team members—Graham, Mauskopf, and Zucker—received bronze medals. But regardless of medals, every participant, like Olympians throughout history, had gained a moment of personal triumph. [David A. Kalson]

Science Fair Awards

Winners in the 46th annual Westinghouse Science Talent Search—the oldest and largest science student competition in the United States—were announced by Science Services of Washington, D.C., on March 2, 1987.

The 40 finalists were chosen from 1,295 seniors from high schools throughout the United States. The top 10 finalists received scholarships totaling $110,000; the other 30 finalists each received $1,000 cash awards. Scholarships and awards were provided by the Westinghouse Electric Corporation. The finalists spent five days in Washington as guests of Westinghouse.

First place and a $20,000 scholarship were awarded to 17-year-old Louise Chia Chang, a student at the University of Chicago Laboratory Schools High School. Chang won for her study of the role genes might play in the onset of cancer. Working with mouse cells, Chang isolated three genes that are more active in cancerous cells than in normal cells. She found that one of the genes produces a protein-destroying chemical that may contribute to the cancer process.

Second place and a $15,000 scholarship went to Elizabeth Lee Wilmer, 16, of Stuyvesant High School in New York City. Her winning project was a mathematical investigation of the properties a map must have so that only three colors need to be used to indicate separate regions on the map, with no two bordering regions having the same color.

Third place and a $15,000 scholarship went to Albert Jun-Wei Wong, 16, of Oak Ridge High School in Oak Ridge, Tenn., for a computerized system that shows how networks of brain cells may recognize patterns. Research such as this could have a bearing on the development of computers that mimic human intelligence.

Fourth place and a $10,000 scholarship were awarded to Joseph Chen-Yu Wang, 17, of Forest High School in Ocala, Fla. Wang used the facilities of the University of Florida radio observatory to study radio waves from Jupiter in an effort to learn how the *polar-*

Louise Chia Chang of Chicago, center, won first place in the 46th annual Westinghouse Science Talent Search competition in March 1987. Second place went to Elizabeth Lee Wilmer of New York City, right, and third place to Albert Jun-Wei Wong of Oak Ridge, Tenn.

Science Fair Awards

Continued

ization of the waves—their plane of vibration—is changed by such influences as Earth's upper atmosphere.

Fifth place and a $10,000 scholarship went to David J. Bernstein, 15, of Bellport High School in Brookhaven, N.Y. He worked out new procedures for computing irrational numbers to an unlimited number of decimal places. One such number is *pi* (approximately 3.14159), the ratio of the circumference of a circle to its diameter.

Sixth place and a $10,000 scholarship were awarded to Stephen A. Racunas, 16, of Valley High School in New Kensington, Pa. Racunas won for developing a computerized simulation of "optical molasses"—atoms that have been slowed by laser beams.

Seventh place and a $7,500 scholarship went to Maxwell V. Meng, 16, of Centennial High School in Ellicott City, Md., for his study of an inherited disorder linked to the development of premature heart disease. He found that victims of the disorder have reduced levels of two important proteins in their bodies.

Eighth place and a $7,500 scholarship were awarded to Todd A. Waldman, 17, of Walt Whitman High School in Bethesda, Md. Waldman studied the genetics of *Escherichia coli* bacteria and caused one of the organism's genes to undergo a *mutation* (genetic change).

Ninth place and a $7,500 scholarship went to Maria J. Silveira, 17, of Bronx High School of Science in New York City. Silviera conducted research on a parasite, *Toxoplasma gondii*, which causes severe brain damage resulting in blindness, seizures, or mental retardation. Her study may help explain how the parasite affects the cells of the host it infects.

Tenth place and a $7,500 scholarship were awarded to Michael P. Mossey, 18, of Greenhills High School in Cincinnati, Ohio. Mossey refined a computer program for finding sets of numbers in which no two pairs of numbers within the set produce the same *difference* (the number resulting when one number is subtracted from another). [David L. Dreier]

Space Technology

The Soviet Union took the lead in manned space flight in 1986 and 1987 with its new *Mir* space station and the launch in May 1987 of the massive new Energia booster. Japan and China also made significant strides with their space programs. European and United States space programs, meanwhile, were seriously disrupted because of the January 1986 explosion of the U.S. space shuttle *Challenger* and the failure in May 1986 of a European Space Agency (ESA) rocket.

Turmoil continued in the U.S. space program as the National Aeronautics and Space Administration (NASA) worked to redesign the flawed solid-fuel booster rocket made by Morton Thiokol that caused the *Challenger* accident. A leak of hot gases from the right solid-fuel booster rocket led to the explosion of the main fuel tank. In the fiery tragedy, all seven crew members died.

The leak occurred because two rubber O-ring seals failed to plug a gap in the joint between two rocket sections. NASA officials in July 1986 announced that the booster rocket would be redesigned with more insulation on the inside of the booster walls to prevent hot gases from reaching the joint and the O-ring seals. NASA said that the addition of a third O-ring seal would provide even greater safety.

Future shuttle missions. NASA set Feb. 18, 1988, as its target date to resume shuttle launches, though space officials privately said that no launch would occur before mid-1988. President Ronald Reagan on Aug. 15, 1986, approved construction of a new $2.8-billion shuttle orbiter to replace the *Challenger*, keeping the number of orbiters in the shuttle fleet at four. NASA had determined that a four-orbiter shuttle fleet was necessary to help build and supply the planned U.S. space station and carry out scientific and military space missions.

Return to rockets. The *Challenger* accident also caused space officials to place a greater emphasis on the use of unmanned nonreusable rockets for space launches. Previously, they relied almost exclusively on the shuttle.

During 1986 and 1987, the United States was able to recover from serious accidents involving the Titan and Delta rockets, which occurred just after the *Challenger* disaster in 1986. From late 1986 to mid-1987, the United States launched 10 unmanned missions, starting with two military satellites orbited on Sept. 5, 1986.

On September 17, a National Oceanic and Atmospheric Administration (NOAA) polar-orbit weather satellite was placed in orbit by an Atlas rocket launched from Vandenberg Air Force Base in California. On November 13, a Scout rocket launched from Vandenberg placed an Air Force scientific satellite into orbit, and on December 4, an Atlas rocket launched from Cape Canaveral carried a Navy communications satellite into orbit.

An Air Force Titan rocket was used to launch a secret military payload from Vandenberg on Feb. 11, 1987. It was the first successful launching of a Titan rocket since a different type of Titan exploded on liftoff in April 1986. A NASA Delta rocket launched from Cape Canaveral on Feb. 26, 1987, carried a NOAA weather satellite.

On March 20, NASA launched a Delta rocket carrying a communications satellite for Indonesia. NASA suffered a setback, however, on March 26 when an Atlas Centaur rocket carrying a Navy communications satellite was struck by lightning and destroyed.

European troubles. Many commercial satellites whose missions were delayed by the shuttle disaster sought to have their payloads launched by the ESA's Ariane rocket. But the ESA had suffered a serious failure of its own in May 1986 when ground controllers had to destroy an Ariane-2 rocket carrying a communications satellite because the rocket's third stage failed to ignite. Ariane launches were delayed until at least August 1987 as a result of the accident.

Space station. Plans for a U.S. space station, which would include participation by the ESA, Canada, and Japan, ran into trouble during 1986 and 1987. In mid-1986, concerns of NASA astronauts and managers led to key engineering changes in the station's design. Then, in late 1986, ESA and Japanese space officials complained that they were not being guaranteed a significant role in the space station's design and operation, though their contributions to the station's construction were expected to exceed $3 billion. At one point the ESA even threatened to quit the program.

In December 1986, the U.S. Department of Defense revealed that it was interested in using the space station to perform military research. This threatened to jeopardize NASA's preliminary agreements with its foreign partners, which stated that the station would be used only for peaceful purposes. A reaffirmation of the partnership came out of a February 1987 meeting in Washington, D.C., but the defense issue continued to be a problem into mid-1987.

In December 1986, NASA announced that its earlier estimate of $8-billion for the U.S. share of space-station funding was far too low. NASA revised its estimate to $15 billion for the basic station structure and $20 billion for other equipment. As a result, the White House began to reexamine the practicality of the space station. In April, President Reagan announced that he supported a scaled-back version of the space station that would cost $13.5 billion.

The Soviet space program continued at an impressive pace, though it too was affected by at least four accidents. The Soviet Union usually does not announce its failures, but most can be detected by Western satellites.

On Oct. 3, 1986, a Soviet satellite intended to warn against an enemy missile attack failed to achieve the correct altitude and went into a useless orbit. This apparently was caused by a malfunctioning booster rocket. Within 12 days of the failure, however, the Soviets used the same type of rocket to launch another missile-warning satellite that succeeded in achieving the proper orbit. Three other Soviet space failures occurred in early 1987. Two of the accidents involved failures with the Proton, the world's largest rocket.

The Soviet failures were minimal, however, compared with their successes. During 1986, for example, the Soviets launched 91 space missions, carrying a total of 114 different satellites, including several satellites atop a

"This is Iowa. Io is a moon of Jupiter."

single rocket. From January to May 1987, they launched 34 missions.

One of the most important Soviet achievements was the launch on May 15 of the Energia booster, which can carry payloads weighing 100,000 kilograms (220,000 pounds) into orbit and will be used to launch a new Soviet space shuttle.

The Soviets also carried out extensive manned operations on the *Mir* space station. Soviet cosmonauts Leonid D. Kizim and Vladimir Solovyov—who were launched to the *Mir* on March 13, 1986—returned to Earth on July 16 after spending 125 days in orbit.

Cosmonauts Yuri Romanenko and Aleksandr Laveikin docked with the station on Feb. 8, 1987, to begin a mission that was expected to last until August 1987. On April 5, an 18,000-kilogram (40,000-pound) module, carrying West European astrophysics instruments, failed to dock properly with the station. The module approached within 180 meters (600 feet) of the *Mir*—then its control system failed. A second attempt failed on April 9, but on April 12, the module successfully docked with the space station after the two cosmonauts made an unscheduled spacewalk to aid the link-up.

The Japanese space program also matured in late 1986. On Aug. 13, 1986, Japan achieved the first successful launch of its new H-1 rocket.

On Feb. 4, 1987, Japan launched an X-ray astronomy satellite, followed by the launch on February 19 of that nation's first remote-sensing spacecraft, the *Marine Observation Satellite* (*MOS-1*). *MOS-1* immediately began returning images of Earth comparable to those provided by the U.S. *Landsat* spacecraft in the 1970's.

China's space program. China also made progress in space in 1986 and 1987. In November 1986, three U.S. companies—Pan Am Pacific Satellite Corporation; Dominion Video Satellite, Incorporated; and Teresat, Incorporated, of Houston—all signed agreements with the Chinese to launch satellites. [Craig P. Covault]

In WORLD BOOK, see SPACE TRAVEL.

Zoology

New findings reported in 1987 may shed light on how animals use Earth's magnetic field to find their way around. Many animals have an internal "compass" that enables them to orient themselves in relation to Earth's magnetic lines of force. But scientists do not know how animals sense and interpret Earth's magnetic field.

Compass in a mollusk. Zoologists Kenneth J. Lohmann and A. O. Dennis Willows of the University of Washington in Seattle reported in January that a Pacific Ocean *nudibranch*—a mollusk without a shell—changes direction according to its position within Earth's magnetic field. The researchers also discovered that the phases of the moon influence the mollusk's response to the field.

Lohmann and Willows first tested nudibranchs in a darkened laboratory. The scientists manipulated the magnetic field around the animals, sometimes canceling out the effects of Earth's magnetism with magnetic coils. During four days of testing, they discovered that the mollusks tended to face east when allowed to sense Earth's natural magnetic field.

Over the next four months, however, Lohmann and Willows found that there were periods when the nudibranchs were *not* inclined to face east. Searching for a reason for this apparent contradiction, the investigators wondered whether the mollusks might be influenced by the changing phases of the moon. Other researchers had reported finding evidence that fruit flies, flatworms, and homing pigeons—other animals with internal compasses—are affected by the moon.

To test their hypothesis, Lohmann and Willows placed nudibranchs in a simple maze at various times of the month. During a full moon, three-fourths of the mollusks released into the maze turned east. But the animals showed no preference for the east during a new moon. These findings showed that the nudibranch's movements with relation to Earth's magnetic field are probably related, in some unknown way, to the cycles of the moon.

A pair of black-footed ferrets emerge warily from their burrow in Wyoming. Found only in the western Great Plains of the United States, black-footed ferrets are among the rarest animals of North America. An outbreak of canine distemper has reduced their number to no more than 20. Conservationists captured 6 ferrets in 1986 to breed in captivity.

Finny stowaways. The first authentic example of brood parasitic behavior among fishes was reported in September 1986 by zoologist Tetsu Sato of Kyoto University in Japan. Brood parasitic behavior is a form of trickery by which an animal gets other animals to raise its young. Certain kinds of birds, such as the cowbird, are best known for this type of behavior. Such birds lay their eggs in other birds' nests.

Sato studied the mochokid catfish, found in Lake Tanganyika in Africa. The mochokid mixes its eggs with those of a small, colorful fish called the cichlid, apparently by depositing eggs where cichlids are spawning.

Female cichlids incubate their eggs in their mouth. Once the eggs hatch, the baby fish move in and out of the mother's mouth to feed and use the mouth as a refuge until they mature. But stowaway catfish eggs, accidentally scooped up by the mother cichlid along with her own eggs, spell disaster for the cichlid young.

Sato found that catfish eggs hatch before cichlid eggs. The intruders remain in the host mother's mouth and feast on the newly hatched baby cichlids. The young catfish thus obtain both protection and food from the mother cichlids.

Foul fish foil fertility. Water pollution caused a decline in the birth rate of seals in the waters off the Netherlands, Dutch marine ecologist Peter J. H. Reijnders reported in December 1986.

The seals live in the western part of the Wadden Sea, an inlet in the North Sea. Their numbers dropped from more than 3,000 to fewer than 500 from 1950 to 1975. During that time, scientists performing autopsies on the seals found the animals' tissues were becoming increasingly contaminated with such pollutants as heavy metals and *polychlorinated biphenyl compounds* (PCB's). These pollutants flow into the Wadden Sea from the Rhine River.

Reijnders tested the effects of PCB's on 24 female seals. Twelve seals were fed PCB-laden fish from the Wadden Sea, and the other 12 seals were fed untainted fish from the northeast Atlantic Ocean. The seals' hormone lev-

Zoology

Continued

Zoologists count Asian cockroaches caught in traps in a Florida yard, *above left*. The Asian cockroach, *above right,* a newcomer to the United States, apparently arrived in Tampa, Fla., by plane or ship sometime between 1984 and 1986. Unlike other common cockroaches, the Asian roach is attracted to light and thrives outdoors. It also invades homes and offices, moving from room to room as lights are turned on. These super-roaches were expected to spread to other Southeastern and Southwestern states.

els were monitored, and the animals were allowed to mate.

Among the seals fed tainted fish from the Wadden Sea, there were only 4 pregnancies, compared with 10 pregnancies in the other group. Reijnders concluded that pollution in the Wadden Sea is responsible for the decline in the seal population.

Reijnders found that hormonal levels in both groups of seals were about the same, indicating that pollutants do not prevent the fertilization of egg cells, an event that is dependent on adequate concentrations of hormones. Reijnders thus concluded that pollutants cause the destruction of eggs after the eggs are fertilized.

A six-legged immigrant that no one would like to see had established permanent residence in the United States by 1987. The newcomer, the Asian cockroach, apparently arrived in Tampa, Fla., by ship or plane sometime between 1984 and 1986.

This roach has an odd quirk: It likes the limelight. Unlike its close cousin, the reclusive German cockroach, the Asian cockroach thrives outdoors and is attracted to lights, turned-on TV sets, and white walls. It also settles on damp towels.

The Asian cockroach also flies. "This could be the roach that ruins the backyard barbecue," a Florida entomologist remarked. The roach is expected to spread north and west to other warm areas.

Lobsters' workout. Every lobster lover knows that the two claws of this delectable shellfish are not equal. One claw is the slender "cutter"—it has nimble muscle fibers for nabbing slow-moving fish. The other claw, the large "crusher," comes in handy when the lobster feasts on clams or oysters. The lobster uses its crusher claw to break the tightly closed shells and get at the soft-bodied mollusks inside.

But a lobster's claws do not start out so specialized, marine biologists C. K. Govind and Fred Lang of Toronto University in Canada reported in July 1986. Govind and Lang found that it takes about two months for one claw to become the crusher—and then only if the lobster has a chance to exercise it.

At Woods Hole Marine Biological Laboratory in Massachusetts, Govind and Lang raised 26 lobsters, isolated on smooth plastic trays, and found that only 6 developed crushers. But 17 of another 23 captive lobsters that were given buttons or pieces of oyster shells to "play with" formed crushers.

In their natural environment, the researchers concluded, lobsters exercise their claws by poking and pinching one another. Even when both claws are used, however, only one—and sometimes neither—becomes the crusher. Thus, the key to crusher development apparently lies in working one claw much more than the other.

Romantic reptiles. Even all-female species benefit from a little romance in their lives, University of Texas zoologist David Crews reported in December 1986. According to Crews, courtship helps the desert grassland whiptail lizard lay more eggs.

Found in the United States Southwest and northern Mexico, this lizard species is made up entirely of females. They reproduce by *parthenogenesis*, a kind of virgin birth common in insects but rare among higher forms of life. In the lizard's reproductive organ, the egg cell makes a second set of *chromosomes* (structures that carry an organism's genes) just before it begins to divide and develop. As a result, eggs are genetically complete without the need for fertilization by a male sperm cell—the event that usually provides the second set of chromosomes. Thus, every individual whiptail lizard adds new members to the population.

Crews discovered, however, that when he put two whiptail lizards in a laboratory cage, they courted just as a male and female might. They "flirted," then the lizard playing the male curled around and grasped the other.

Crews discovered that each lizard goes through a cycle of male-female behavior as a result of changes in hormone levels. He observed that a pair of whiptails switched roles about every 10 to 14 days. Crews learned that after about two weeks, *ovulation* (the release of eggs) occurs in the partner taking

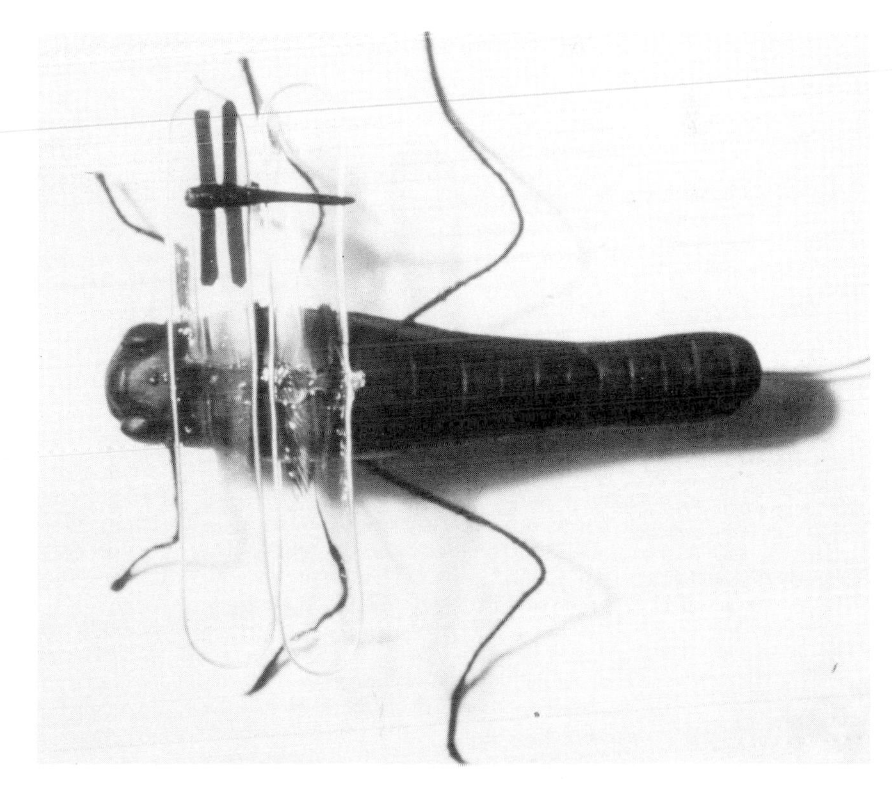

Artificial models of ancient insects, constructed by zoologists at the University of California in Berkeley, may explain why insects began to develop wings some 400 million years ago. Using thermometers, the scientists found that the wings act like solar panels, transferring heat absorbed from a sun lamp to the insects' bodies. So they theorized that wings began to evolve to regulate body temperature. As wings got larger, insects used them to fly.

Zoology

Continued

A giant panda feeds on a bamboo plant. A shortage of such plants—a major part of the animal's diet—may threaten the existence of pandas. According to reports in 1987, only about 700 pandas were left in the wild, all of them in the Sichuan (Szechwan) Province of China. The expanding population of China has forced pandas into smaller and more isolated areas. When a bamboo species flowers and dies out in such an area, pandas find it difficult to obtain food. Some 200 pandas are thought to have starved to death since the late 1970's.

the female role. Ovulation then causes the "female" to produce extra progesterone, a hormone that makes her act out the male role. During the male phase, the lizard's ovaries progressively enlarge, producing more estrogen. This eventually causes the lizard to again assume the female role. As the partners go through the same cycle together, they promote each other's fertility.

Thirsty tree frogs. Amphibians that live in deserts usually must work hard to prevent dehydration. But one desert-dwelling frog from Paraguay survives by sipping raindrops, according to a January 1987 report. "The drinking behavior observed . . . has not been described in any other amphibian," said biologist Lon L. McClanahan of California State University in Fullerton and zoologist Vaughan H. Shoemaker of the University of California in Riverside.

Some desert amphibians survive dry conditions by burrowing deep into the soil, emerging briefly when it rains to mate and eat. But the Paraguayan frog lives in trees and lacks the ability to dig.

Like other amphibians, this tree dweller stores water in its bladder. But while many of its relatives sit in puddles and absorb water through their porous skin, the Paraguayan frog does not need to wait for a heavy rain to stock up.

In both the field and laboratory, McClanahan and Shoemaker dripped water onto the heads of tree frogs and observed their response. The animals always raised their head and "swallowed" the waterdrops through their nostrils.

The scientists also learned that the Paraguayan frogs have special glands that produce a waxy substance. The animals wipe the wax over their bodies to slow the loss of water and other body fluids. This adaptation, plus their drinking behavior, enables the frogs to spend more time feeding and less time hiding from the hot sun.

[William J. Bell and Elizabeth Pennisi]

In World Book, see Animal; Zoology.

Science You Can Use

In areas selected for their current interest, *Science Year* presents information that the reader as a consumer can use in making decisions—from buying products to caring for personal health and well-being.

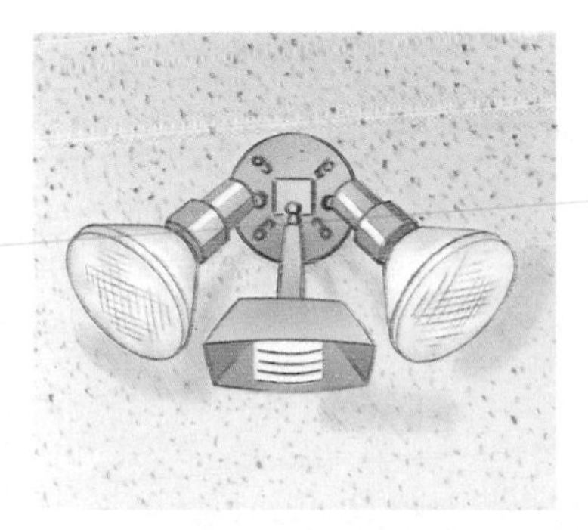

Automatic 35-mm Cameras: More Help for the Amateur

Automatic 35-mm cameras take the guesswork out of picture-taking by setting the correct exposure and focusing automatically.

A revolution in electronics and camera design has created a new class of camera—the totally automatic 35-millimeter (mm) camera. Often advertised as "auto-everything," these cameras take the difficulties out of 35-mm photography. They automatically set the correct exposure and focusing distance, determine the film's speed, advance and rewind the film, and activate the flash if necessary, to mention just a few features.

Not all automatic 35-mm cameras are alike, however. The various models offer a wide—and sometimes confusing—array of features. And since the cost of an automatic 35-mm camera is tied to its features, you will want to determine how many of them you really need before you purchase your new camera.

The original 35-mm camera was designed for relatively sophisticated photographers. The 35-mm (1½-inch) film format is excellent for producing quality prints and slides in a wide range of light conditions. With the advent of the single-lens reflex (SLR) 35-mm camera after World War II, professional photographers or serious amateurs could exercise more precise control over their pictures, because they could use a variety of lenses—such as wide-angle or telephoto—for different types of situations. But to use an SLR 35-mm camera properly requires the mastery of some rather complex photographic skills. So the casual photographer was often discouraged by the difficulties of using these cameras.

Today's non-SLR automatic 35-mm cameras retain the advantages of using 35-mm film, yet are simple to use and relatively inexpensive. These cameras are made possible by two technological advances—the microchip and the combination of the shutter and lens into a single component. A preprogrammed microchip, a silicon chip no bigger than a thumbnail that holds tiny electronic circuits, is the "brain" of the automatic 35-mm camera (in the Special Reports section, see THE MICROCHIP—A MINIATURE MARVEL).

The shutter controls the length of time that light reaches the film. The lens focuses rays of light onto the film, and its opening controls the amount of light falling on the film. By combining the shutter and lens into a single component, manufacturers were able to make cameras smaller. The microchip—along with the shutter and lens component—enabled manufacturers to produce automatic 35-mm cameras that are small in size, inexpensive to manufacture, and simple to use.

As a rule, the more features a camera has, the higher the price. List prices for automatic 35-mm cameras range from approximately $150 to $300, but big discounts—as much as 50 per cent—are usually available.

Before you select your automatic 35-mm camera, it is important to understand the contribution each automatic feature can make to your picture-taking pleasure.

All automatic 35-mm cameras have automatic *exposure* (the length of time that light is allowed to strike the film to produce an image). They contain a meter that registers the amount of light reflected from the object in the center of the viewfinder. A preprogrammed microchip uses this information to set the proper exposure.

All automatic 35-mm cameras have a built-in flash, but how they operate varies from model to model. A flash is used to supplement existing light. In some automatic cameras, you must turn on the flash yourself if a warning light shows in the viewfinder. In others, the flash goes off automatically when the camera's exposure meter registers insufficient light. A few cameras have a "fill-flash" switch that allows you, for example, to use the flash to lighten areas of a person's face that would otherwise be obscured by deep shadows.

In order for any camera's automatic exposure system to work, it must know the speed of the film you are using—in other words, the film's sensitivity to light. With some 35-mm cameras, you must set this speed yourself, but many automatic 35-mm cameras take advantage of a new film-speed coding system to do the job automatically. The system is called DX coding, and consists of a checkerboard image on the film *cassette*, or casing. This checkerboard tells the camera not only the film speed—expressed as either an ISO (*I*nternational *S*tandards *O*rganization) number or an ASA (*A*merican *S*tandards *A*ssociation) number—but also the number of frames on the roll of film. A sensor in the camera "reads" the code and sets the film speed and number of shots automatically.

Some cameras with DX coding can adjust to only a limited number of speeds. Make sure that the camera you buy can accommodate all the film speeds you may want to use.

The autofocus feature of these automatic 35-mm cameras uses a beam of invisible infrared light to determine the distance between the camera and the subject. The beam is shot out to the subject through a special opening in the camera and then reflected back to a sensor in the camera. The camera's ingenious electronic circuitry calculates the distance to the subject from the time it takes for the beam to bounce back, and then adjusts the focus accordingly.

The autofocus feature, however, does have drawbacks. Instead of focusing the subject according to its exact distance, the camera chooses one of several preset *focus zones*. If a subject is 3.6 meters (12 feet) away from the photographer, for example, an autofocus camera may set the focus at the focus zone between 3 meters (10 feet) and 6 meters (20 feet). If the range is too broad, some of your pictures may be out of focus. The more focus zones a camera has, the sharper the focus. The number of focus zones available varies from 2 to 13, depending on the camera model. As you might expect, higher-priced models, in general, have more focus zones than do less expensive models. The autofocus system can also be fooled in some unusual situations—such as shooting through a window or photographing very shiny objects that scatter the infrared beam.

Another potential drawback is that autofocusing measures the distance to whatever happens to be centered in

Features of Automatic 35-mm Cameras
Most automatic 35-mm cameras have such standard features as automatic exposure setting; automatic focusing; and automatic film loading, advance, and rewind. Some models offer other options, such as a lens selector to switch from a wide-angle to a telephoto lens, and a DX-coding sensor for automatic film speed setting.

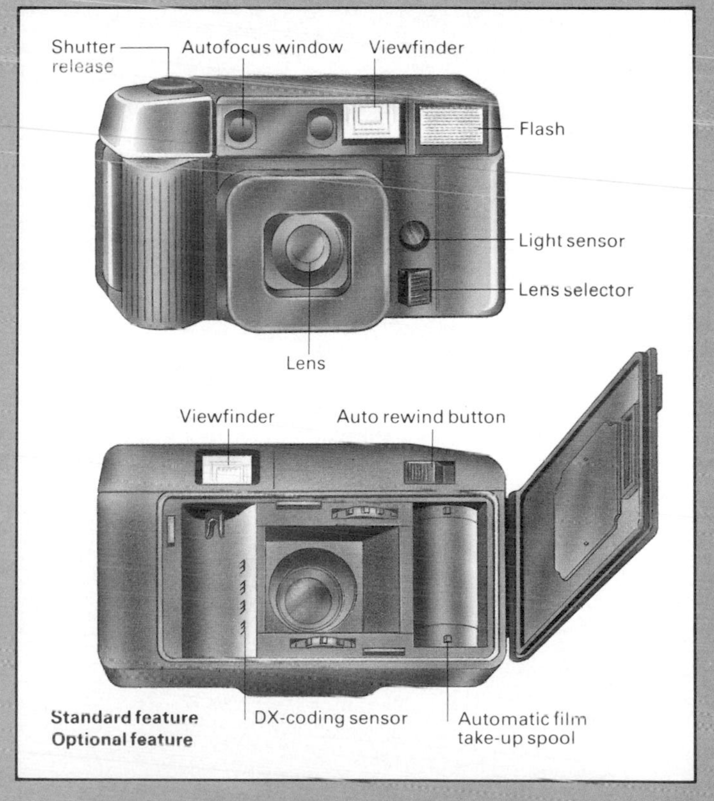

Most automatic 35-mm cameras will focus on whatever object is in the center of the viewfinder, *top.* To create compositions with the main subject not in the center of the photo, *above,* some cameras have a feature called *focus hold.* To use focus hold, first center the subject in the viewfinder and press the shutter button halfway down. This locks the focus in place. Then, move the camera to position the subject off to one side and shoot the picture.

the viewfinder. Unless your main subject is right in the middle of the picture, the camera will focus on something else, and your main subject will be out of focus. If you wish your main subject to be off-center, however, in order to make a more interesting composition, you will need a feature called *focus hold.* This feature allows you to first center your subject—a cat, for example—in the viewfinder, lock in the focus by holding the shutter button halfway down, then move the camera to position the cat off to one side. When you take the photograph, the cat will be in focus even though it is not in the center of the picture.

Most automatic 35-mm models are equipped with an auto winder, which does away with the need to manually wind the film through the camera. You simply drop in the film, pull out the film's leader, and close the camera. A motor automatically advances the film to the first frame. Every time you shoot, the motor winds the film to the next frame. After the last shot on the roll, the motor rewinds the film automatically. The top-of-the-line Fuji TW-300 adds an extra twist. After you drop in the film, the camera winds the film to the end of the roll, then rewinds it back into the cassette frame by frame as you shoot. This prevents photos from being spoiled if you accidentally open the camera and let in light.

All these functions are battery-powered. Most of these cameras operate on two AA or AAA batteries. Several models use ultralong-life lithium batteries, which last about five years in normal use. A few cameras can run on either kind of battery. Some models have a signal that tells you when batteries are running down.

Apart from these automatic features, a particularly valuable option is a telephoto focal length for taking portraits of people. *Focal length* is the distance from the center of a lens to the point at which light focuses on the film. The standard lens for most automatic 35-mm cameras has a wide-angle focal length of 35 mm to 38 mm, which is fine for outdoor scenes and group pictures. Head-and-shoulder portraits, however, require a focal length of 60 mm or 70 mm. Several camera makers now offer models with the ability to take both kinds of pictures. At the press of a button, the focal length can be changed from wide angle to telephoto and back. The view in the viewfinder also shifts to show you what the shot will look like. One model—the Pentax IQZoom—zooms between wide angle and telephoto, allowing for any focal length between the two.

If you want a fairly rugged camera for outdoor use, several manufacturers offer 35-mm automatic cameras sealed against dust and moisture.

Although new automatic cameras cannot duplicate the complete control over the final picture that is the hallmark of the advanced SLR, most amateur photographers welcome the ease, compact size, and versatility of this new breed of camera, which almost guarantees a perfect picture every time. [Arthur Fisher]

Choosing the Right Insecticide

With more than 800,000 kinds of insects in the world, chances are that some will crawl or fly their way into your home. Unwanted insects carry diseases and can destroy food, clothing, and property.

People spend millions of dollars each year on *insecticides* (chemicals that kill insects). But how do you know which one is best for eliminating the pest bothering you? Some insecticides are called *stomach poisons* because they kill insects that eat them. Others are called *contact poisons* because they kill insects that touch them by penetrating the insect's skin and destroying their nervous systems. There are also *chemosterilants* that make insects unable to reproduce.

Choosing an insecticide for your particular unwanted houseguest can be confusing. Insecticides come in a variety of forms, from sprays to powders. You must also decide whether to apply the insecticide yourself or hire an *exterminator* (pest-control expert) to do the job for you.

The type of insecticide you need depends on the type of pest you want to kill. Surface sprays, used on floors and other surfaces, are good for getting rid of ants, cockroaches, and other crawling insects. These sprays have coarse particles that leave a chemical film on surfaces. This film kills insects when they crawl over it.

Space sprays, or aerosols, are designed to kill flying insects, such as wasps or houseflies. These sprays have fine particles that float in the air. Sometimes space sprays can be used to force nonflying insects out of their hiding places—from behind baseboards, for example—so that they can be killed. Most surface and space sprays have one drawback, however—they give some insects only a weak dose because they spread unevenly in the air and on various surfaces. Some insects survive repeated weak doses of the spraying because they happen to have genes that make them resistant. They produce offspring that are also resistant. This resistance grows stronger in succeeding generations. New synthetic repellents—as well as new poisons—offer a solution to this problem because insects have not yet developed a resistance to them.

Insecticide dusts and powders work by clinging to crawling insects. When the insect cleans itself—by drawing its legs or feelers through its mouth—it ingests the chemical and dies. In addition, the insect often carries the powder back to the nest, where it may kill others.

Baits and traps are available for insects that live in hard-to-reach places and venture out at night, such as roaches. Some traps contain an *attractant*, a substance that draws the insect into the trap. Once inside, the insect eats a poison, which either kills it immediately while it is in the trap or kills the insect after it has left the trap and returned to its nest. Some traps on the market do not contain insecticides. Instead, they trap insects with a glue strip. Because insects usually do not survive a visit to a trap, this form of

Insecticides are packaged in a variety of forms, including sprays, powders, and traps.

pest control is sometimes more effective than sprays or powders.

A new development in insecticide application was introduced in 1985 by the d-Con Company of Montvale, N.J. The company developed a roach and ant killer in a large "magic marker." You simply "draw" the insecticide exactly where you want it.

Of all household pests, the cockroach has proved to be the most resistant to pest-control efforts. Roaches are rapid breeders, producing large numbers of offspring very quickly. They are also adept at developing a resistance to many insecticides. To overcome this problem, scientists have been working on a different roach-control strategy. They are developing chemicals that stop roaches from reproducing.

In 1985, the first birth control spray for cockroaches—Black Flag Roach Ender—was introduced in the United States by Boyle-Midway, Incorporated, of New York City.

Roach Ender is a spray that contains hydroprene, a chemical that makes adult roaches *sterile* (unable to reproduce) and prevents young roaches from reaching maturity. Roach Ender also contains an insecticide that will

How Insecticides Eliminate Pests

	Insect	Examples of Insecticides: Brand Names	Form	How It Works
Ants	Ants are biting insects that build their nests in soil, brick, and wood. They enter buildings through cracks in walls and under outside doors. Once inside, they feed on food and plants.	Black Flag Ant Traps Formula II	Trap	Destroys nervous system
		Raid Ant & Roach Killer	Spray	Destroys nervous system
		Pic Ant Traps	Trap	Stomach poison
Cockroaches	Cockroaches like warm, moist places. They hide behind baseboards and in cabinets in kitchens and bathrooms. They come out at night to feed on food and garbage. Roaches carry diseases.	Black Flag Roach Ender	Spray	Destroys nervous system; prevents reproduction
		Combat Roach Control System	Tray	Prevents digestion; roach starves
		Roach Prufe	Powder	Stomach poison
		Rid-A-Roach	Powder	Stomach poison
Fleas	Fleas often enter homes on the fur of pets, such as cats and dogs. They lay their eggs in carpets, beds, and cracks in floors. Fleas suck the blood of animals and humans.	Raid Flea Killer Plus	Spray	Destroys nervous system; prevents reproduction
		d-Con Flea Kill Fogger	Spray	Destroys nervous system; prevents reproduction
		Ortho Flea-B-Gone Flea Killer Formula II	Spray	Destroys nervous system
Flies, Mosquitoes	Flies and mosquitoes enter homes through open doors or windows. Both insects carry diseases. Flies contaminate food by landing on it; mosquitoes bite animals and humans.	d-Con Flying Insect Killer	Spray	Destroys nervous system
		Ortho Flying & Crawling Insect Killer	Spray	Destroys nervous system
Wasps, Hornets	Wasps build their nests in the ground and in tree hollows. Hornets build their nests outside homes—often under eaves, on porches, or in nearby trees. Both insects give painful stings when disturbed. Wasp stings may be fatal.	Ortho Hornet & Wasp Killer	Spray	Destroys nervous system
		d-Con Jet Stream Wasp & Hornet Killer	Spray	Destroys nervous system

quickly kill roaches that are not resistant to it. Roach Ender should be sprayed around kitchen sinks and cabinets; behind window frames, door frames, and baseboards; under tables and chairs; and in closets. It may take Roach Ender six to seven months to eliminate the cockroach population in the average-sized house or apartment. But within two weeks you can see one of its effects—hydroprene causes the wings of affected roaches to crinkle.

Two similar sterilizing sprays for fleas are Raid Flea Killer Plus, manufactured by S. C. Johnson & Company of Racine, Wis., and Flea Kill Fogger, made by the d-Con Company. These sprays also contain insecticides that kill some of the fleas.

Other new chemicals for killing roaches include Combat Roach Control System, a poison bait in a child-proof plastic tray. Its key ingredient is a chemical that disrupts the roach's ability to convert food to energy so that the insect starves to death even though it continues to eat. Cockroaches walk through the tray and carry the chemical back to the nest. Combat, made by American Cyanamid Company of Wayne, N.J., is sold in packs of 8 or 12 trays for under $4. The trays come in two sizes: a small size for German and brown-banded roaches and a large size for American and Oriental roaches. The company recommends using all the trays in the pack at the same time.

An alternative to roach sprays or roach bait is roach powder. One such powder, called Roach Prufe, contains boric acid, a deadly poison. Less than 5 grams (1 teaspoon) of boric acid can kill an infant. To prevent accidental poisoning, the manufacturer—Copper Brite, Incorporated, of Los Angeles—has added blue coloring to the white powder to distinguish it from flour or sugar. Copper Brite also has added a bitter substance to the powder to discourage children and pets from eating it. Copper Brite claims that the boric acid in Roach Prufe is ground finer than other roach powders and has been given an electrostatic charge to help it cling better to roaches.

Some insecticides kill several types of insects. S. C. Johnson's Raid Ant & Roach Killer also kills silverfish, spiders, and crickets. Many all-purpose insecticides contain chemicals called *organophosphates*. Many roaches, however, are resistant to these chemicals.

Another group of chemicals that kill a variety of insects are called *pyrethroids*. These chemicals kill quickly when sprayed directly on the insect. They also flush insects from hiding places, bringing them into further contact with the spray.

After you have determined which type of insecticide you need, you must decide whether to apply it yourself or hire an exterminator. If you apply it yourself, you must take certain precautions. You should wear protective clothing, such as a mask and rubber or plastic gloves. Make sure that you have enough ventilation and follow directions carefully. And always store insecticides away from children and pets to prevent accidental poisoning.

For particularly destructive insects, such as wood-eating termites, or for a major infestation of other pests, it might be a good idea to hire an exterminator. One way to determine the extent of an infestation is to count the number of insects caught in a trap.

For example, experts estimate that for every roach caught in a glue trap daily, there are 600 to 800 roaches roaming free in your home. If you find one roach in a trap you set out a few weeks ago, you probably do not have a major infestation. But if the traps consistently catch roaches day after day, you have a problem.

Before hiring a pest-control company, ask if the person to be sent to your home is licensed by the state as a pest-control specialist and able to discuss with you the types of chemicals to be used. It is also important to ask the firm if you can request alternative chemicals if you don't trust the ones to be applied by the pest-control specialist. If a pest-control company says no to any one of these questions, it is best to contact another firm.

To prevent insects from entering your home in the first place, keep your home clean and free of garbage. Keep foods covered in containers. Stuff gaps between pipes and walls with fine steel wool, and seal up holes in walls. In the end, the best type of pest control is prevention. [Donald L. Comis]

How Much Home Security Do You Need?

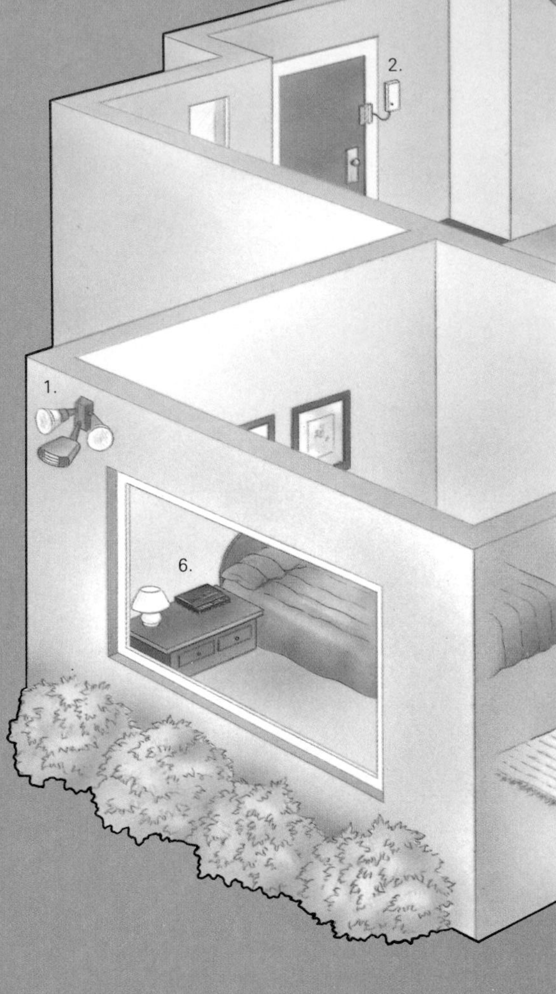

One benefit of modern technology has been the development of new devices and systems to protect houses from burglaries. Police officials say that houses and apartments equipped with home-security systems are 15 times less likely to be burglarized. Yet only 8 per cent of homes in the United States have security systems.

Security devices can be as simple as a dead-bolt door lock or as complex as a computer-controlled alarm system linked to a security agency or a police station. Some alarm systems set off sirens to scare intruders or alert residents to call the police, while others use a combination of sirens and floodlights. Some systems are turned on and off manually; others are controlled by computers.

To determine what type of security system is best for you, it is important to carefully assess your needs. Take a close look at your house or apartment. Are the outside doors solid and strong? Does every door and window lock securely? Is the yard well lit at night, with no dark shadows for a burglar to hide in? How much are you willing to spend? Choosing the right security system is a matter of matching your needs with the various systems available.

Security systems work in different ways, depending on the type of *sensor* they have. A sensor picks up the presence of an intruder and relays this information to set off lights and sirens or to signal police or security guards. Some sensors operate on electric current that passes through wires placed around doors and windows. If an intruder forces open a door or window, the circuit is broken and the alarm goes off. Another type of sensor sets off an alarm if it is jarred or moved. And some sensors are *infrared* (heat wave) detectors. These sensors can detect the body heat of a nearby human—or animal.

There are also sensors that work by detecting sound. Some of these sound sensors can detect more than just the noise of your back door opening or the rustle of bushes outside your window—they can pick up sounds that are above or below the range of human hearing.

Other systems have *microwave* (short-radio wave) sensors. These sensors work somewhat like radar. They send out radio waves that pick up changes caused by moving objects. Ultrasonic sensors use inaudible sound waves to sense movement in a similar fashion.

No sensing method is perfect, however. Infrared sensors can be fooled into sounding an alarm when a room heater turns on or when sunlight suddenly comes through a window. Sound sensors can be tricked when a fan or air conditioner causes an air pressure change in the room. Several companies, however, offer dual sensors that combine, for example, an infrared and an ultrasonic sensor. These dual sensors, which cost a bit more than single sensors, are very hard to fool. They

Types of Security Devices
Security devices vary widely in price and complexity. Outdoor security lights (1) with sound or movement detectors turn on to scare off prowlers. Electronic sensors mounted on doors (2) detect when doors are opened. Infrared sensors mounted on ceilings (3) or indoor walls (4) monitor rooms for the movement of an intruder. Electronic sensors mounted on windows (5) detect the breaking of glass and sound an alarm. In the most sophisticated security systems, sensors are hooked up to a computerized control unit (6), which is usually placed in a bedroom.

sound alarms only when the body heat and changes in sound waves caused by the movement of an intruder are detected simultaneously, preventing most false alarms. There are also dual systems that combine microwave and infrared sensors. But even a dual sensor does not eliminate all problems. A dual sensor, for instance, has no way of knowing whether a body moving through a room belongs to a human intruder or a family pet.

A major consideration in choosing a security system is price. Many security systems are relatively inexpensive and easy to operate. You can purchase an outside security light for less than $200. A security light may have an infrared or an ultrasonic sensor that turns the light on when it detects what may be an intruder. Some security lights have two floodlights and an infrared sensor mounted on a small control box. When the sensor detects a change in heat—from as far away as 15 meters (50 feet) in some cases—it turns on the outdoor floodlights by triggering an electric current inside the control unit.

Systems that can detect an intruder opening a door or window can be found for under $150. These systems involve a complex network of wires that run around windows and doors and through walls to connect with a central control unit. The control unit is usually mounted near the door, and you turn the system on and off with a

key that fits in the control box. Some systems have sensors that are wired to windows and set off a siren if someone tampers with the window. Most wired systems are compatible with one another so you can add other sensors, such as infrared or sound, to window-movement sensors. Wired systems, however, can be difficult to install, especially if you run the connecting wires within your walls.

There are also wireless home security systems. These systems use microwaves to send alarms from sensors to the system's central processing unit. The sensors stay on all the time, but if you turn the central unit off, alarm signals from the sensors are not processed. Wireless systems are easier and less expensive to install because the microwave sensors run on batteries and you do not have to string wires to connect each sensor to the control unit. The lowest-priced wireless systems cost less than $200.

Until recently, wireless systems were plagued with radio or static interference. Such interference either triggers false alarms or blocks the microwave signals from the sensors. The newest wireless systems, however, use coded signals and special frequencies to prevent false alarms.

The most sophisticated wired or wireless security systems are operated by computerized control units. Computerized home-security systems can be operated by a personal computer or by their own control unit, which usually consists of a keypad and a small screen mounted on a wall inside the home. Several manufacturers offer systems that can be wired to a touch-tone phone. To turn on the system, you simply punch a code into the phone. Some systems even allow you to call from outside the home, punch in an identification code, and activate the system.

Computerized systems can control each sensor individually. For instance, if the house is empty during the day, you can program the computer to turn on your entire system. At night, when the family is home, you might instruct the computer to turn on only the sensors that monitor outside doors, leaving inside sensors off so that the family can move around freely.

Many systems have timers that let you predetermine when you want the system to turn itself on and off. Such a system could turn itself on every night after you are in bed and turn itself off in the morning just before you get up. In addition, many computerized control units constantly monitor the entire security system and alert you if there is a problem—with a faulty sensor, for example—thus preventing many false alarms.

Among the most sophisticated home-security systems are those that employ television cameras and monitors. This type of system is expensive, costing at least $1,000. It uses small TV cameras and microphones connected to a TV monitor inside your home to show you who is at your door, or to look at any area around your home within view of a camera. TV systems have motion detectors so that whenever anything moves in front of the camera, the system automatically switches on and sounds a buzzer inside the house. Although designed as front door security devices, TV systems can also be used to watch over backyard pools to prevent unsupervised children from getting into trouble. In addition, parents can install a camera in the nursery so that they can keep an eye on the baby from other rooms in the house.

One of the most sophisticated and expensive TV-based systems is an eight-camera model that costs several thousand dollars and is designed for use in large homes or businesses. The system features automatic switching from one camera position to another to track any action. It also prints pictures of any scene that triggers an alarm. If someone breaks in, this system can provide police with actual pictures of the intruder.

Whether you install your own system or have a professional do it, be sure to learn how to use it properly. Police experts say that half of all false alarms occur because people do not know how to use their security systems properly.

There are home-security systems available for every type of budget. But whether you choose a simple or complex system, be sure it fits your needs and habits. [G. Berton Latamore]

Sorting Out Trends in Oral Health Care

The war against tooth decay and gum disease is being fought more fiercely than ever. People in the United States spend more than $3 billion annually on oral health care products, from toothbrushes to mouthwashes. Tooth decay is being conquered—more than one-third of all children in the United States between the ages of 5 and 17 have no cavities. Gum disease, however, remains a serious oral health problem.

It is not surprising, therefore, that manufacturers of oral health care products have begun introducing new weapons against gum disease and tooth decay. Every drugstore carries toothbrushes in a bewildering variety of shapes and sizes, toothpastes and mouthwashes that claim to fight early gum disease, and even different types and flavors of dental floss. But how do you know which oral health care products are best for you? Understanding the causes of tooth decay and gum disease may help you choose.

Shoppers are confronted with a bewildering array of oral health care products that claim to fight tooth decay and even gum disease.

The culprit behind both tooth decay and gum disease is *plaque*—a colorless, sticky coating of bacteria. It forms constantly on tooth surfaces, and its bacteria react with sugar and other carbohydrates in food to produce harmful acids. These acids can break down a tooth's *enamel* (hard outer covering) and can cause *gingivitis* (irritated and inflamed gums). Plaque contributes to the formation of a mineralized deposit called *tartar*. Tartar below the gum line makes plaque removal more difficult because tartar has a rough and porous surface to which harmful bacteria cling. For this reason, tartar promotes *periodontal disease*, a serious gum problem.

When gum disease reaches advanced stages, it affects the *periodontal ligaments* that anchor the tooth to the gum. It begins when irritants in plaque attack the gums. Eventually, the gums pull away from the teeth and form "pockets" between the tooth and the gum. More bacteria accumulate in these pockets. If left untreated, this gum disease eventually destroys both the periodontal ligaments and the bone supporting the tooth. As a result, the tooth falls out or has to be extracted. Periodontal disease causes 70 per cent of all tooth loss in adults in the United States.

Diligent removal of plaque is the key to preventing gum disease as well as tooth decay. Despite the claims of all the new products on the market, dentists say the best way to get rid of plaque is by daily brushing and flossing. Visiting your dentist regularly—usually twice a year—is also important. Only the dentist, or the dental hygienist, can remove the tartar.

According to the experts, the key weapon in removing plaque is your toothbrush. Toothbrushes come with soft, medium, or hard nylon bristles. Most dentists recommend brushes with

How to Brush Properly
Hold the toothbrush at a 45-degree angle with the bristles pointed toward the gums (1). Move the brush back and forth, gently scrubbing the sides (2) and the biting surfaces of the teeth (3). Also brush the backs of the front teeth (4) and the top of the tongue (5).

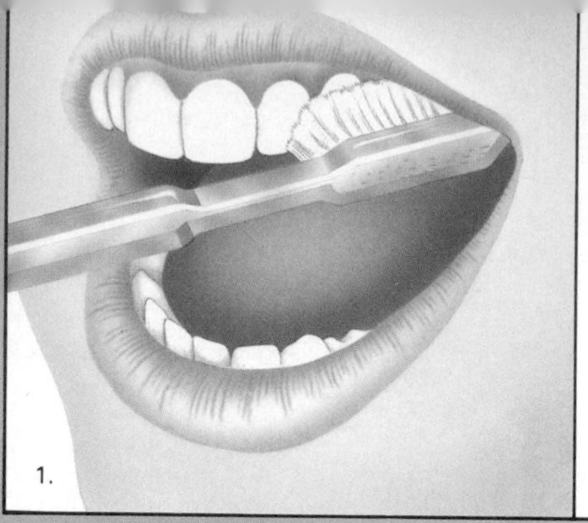
1.

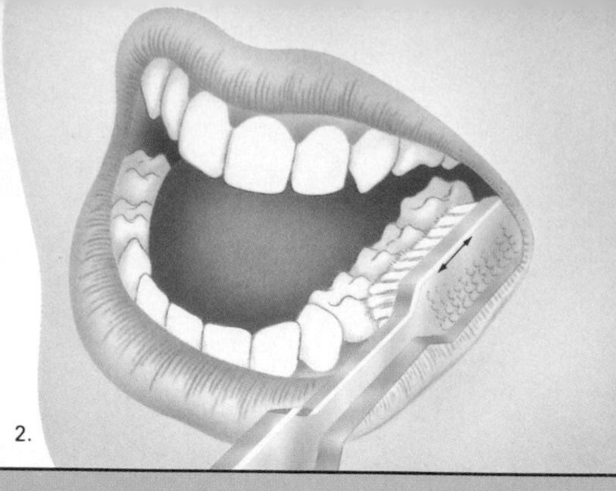
2.

soft or medium-soft bristles that have polished, rounded ends, because these bristles are less likely to injure the gums.

When choosing a toothbrush, make sure that its size and shape allow you to reach every tooth. Manufacturers have developed toothbrushes with unusually shaped handles or bristles, claiming that these brushes make cleaning of hard-to-reach teeth easier. Some toothbrushes have curved handles, small brush heads, or diamond-shaped heads designed to reach back teeth and gums more easily.

More important than the size or shape of your toothbrush is using it properly and brushing long enough. The American Dental Association (ADA) in Chicago recommends brushing teeth in a back-and-forth motion with the brush tilted at an angle of 45 degrees against the gum line. For the best results, research studies have indicated that brushing for approximately three to five minutes after meals will adequately remove plaque from all tooth surfaces.

Conventional electric toothbrushes have vibrating or rotating bristles, but studies have not shown that they do a better job than manual toothbrushes. Handicapped people or young children, however, may find electric toothbrushes easier to use.

Two new electric toothbrushes—Interplak and Rota-Dent—are reportedly effective in removing plaque. Interplak, manufactured by Dental Research Corporation of Tucker, Ga., has several sets of bristles that rotate at high speed, with each set turning in the opposite direction from the set next to it. One study found that this toothbrush removed 98 per cent of plaque, compared with 49 per cent for a conventional toothbrush. Interplak costs about $100.

Rota-Dent, distributed by Pro-Dentik, Incorporated, of Batesville, Ark., resembles the powered instrument dentists and hygienists use to clean teeth. This toothbrush has more than 5,000 fine, rotating bristles and costs approximately $70. If you are uncertain about what type of toothbrush is best for you, ask your dentist.

Although a toothbrush alone will remove plaque from most easy-to-reach tooth surfaces, most people prefer to use toothpaste when brushing. Toothpastes have become more colorful and flavorful to make them more attractive to consumers. In choosing a toothpaste, however, you should check to make sure it contains *fluoride* and mild *abrasives*. Fluoride is a naturally occurring mineral that helps strengthen teeth against decay.

Abrasives in toothpastes are usually

How to Floss Properly
Take a piece of floss about 45 centimeters (18 inches) long and wind it around the middle fingers of each hand (1). Gently guide the floss between two teeth and rub it up and down against each tooth's surface and under the gum (2). Also rub it against the back teeth (3).

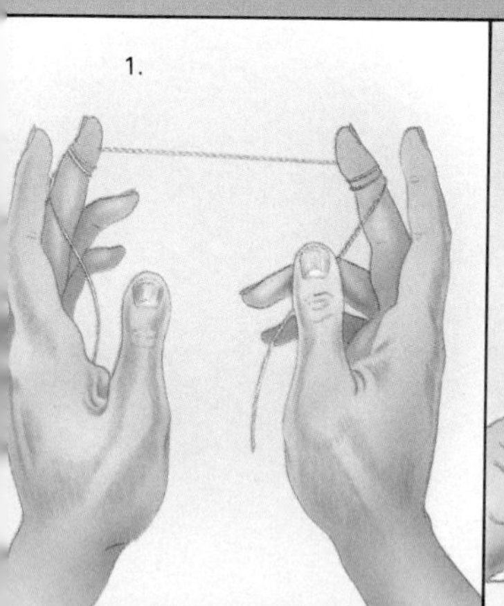
1.

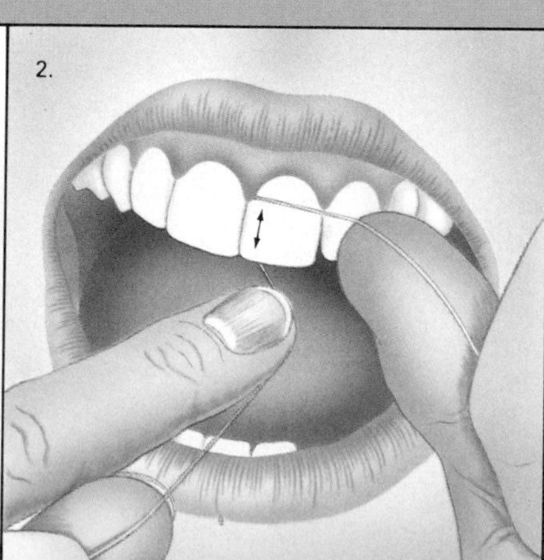
2.

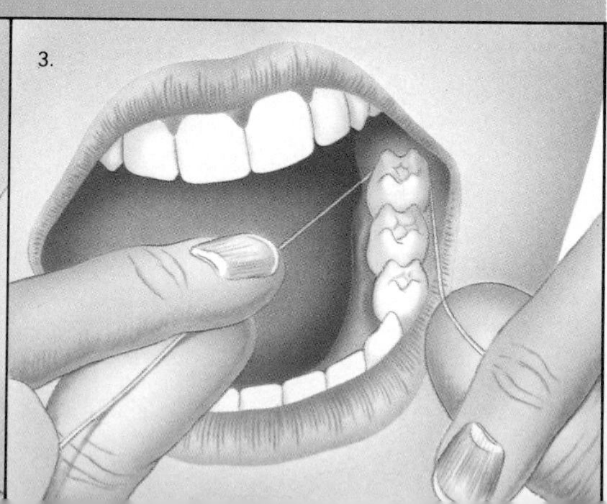
3.

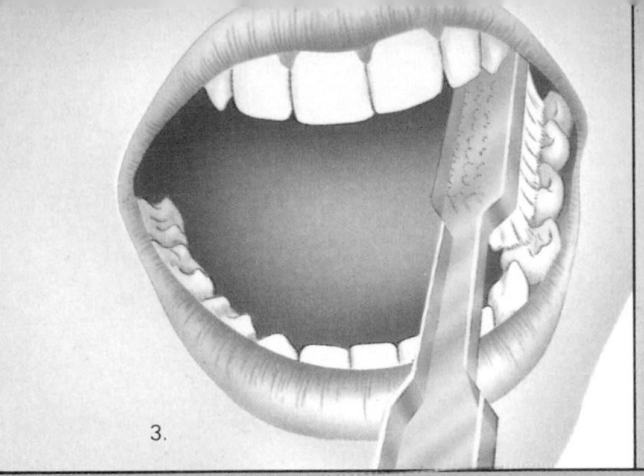

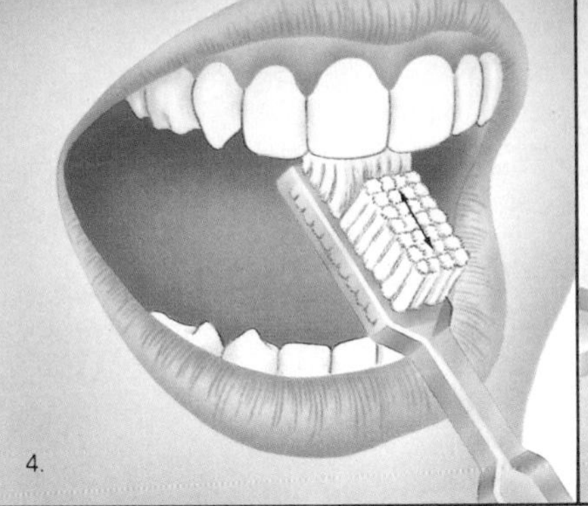

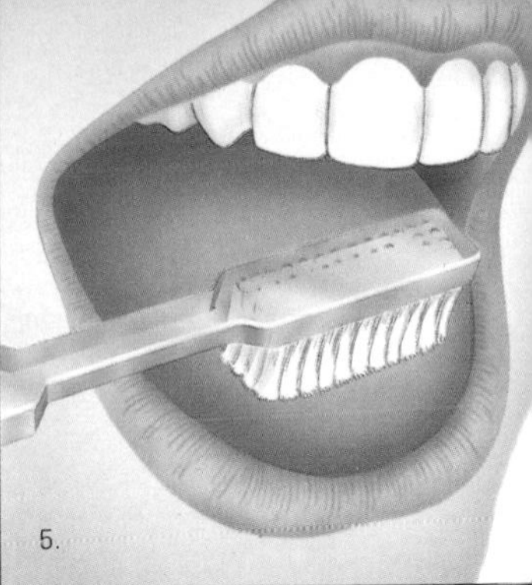

inorganic salts that help remove plaque, food particles, and stains from teeth. Studies have shown that the abrasive ingredients of a toothpaste enhance the cleaning power of a toothbrush. According to the ADA, mild to moderate abrasives will not hurt tooth enamel, but toothpastes with harsher abrasives could damage teeth and gums. Ask your dentist which toothpaste to use.

Some toothpastes on the market today are reputed to have chemical additives that fight plaque and control tartar. Most of these products, however, do not contain any chemical additives or special ingredients. The ADA has not yet put its seal of acceptance on any toothpaste that claims to chemically fight plaque. It has accepted two tartar-control toothpastes made by Crest and Colgate—but only for their fluoride content. The ADA considers their tartar-control benefits to be cosmetic because though they contain chemicals that inhibit the unsightly formation of tartar on the teeth, they do not affect the disease-causing tartar under the gums.

Flossing is as essential to dental hygiene as brushing. Flossing is done with a special type of thread called *dental floss.* Floss reaches between teeth and under the gum line to remove plaque and food debris that toothbrushes miss. It is very important to learn how to floss properly so you do not injure your gums. Your dentist or hygienist can teach you the right method.

Flosses come waxed or unwaxed. There is no significant difference in the cleaning ability of these flosses, however. Some experts say waxed floss may make it easier to clean between tightly spaced teeth and around dental work. Other experts say unwaxed floss is better because its strands spread out to pick up more plaque.

Mouthwashes have also joined the antiplaque bandwagon. Some mouthwash manufacturers say that their new products contain chemicals that "loosen" plaque. Manufacturers of mouthwashes that have been on the market for many years have also begun to make antiplaque claims. Not enough scientific information is available to prove that these claims are true, however. The one exception is Peridex, a prescription mouthwash manufactured by Procter & Gamble Company of Cincinnati, Ohio. Peridex, which has been accepted by the ADA, contains chlorahexadine, a chemical that studies have shown is an effective antiplaque agent.

Experts do not know whether reducing plaque with chemicals is helpful. In any case, the ADA warns that mouthwashes should not be substituted for brushing and flossing. It does recommend, however, that people supplement brushing and flossing with rinses that contain fluoride. Studies have shown that fluoride rinses reduce tooth decay by 40 per cent when used daily.

Besides toothbrushes, toothpastes, flosses, and rinses, there are other dental aids you can buy to use at home. An irrigator, for example, sends out jets of water that help remove food particles from between teeth. Irrigators may be useful for people with braces or dentures, but they have little effect on plaque.

More research is needed to determine the value of all the new oral health care products. In the meantime, proper brushing and flossing and regular visits to the dentist remain our best weapons against tooth decay and gum disease. [Richard W. Asa]

Facts About Acne Treatment

Gently washing the face with water and a nonoily soap will help keep skin clear. For blemishes, dermatologists recommend acne medications, *inset,* that contain benzoyl peroxide and other drugs approved by the U.S. Food and Drug Administration for acne treatment.

Pimples. Zits. Blemishes. No matter what we call them, they are all acne, a skin disorder that plagues most teen-agers and many adults. Although acne cannot be cured or prevented, those much-lamented "unsightly blemishes" can be kept to a minimum with do-it-yourself home care and over-the-counter or prescribed medications.

Common acne, *acne vulgaris,* most often occurs in teen-agers. It usually disappears about five years after it starts. But in approximately 5 per cent of the United States population, outbreaks continue into adulthood. Another type of acne, *acne cosmetica,* occurs in adults who use heavy moisturizing lotions, makeup foundations, and oily soaps.

Myths about acne abound. Contrary to common belief, researchers have found that eating chocolate, pizza, fried foods, and cola drinks does not cause acne and, in most people, does not worsen it either. Stress, however, probably does play a role. Scientists note that events such as exams, a big date, or an argument with parents or friends seem to trigger skin eruptions.

The exact cause of *acne vulgaris* is unknown. Because it appears during early adolescence, *dermatologists* (physicians who specialize in skin care) believe that the body's increased production of hormones may play some role. The sex hormone *androgen* stimulates the *sebaceous glands,* oil glands located underneath the skin. The oil then empties out through ducts below the skin. The most dense concentrations of oil glands and ducts occur on the face and the upper back and chest, the most common sites of acne eruptions.

Acne is caused when the oil ducts develop plugs of excess oil and dead skin cells. A small plugged-up duct is called a *whitehead* or a *blackhead.* The blocked-up gland provides an ideal breeding ground for bacteria, sometimes resulting in a leak or rupture of the wall of the gland or duct. A leak in the wall can produce a small red bump or a larger bump filled with yellowish pus. A leak or rupture that does not drain completely may cause an infection, forming a red lump called a *cyst.* Cysts can scar skin tissue as they heal, and some cysts leave lumps that may last for months or years.

Not much more is known about the cause of *acne cosmetica,* but it appears that heavy use of makeup, creams, or after-shave moisturizers can block glands and ducts—leading to a build-up of oil, dead cells, and bacteria.

Acne flares up in some women just before menstruation or as a result of a change in their use of birth control pills. Presumably, changes in hormone production are responsible for these outbreaks. And some medications can trigger acne. These include some vitamin and mineral preparations, *steroids* (chemical compounds that influence the body's metabolism), and certain drugs that affect the nervous system.

Because acne can generate emotional distress and potentially lasting physical scars, it is important to treat this skin disorder properly. Do not squeeze or pick pimples; this may spread the infection and cause scarring. And try to keep your hands, which carry oils and bacteria, off your face.

Dermatologists advise people with acne to avoid foundation makeup and oily moisturizers and makeup remover. They also suggest washing the affected areas three times a day with a mild, nonoily soap. Wash your skin gently to remove surface oils. Vigorous scrubbing or use of abrasives may rupture plugged ducts. It is also important to wash your hair frequently and to keep your hair combed up off your face and forehead.

According to dermatologists, antibacterial soaps and astringents do not clear up acne. Medicated soaps, which are rinsed off the skin are less effective than medicated lotions that are left on the skin.

The U.S. Food and Drug Administration (FDA) has approved many drugs for acne. For treating and preventing mild acne, these include over-the-counter medications containing benzoyl peroxide, sulfur, resorcinol with sulfur, or salicylic acid, all of which cause the skin to dry and peel.

Dermatologists advise people with *acne cosmetica* to avoid using creams, lotions, and foundation makeup. An outbreak of this type of acne can take at least two months to clear up. Those who start wearing makeup again should choose cosmetics labeled "oil-free" or "less-acnegenic."

If acne persists—or becomes severe—you should consult a dermatologist. The doctor may prescribe lotions, perhaps containing antibiotics, that can be applied to the skin. For severe acne, a dermatologist may prescribe antibiotics to be taken orally to kill the bacteria in the glands under the skin.

Tetracycline is the antibiotic that is most commonly prescribed. While it is regarded as extremely safe for long-term use, it is not recommended for pregnant women or young children as it may stain developing bones and teeth. This antibiotic also increases the risk of sunburn.

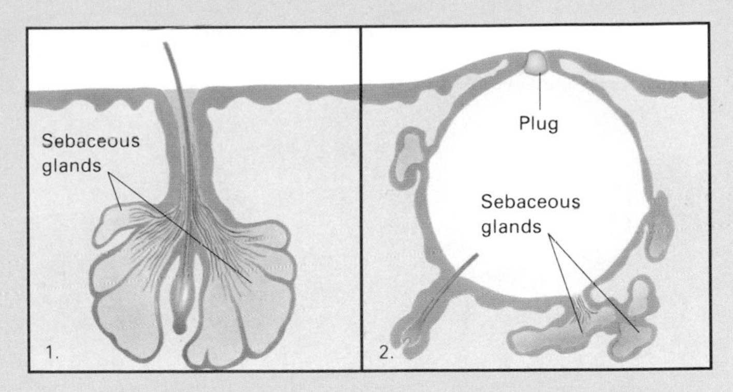

How a Pimple Develops
Normally, oil secreted by the sebaceous glands, along with dead skin cells, travels through ducts leading from the glands and empties out through a channel that opens on the skin's surface (1). A pimple forms when excess oil and dead skin cells plug the ducts (2). This causes a build-up of oil, dead cells, and bacteria, which expands the ducts' walls.

Dermatologists also may prescribe medication containing vitamin A acid, which helps unplug the ducts. Vitamin A acid may darken black skin, however, and it should not be used by anyone exposed to continuing high levels of sunlight.

Since 1982, a highly effective drug called *13-cis-retinoic acid* has been available in the United States to treat cystic, or deep-scarring, acne. The drug reduces the size of sebaceous glands, as well as their oil production. But it has some serious side effects, including birth defects in children conceived while the mother is taking the drug. Therefore, 13-cis-retinoic acid is recommended only for cystic forms of acne or extremely severe cases of acne that have not responded to other types of treatments.

New and improved techniques for the removal of acne scars also are available today. Such techniques include *dermabrasion*, in which motor-driven brushes or diamond wheels remove scarred skin, and chemical face peels, which use special acids to remove outer skin layers. Injections of collagen, a purified animal protein, are sometimes used to fill in sunken scars. Since the collagen is gradually reabsorbed, the treatment has to be repeated at 18- to 24-month intervals.

Punch grafts, in which small plugs of skin are transplanted from the back of the neck to the face is another alternative. A doctor may also recommend surgery to remove deep scars.

Acne is not a pleasant experience. But with proper care and treatment, it is possible to keep outbreaks to a minimum and improve even the most severe cases. [Lynne Lamberg]

People in Science

A love of science and what it can teach us about the world we live in often springs up early in life; whether it flourishes depends on education. This section deals with both the learning and teaching of science. It begins with the story of a man who, after early training in a special science high school, went on to learn how cancer-causing substances can reveal themselves. It continues with a look at several high schools of science and what they are doing to encourage a love of science in both gifted and more ordinary students.

SOCIAL
STUDIES
MATH

This expert on cancer-causing substances applies his research to products we encounter in our everyday lives.

Bruce N. Ames

BY LUCILLE DAY

One of the first things visitors notice about the office of biochemist Bruce N. Ames is a collection of cartoons taped to a file cabinet. One of his favorites shows a lion cautioning another as they stare at a hunter: "Don't eat one of them. They're loaded with additives and preservatives." Cartoons that satirize the dangers of chemicals added to foods are particularly appreciated by Ames, who developed the world's most widely used test for identifying food additives and other substances likely to cause cancer.

Another cartoon shows an elderly couple grocery shopping. The woman confides to her husband, "I don't trust natural. People die all the time from natural causes." This, too, is an especially appropriate sentiment to find on Ames's file cabinet, because Ames—an expert on diet and cancer—has been instrumental in alerting the public about cancer-causing substances that occur naturally in food.

Ames, chairman of the Biochemistry Department at the University of California at Berkeley, is a multitalented scientist. As a biochemist, he studies the chemistry of living organisms. He is also a geneticist, concerned with the workings of *genes*—the complex chains of *deoxyribonucleic acid* (DNA) that are found in all living cells and that determine all inherited traits.

But Ames's special gift as a scientist is his ability to build bridges between his work in the laboratory and everyday life—to find practical applications for his research. Because of his interest in DNA and the connection between DNA damage and cancer, he invented a test to detect *mutations*—changes in the structure of genes. He then started looking for *mutagens*—agents that cause mutations—in common substances.

And Ames found such mutagens, lots of them, including hair dyes and chemicals used to make children's sleepwear fire-resistant. These substances were subsequently shown to be *carcinogenic*—cancer-causing—as well. Today, Ames's test is an important tool of *genetic toxicology*—the study of substances in the environment that damage genes.

Bruce Ames was born on Dec. 16, 1928, in New York City. Thanks to his father—who was the chairman of a high school chemistry department, and later supervisor of science for all the New York City public schools—Ames and his two younger sisters were always surrounded by science. As a youngster, he was particularly interested in chemistry and biology, eagerly reading the books his father left lying around the house. During the summers, which the Ames family spent at a lakeside cabin in the Adirondack Mountains of New York, he explored the natural world and collected frogs, salamanders, and snakes.

He also had a voracious appetite for books. "I always read enormous amounts," Ames says. "I would come back from the library with a whole stack of books."

As a teen-ager, Ames attended the Bronx High School of Science, where, as the name implies, special attention is given to the study of science (in the People in Science section, see HIGH SCHOOLS OF SCIENCE). There, he conducted experiments to learn how plant hormones affect the growth of tomato root tips grown in laboratory glassware. Propelled by that first taste of the excitement of scientific research, Ames enrolled at Cornell University in Ithaca, N.Y., to study the subjects that had appealed to him since childhood—chemistry and biology.

Ames, a slender, agile-looking man with lively hazel eyes, discovered another talent during his college years—dancing. "It was the first thing I really did well in my life," he declares. He still dances, and his specialties include Scottish country dancing and ballroom dancing. "I'm always happy to waltz all night," he says.

Although Ames was extremely imaginative and adept at problem solving, he was not an "A" student at Cornell. He liked to follow his own interests—which did not always coincide with the assigned reading for his courses. "I just went off my own way in whatever I was doing," he recalls. "For example, I got tremendously excited about Jane Austen when I was in college, and one semester I read all of Jane Austen instead of the books for the courses I was taking."

Nevertheless, Ames managed to do well enough in his courses to be accepted at a prestigious graduate school. He graduated from Cornell in 1950 and headed west to the California Institute of Technology (Caltech) in Pasadena.

Ames found Caltech exhilarating. "Those were marvelous years,"

The author:
Lucille Day is a science writer and director of Precollege Education at Lawrence Berkeley Laboratory in Berkeley, Calif.

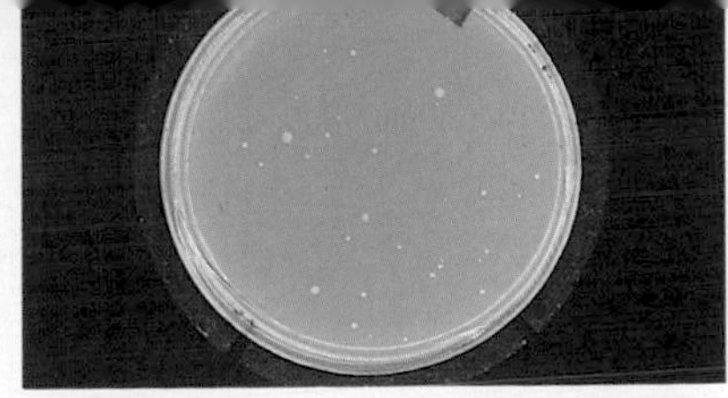

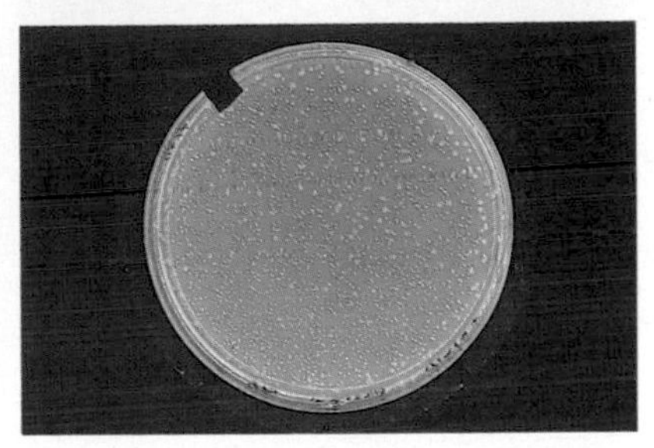

Bruce Ames, *above left,* examines a laboratory dish containing bacteria used in the Ames test to identify *mutagens*—substances that cause *mutations* (changes) in genes. Such mutagens are also likely to be *carcinogens*— cancer-causing substances. In this test, mutagens cause a change in the bacteria that enables the bacteria to grow. Only a few bacteria grow on one dish, *upper right,* indicating it does not contain a mutagen. Many bacteria grow in another dish, *above right,* showing that the substance being tested is a mutagen.

he says. "The students were very high caliber, and many of them are now in the National Academy of Sciences"—a highly distinguished society of accomplished scientists to which Ames himself was elected in 1972.

There were also exceptional professors at Caltech working in Ames's field. During the early 1950's, when Ames was a graduate student, not much was known about how genes worked. Even the molecular structure of DNA was a mystery until American biologist James D. Watson and British biologist Francis H. C. Crick devised their model of DNA in 1953. But some people learned about genes by studying the products of genes—enzymes and other proteins—a field known as *biochemical genetics*. Several of these pioneers in biochemical genetics, including George W. Beadle, Norman H. Horowitz, and Herschel K. Mitchell, were professors at Caltech.

Ames joined Mitchell's laboratory, and decided to tackle the problem of figuring out the step-by-step process used by a cell to make an amino acid called *histidine*. Amino acids are the building blocks of proteins. He chose to study the histidine-making process in *Neurospora crassa*—an orange-colored bread mold often used in genetic research.

A cell of this bread mold contains a set of genes that are needed by the cell to produce histidine. Each gene contains instructions for making an enzyme. When a mutation occurs in one of the genes, the mutant mold cell makes a defective enzyme, and no histidine can be made. Using a number of such mutants, Ames was able to unravel the series of chemical reactions the cell carries out to make histidine—reactions that are controlled by the mold cell's genes and that reflect the ordered activity of those genes.

In 1953, having completed this ambitious research project after only three years at Caltech, Ames received his Ph.D. degree and headed back east to do research at the National Institutes of Health (NIH) in Bethesda, Md. He decided to attack the important question of how the activity of genes is controlled in a cell. This puzzle of gene regulation—how genes are turned "on" and "off" so that enzymes and other proteins are produced at the right time—is crucial to understanding fundamental issues of biology, such as how a single fertilized egg develops into a human being.

Ames continued to work with the bread mold for a while, but soon switched to using a bacterium called *Salmonella typhimurium*, because it is a simpler organism and it grows faster—making it possible to complete experiments more quickly. Even so, it took about 15 years of research before Ames discovered the chemical signals that turn on and off the genes involved in making histidine.

While working at the forefront of research in his field, Ames revealed another side of his creativity—inventiveness. He invented many of the techniques required for his research. One of the most important was a process for separating out certain proteins, which he developed with cell biologist Robert G. Martin. This technique uses a *centrifuge*—a device that holds a ring of test tubes and spins them around at very high speed—to separate certain proteins in a sugar solution from one another. Heavier proteins wind up at the bottom of the test tubes, while lighter ones are suspended higher up. Being able to separate one protein from others makes it possible to study its structure and function. Ames first used this technique to separate the enzymes required by a cell to make histidine. Since then, thousands of researchers have used the method to separate many different kinds of proteins.

As a boy, Ames spent family vacations at a lakeside cabin, exploring the world of nature, *top.* His interest in science eventually led him to the California Institute of Technology in Pasadena, where as a graduate student in the early 1950's, *above,* he did research in genetics and biochemistry.

In 1958, Giovanna Ferro-Luzzi, a young Italian biochemist working at Johns Hopkins University in Baltimore, attended an NIH seminar given by Ames. Shortly after that, the two met at another scientific gathering. Then, Ames says, "I started chasing her—or vice versa." He invited her to help him plant dahlias in his garden, and soon afterward took her to see the celebrated Bolshoi Theater Ballet on tour from the Soviet Union. In 1960, they married.

Ferro-Luzzi, an energetic, enthusiastic woman—and a highly acclaimed scientist in her own right—says that her husband brings the same creativity to his family life as he does to his research. She describes him as "very understanding, persistent, and imaginative—with sort of an artistic flair thrown in."

"He'll come into the kitchen," she explains, "and think how some gadget could be designed a little better—that's very characteristic of him." She adds, good-humoredly, "He's also very dreamy—most of the time he's not listening to what I ask him or tell him to do."

Shortly after they were married, Ames and Ferro-Luzzi spent a year abroad, doing research in the labs of two Nobel laureates—Crick at Cambridge University in England and biologist François Jacob of the Pasteur Institute in Paris. At Cambridge, Ames studied *polyamines*—small molecules that bind to DNA and, he believes, protect it against mutagenic substances. In Paris, he continued his work on the genes involved in making histidine.

Ames returned to the NIH in 1962, and a year later, two important events took place. First, Ames's daughter, Sofia, was born. Then, the day after her birth, the American Chemical Society (ACS) told Ames he had been chosen to receive the Eli Lilly Award for his research—his first major scientific award.

When the editors of *Chemical and Engineering News* magazine asked for a photograph to accompany the announcement of Ames's award, Ferro-Luzzi suggested that he send a picture she had recently taken. It showed Ames, who was recovering from a cold, sitting under an avocado plant looking dejected, unshaven, and in need of a haircut. Amused, he sent it off—and when the magazine appeared, he received an avalanche of fan mail from people delighted to see "a scientist the way he normally looks."

When asked to provide a portrait to accompany a news story about his winning an award, Ames, as a joke, sent a photo taken while he was recovering from a cold—and looking rather disheveled and dejected. Ames received a flood of mail from readers delighted to see a picture of "a scientist the way he normally looks."

The following year was also eventful for Ames. That year, his son, Matteo, was born. Characteristically, Ames brought the same inventiveness that served him in the laboratory to raising his children—especially as they grew into teen-agers. "All his sympathy, understanding, and flexibility came out," says Ferro-Luzzi. "He could really converse with them, I think, much better than I could."

But 1964 turned out to be memorable for another reason besides the birth of Ames's son. During that year, he happened to read the list of ingredients on a package of potato chips. Mulling over the list, he began to think about all the new synthetic chemicals entering the environment and the damage they might cause in human cells. A natural problem-solver, Ames began to wonder if a simple test could be developed to screen these chemicals for their ability to cause mutations in genes. As a hobby, he started looking for mutagens, using the *Salmonella* bacteria strains he had developed for his research on histidine.

In the *Salmonella* test—which everyone but the modest Ames refers to as "the Ames test"—certain strains of *Salmonella* bacteria are placed in laboratory dishes along with some nutrients and the chemical to be tested. Normally, these special bacterial strains will not grow in the laboratory dishes, because they contain a mutation that makes them unable to produce histidine. But when a substance that causes mutations is added, such as cigarette smoke, some of the bacteria will acquire a second mutation. And because some of these newly mutated bacteria are able to make histidine, they grow and form small clumps of cells called colonies. A chemical that causes mutations in the genes of bacteria may cause mutations in human genes, too. So a mutagen identified by the Ames test is well worth further scrutiny.

Ames became convinced that one important property of cancer-causing substances is their ability to damage DNA—in other words, to act as mutagens. "We kept lists of commonly studied carcinogens and kept working to make improvements in the test that could detect them as mutagens," says Ames.

But at first, the *Salmonella* test failed to detect some important

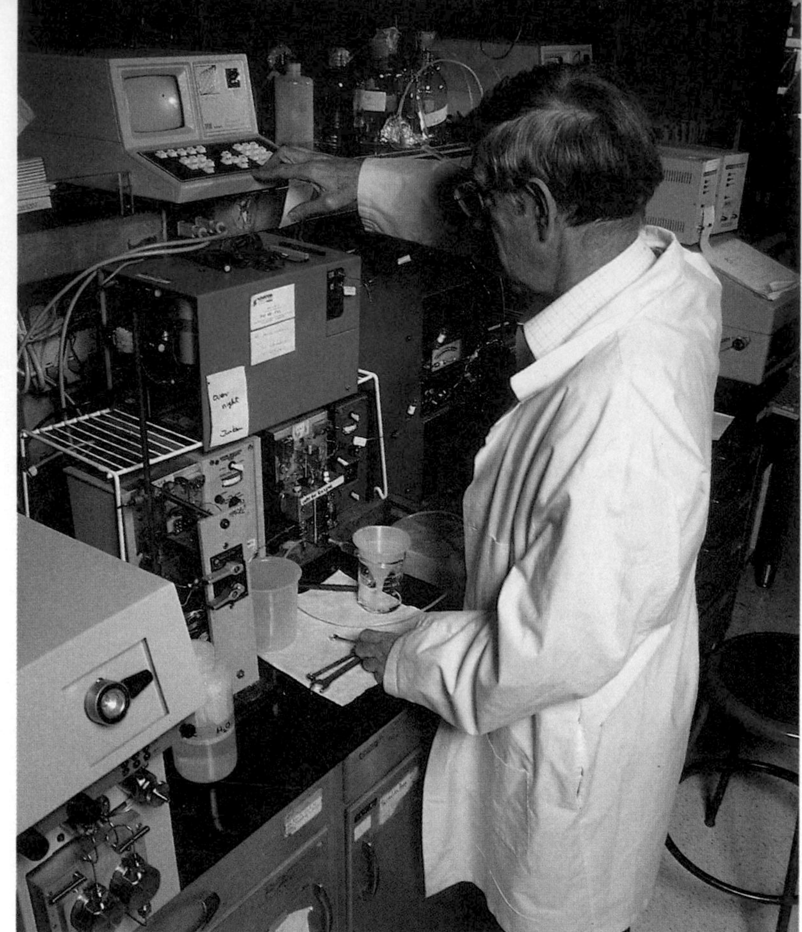

In his kitchen, *above,* Ames brews a morning cup of coffee, even though his research has shown that coffee and other foods contain carcinogens. Ames says the amount of carcinogens in these foods is much too small to worry about if the foods are consumed in moderation. In his laboratory at the University of California at Berkeley, Ames continues to look for links between diet and cancer, *right.*

mutagens that had already been identified as carcinogens. So Ames and his colleagues kept tinkering with the test, hoping to improve its sensitivity.

Ames then acted on a hunch. He knew that one of the functions of an animal's liver is to break down *toxic* (poisonous) substances taken into the body. Perhaps, Ames reasoned, the liver converts some substances into mutagens. He added ground-up rat liver to the test dishes in an attempt to imitate what happens to chemicals in a living animal's body.

The hunch paid off. Ames retested powerful carcinogens—such as benzpyrene, a chemical found in cigarette smoke and many other substances—that the test had previously failed to identify as mutagens. "This time," says Ames with satisfaction, "the test worked." With the added rat liver and other improvements, Ames and his group were able to show that more than 80 per cent of carcinogens are also mutagens.

Ames says that his most important talent in science is his imagination. And his hunch about how to improve his test for environmental mutagens is a typical illustration of his creative process. "When I come across a new fact, it's like trying to fit a piece into a jigsaw puzzle. I compare it to all the other facts in my mind to see

Ames has many activities besides his own research. For example, he gives advice and monitors the work of one of his graduate students, *above left,* then spends hours in his office, *above,* reading reports and evaluating research grant proposals for the government and for private foundations.

what connects," he explains. "I'm always making odd connections and asking what they might explain."

Today, the Ames test is used in more than 3,000 laboratories—including those of most major drug and chemical companies. It is widely regarded as an important first step in identifying substances that might cause cancer. After a substance is shown to be mutagenic by the Ames test, it is subjected to animal cancer tests, in which rats or mice are fed the substance over a long period of time to see if they develop cancer. The Ames test is performed first because it is inexpensive, and because it allows a chemical to be screened in a matter of days and at a cost of a few hundred dollars. Animal cancer tests, in contrast, are extremely expensive—about $500,000 per substance tested—and take years to produce results. With hundreds of thousands of new chemicals being developed each year, it would not be feasible for companies to conduct animal tests on all of them.

Ames, who is not particularly motivated by money, never patented his widely used test. "I figured I didn't need the money," he says. Besides, he adds, characteristically, "I was too busy with other things anyhow."

In 1968, Ames left the NIH and headed west again—this time to join the faculty at the University of California, Berkeley, as a full professor. Ferro-Luzzi, an expert on the proteins involved in transporting amino acids into cells, also became a researcher—and later a professor—in the Biochemistry Department. Ames continued to make a number of important improvements in the *Salmonella* test, taking advantage of recent developments in genetic engineering technology. Using genetic engineering techniques, Ames and his co-workers created bacteria with custom-made mutations—that is, they knew both the exact location and the type of chemical changes in the DNA of the mutant strain of bacteria. These customized bacteria became useful tools for learning precisely what physical changes occur in the DNA molecule when it is exposed to a mutagen.

Ames receives the Charles S. Mott Prize for cancer research from General Motors (GM) Chairman Roger B. Smith in 1983. The GM Research Foundation award is one of the many honors that have been bestowed on Ames for his research.

Since 1981, Ames has held an additional post at the Lawrence Berkeley Laboratory (LBL), a national scientific laboratory operated by the University of California for the United States Department of Energy. He is currently collaborating with Lois S. Gold of LBL to develop a computer database—a computerized "library"—designed to help people evaluate the hazards of both artificial and natural cancer-causing substances.

Ames and Gold's database includes the results of thousands of animal cancer tests conducted on hundreds of carcinogens. Using this information—and the estimated amount of each of the substances that a human being will be exposed to during a lifetime—they have created a "possible hazard scale." This scale rates substances according to the strength of the carcinogen and the typical human exposure. Ames hopes the scale will serve as a kind of yardstick to help people estimate the relative risks of different substances—and to pinpoint the most hazardous substances.

But Ames does not consider the hazard scale the final word on these carcinogens. The ratings are based on animal cancer tests, and, as he is quick to point out, "A substance that causes cancer in rats doesn't necessarily cause cancer in mice, let alone people."

Ames is ready to speak up when he sees a genuine hazard to the public's health. He has served on a number of government advisory panels and is frequently asked to testify before government committees concerned with toxic chemicals in the environment. In 1981, for example, concerned about factory workers exposed to the pesticide ethylene dibromide (EDB), he testified before the California Senate Committee on Toxics and Public Safety Management. As a result of the testimony of Ames and others, the standard for EDB exposure in California is now 100 times more stringent than the national standard.

On the other hand, Ames believes that in certain cases, people are unduly concerned about tiny amounts of synthetic chemicals in the environment. For example, in California's "Silicon Valley"—an area where many computer-related industries are located—well water contains small amounts of trichloroethylene (TCE), a carcinogenic industrial chemical. But, says Ames, Silicon Valley well water is ac-

Imitating the figures in American artist Grant Wood's famous painting *American Gothic,* Ames and his wife, biochemist Giovanna Ferro-Luzzi, strike a pose in their garden.

tually safer to drink than a glass of ordinary chlorinated tap water. The tap water contains trace amounts of chloroform, a carcinogen used to sterilize and disinfect drinking water. But neither kind of water, says Ames, poses very much risk to the public.

The money and other resources currently devoted to tracking down and removing tiny amounts of weak carcinogens could be channeled more usefully, Ames believes. "We should try to determine what's really important in causing cancer and zero in on that," he says. Useful efforts, in his opinion, would be studies to identify significant artificial and natural carcinogens, programs geared to help people stop smoking—a major cancer risk—and research into the relationship between diet and cancer.

Ames himself has given a great deal of thought to diet and cancer. In 1983, he wrote an article for *Science* magazine that thrust the issue into the limelight. In the article, Ames discussed common foods that naturally contain mutagens and carcinogens. Since the early 1980's, he and others have used the Ames test to detect naturally occurring mutagens in a variety of foods, including coffee, mushrooms, celery, parsley, peanut butter, black pepper, comfrey herb tea, and mustard.

Many plant foods contain natural toxic chemicals because these chemicals protect the plants from hungry insects, explains Ames. He and others are now showing that many of these substances are both mutagenic and carcinogenic.

"The amount of nature's pesticides we are ingesting is at least 10,000 times the level of artificial pesticides," he explains. That is one reason why foods such as peanut butter and beer, which contain natural carcinogens and which people routinely eat and drink, score much higher on Ames's hazard scale than do artificial pesticides, such as DDT.

Fortunately, Ames points out, human beings and other animals have evolved many defenses against dangerous natural substances in plants, defenses that also help to protect them against artificial

Ames keeps informed about news involving cancer-causing substances and public health by clipping newspaper stories about chemicals present in the environment.

carcinogens. For example, the liver sometimes converts mutagenic substances to harmless ones. Also, the digestive tract has a protective lining and regularly sheds cells that have been exposed to mutagens. And finally, cells throughout the body contain enzymes and other chemicals that help protect against DNA damage.

"But none of these defenses is perfect," Ames cautions. "There are always little bits of harmful chemicals getting by."

Ames is not suggesting that we give up peanut butter and other well-loved foods, however. The hazards are too small to worry about, he says—and besides, there is some good news about diet and cancer. As Ames pointed out in his *Science* article, foods are also full of natural anticarcinogens—chemicals that protect against cancer. These include vitamin E; carotenoids, substances found in green and yellow vegetables; selenium, which abounds in meat and seafood; vitamin C; and uric acid.

Rather than trying to eliminate every carcinogen from our diet—an impossible task—Ames suggests that we should try to identify the more important ones. He also thinks it is useful to eat a balanced diet that includes protective anticarcinogens. Large amounts of some of these substances, however, can cause health problems. High doses of selenium are toxic, for example, and too much uric acid can cause gout. More research is needed before health experts can specify ideal amounts of anticarcinogens in the diet.

An important new area of research in Ames's laboratory that ties in to the diet-cancer connection is the study of damage caused when the element oxygen combines with DNA. Ames and other scientists believe that such damage to DNA could play a major role in both cancer and aging. Many of the mutagens and carcinogens in the diet may act in the same way to damage DNA—by generating oxygen-containing molecules called *free radicals* that combine readily with the DNA and change its chemistry and function. This hypothesis suggests that substances that prevent free radicals from forming should help prevent DNA damage. As it turns out, many of the anticarcinogens in food are chemicals that combine with free radicals before they damage the DNA.

Ames runs a large laboratory at Berkeley, and meets twice a week with his research team to discuss the work. Because of his eminence as a scientist, outstanding researchers and students from all over the world apply for positions with him. On the wall outside Ames's laboratory are pictures of his extended scientific family—all the students and researchers who have ever worked with him. The gallery of Ames's alumni was started, he says, "so that new people can attach faces to the names they read in the research papers." Ames himself has published more than 200 research papers.

Ames also teaches a laboratory course for biochemistry students at Berkeley, which brings him into contact with another 100 or so students every year. His enthusiasm and sense of humor, as well as his scientific reputation, make him a popular teacher and a sought-after speaker for universities, government agencies, companies, and the general public.

In addition to overseeing the work in his laboratory, Ames does a lot of reading. Besides keeping up with new research in his field, he is often asked to evaluate papers for publication in various scientific journals. Government agencies and private foundations also seek his advice on funding research projects—which means reading and evaluating lengthy proposals.

But in spite of these demands on his time, Ames's research continues to thrive, as his dozen or so major scientific awards testify. These include the Charles S. Mott Prize of the General Motors Cancer Research Foundation and the John and Alice Tyler Ecology Award. The fact that Ames has won the highest awards for both cancer research and environmental achievement reveals his unusual versatility as a scientist.

Recently, Ames again applied his practical mind to his favorite theoretical issue—the role of DNA damage in aging and in such illnesses as heart disease and cancer. He developed a test to measure chemicals in the urine that indicate DNA damage associated with aging and these ills. The test will help determine what factors are related to DNA damage.

"The more we can understand about what's really damaging DNA, the more we can study defenses that might prevent the damage," says Ames. So, for example, the test will help show differences in DNA damage between smokers and nonsmokers and will help determine whether such anticarcinogens as vitamin E and selenium reduce the amount of DNA damage.

Ames is also developing a blood test that uses levels of oxidized fat—fat molecules that have combined with oxygen to form chemicals called *peroxides*—as a measure of abnormal processes in the body that could cause heart disease and DNA damage. He hopes that both these new tests—or similar ones—will ultimately prove useful as diagnostic tools in medicine.

Daniel E. Koshland, Jr., a biochemistry professor and the editor of *Science* magazine, says that Ames is that rare individual who can come up with new theories about a basic scientific problem and then put his ideas to practical use. "A lot of people are very good theorists but can't translate theory into what it means for industry or people or the environment," says Koshland. Ames, he says, is one of the few theorists who can develop practical applications for his ideas.

Although Bruce Ames's reputation as an innovative scientist is firmly established, he has no intention of resting on his accomplishments. He intends to continue his lifelong commitment to unraveling the secrets of DNA, and applying that knowledge to our everyday lives.

Science Year highlights four public high schools whose methods of teaching science and math are setting examples for other U.S. schools.

High Schools of Science

BY SANDY FRITZ

A teen-aged girl peers into a small bottle stuffed with rotten banana, then sharply raps it against the edge of a table. A host of tiny bugs fall out, and she sweeps them into a clean jar.

"Fruit fly duty," comments a 15-year-old observer, moving a safe distance away. "It's the pits."

What's fun, on the other hand, is examining parts of those flies under an electron microscope or weighing them on a scale sensitive enough to detect a fingerprint. Many high school science students only dream about using this kind of equipment, which is usually found only in top-flight research laboratories. But these teen-agers have the sophisticated microscope and scale available to them at a public high school—The North Carolina School of Science and Mathematics in Durham.

This academy is one of a few public high schools in the United States devoted to intensive training in math and science; others in-

Comparing U.S. and Foreign High Schools

The charts below compare the percentages of students who take three years of high school science and math, *left,* and the percentages of correct answers by 12th-grade physics students on an international science test, *right*. United States students lag behind their counterparts in both areas.

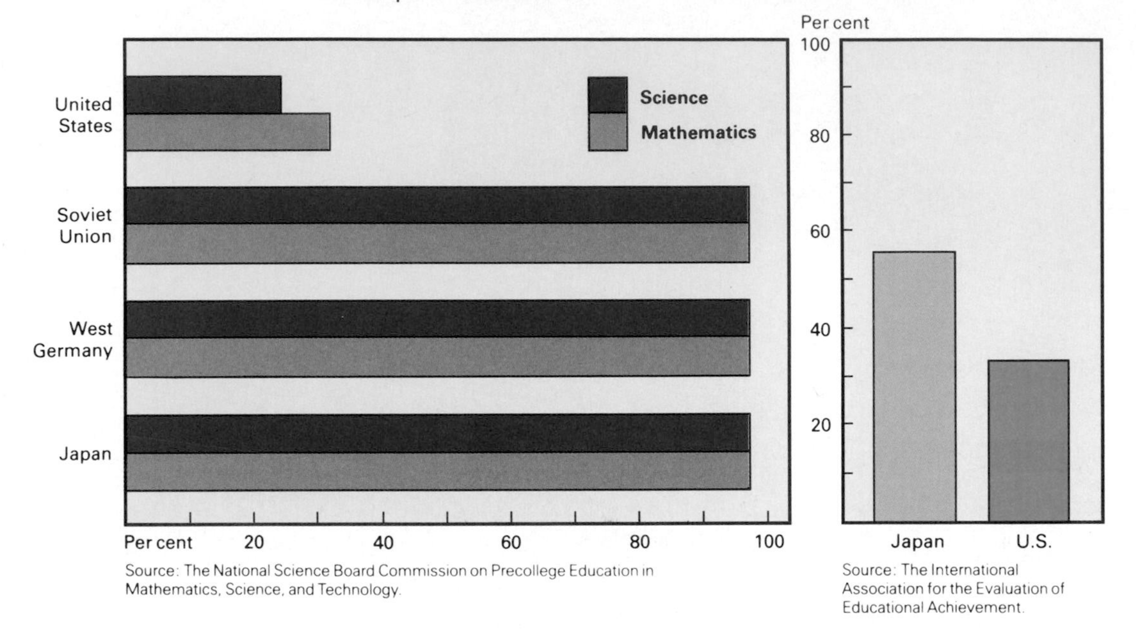

clude New York City's Bronx High School of Science; the Louisiana School for Math, Science, and the Arts in Natchitoches; and the Illinois Mathematics and Science Academy in Aurora.

These schools were created because their founders believed existing schools did not adequately prepare students in math and the sciences. They considered most school facilities, equipment, and textbooks outdated or unsuitable. More important, many educators believed that ordinary public schools suffered from a serious shortage of good teachers. The National Science Teachers Association reports that fully half of the U.S. high school science and math teachers hired in 1981 were unqualified for some of the subjects they were required to teach. Geophysicist Frank Press, president of the National Academy of Sciences (NAS), agrees that the shortage of qualified teachers is a major problem. "Fewer and fewer superior college students are entering the teaching profession," he says, "and many qualified teachers of mathematics and science are seeking new and more attractive careers in business."

Founders of the science academies also believed that course requirements at U.S. schools were far too lax. In 1982, the NAS and the National Academy of Engineering reported that two-thirds of U.S. school districts did not require high school students to complete more than one year of math and one year of science in order

The author:
Sandy Fritz is a free-lance science writer.

Special math and science high schools that mix innovation and tradition have been set up to encourage gifted U.S. students. At the Louisiana School for Math, Science, and the Arts, *top left,* a chemistry student learns the ancient art of papermaking. A Bronx High School of Science student, *above,* explores electronic technology in a robotics class. The one-on-one instruction provided by a skilled math teacher at The North Carolina School of Science and Mathematics, *above left,* is typical of the quality teaching found at all of the schools.

to graduate. By contrast, Japan, the Soviet Union, and West Germany require high school students to pass at least one course of science and one of math every year. According to Press, in 1984 all secondary school students in Japan generally spent three times more hours in science classes than did U.S. students who planned to major in science.

Inadequate instruction and lax requirements produce students with little interest in science and math. A National Science Board survey reveals that only 16 per cent of U.S. high school students enroll in physics classes and only 35 per cent take chemistry. According to the National Science Foundation, few students enroll in any science courses after the 10th grade.

Not surprisingly, test scores reflect the students' poor preparation. From 1963 to 1980, high school achievement as measured by math scores on the College Entrance Examination Board's Scholastic Aptitude Test (SAT) showed a steady decline that was only partially caused by an increase in the number of students taking the test. Although SAT scores improved slightly in the early 1980's, science and math skills still suffer. International tests show that students in the United States are still less scientifically knowledgeable than their counterparts in Japan.

The public became more aware of the problems in science education after the National Commission on Excellence in Education published its 1983 study, *A Nation at Risk.* The media widely publi-

All is not math and science at the exemplary schools. Courses in the arts and humanities, such as this orchestra and choir class at the Illinois Mathematics and Science Academy, round out the students' education.

cized the report, which suggested that the public school systems in the United States were not keeping up with industry's demand for technologically "literate" workers—one reason some U.S. businesses found it difficult to compete with nations such as Japan. Legislators, professional science organizations, and educators soon began to propose remedies, which included higher salaries for math and science teachers and more intensive preparation for students in elementary schools. Many experts also suggested the establishment of *exemplary* high schools for math and science, which would serve as examples for other schools. The National Science Board recommended that 1,000 exemplary high schools be established to serve as testing grounds for math and science teaching methods.

Today, the science academies in Illinois, Louisiana, New York, and North Carolina—some of which were founded before the current concerns about education arose—are serving this purpose. Indiana, New Jersey, Texas, and Virginia are considering founding their own exemplary schools as well. These academies have proved that science and math education can be exciting. The schools' students, many of whom are considered gifted, are motivated to learn, their morale is high, and they achieve high academic goals.

Physicist Leon M. Lederman, a creator of the Illinois academy and director of the Fermi National Accelerator Laboratory in Batavia, Ill., sees the science academies as fertile ground for cultivating the next generation of scientific leaders. "When you see one of these kids walking down the hall or chatting with somebody, sometimes

Programs developed at the high schools of science can benefit students enrolled at ordinary public schools. Using computers, microphones, and telephone hookups, a teacher at the Louisiana School for Math, Science, and the Arts, *below left,* provides instruction for students in a calculus class at another school miles away, *below right*.

Although the homework load is heavy, the gifted math and science scholars still find time for fun. Roommates at the Louisiana School for Math, Science, and the Arts take a break from studying in their dorm room, *above left*. The annual prom, *top*, provides diversion for students at The North Carolina School of Science and Mathematics, as do extramural sports such as softball, *above*.

you think, 'This kid might be another Einstein; he might do something that will change the way people live on the planet.' "

Cultivating gifted and talented students may indeed produce leaders in the sciences, but only a few students are able to attend exemplary schools. Will these isolated centers of excellence influence education in nonspecialized public high schools, where the majority of students are taught? Administrators of the science academies say yes. For example, the academies offer outreach programs that enable students enrolled in public high schools to be taught by the academies' teachers. The Louisiana School for Math, Science, and the Arts is literally linked with other schools in the state's public school system—via computer and telephone. Students in two schools many miles away from the academy watch calculus problems appear on their computer screens while their unseen teacher remains on the academy's campus. Every summer, the Louisiana school hosts a five-week science and math program elsewhere in the state that attracts about 600 teen-agers who are unable to attend the academy.

In addition, teachers and administrators at most of these exemplary schools give workshops for teachers at ordinary schools. Teaching methods found to be effective at an exemplary school are in this way spread throughout a school system. Nearly 3,000 teachers have attended workshops at the North Carolina school. The Illinois Mathematics and Science Academy plans a similar program.

"The goal is that every kid will have the opportunity to achieve his or her full potential to live and to learn and to grow," says F. Borden Mace, the North Carolina school's first dean of operations.

If the academies are true to this mission, we can expect some dramatic changes in U.S. science education in the next 20 years.

Bronx High School of Science

Mark Hochberg, a senior at Bronx High School of Science in New York City, prepares for his advanced computer programming class.

Beneath towering mosaic murals depicting science greats—British mathematician Sir Isaac Newton, French physicist Marie Curie, and American inventor Thomas Edison—limps 16-year-old Mark Hochberg. It is the break between classes, and some of Mark's 2,800 fellow students at the Bronx High School of Science flood the halls and rush past him.

"I hurt my foot last night racing around the house after my brother," he explains. This senior, who is worried about missing a week of volleyball practice, also happens to be well versed in five computer-programming languages.

Although able to pass an entrance examination that 6 out of 7 applicants to Bronx Science fail, Mark started his four years at the school without any particular interest in science or math. "Then," he says, "something happened." Mark credits his teachers' interesting classroom presentations for sparking his desire to learn more about math and science.

This year, Mark is pursuing his new-found interest by taking courses in electronics and advanced computer programming—in addition to English, political science, and history classes. Mark's after-school time is divided among the backgammon club, the engineering club, the math team, and—feet permitting—the volleyball team, one of many interschool sports offered at Bronx Science. Like most of his fellow students, Mark spends three to four hours a night on homework. "Where do I find the time?" he says, laughing. "Simple. I just don't sleep."

Bronx High School of Science, founded in 1938, is situated in a quiet New York City neighborhood. Every school day, Mark catches a public bus for a 20-minute ride from his home to school. Mark's classmates, who are drawn from all five of the boroughs that make up New York City, also commute. Forty per cent of the student body come in from far-off Queens or Brooklyn—a trip that can take two hours and three transfers between buses and subways each way.

The inconvenience is far outweighed by the education the students receive. Facilities at Bronx Science are not lavish, but the energy of its students and faculty makes up for any lack. The bright, motivated students inspire the school's 160 teachers. The faculty challenges the students to discover things for themselves. Instead of giving lectures, many teachers use the Socratic method, leading students to discovery by asking thought-provoking questions instead of providing ready answers.

The school's principal, Milton Kopelman, explains that teaching problem-solving skills is a major goal at Bronx Science. "We want to produce kids who can identify problems and work through to solutions," he says, "not kids who are good at memorizing."

Bronx Science also challenges students to pursue their studies beyond the classroom. Students interested in independent research can work with professional scientists in places such as Rockefeller University, the Memorial Sloan-Kettering Cancer Center, and the American Museum of Natural History, all in New York City. The professional scientists act as *mentors*, advising and encouraging the students. Some student research is published in the school's own science publication, *The Journal of Biology*, which is the only high school publication so respected by professionals that it is indexed in *Biological Abstracts*, a worldwide reference work.

Even students who do not choose to do independent research leave the school with an excellent education. The school requires all students to take four years of science and three years of math. Humanities are also emphasized. Students must complete four years of English and social studies, three years of a foreign language, and a music course. In addition, before they graduate, students must pass a computer literacy course and a technical drafting course.

The curriculum has proved to be a formula for excellence. Many former students have become successful—whether their careers include science or not. For example, former U.S. Secretary of Defense Harold Brown and *Ragtime* author E. L. Doctorow are Bronx Science graduates.

The school is nonetheless particularly proud of its alumni scientists, some of whom have become leaders of international renown. In 1972, Bronx Science alumnus Leon N. Cooper shared the Nobel Prize in physics for his work on *superconductivity* (the ability of a material to conduct electricity without resistance). Bronx Science graduates Sheldon L. Glashow and Steven Weinberg shared the Nobel Prize in physics in 1979 for developing a principle that unifies two fundamental forces of nature—electromagnetism and the weak nuclear force. Among other notable graduates is biochemist Bruce N. Ames, inventor of a test used to detect cancer-causing substances (in the People in Science section, see BRUCE N. AMES).

Bronx Science's reputation encourages students such as Mark Hochberg to excel. "You come here knowing the work will be hard," Mark says. "But you also know you'll be getting one of the best educations in the world."

Before commuting back to their homes, three Bronx Science students linger in this urban school's dramatic foyer. The engraved motto under the mural challenges scholars to be creative: "Every great advance in science has issued from a new audacity."

The North Carolina School of Science and Mathematics

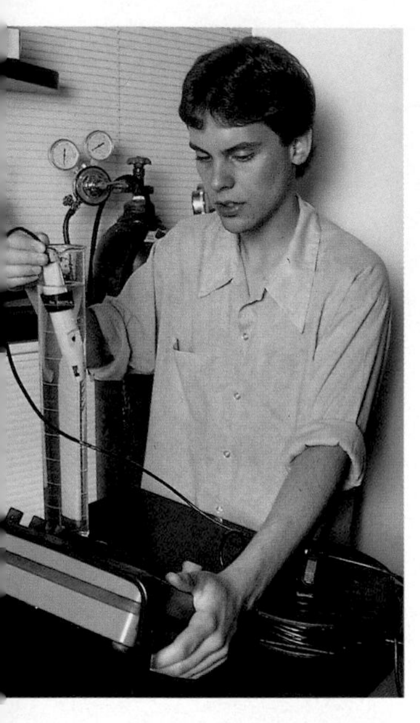

John Edgerton measures the oxygen content of a sample of lake water for a biology project at The North Carolina School of Science and Mathematics in Durham.

If 15-year-old John Edgerton had his way, he would be making an ammonium laser. "I was really excited about the idea," the high school junior said. "But I would have had only a week to make it, and my teacher said it would take longer. So I'm thinking about making an electric guitar instead."

It is "Special Projects Week" at the North Carolina School of Science and Mathematics. Classes are suspended, and the school's 480 students have jumped into faculty-approved science projects. Later, the students will present their findings. "The teachers list their specialties," a student explains. "You can pick an adviser and discuss a project. Some of the stuff they're interested in is just amazing."

Some of the students' projects are amazing as well. Those interested in physics can make *holograms*, three-dimensional pictures produced by laser light. Future aeronautical engineers can assemble rockets. Those interested in electronics can create their own robots. Chemistry "magic" is taught to some students, who learn flashy demonstrations—combining chemicals to make a dazzling fluorescent liquid, for example. The students then perform chemistry magic for middle school students in the Durham area. John Edgerton's electric guitar project, though less complicated than an ammonium laser, will allow him to dabble in electrical engineering and study the physics of sound production.

The chance to work in such an atmosphere encourages 800 top 10th-grade students from all parts of North Carolina to compete every year for 200 places at the two-year school for math and science. SAT scores, academic standings, outside-school activities, and interviews are among the criteria used to select students.

The academy is a boarding school. Once admitted, students from all parts of the state move to the school's campus in the city of Durham, which is also the home of three universities and Research Triangle Park, a development comprising more than 40 high-technology and scientific-research organizations. The school, founded in 1980 as part of a plan to move North Carolina from an agricultural to a high-technology economy, is tuition-free. Costs for the students' housing and food are also borne by the state.

Unlike the Bronx High School of Science, the North Carolina school is virtually filled with modern equipment. Every science course includes a hands-on laboratory class. Entire rooms are devoted to computers, while expensive spectrophotometers and gas and liquid chromatographs, instruments used to analyze compounds, fill the chemistry laboratories.

The founders of the school designed a mentorship program like that at Bronx Science, and it has proved enormously popular. In nearby high-technology research centers operated by major corporations and Duke University, students work alongside mentors, some of whom later hire the students for summer jobs. Steve War-

Math and computer science classes are held in the oldest building on the North Carolina campus, a former hospital, built in 1908 and now on the National Historic Register.

shaw, who directs the school's science department, says the program helps students find out about the workaday world. "The things that we talk about in class here, they can see being applied by people working in laboratories and offices."

Senior Gretchen Case, for example, does cell pathology research with her mentor, zoologist David R. McClay of Duke University. Her chemistry background at the school prepared her for the project, which involves *antibodies* (immune system molecules that help the body fight bacteria, viruses, and other invaders).

The core of the North Carolina program is not the school's equipment or special electives but its strict requirements for daily course work. In two years, every student must become proficient in computer use and must pass biology, chemistry, and physics courses and two courses each of math and English. Courses in history, physical education, and foreign language are also required. Students work a few hours a week for the school, tending the yards, cleaning the cafeteria, or collecting those unpleasant fruit flies. A total of 45 hours of community service work is also required.

Such a workload, combined with living away from home, could overwhelm the teen-agers at the academy. To minimize stress, administrators strive to create a friendly, noncompetitive atmosphere. Students are not told of their class standings. No one is named valedictorian, and students who "beat" their classmates academically are not rewarded. Instead, students are encouraged to compete with the teachers' and their own expectations—which can be very high.

"I don't know about anyone else around here," says Edgerton, "but I've never had to work so hard in my whole life."

Louisiana School for Math, Science, and the Arts

Greg Griffin uses crossed sticks and a wire to illustrate the curve of a trigonometry concept called a sine wave at the Louisiana School for Math, Science, and the Arts in Natchitoches.

It is lunchtime at the Louisiana School for Math, Science, and the Arts, and the conversation does not at the moment center on sports, rock groups, or cars—but on a type of *vector* (a term used in math and physics to describe the strength and direction of a force). Fifteen-year-old Greg Griffin and his lab partner are enthusiastically hashing out the simplest way to determine a linearly independent vector. "Boy," Greg says after finding the solution, "I can't wait to take differential equations class next semester."

Outside, the smell of freshly cut grass rises from the broad green lawns of the southern city of Natchitoches. But, as demonstrated by the students' lunchroom conversation, there is more than grass growing on the campus of this Louisiana school. The dorms, cafeteria, and laboratories are sprouting young physicists, biologists, chemists, and mathematicians.

This two-year public boarding school, which opened in 1983, is modeled on The North Carolina School of Science and Mathematics. It differs chiefly in that it also includes artistically gifted students. Physics students like Greg go to class alongside painters, singers, and dancers. Required courses such as English, economics, and history bring the two groups together.

For all students, criteria for admission include grades, SAT and other test scores, and interviews. Every year about 650 10th-graders apply for 200 openings.

If students find it hard to be admitted to the school, prospective teachers face even greater competition. According to founder Robert A. Alost, more than 1,200 teachers applied for the school's original 17 positions. Because the academy does not require its teachers to be certified by the state of Louisiana, they are drawn from throughout the United States. Many hold advanced degrees; most gave up university teaching positions to come to Natchitoches.

All teachers at the academy use college textbooks in their classes and move through the course work in much less time than would be spent in an ordinary high school. Students able to exceed their teachers' goals can work at an even faster pace. "We had a youngster who finished trigonometry in six weeks," says Alost. "Our kids are not hampered by rules limiting the number of classes they can take once they prove they can handle the academics here."

The school also educates students to be information givers instead of information consumers. One way to accomplish this, according to Richard G. Brown, the school's director, is to publicly televise students in learning situations. In these educational teleconferences, students might be seen communicating with experts in a distant city through a satellite television link. While students benefit from the opportunity to speak directly to experts and from the challenge of being fully prepared for the event, television viewers can share in the information such a conference produces.

The school's first teleconference was broadcast on cable television in March 1987. Fifteen students who had studied agriculture led an auditorium full of students in a discussion with Louisiana state farming experts and federal officials from the U.S. Department of Agriculture in Washington, D.C. The conference was a success for the students—and nearly 3 million cable TV viewers tuned in.

Greg Griffin was one of many students who came to the academy in pursuit of such opportunities—and the chance to take advanced courses not available at their hometown schools. By studying high-level math at the Louisiana academy, Greg became fascinated with physics. "I like physics because it's the most fundamental of the sciences," he says. "Biology is chemistry, and chemistry is physics. Other sciences really hinge on physics. I'm a 'why' person and I like to get down to the essence of things."

Greg likes balancing his physics courses with Latin, one of six languages offered at the school, and piano class. In addition to taking other required humanities and science courses, Greg and his fellow students are expected to perform three hours of community service and seven hours of work on campus every week.

With a schedule like this, there is little time for lunchroom banter about linearly independent vectors. Greg glances at his watch, sees it is time for piano class, and rummages through his bag for his sheet music. He looks up suddenly, and adds, "Someday I'd like to have the freedom to do research on whatever I want. I'd like to run my own show."

Students relax after school on the campus of the Louisiana School for Math, Science, and the Arts, which is located in a rural part of the state.

Illinois Mathematics and Science Academy

In a laboratory at the Illinois Mathematics and Science Academy in Aurora, Marcie Edwards, right, and her lab partner conduct a chemistry experiment.

Frogs, a pig fetus, a sheep's eye, a grasshopper. Some kind of witch's brew? No, a list of dissection projects that drew 15-year-old Marcie Edwards to science. Her favorite project was the sheep's eye. "The teacher gave you the eye," Marcie says as she adjusts the flame under a beaker. "It looked just like an eye—it *was* an eye—and you had to cut it open. I thought, 'Wow! This is intriguing, to see what's actually inside an eye,' " she says, grinning widely. "Everyone else was gagging."

A strong sense of curiosity combined with plain old hard work has made Marcie an exceptional student. The Illinois Mathematics and Science Academy, 56 kilometers (35 miles) west of Chicago in Aurora, is now helping her attain her goal of becoming an exceptional scientist.

A new kid on the science-education block, the Illinois school opened in 1986. Leon M. Lederman, who proposed the concept of the school, especially enjoyed the experience of creating an exemplary school. "It is very rare, I think, in history that somebody hands you a blank pad of paper and tells you to design a new educational strategy for bright kids, with no constraints," he says. "[You can] forget about boards of education, forget about all of the rules, and sit down with the best possible advice you can get and design a school for bright kids."

The new public school was designed to combine the most successful features of math and science schools in other states. When full enrollment is reached in 1989, about 900 students will attend 10th, 11th, and 12th grade at the tuition-free boarding school. Like the Louisiana School for Math, Science, and the Arts, the Illinois academy was designed to accommodate students who are gifted in nontechnical subjects, as well as scientifically talented students such as Marcie.

Along with providing a model for high-quality science education, the school hopes to reach gifted students living in the state's isolated rural areas. Finding these students is a challenge, according to Lederman. "If a genius is born in the middle of a jungle, you've probably lost him," he says. "If he's born in a town 20 miles north of Carbondale, Illinois, you still might lose him as a scientist if the educational opportunity isn't there."

Students brought to the school find a huge, 30,500-square-meter (330,000-square-foot) building. Although laboratory equipment is not as abundant as in The North Carolina School of Science and Mathematics, the building itself is well appointed, containing an indoor swimming pool and three gymnasiums. Dormitories large enough to house 900 boys and girls have been built in a horseshoe arrangement around the back of the circular school building.

Most faculty members find that working at the Illinois academy is more like teaching at a college than teaching at an ordinary high

Classrooms inside the Illinois Academy, which opened in 1986, are spacious and modern.

school. Teachers use college-level textbooks in their classes and, like professors, hold evening office hours in order to be available for tutoring. Many of the school's full-time teachers—all of whom have advanced degrees—say they have never worked in such a challenging atmosphere.

Teachers at the school approach science and math in a somewhat unusual way. Academy director Stephanie Marshall says, "My philosophy is that the kids have to have fun with math and science. Having fun is really the joy of discovery. We want students to embrace math and science within the context of ethical problem solving."

To help the students relate their technical studies to human concerns, one Saturday every month is devoted to special classes, called "Saturday seminars." Once, students listened to a visiting poet describe his view of the relationship between poetry and science. On another Saturday, the entire school met to play "Congress," the students acting as members of the House of Representatives and the Senate. Marshall also hopes to institute required courses in the history of science.

Providing student access to mentors, which has been very successful in the other science academies, is also planned for Illinois. Students may soon be able to work closely with research scientists at Argonne National Laboratory, a nuclear energy research center southwest of Chicago. The Fermi National Accelerator Laboratory, for high-energy physics research, already sponsors Saturday morning classes for high school students and has also shown interest in providing mentors.

Aside from such enrichment programs, the school has a rigidly structured curriculum. All first-year students must take classes in chemistry, physics, math, foreign language, physical education, English, and social science. Juniors and seniors can take some elective courses in addition to one math and two science courses each year. Even though the students at the Illinois academy are gifted, they do not always find the work easy.

"These students don't come in here and blaze through a college chemistry course at all," says chemistry teacher Steve Gramsch. "They have to work at it if they want to succeed. Many of them are learning that for the first time." Administrators expect the intense course work to prepare most of the first class of graduates in 1988 to enter college as sophomores—skipping the first year of undergraduate work altogether.

For Marcie Edwards, Monday morning starts at 7:45 A.M. with physics, followed by chemistry and then algebra. Next, she studies German, history, or English. After-school time is reserved for swim team practice. On top of all this, Marcie is on 24-hour call as student council secretary. "It takes a lot of time, it takes a lot of effort," she says.

The efforts of students like Marcie and their teachers are expected to be well spent. These future scientists will not be the only ones to benefit from their special training; the science academies should eventually benefit not only public education, but business and government as well by raising standards for science education across the United States.

The Illinois Academy's main building, which covers 3 hectares (7½ acres), blends into the prairie landscape.

World Book Supplement

Revised articles reprinted from the 1987 edition of *The World Book Encyclopedia.*

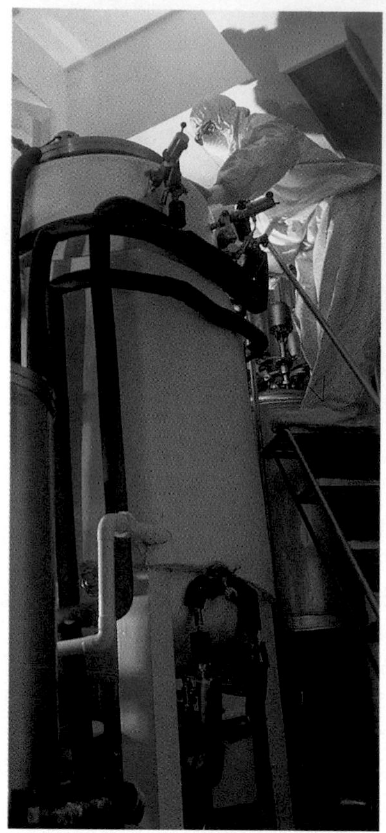

© Hank Morgan, Photo Researchers

Producing an Experimental Vaccine

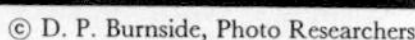

© D. P. Burnside, Photo Researchers

Tagging a Bear Cub for Study in the Wild

© Tim McCabe, Taurus

Developing New Plant Varieties

© Dave Woodward, Taurus

Photographing a File Clam

The Work of Biologists includes a wide range of activities and takes place in many different settings. For example, some biologists work outdoors, studying wildlife in the forests or the ocean. Others carry on laboratory research in such areas as plant breeding or the production of drugs.

BIOLOGY

BIOLOGY is the scientific study of living things. There are more than 2 million species of living things on the earth. They range in size from microscopic bacteria to huge blue whales and towering redwood trees. Living things also differ greatly in where and how they live. However, all forms of life share certain characteristics that set them apart from nonliving things. These characteristics include the ability to reproduce, to grow, and to respond to changes in the environment.

Traditionally, biology has been divided into two major fields. *Botany* deals with plants, and *zoology* with animals. Botany and zoology are further divided into various branches and specialized areas of study. But most branches of biology—for example, *anatomy* (the study of the structure of living things) and *genetics* (the study of heredity)—apply to both plants and animals.

Biology may also be divided into *ecology*, *physiology* and *systematics*. Ecology deals with the relationships among living things and between organisms and their environment. Physiology concerns life functions, such as digestion and respiration. Systematics, also called *taxonomy*, is the scientific classification of plants and animals.

Biologists often make use of the methods and findings of other sciences. For instance, they rely on physics and chemistry to help them understand the processes that occur in living plants and animals. They use statistics in studying changes in the size of an animal or plant *population*—that is, the number of organisms of a particular species in an area. *Exobiologists* work with astronomers in searching for life elsewhere in the universe.

Achievements in biological research have greatly affected people's lives. For example, farm production has soared as biologists have helped develop better varieties of plants and new agricultural techniques. Discoveries in biology have enabled physicians to prevent, treat, or cure many diseases. Research on the relationships between living things and their environment has helped in the management of wildlife and other natural resources.

Garland E. Allen, the contributor of this article, is Professor of Biology at Washington University and coauthor of The Study of Biology.

What Biologists Study

Biology is such a broad subject that most biologists specialize in some area of study. But in whatever area they work, all biologists are interested in both the parts of living things and how the parts work together.

Certain biologists study organisms that live in a specific environment. *Marine biologists*, for example, investigate life in the ocean. Some biologists concentrate on a particular type of organism. *Ornithologists*, for instance, study birds. Many biologists examine the parts of living things. *Cytologists*, for example, deal with the structure, composition, and functions of cells. Other biologists analyze life processes. *Embryologists*, for example, investigate the formation and development of animals and plants before they become independent organisms.

The techniques and tools that biologists use depend on what they are investigating. Many biologists conduct experiments to gain information and to develop and test theories. Their experiments may involve making a change in an organism's way of life or its environment and then observing the effects of that change. For example, a biologist may change the diet of an animal and study how the animal's growth and functioning are thereby affected. The microscope has long been one of the biologist's most useful tools. An entire branch of biology, called *microbiology*, is devoted to the study of organisms that can be seen only with a microscope. Other techniques and tools used by biologists range from aerial surveys of plant and animal populations to techniques that isolate the molecules of living cells.

© Runk/Schoenberger from Grant Heilman

An Ecologist inspects soybean plants that have been exposed to various pollutants in a specially designed tent. Many ecologists study the effects of pollution on plant and animal life.

History of Biology

Beginnings. In prehistoric times, people gradually developed a great deal of practical biological knowledge. They learned to grow many kinds of plants and to tame and raise certain animals. In ancient times, people of China, India, and the Middle East accumulated further knowledge of plants and animals. For instance, they knew how to use numerous plants as medicines or poisons. The Egyptians learned some anatomy and physiology through embalming their dead.

The ancient Greeks made major advances in biology. Unlike most other people of the time, some Greek thinkers did not believe that gods or spirits caused natural events. Instead, they saw nature as operating according to laws that people could discover. About 400 B.C., a Greek physician named Hippocrates taught that diseases have only natural causes. He also emphasized the relationships among the parts of an organism and between an organism and its environment. Hippocrates is often called the father of modern medicine.

During the 300's B.C., the Greek philosopher Aristotle gathered a vast amount of information about plants and

Major Fields of Biology

Anatomy deals with the structure of living things.

Bacteriology is the study of bacteria.

Biochemistry examines the chemical processes and substances that occur in living things.

Biophysics applies the tools and techniques of physics to the study of living things.

Botany is the study of plants.

Cryobiology, *KRY oh by AHL uh jee,* analyzes how extremely low temperatures affect living things.

Cytology, *sy TAHL uh jee,* studies the structure, composition, and functions of cells.

Ecology concerns the relationships living things have with one another and their environment.

Embryology deals with the formation and development of plants and animals from fertilization until they become independent organisms.

Entomology, *EHN tuh MAHL uh jee,* is the study of insects.

Ethology, *ih THAHL uh jee,* concerns animal behavior under natural conditions.

Evolutionary Biology is the study of the evidence supporting the theory of evolution.

Genetics is the study of heredity.

Ichthyology, *IHK thee AHL uh jee,* is the study of fishes.

Immunology concerns the body's defenses against disease and foreign substances.

Limnology, *lihm NAHL uh jee,* studies bodies of fresh water and the organisms that live in them.

Marine Biology investigates life in the ocean.

Medicine is the science and art of treating and healing.

Microbiology deals with microscopic organisms.

Molecular Biology analyzes molecular processes in cells.

Neurobiology deals with the nervous system of animals.

Ornithology, *AWR nuh THAHL uh jee,* is the study of birds.

Paleontology, *PAY lee ahn TAHL uh jee,* is the study of fossils.

Pathology examines the changes in the body that can cause disease or are caused by disease.

Physiology deals with the functions of living things.

Sociobiology focuses on the biological basis for social behavior in human beings and other animals.

Systematics, or *taxonomy,* is the scientific classification of plants and animals.

Virology, *vy RAHL uh jee,* concerns viruses and virus diseases.

Zoology, *zoh AHL uh jee,* is the study of animals.

animals. He was one of the first thinkers to classify animals according to their own characteristics rather than according to their usefulness to people. Pliny the Elder, a Roman naturalist who lived during the first 100 years after Christ's birth, also collected many facts about plants and animals. He included the information in his 37-volume *Natural History*.

During the A.D. 100's, Galen, a Greek physician who practiced medicine in Rome, contributed greatly to advances in anatomy and physiology. He gained much of his knowledge from treating injured gladiators and dissecting apes and pigs.

The growth of biological knowledge slowed during the Middle Ages, a 1,000-year period in European history that began in the 400's. However, works by Hippocrates, Aristotle, Galen, and other ancient authorities were collected, preserved, and translated by Arab scholars in the Middle East. The Arabs also made major contributions of their own in biology. The works of the ancient Greek and Arab scientists eventually made their way to Europe. During the Middle Ages, the authority of the ancient writers was unquestioned, though their works contained many errors.

The Renaissance. From the early 1300's to about 1600, a new spirit of inquiry spread across western Europe. During this period, called the Renaissance, many anatomists and physiologists began to challenge the authority of the ancient writers. They believed that people should rely on experimentation and observation rather than accept without question the ideas of the ancients.

The emphasis on observation stimulated the development of a high degree of naturalism and accuracy in biological illustration. During the late 1400's and early 1500's, the great Italian artist Leonardo da Vinci made hundreds of drawings of the human body in which he paid careful attention to detail and proportion. Leonardo based his work on dissections of human corpses. The first scientific textbook on human anatomy was published in 1543. This work, titled *On the Fabric of the Human Body*, was written by Andreas Vesalius, an anatomist born in what is now Belgium. Like Leonardo, Vesalius based his work on dissections he had made of human corpses. The book, richly illustrated with exceptionally lifelike drawings of human anatomy, corrected many of Galen's mistaken ideas.

One of the most important discoveries in physiology in the 1600's was made by William Harvey, an English physician. In 1628, Harvey published the results of his experiments showing how blood, pumped by the heart, circulates through the body.

Early Discoveries with the Microscope. The introduction of the microscope led to great discoveries in biology during the middle and late 1600's. About 1660, an Italian anatomist named Marcello Malpighi, with the aid of a microscope, became the first person to observe the movement of blood through the capillaries. In 1665, Robert Hooke, an English experimental scientist, published *Micrographia*, a book containing detailed drawings of many biological specimens as seen with a microscope. The book included the first drawings of cells. In the mid-1670's, Anton van Leeuwenhoek, a Dutch amateur scientist, discovered microscopic life forms, thus opening up a new world for investigation.

The Origins of Scientific Classification. During the 1700's, Europeans came into increasing contact with distant parts of the world and thereby learned of many unfamiliar plants and animals. Naturalists realized that they needed a classification system that could include those plants and animals. In 1735, the Swedish naturalist Carolus Linnaeus (also called Karl von Linné) published a system of classification in which he grouped organisms according to structural similarities. His system forms the basis of scientific classification used today.

Classifying organisms according to structural similarities stimulated interest in *comparative anatomy*—the comparison of the anatomical structures of different organisms. The leading comparative anatomist of the late 1700's and early 1800's was Baron Cuvier of France.

SCALA/Art Resource

The Cultivation of a Date Palm is shown in this ancient Mesopotamian carving. Much early biological knowledge dealt with farming and other practical matters.

S. Champier, *Symphonia Platonis*, 1516

Dissections of Animals were carried out by the Greek physician Galen during the A.D. 100's. Galen's studies greatly advanced the knowledge of anatomy.

Metropolitan Museum of Art, Rogers Fund, 1913

A Recipe for Cough Syrup made from plants appears in an Arabic manuscript from the 1220's. The Arabs made major contributions in botany and medicine.

Cuvier noticed that most kinds of animals have one or another of a very few basic body types. He devised a system of classifying animals according to basic body types that is still used in modified form. Cuvier also applied the methods of comparative anatomy to another field he helped establish, *paleontology*—the study of fossils.

The Theory of Evolution. Most biologists had long believed that each species of life had remained unchanged and no new species had appeared since the world began. However, biologists began to question those beliefs during the late 1700's. They noted that farmers had produced new varieties of plants and animals by selective breeding. In addition, voyages of exploration had revealed isolated groups of plants and animals that contained many species which varied only slightly from one another. Biologists wondered why there should be so many species with little variation. Such observations led many biologists to believe that species change over time and that some species had *evolved* (gradually developed) from others.

During the early 1800's, several biologists proposed explanations of how species evolve. The most convincing theory was eventually reached independently by two British naturalists—Charles Darwin and Alfred Russel Wallace. However, Darwin presented his ideas in a widely read book, and his work became better known.

Darwin detailed his theory of evolution in *The Origin of Species* (1859). According to Darwin, some organisms are born with traits that help them survive and reproduce. They pass the favorable traits on to their offspring. Other members of the same species that have unfavorable traits are less likely to survive and reproduce. The unfavorable traits eventually die out. Darwin proposed that species evolve as more and more favorable traits appear and are passed from generation to generation. He called the process *natural selection*.

Materialistic Physiology and the Cell Theory. Many physiologists of the late 1700's had come to think of life as the total of the physical and chemical processes occurring in an organism. Unlike some other biologists, they did not believe that living things are guided in their functioning by any spiritual or supernatural forces. Instead, they felt that living things are nothing more than special combinations of materials and function like machines. Such views are called *materialistic physiology* or *mechanistic materialist physiology*.

Antoine Lavoisier, a French chemist, applied the techniques of chemistry to physiology in the late 1700's. He compared respiration to the burning of a candle because both processes use oxygen and produce heat and carbon dioxide. Beginning in the mid-1800's, the French physiologist Claude Bernard introduced a new approach to materialistic physiology. Bernard saw living things as highly organized sets of control mechanisms that work to maintain the internal conditions necessary for life. He pointed out that in a mammal, for example, such mechanisms keep body temperature constant in spite of variations in the temperature outside the organism.

Paralleling developments in physiology was a growing understanding of the cell. In the late 1830's, two Germans—the botanist Matthias Schleiden and the physiologist Theodor Schwann—proposed that the cell was the basic structural and functional unit of all plants and animals. In 1858, Rudolf Virchow, another German scientist, published his theory that all diseases were diseases of the cell. In combination, these ideas are called the *cell theory*.

Building on materialistic physiology and the cell theory, Louis Pasteur, a French chemist, and Robert Koch, a German physician, firmly established a new theory of disease during the middle and late 1800's. Through their studies, Pasteur and Koch proved what was called the germ theory. According to the theory, many diseases are caused by microscopic organisms.

The Growth of Modern Biology. During the late 1800's, Darwin's theory of evolution had stimulated much speculation among biologists about the origin,

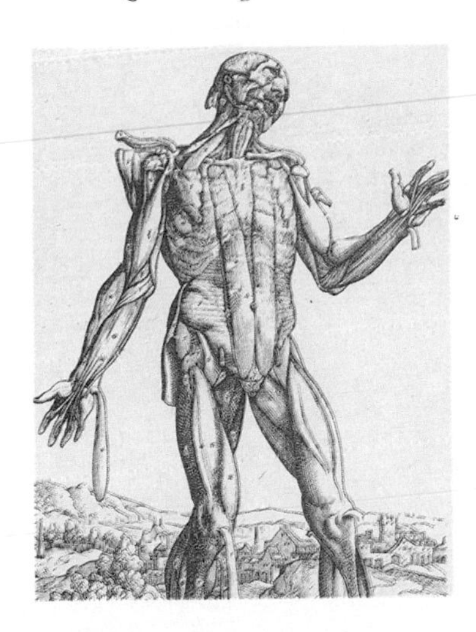

The Human Muscular System is shown in this illustration from Andreas Vesalius' *On the Fabric of the Human Body* (1543), the first scientific text on human anatomy.

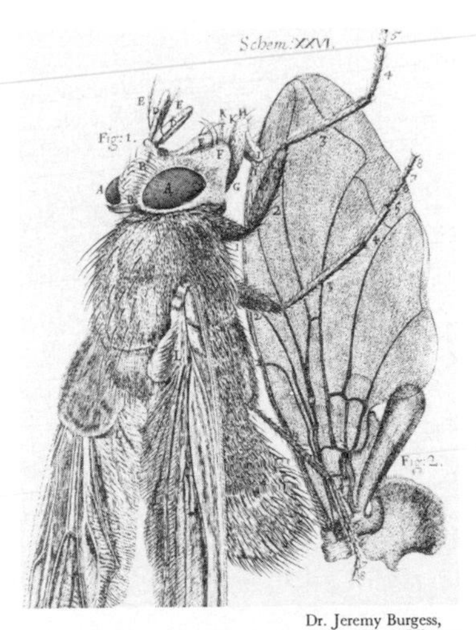

Dr. Jeremy Burgess, Science Photo Library

A Detailed Drawing of a Fly was published in *Micrographia* (1665) by Robert Hooke of England. Hooke pioneered in studying specimens with the microscope.

Drawing by U. Schleicher-Benz in "Lindauer Bilderbogen"; © Jan Thorbeck Verlag, Sigmaringen, West Germany

A Flower "Clock" proposed by the Swedish naturalist Carolus Linnaeus in 1745 arranges species by the times of the opening and closing of the blooms.

nature, and development of organisms. By the early 1900's, however, many biologists strongly rejected the emphasis on theory and speculation. Instead, they stressed the value of carefully controlled experiments and the application of mathematical techniques to biology. This helped lead to an enormous expansion of biological knowledge, particularly in the understanding of the chemical and molecular basis of life.

Genetics was established as a branch of biology in the early 1900's. It developed chiefly from experiments conducted during the mid-1800's by Gregor Mendel, an Austrian monk. On the basis of his experiments, Mendel discovered that physical characteristics are produced by basic hereditary units that transmit traits from generation to generation. About 1910, Thomas Hunt Morgan, an American biologist, found that Mendel's hereditary units—later called *genes*—are located on structures called *chromosomes* within cells. Biologists at the time also noted that changes in hereditary traits correspond to visible changes in chromosome structure.

During the 1940's, geneticists found that genes guide the manufacture of the proteins by which cells regulate their chemical processes. In 1953, biologist James D. Watson of the United States and physicist Francis H. C. Crick of Great Britain proposed a model of the molecular structure of *deoxyribonucleic acid* (DNA), the material in chromosomes that controls heredity. Knowing the structure of DNA enabled biologists to understand the molecular basis of many life processes.

Breakthroughs in genetics helped alter biologists' approach to the study of evolution. By the 1960's, many biologists were studying evolution in terms of changes in the kinds and numbers of genes in a population.

The field of ecology began to develop dramatically in the early 1900's. Scientists had long recognized the importance of the relationships among organisms and between organisms and their environment. But the development of ecology as a separate branch of biology occurred only after the introduction of such techniques as the statistical analysis of complex systems of relationships. Since the 1960's, concern about environmental effects of pollution has stimulated research in ecology.

Great advances have also been made during the 1900's in *neurobiology*—the study of the nervous system. Neurobiologists have learned much about how nerves function individually and in organized groups.

Current Research and Issues. The study of the human *immune system*—that is, the body's defense system against disease and foreign substances—is one area at the frontier of biological research. Scientists are learning how our bodies produce a seemingly endless variety of disease-fighting proteins called *antibodies*. Each antibody is tailored to combat one of many foreign substances called *antigens*. Biologists have discovered that the body can produce a great number of different antibodies because certain genes rearrange themselves to produce antibodies that attack specific antigens.

Since the 1950's, biologists have been collecting evidence for the theory that life began in a series of chemical reactions early in the earth's history. They have produced complex biological molecules in chemical experiments that reproduce conditions thought to have existed on the earth billions of years ago.

Since the 1970's, a growing number of biologists have questioned the idea that evolutionary change occurs only as a result of a gradual process. Instead, they accept the idea that evolution may proceed at times by abrupt changes leading to the replacement of one species by another. Although there is questioning over details, most biologists believe in the general outlines of the theory of evolution. However, some other people reject the theory because of the many gaps in our understanding of how particular species evolved. Still other people object to the idea of evolution because it conflicts with their religious beliefs about the creation of life. See EVOLUTION.

By the late 1970's, scientists had learned how to remove genes from one species and insert them into another. The process is called *genetic engineering*. Genetic

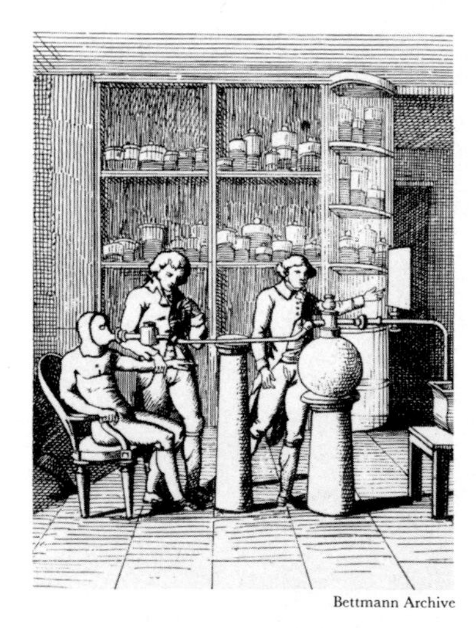

Bettmann Archive

An Experiment on Respiration is shown in this engraving of Antoine Lavoisier's laboratory. The French chemist studied physiological processes in the late 1700's.

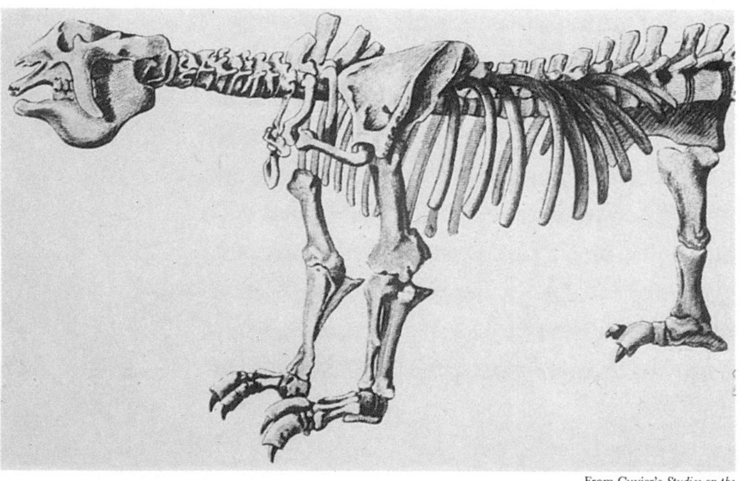

From Cuvier's *Studies on the Bones of Fossil Vertebrates*, 1812

The Skeleton of an Extinct Giant Sloth was drawn by Baron Cuvier of France, the leading comparative anatomist of the late 1700's and early 1800's. Cuvier pioneered in *paleontology*, the study of fossils. He used the methods of comparative anatomy to determine the structures of prehistoric animals from their fossil remains.

Important Dates in Biology

c. 400 B.C. Hippocrates established the principles of modern medical practice based on the idea that diseases have only natural causes.

A.D. 100's Galen greatly extended knowledge of anatomy and physiology through his treatment of injured gladiators and dissections of apes and pigs.

1543 Andreas Vesalius' *On the Fabric of the Human Body,* the first scientific text on human anatomy, was published.

1628 William Harvey published his discovery of how blood circulates through the body.

1665 The first drawings of cells appeared in Robert Hooke's book *Micrographia.*

Mid-1670's Anton van Leeuwenhoek discovered microscopic forms of life.

1735 Carolus Linnaeus classified organisms according to their structural similarities, laying the foundation for modern scientific classification.

Late 1700's Antoine Lavoisier conducted chemical studies of such physiological processes as respiration and the conversion of food to energy.

c. 1800 Baron Cuvier made major contributions in *comparative anatomy* (the comparison of the structures of different species) and *paleontology* (the study of fossils).

1838-1839 Matthias Schleiden and Theodor Schwann proposed that the cell is the basic unit of life.

Mid-1800's Gregor Mendel discovered the basic laws of heredity.

1859 Charles Darwin set forth his theory of evolution in *The Origin of Species.*

Middle and late 1800's Louis Pasteur and Robert Koch firmly established the germ theory of disease.

1953 James D. Watson and Francis H. C. Crick proposed a model of the molecular structure of *deoxyribonucleic acid* (DNA), the hereditary material in chromosomes.

Late 1970's Researchers used genetically engineered bacteria to produce human *insulin*—a hormone used to treat people with diabetes.

1983 Researchers used genetic engineering to transfer human growth hormone genes into mice, causing the mice to grow to about twice their normal size.

engineering offers many potential benefits in medicine, industry, and agriculture. For example, scientists have transferred to bacteria the human gene that produces *insulin*—a hormone that regulates the body's use of sugar. The bacteria then produce insulin, which can be used to treat people who have diabetes. However, some people question the morality of interfering with the hereditary makeup of living things through genetic engineering. Genetic engineering has also caused concern that the release of genetically engineered organisms into the environment may have harmful effects. For this reason, scientists involved in genetic engineering have developed safety guidelines.

Careers in Biology

Biology offers a wide variety of careers and a broad range of work settings, from research laboratories to national parks. High school and college courses helpful to students preparing for a career in biology include chemistry, mathematics, and physics as well as courses in biology. A bachelor's degree is sufficient for some careers in biology, but many positions require a graduate degree. Some people with a bachelor's degree teach in junior high and high schools. Others work as technicians in research laboratories. Many biologists with advanced degrees teach and conduct research at universities.

Job opportunities for biologists in agricultural research and in industry are increasing, especially in the areas of genetic engineering and ecology. Such biologists may work to develop new varieties of food crops or to create organisms capable of producing drugs.

Many government agencies responsible for public health, sanitation, and water quality employ biologists. Careers in biology also include work in zoos and botanical gardens. In addition, some companies and government agencies hire biologists to study the environmental effects of proposed construction projects and problems caused by environmental pollution. GARLAND E. ALLEN

From Schwann's *Microscopical Researches*, 1839

Drawings of Cells made by Theodor Schwann of Germany in the late 1830's helped convince scientists that all plants and animals are made up of cells.

SEF/Art Resource

Charles Darwin, a British naturalist, set forth his theory of evolution in *The Origin of Species* (1859). Darwin's ideas revolutionized biological thought.

© Hank Morgan

Genetically Engineered Yeast, shown on the screen, produce a hepatitis vaccine. Many new uses of genetic engineering are predicted for the future.

John Zoinar from Peter Arnold

Monitoring a Fermentation Process

A. G. E. Fotostock from Peter Arnold

Building a Three-Dimensional Model of a Molecule

Dan McCoy, Rainbow

Measuring Molecular Weights with a Mass Spectrometer

Research in Chemistry attempts to answer questions about the nature of substances. Some chemists, for example, try to understand the chemical changes that substances go through. Others may use models or highly advanced instruments to explore the structure and composition of substances.

CHEMISTRY

CHEMISTRY is the scientific study of substances. Chemists investigate the *properties* (characteristics) of the substances that make up the universe. They study how those substances behave under different conditions. They attempt to explain the behavior of a substance in terms of the substance's structure and composition. Chemists also seek to understand chemical changes. Chemical changes involve alterations in a substance's chemical makeup. The combination of iron with oxygen from the air to form rust is a chemical change. Substances may also go through physical change without altering their chemical makeup. Water changes physically but not chemically when it freezes.

Chemical changes occur constantly in nature and make life on the earth possible. During a thunderstorm, for instance, lightning causes a chemical change in the air. The electrical energy and heat of a lightning bolt cause some of the nitrogen and oxygen in the atmosphere to combine and form gases called *nitrogen oxides*. The nitrogen oxides dissolve in raindrops that fall to the ground. In the soil, they are chemically changed into *nitrates*, substances that serve as fertilizer.

Chemical changes also occur as wood burns and becomes ashes and gases. The food we eat goes through many chemical changes in our bodies. Chemical changes that take place when gasoline burns supply power for automobiles.

Chemists have learned much about the chemical substances and processes that occur in nature. In addition, chemical researchers have created many useful substances that do not occur naturally. Products resulting from chemical research include many artificial fibers, drugs, dyes, fertilizers, and plastics. The knowledge gained by chemists and the materials they have produced have greatly improved people's lives.

The Work of Chemists

Chemistry involves the study of many different substances. Substances differ greatly in their properties, structure, and composition. The methods that chemists use and the questions that they attempt to answer also differ greatly. However, all chemists share certain ideas that are fundamental to their work.

Fundamental Ideas of Chemistry. The most basic chemical substances are the chemical elements. They

Melvyn C. Usselman, the contributor of this article, is Associate Professor of Chemistry at the University of Western Ontario.

Ray Pfortner from Peter Arnold

Collecting Soil Samples at a Hazardous Waste Site

Dick Luria, Photo Researchers

Perfecting a Formula for a New Perfume

Dick Luria, Photo Researchers

Inspecting Drug Purification Equipment

The Practical Applications of Chemistry range from the development of new methods of disposing of hazardous wastes to the discovery of new formulas for perfumes. Cosmetics, drugs, dyes, fertilizers, and synthetic fibers are only a few of the products resulting from chemical research.

are the building blocks of all other substances. Each chemical element is made up of only one kind of atom. The atoms of one element differ from those of all other elements. Chemists use letters of the alphabet as symbols for the elements. The symbols for the elements carbon, hydrogen, oxygen, and iron, for example, are C, H, O, and Fe. There are 92 elements known to exist in nature. About 15 more have been produced artificially. See Element, Chemical.

Electrical forces at the atomic level create chemical bonds that join two or more atoms together, forming molecules. Some molecules consist of atoms of a single element. Oxygen molecules, for example, are made up of two oxygen atoms. Chemists represent the oxygen molecule by the chemical formula O_2. The *2* indicates the number of atoms in the molecule. See Molecule.

When atoms of two or more different elements bond together, they form a chemical compound. Water is a compound made up of two hydrogen atoms and one oxygen atom. The chemical formula for a water molecule is H_2O. See Compound.

Compounds are formed or broken down by means of chemical reactions. All chemical reactions involve the formation or destruction of chemical bonds. Chemists use *chemical equations* to express what occurs in chemical reactions. Chemical equations consist of chemical formulas and symbols that show the substances involved in chemical changes. For example, the equation

$$C + O_2 \rightarrow CO_2$$

expresses the chemical change that occurs when one carbon atom reacts, or bonds, with an oxygen molecule. The reaction produces one molecule of carbon dioxide, which has the formula CO_2.

The Broad Range of Study. Chemists study substances according to questions they want to answer. Many chemists study special groups of substances, such as compounds containing carbon-to-carbon bonds. Some chemists specialize in techniques that enable them to analyze any substance and identify the elements and compounds it consists of. Other chemists study the forces involved in chemical changes. Much chemical research deals with the atomic and molecular structures of substances. Certain chemists try to predict chemical behavior from theories about the forces at work within the atom. Chemists also work to create new substances and to make synthetic forms of rare but useful natural materials. Their field is called *synthetic chemistry*. A number of chemists apply their knowledge to finding ways of using substances and chemical processes in agriculture, industry, medicine, and other fields.

In some cases, chemistry overlaps such sciences as biology, geology, mathematics, and physics to such an extent that *interdisciplinary sciences* have been established. *Biochemistry*, for example, combines biology and chemistry in studying the chemical processes of living things.

Tools and Techniques. Chemists use a wide variety of tools and techniques. Specialized instruments and electronic computers help chemists make accurate measurements. A device called a *mass spectrometer*, for example, enables chemists to determine the *mass* and atomic composition of molecules. Mass is the total quantity of matter that anything contains. Chemists can identify how atoms are arranged in molecules by using instruments that measure the radiation absorbed and given off by the molecules. A technique called *chromatography* enables chemists to separate complicated mixtures into their parts and to detect and measure low concentrations of substances, such as pollutants in air and water.

History of Chemistry

Beginnings. In prehistoric times, people made many useful discoveries by observing the properties of natural substances and the changes those substances go through. About $1\frac{1}{2}$ million years ago, people began to use fire. Fire was the first chemical reaction that human beings learned to produce and control. The use of fire enabled people to change the properties of substances. They used fire for such purposes as cooking, hardening pottery, and smelting metal ores. Fire also enabled them to create new materials. About 3500 B.C., for example, people learned to make bronze by melting copper and tin together.

The people of many ancient cultures believed that gods or spirits caused natural events. About 600 B.C., however, certain Greek philosophers began to regard nature in a different way. They believed that nature worked according to laws that people could discover by observation and logic.

Several ancient Greek philosophers developed theories about the basic substances that make up the world. Empedocles, who lived during the 400's B.C., argued that there were four primary elements—air, earth, fire, and water—and that they combined in various proportions to form all other substances.

About 400 B.C., a Greek philosopher named Democritus taught that all matter was composed of a single material that existed in the form of tiny, indestructible units called atoms. According to his theory, differences among substances were caused only by differences in the size, shape, and position of their atoms.

The Greek philosopher Aristotle, who lived during the 300's B.C., claimed that each of the four primary elements proposed by Empedocles could be changed into any of the other elements by adding or removing heat and moisture. He stated that such a change—called *transmutation*—occurred whenever a substance was involved in a chemical reaction or changed from one physical state—solid, liquid, or gas—to another. Aristotle believed that water, for example, changed to air when it was heated.

Alchemy. During the first 300 years after the birth of Christ, scholars and craftworkers in Egypt developed a chemical practice that came to be called *alchemy*. They based their work on Aristotle's theory of the transmutation of elements and tried to change lead and other metals into gold. Alchemy began to spread to the Arabian Peninsula in the A.D. 600's and to much of western Europe in the 1100's. Until the 1600's, alchemy was a major source of chemical knowledge.

Despite centuries of experimentation, alchemists failed to produce gold from other materials. They did gain wide knowledge of chemical substances, however,

Major Branches of Chemistry

Analytical Chemistry determines the properties of chemical substances and the structure and composition of compounds and mixtures.

Qualitative Analysis identifies the types of elements and compounds that make up substances.

Quantitative Analysis measures the amounts of the different chemicals that make up substances.

Radiochemistry involves the identification and production of radioactive elements and their use in the study of chemical processes.

Applied Chemistry refers to the practical use of the knowledge of chemical substances and processes.

Agricultural Chemistry develops fertilizers and pesticides and studies the chemical processes that occur in the soil and that are involved in crop growth.

Environmental Chemistry studies, monitors, and controls chemical processes and other factors in the environment and their relationships to living things.

Industrial Chemistry involves the chemical production of raw materials and the development, study, and control of industrial chemical processes and products.

Biochemistry deals with the chemical processes of living organisms.

Inorganic Chemistry concerns chemical substances that do not contain carbon-to-carbon bonds.

Organic Chemistry is the study of chemical substances that contain carbon-to-carbon bonds.

Physical Chemistry interprets chemical processes in terms of physical properties of matter, such as mass, motion, heat, electricity, and radiation.

Chemical Kinetics studies the sequence of steps in chemical reactions and the factors that affect the rates at which chemical reactions proceed.

Chemical Thermodynamics deals with the energy changes that occur during chemical reactions and how temperature and pressure differences affect reactions.

Nuclear Chemistry is the use of chemical techniques in the study of nuclear reactions.

Quantum Chemistry analyzes the distribution of electrons in molecules and interprets the chemical behavior of molecules in terms of their electron structure.

Radiation Chemistry concerns the chemical effects of high-energy radiation on substances.

Solid-State Chemistry deals with the composition of solids and the changes that occur within and between solids.

Stereochemistry studies the arrangement of atoms in molecules and the properties that follow from such arrangements.

Surface Chemistry examines the surface characteristics of chemical substances.

Polymer Chemistry deals with plastics and other chainlike molecules formed by linking many smaller molecules.

Synthetic Chemistry involves combining chemical elements and compounds to duplicate naturally occurring substances or to produce compounds that do not occur naturally.

and invented many tools and techniques still used by chemists. Alchemists used such laboratory equipment as funnels, strainers, balance scales for weighing chemicals, and *crucibles* (pots for melting metals). They also discovered new ways of producing chemical changes and learned to make and use various acids and alcohols.

Alchemists also searched for a substance that could cure disease and lengthen life. During the 1500's, some alchemists and physicians began to apply their knowledge of chemistry to the treatment of disease. The medical chemistry of the 1500's and 1600's is called *iatrochemistry* (pronounced *eye* AT *roh KEHM uh stree*). The prefix comes from *iatros*, the Greek word for *physician*. Iatrochemists made the first studies of the chemical effects of medicines on the human body.

Robert Boyle, an Irish scientist of the 1600's, was one of the first modern chemists. He taught that theories must be supported by careful experiments. Boyle conducted many experiments that showed that air, earth, fire, and water are not true elements. He believed that the best explanation of the properties of matter was provided by an atomistic theory that described substances as composed of tiny particles in motion.

The Phlogiston Theory (pronounced *floh JIHS tuhn*) was a very successful chemical theory, though it was eventually replaced by a better one. The theory was developed in the early 1700's by a German chemist and physician named Georg Ernst Stahl. Stahl wrote that all flammable materials contained a substance called *phlogiston*. According to his theory, materials gave off phlogiston as they burned. Air was necessary for combustion because it absorbed the phlogiston that was released. Plants, in turn, removed phlogiston from the air. They therefore became rich in the substance and burned when dry. Like all other good chemical theories, the phlogiston theory provided an explanation for the results of a variety of experiments and offered clues to areas of study in which new discoveries could be made. For that reason, the theory was widely accepted in the 1700's and led to many findings in chemistry.

Chemists of the middle and late 1700's developed techniques for isolating and studying gases. They based their work on the phlogiston theory and made numerous discoveries. In the 1750's, a Scottish chemist and physician named Joseph Black identified carbon dioxide. It was the first gas recognized to have properties different from those of air. In 1766, Henry Cavendish, an English chemist and physicist, discovered hydrogen. Because hydrogen is very flammable, Cavendish believed it was pure phlogiston. Oxygen was discovered independently by a Swedish chemist named Carl Scheele in the early 1770's and an English chemist named Joseph Priestley in 1774. Wood burns more vigorously in oxygen than in air. Priestley therefore believed that oxygen could absorb great quantities of phlogiston. He called oxygen *dephlogisticated air* (air without phlogiston).

Lavoisier's Contributions. Antoine Lavoisier, a French chemist, revolutionized chemistry in the late 1700's. He repeated many of the experiments of earlier chemists but interpreted the results far differently. Unlike earlier chemists, Lavoisier paid particular attention to the weight of the ingredients involved in chemical reactions and of the products that resulted. He found that the weight of the products of combustion equals that of the original ingredients. His discovery came to be known as the *law of the conservation of mass* (or *matter*).

Lavoisier noted that the weight of the air in which combustion occurred decreases. He found that the weight loss results from the burning material combining with and removing a substance in the air. That substance, which was the same as dephlogisticated air, Lavoisier called oxygen. Lavoisier's oxygen theory of combustion came to replace the phlogiston theory.

Lavoisier and the Marquis de Laplace, a French astronomer and mathematician, also carried out experiments demonstrating that respiration in animals is chemically similar to combustion. Their studies of the chemical processes of living organisms were among the first experiments in biochemistry. Lavoisier also helped work out the present-day system of chemical names. He

The Metropolitan Museum of Art, Rogers Fund, 1931

Metal Smelting and Casting are shown in this Egyptian wall painting from about 1474 B.C. Ancient peoples knew how to use various substances to make many things they needed.

Granger Collection

The Alchemist's Workshop was the forerunner of the modern chemical laboratory. Alchemists used such laboratory equipment as funnels, strainers, and balance scales

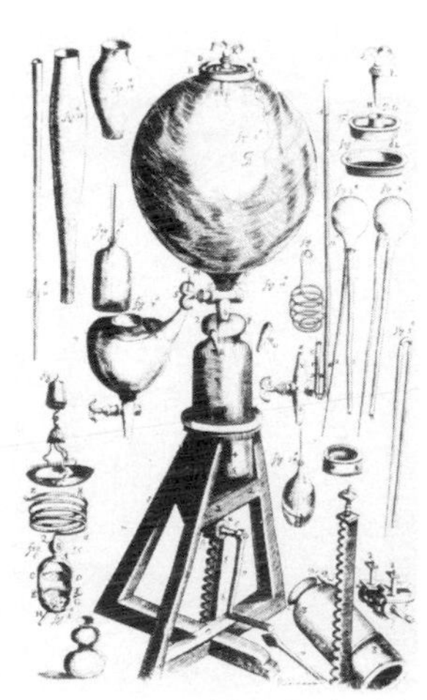

From Boyle's *New Experiments* 1660

An Air Pump built by Robert Boyle and Robert Hooke in the mid-1600's was used to investigate the nature of vacuums.

Important Dates in Chemistry

c. 3500 B.C. People learned to make bronze.

c. 400 B.C. Democritus proposed an atomic theory.

A.D. 600's Alchemy began to spread from Egypt to the Arabian Peninsula and reached western Europe in the 1100's.

1600's Robert Boyle taught that theories must be supported by careful experiments.

Early 1700's Georg Ernst Stahl developed the phlogiston theory.

1750's Joseph Black identified carbon dioxide.

1766 Henry Cavendish discovered hydrogen.

1770's Carl Scheele and Joseph Priestley discovered oxygen.

Late 1700's Antoine Lavoisier stated the law of the conservation of mass and proposed the oxygen theory of combustion.

1803 John Dalton proposed his atomic theory.

1811 Amedeo Avogadro suggested that equal volumes of all gases at the same temperature and pressure contain equal numbers of particles.

Early 1800's Jöns J. Berzelius calculated accurate atomic weights for a number of elements.

1828 Friedrich Wöhler made the first synthetic organic substance from inorganic compounds.

1856 Sir William H. Perkin made the first synthetic dye.

1869 Dmitri Mendeleev and Julius Lothar Meyer discovered the periodic law.

1910 Fritz Haber patented a process to produce synthetic ammonia.

1913 Niels Bohr proposed his model of the atom.

1916 Gilbert N. Lewis described electron bonding between atoms.

1950's Biochemists began to discover how such chemicals as *deoxyribonucleic acid* (DNA) and *ribonucleic acid* (RNA) affect heredity.

Early 1980's Chemists began working to develop a solar-powered device that produces hydrogen fuel by means of the chemical breakdown of water.

published his ideas on combustion, respiration, and the naming of compounds in *Elementary Treatise on Chemistry* (1789), the first modern textbook of chemistry.

Dalton's Atomic Theory. In 1803, an English chemist named John Dalton developed an atomic theory based on the idea that each chemical element has its own kind of atoms. He believed that all the atoms of a particular element had the same mass and chemical properties. The theory could explain and predict the results of various experiments and was gradually accepted.

According to Dalton's theory, a fixed number of atoms of one substance always combined with a fixed number of atoms of another substance in forming a compound. Dalton realized that substances must combine in the same proportions by weight as the weight proportions of their atoms. Chemists had already observed that pure substances do combine in fixed proportions. They called that finding the *law of definite* (or *constant*) *proportions*. Dalton's theory explained the law.

Dalton was the first to calculate the weights of the atoms of several elements. By 1814, Jöns J. Berzelius, a Swedish chemist, had obtained accurate atomic weights for a number of elements. He also began the system of using letters of the alphabet as symbols for elements.

Formation of the Periodic Table. In 1869, a Russian chemist named Dmitri Mendeleev and a German chemist named Julius Lothar Meyer independently announced their discovery of the *periodic law*. The law is based on their observation that when elements are arranged in a table according to their atomic weights, elements with similar properties appear at regular intervals, or *periods*, in the table. The two chemists rearranged the table in columns so that elements with similar properties were grouped together. Such an arrangement became known as the *periodic table*. Both men left gaps in the table, and Mendeleev correctly predicted that elements with certain properties would be discovered to fill the gaps. The modern periodic table summarizes the chemistry of all the known elements. See ELEMENT, CHEMICAL (Periodic Table of the Elements).

Granger Collection

Antoine Lavoisier studied many chemical processes in the late 1700's. This engraving shows his experiment proving that water consists of hydrogen and oxygen.

Granger Collection

John Dalton developed an atomic theory in 1803. His theory, based on the idea that each chemical element has its own kind of atoms, gradually won acceptance.

Granger Collection

Friedrich Wöhler made the first organic substance from inorganic chemicals in 1828, showing that living things were not the only source of organic compounds.

A WORLD BOOK SCIENCE PROJECT

A SIMPLE CHEMICAL EXPERIMENT

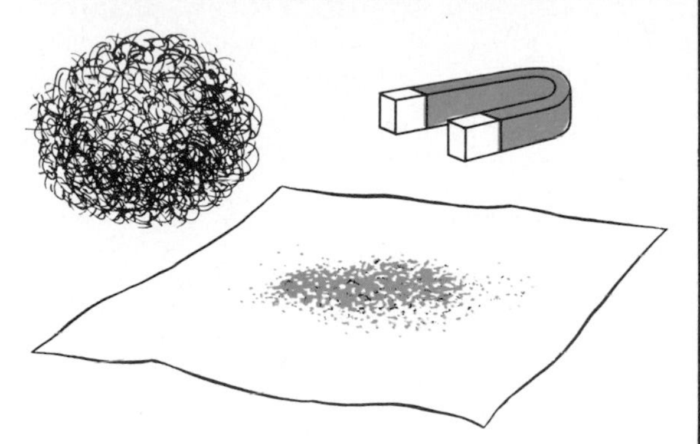

This experiment shows that iron and oxygen can combine in different ways to form two compounds. One compound is a reddish, nonmagnetic powder, and the other is a blue-black magnetic powder.

MATERIALS

Steel wool rusts slowly in water, but quickly in a solution containing hypochlorous acid. You can make such a solution by mixing bleach and vinegar. The hypochlorous acid ($HClO$) in the solution reacts with the iron (Fe) in steel wool to form hydrated ferric oxide ($Fe_2O_3 \cdot H_2O$). By heating this oxide, you can change it to magnetic oxide of iron (Fe_3O_4).

WORLD BOOK illustration by Raymond Perlman

PROCEDURE

Place the ball of steel wool into one of the jars and add enough water to cover the ball. You see no change in the steel wool because water affects iron slowly.

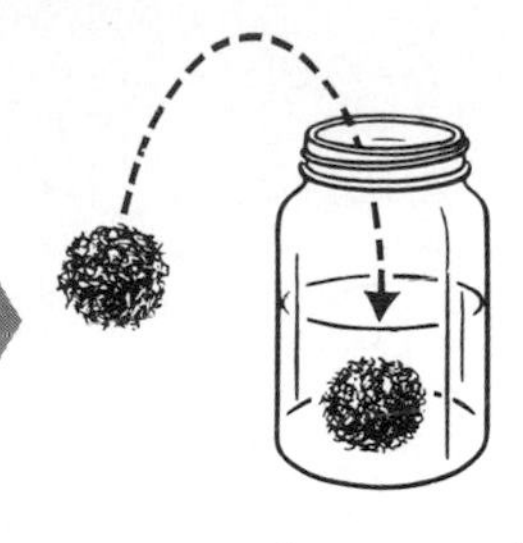

Add 4 teaspoons (20 milliliters) of bleach and 2 teaspoons (10 milliliters) of vinegar to the water and stir. The steel immediately begins to turn red as the iron in it reacts with hypochlorous acid.

In about 5 minutes, the liquid is full of red powder. This powder is hydrated ferric oxide (rust). Remove the steel wool from the jar and wait for the powder to settle.

After the powder has settled, carefully pour off the clear solution. The powder and some liquid will remain at the bottom of the jar.

Wash the powder by filling the jar with water and stirring. Once again, allow the powder to settle and pour off most of the water.

Place a paper napkin over the mouth of another jar and pour the mixture of powder and liquid into the napkin. Wait for all the liquid to filter through the paper.

Spread out the napkin and allow the powder to dry. Then test it by touching a magnet to it. Hydrated ferric oxide is not magnetic.

Place the powder on an old spoon and heat it in the flame of a candle. The red powder slowly turns blue-black as it changes to magnetic oxide of iron.

After the color of the powder has changed completely, test it again for magnetism. The blue-black grains will cling to the magnet.

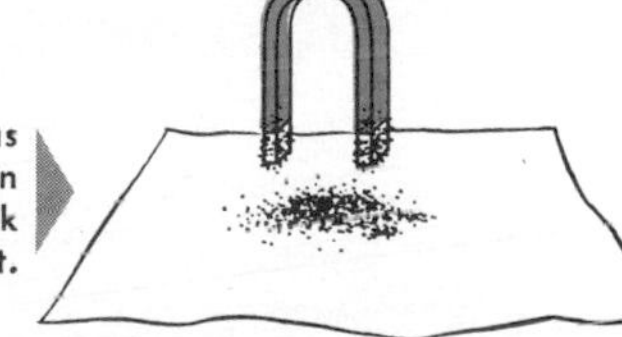

CHEMISTRY

Development of Organic Chemistry. From the time of the alchemists, researchers had investigated various substances found in plants and animals. Such organic substances, however, proved especially difficult to analyze. As a result, nearly all early chemical knowledge was obtained by studying simpler inorganic substances.

Most chemists of the early 1800's believed that organic compounds could be produced only with the aid of a *vital force*, a life force present in plants and animals. That belief is called *vitalism*. In 1828, a German chemist named Friedrich Wöhler mixed two inorganic substances, heated them, and obtained *urea*—an organic compound found in urine. Wöhler thus made the first synthetic organic substance from inorganic materials and proved that a vital force is not necessary for the production of an organic compound.

During the 1800's, chemists isolated many organic substances. They discovered that most organic compounds consist mainly of carbon combined with hydrogen, nitrogen, and oxygen in various proportions. Chemists found that, in certain cases, two separate organic compounds with different properties are composed of the same elements in the same proportions. Berzelius called such compounds *isomers*. Isomers have the same kinds and numbers of atoms but differ in the ways the atoms are joined together.

In the mid-1800's, chemists began to include ideas about molecular structure in their theories. They developed the *valence theory* to explain how atoms combine and form molecules. Valence refers to the normal number of bonds one atom can form with other atoms. In 1858, a German chemist named Friedrich Kekulé von Stradonitz proposed that carbon atoms can bond to four other atoms and that carbon atoms can link together to form chains. As a result of his ideas, chemists quickly recognized organic compounds as molecules based on a framework of carbon-to-carbon bonds along with other bonds from carbon atoms to hydrogen, nitrogen, oxygen, and a few other atoms.

By 1900, the study of organic substances had become a major branch of chemistry. Chemists have since learned how to produce numerous complex organic molecules. In the mid-1900's, Melvin Calvin, an American chemist, solved many long-standing mysteries of *photosynthesis*, the chemical process by which plants make food. Since the mid-1900's, biochemists have discovered how such organic substances as *deoxyribonucleic acid* (DNA) and *ribonucleic acid* (RNA) affect heredity (see HEREDITY; NUCLEIC ACID).

Development of Physical Chemistry. During the 1800's, many chemists and physicists investigated the properties of substances and the energy changes that accompany chemical reactions. They based their work on ideas about the structure and behavior of atoms and molecules. Such study is called *physical chemistry*.

One of the first scientists to explore the area of physical chemistry was Amedeo Avogadro, an Italian physicist. In 1811, Avogadro suggested that equal volumes of all gases at the same temperature and pressure contain equal numbers of particles. His idea, known as *Avogadro's law*, enabled chemists to calculate relative atomic weights. Later in the 1800's, physical chemists developed the *kinetic theory of gases*. The theory describes gases as clusters of particles in constant motion and explains how such clusters moving at high speed determine pressure, temperature, and other properties of gases.

During the mid-1800's, physicists formulated the principles involved in the conversion of heat into mechanical energy and vice versa. They thereby laid the foundations for *chemical thermodynamics*, the study of the changes in heat that accompany many reactions.

During the 1870's, an American scientist named Josiah Willard Gibbs developed the *phase rule*. The rule explains the physical relationships among the solid, liquid, and gaseous *phases* (states) of matter. Jacobus van't Hoff, a Dutch chemist, relied on the phase rule in his studies of how crystals form in various solutions. Van't Hoff's work led to the development of *stereochemistry*, the study of the arrangement of atoms in molecules.

In the late 1800's, physical chemists Svante A. Arrhe-

Granger Collection

Sir William H. Perkin discovered the first synthetic dye accidentally in 1856. He produced mauve while trying to make quinine from a coal tar product.

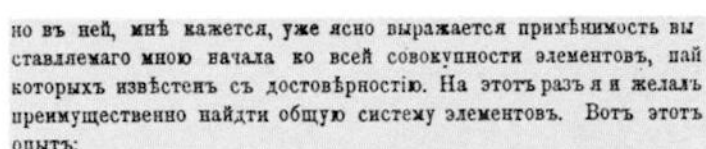

но въ ней, мнѣ кажется, уже ясно выражается примѣнимость выставляемаго мною начала ко всей совокупности элементовъ, пай которыхъ извѣстенъ съ достовѣрностію. На этотъ разъ я и желалъ преимущественно найдти общую систему элементовъ. Вотъ этотъ опытъ:

			Ti=50	Zr=90	?=180.
			V=51	Nb=94	Ta=182.
			Cr=52	Mo=96	W=186.
			Mn=55	Rh=104,4	Pt=197,4
			Fe=56	Ru=104,4	Ir=198.
			Ni=Co=59	Pl=106,6,	Os=199.
H=1			Cu=63,4	Ag=108	Hg=200.
	Be=9,4	Mg=24	Zn=65,2	Cd=112	
	B=11	Al=27,4	?=68	Ur=116	Au=197?
	C=12	Si=28	?=70	Sn=118	
	N=14	P=31	As=75	Sb=122	Bi=210
	O=16	S=32	Se=79,4	Te=128?	
	F=19	Cl=35,5	Br=80	I=127	
Li=7	Na=23	K=39	Rb=85,4	Cs=133	Tl=204
		Ca=40	Sr=87,6	Ba=137	Pb=207.
		?=45	Ce=92		
		?Er=56	La=94		
		?Yt=60	Di=95		
		?In=75,6	Th=118?		

а потому приходится въ разныхъ рядахъ имѣть различное измѣненіе разностей, чего нѣтъ въ главныхъ числахъ предлагаемой таблицы. Или же придется предполагать при составленіи системы очень много недостающихъ членовъ. То и другое мало выгодно. Мнѣ кажется притомъ, наиболѣе естественнымъ составить кубическую систему (предлагаемая есть плоскостная), но и попытки для ея образованія не повели къ надлежащимъ результатамъ. Слѣдующія двѣ попытки могутъ показать то разнообразіе сопоставленій, какое возможно при допущеніи основнаго начала, высказаннаго въ этой статьѣ.

Li	Na	K	Cu	Rb	Ag	Cs	—	Tl
7	23	39	63.4	85.4	108	133		204

Granger Collection

A Periodic Table grouping the elements by atomic weights was proposed by Dmitri Mendeleev in 1869. It appeared in the *Journal of the Russian Chemical Society.*

Lawrence Berkeley Laboratory

Melvin Calvin, winner of the 1961 Nobel Prize in chemistry, used a radioactive tracer to map the chemical reactions that occur during photosynthesis.

nius of Sweden and Wilhelm Ostwald of Germany proposed that electricity is carried through solutions by charged atoms or molecules called *ions*. Ostwald wrote one of the first textbooks in *electrochemistry*, the study of chemical changes associated with electrical forces.

Since the early 1900's, chemists and physicists have devoted much study to the structure of atoms and molecules. In 1913, a Danish physicist named Niels Bohr suggested a model of the atom in which electrons are arranged in successively larger orbits around a small nucleus of protons and neutrons. He believed that many of the properties of an element depend on the number of electrons in the outer orbit of the atoms of that element.

Bohr's model of the atom also helped explain how atoms interact with light and other forms of radiation. Bohr assumed that the absorption and *emission* (giving off) of light by an atom involve a change in the energy state of an electron and a resulting electron jump from one orbit to another. Chemists have gained much information about the structure of molecules by measuring their absorption and emission of radiation.

In 1916, Gilbert N. Lewis, an American chemist, proposed that the bond between atoms in a molecule consists of a pair of electrons that both atoms share. His idea led to the *electron pair theory*, which explains the bonding characteristics of elements in terms of the arrangement of their electrons. See BOND (chemical).

Growth of Industrial Chemistry. The use of chemical knowledge by manufacturers started with the origins of chemistry itself. During the 1700's, however, manufacturers of such products as acids, alkalis, and soap began to use the knowledge of chemists on a broad scale to improve their products and production methods. During the 1800's, factories turned out huge quantities of such chemicals as sulfuric acid, sodium carbonate, and bleaching powder. In 1856, the English chemist Sir William H. Perkin produced *mauve*, also called *aniline purple*—the first synthetic dye. Its popularity soon led to the synthesis of other dyes for the textile industry.

By 1900, Germany had the most advanced chemical industry in the world. In 1910, a German chemist named Fritz Haber patented a process to produce ammonia from hydrogen and nitrogen. His work led to the large-scale manufacture of chemical agricultural fertilizers. During World War I (1914-1918) and World War II (1939-1945), the chemical industry expanded greatly in several countries to meet the demand for such war materials as explosives, medications, and synthetic rubber.

After World War II, the chemical industry continued to produce a great variety of goods for consumers. The development of new materials resulted in the widespread use of plastics and such synthetic fibers as nylon and polyesters. In addition, further discoveries led to the availability of many new drugs, food preservatives, fertilizers, and pesticides.

Current Research. Biochemistry is a particularly active area of scientific research today. New instruments have enabled biochemists to study the action of chemicals within an organism without harming the organism. Biochemists are studying substances suspected of causing cancer or genetic damage in order to determine what molecular features are responsible for the harmful effects. Other chemists are investigating how chemical pollutants affect the environment and how they break down into other substances.

Synthetic chemistry is another area of active research. Chemists synthesize many thousands of new compounds each year. They have discovered chemical agents that can be used in reactions to add special groups of atoms to specific parts of other molecules. Researchers design new molecules and use such agents in a series of reactions to build the new compounds. Their techniques have led to the creation of many drugs.

The study of the surface properties of chemical compounds—called *surface chemistry*—is another promising field of present-day research. Chemists have learned that surface characteristics are responsible for the ability of certain substances—called *catalysts*—to speed up the rate of chemical reactions. Chemists today are also working to develop a chemical cell that would use the energy of sunlight to break up water molecules into oxygen and hydrogen. The hydrogen thus produced could be used as fuel. Such cells may one day provide a valuable new source of energy.

IBM Thomas J. Watson Research Center, IBM

The Scanning Tunneling Microscope, *above left,* provides chemists with new insights into the surface properties of materials. Its image of the surface of silicon, *above right,* shows both the surface atoms and the bonds that hold them in place.

The Chemical Industry

The chemical industry plays a vital role in the production of many manufactured goods. The industry provides a tremendous variety of materials to other manufacturers. It also produces many chemical products that benefit people directly. Major products of the industry include detergents, drugs, dyes, fertilizers, food preservatives and flavorings, glass, metal alloys, paper products, plastics, and synthetic fibers.

Most major chemical products are *basic chemicals* used in the manufacture of other products. Sulfuric acid is the chief basic chemical in the United States and many other countries. It is used to produce fertilizers and numerous other chemicals. Other basic chemicals include chlorine, nitrogen, and oxygen; such alkalis as lime and sodium hydroxide; and chemicals used in plastics.

The production of chemicals has become increasingly concentrated in *multinational companies*, which have plants and offices in a number of countries. To help keep costs low, the companies tend to locate their factories in countries where raw materials are readily available. Many basic chemicals are produced in developing countries by factories of multinational firms. But chemicals requiring advanced production methods are made mainly in developed countries.

Most chemical companies have research and development programs. Chemists in those programs work to develop new substances, new uses for known chemicals, and improvements in production techniques.

The success of the chemical industry has been accompanied by environmental and safety problems. For example, the use of huge amounts of pesticides has resulted in soil and water pollution. In addition, the production of some chemicals results in harmful waste products that must be disposed of safely. Many chemical dumps for the storage of such wastes have leaked, threatening the health of people in nearby areas. Since the mid-1970's, a number of accidents have occurred at chemical plants in several countries and have resulted in the release of harmful substances.

Chemical companies have had to spend much money in efforts to solve environmental and safety problems. For example, they are working to develop insecticides that will quickly break down into harmless substances in the environment. They are also seeking safer methods of disposing of chemical wastes and of cleaning up chemical dumps. In addition, they are increasing safety precautions at chemical plants to guard against accidents.

Careers in Chemistry

Chemistry offers a variety of challenging career opportunities in education, industry, and government. High school and college courses helpful to students preparing for a career in chemistry include mathematics and physics as well as classes in chemistry. Many chemical instruments make use of computer technology, and so classes in computer science are also useful. Writing courses help chemists develop their ability to communicate scientific information to others.

A bachelor's or master's degree in chemistry is sufficient for some careers, including teaching chemistry in junior high and high schools. Some chemists with advanced degrees teach at universities or conduct research. A doctor's degree is important for students who wish to pursue *basic research*—that is, the study of fundamental laws and processes of chemistry.

Many university graduates with specialized knowledge in chemistry find employment in industry. They work as plant superintendents, chemical engineers, quality control personnel, and salespeople. In addition, a large number of chemists are hired by government agencies involved in such areas as trade, environmental protection, and public health.

Melvyn C. Usselman

Related Articles in World Book. Each chemical element has a separate article. For a list, see Element, Chemical. See also the following articles:

Outline

A Female Leaf-Roller uses a leaf as a "nest." This type of beetle makes cuts in the leaf, rolls it up, and lays eggs in the folds.

Thomas Eisner

A Bombardier Beetle, *above,* defends itself by squirting a hot, irritating jet of gas at its attacker. The insect produces this spray by mixing chemicals from two organs located at the end of its body.

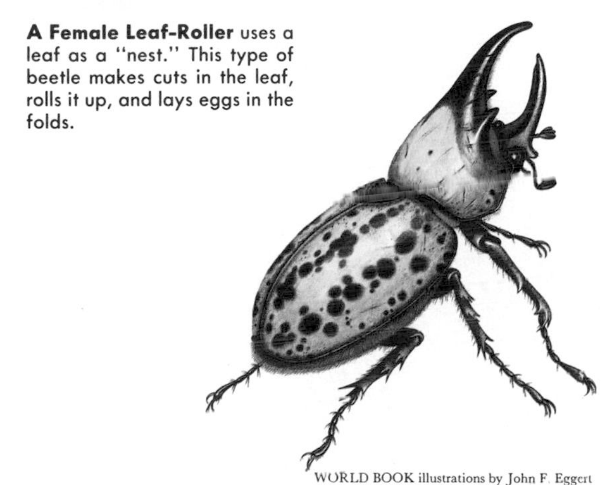

WORLD BOOK illustrations by John F. Eggert

A Male Eastern Hercules Beetle, *above,* has a long horn. This beetle is also one of the largest beetles in North America. It grows to $2\frac{1}{2}$ inches (6 centimeters) long.

Hans Reinhard, Bruce Coleman Inc.

Male Stag Beetles have long jaws that they use to defend themselves. Stag beetles get their name from the jaws, which resemble the antlers of a *stag* (male deer).

BEETLE

BEETLE is one of the most common of all insects. There are about 300,000 *species* (kinds) of beetles. They live everywhere on earth except in the oceans. Beetles are found in rain forests and in deserts. They live in freezing cold areas and in hot springs. They inhabit mountain lakes and can even survive in polluted sewers.

Beetles have typical insect body parts, including antennae, three pairs of legs, and a tough *exoskeleton* (external skeleton). However, unlike other insects, adult beetles have a pair of special front wings called *elytra.* These wings form leathery covers that protect the beetle's body. Because of their shell-like skeleton and hard wing covers, beetles have been called the "armored tanks" of the insect world.

Beetles vary greatly in shape, color, and size. Some, such as click beetles and fireflies, are long and slender. Others, including ladybugs, are round. Most beetles are brown, black, or dark red in color. But some have bright, shiny, rainbow colors. The smallest beetles, called feather-winged beetles, measure less than $\frac{1}{50}$ inch (0.5 millimeter) long. One of the largest is the Goliath, which lives in Africa. It grows about 5 inches (13 centimeters) long and weighs over $1\frac{1}{2}$ ounces (42 grams).

Most species of beetles are *solitary insects*—that is, they live alone and have no family life. The young develop without help from their parents. A few species of beetles are *social insects.* These beetles spend at least part of their life in family groups.

Beetles have many enemies, including birds, reptiles, and other insects. Most beetles protect themselves by hiding or by flying away. A few produce bad-smelling chemicals that discourage attackers. Some beetles bite.

Many beetles are pests because they feed on farm crops, trees, or stored food. But some beetles are helpful to people. For example, ladybugs and certain other beetles save crops by eating insect pests. Other beetles are important because they eat dead plants and animals and thus remove them from the environment.

The Bodies of Beetles

Like other insects, beetles have a body that is divided into three main parts. These parts are: (1) the head, (2) the thorax, and (3) the abdomen.

The Head includes the beetle's mouthparts, eyes, and a pair of antennae. The eyes and antennae are the insect's chief sense organs.

Mouthparts. Beetles have chewing mouthparts. In beetles called *weevils,* the mouthparts are part of a long snout. A beetle's jaws are called *mandibles.* A number of beetles have large, pincerlike mandibles.

The External Anatomy of a Beetle

WORLD BOOK illustrations by John F. Eggert

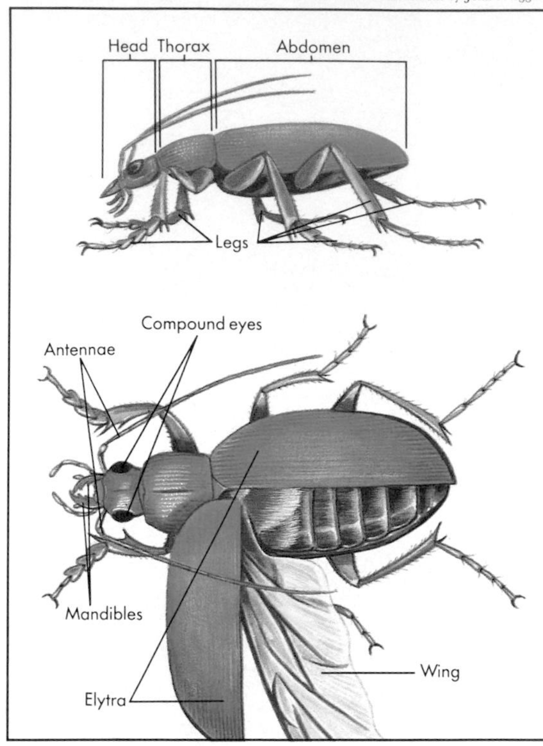

The Internal Anatomy of a Beetle

The diagram below gives an internal view of a typical female beetle. Included are features of the reproductive system, the nervous system, and the systems for circulating blood and digesting food.

WORLD BOOK diagram by Lori L. Grove

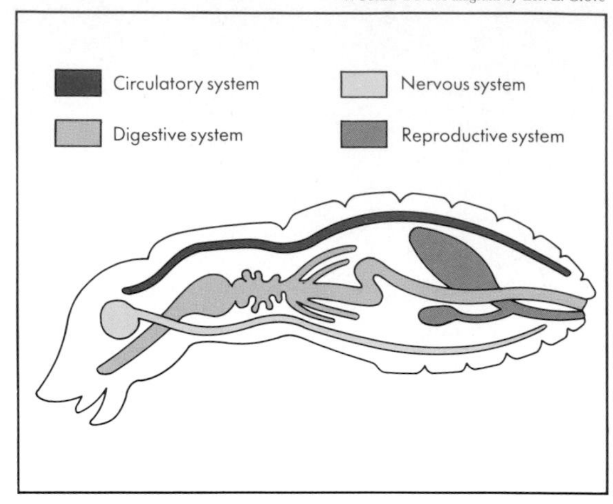

Eyes. Beetles have a *compound eye* on each side of the head. Each eye consists of a bundle of tiny, light-sensitive lenses. Each lens contributes a small bit of the total image that a beetle sees. Most beetles see motion and colors quite well. A few species are blind.

Antennae vary greatly in shape and form among beetles. Many beetles have antennae made up of threadlike or beadlike segments. In many of these beetles, the tip segments of the antennae are club-shaped. Some beetles have elbow-shaped or featherlike antennae. A beetle's antennae are covered with hairs and special organs that can detect specific odors. Some beetles have special sense organs near the base of the antennae that provide a simple type of hearing. These organs send messages to the brain when certain sounds vibrate the antennae.

The Thorax forms the middle of the beetle's body. It consists of three segments, each with a pair of legs. The second and third segments each have a pair of wings.

Legs. Each leg of a beetle has five segments and claws at the end. Most beetles that are fast runners have long, slender legs. Other beetles have short, stout legs, often with flat pads on the bottom. These pads have hundreds of expanded hairs that act like suction cups and enable the beetle to walk upside down on slick surfaces. The legs of digging beetles have toothlike projections that are used to scrape away soil. Most swimming beetles have flattened hind legs. In some species, these are fringed with long hairs to form paddles.

Wings. A beetle's front wings, the elytra, are attached to the second segment of the thorax. The hind wings are attached to the third segment. In most species, the elytra cover the hind wings when the insect is not flying. To fly, a beetle pops open the elytra and holds them upward and outward so that it can move its hind wings freely.

The Abdomen contains the reproductive organs and the chief organs of digestion. It typically consists of 10

The Life Cycle of a Beetle

A beetle goes through four stages of development: (1) egg, (2) larva, (3) pupa, and (4) adult. The illustration below shows the development of a broad-nosed beetle. The egg, which is laid in the ground, hatches into a wormlike larva. As the larva grows, it sheds its outer skin several times before becoming a pupa. The adult organs develop in the pupa. When this process is complete, the adult emerges.

WORLD BOOK illustration by John F. Eggert

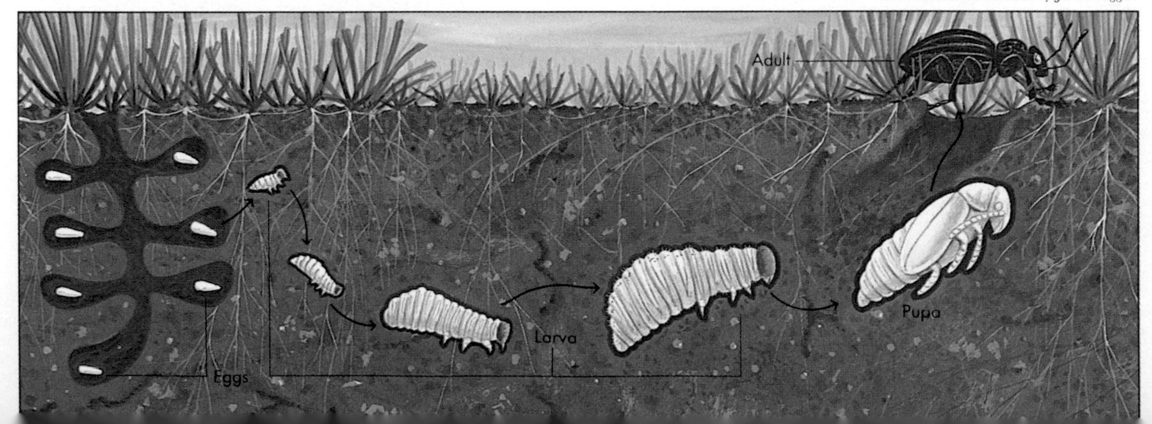

Some Kinds of Beetles

There are more than 300,000 species of beetles. They live in nearly every type of environment on earth except in the oceans. These drawings and photographs provide examples of the great variety of sizes, shapes, and colors of beetles.

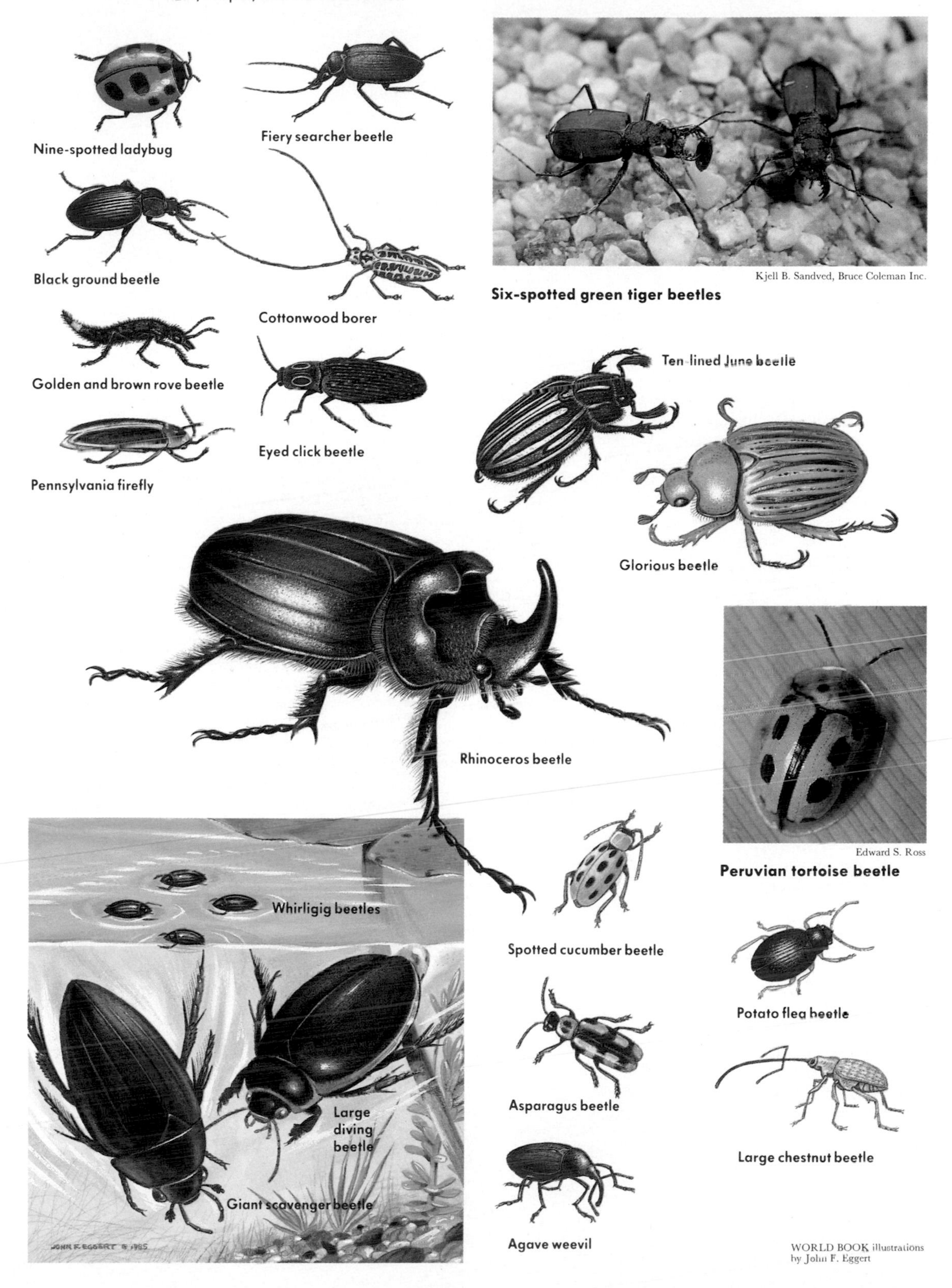

Six-spotted green tiger beetles

Peruvian tortoise beetle

WORLD BOOK illustrations by John F. Eggert

segments, though only 5 to 8 segments may be visible. The segments are usually soft on the upper surface where they are covered by the elytra. The undersurface is harder for protection. Each segment of the abdomen has a pair of tiny holes called *spiracles*. Oxygen enters the beetle's body through the spiracles.

The Life Cycle of Beetles

A beetle passes through four stages of development during its life: (1) egg, (2) larva, (3) pupa, and (4) adult. The beetle changes greatly in appearance and structure from one stage to another.

The Egg. Most female beetles lay eggs with oval shapes and drab colors. A female beetle may lay from a few to several thousand eggs at one time, depending on the species. Most beetles place their eggs on the surface of their food or inside cracks or holes. Eggs laid in the spring or summer may take a week to a month to hatch. Some species lay eggs in the fall. The eggs of these beetles hatch the following spring.

The Larva of a beetle is also called a *grub*. This form of the insect looks much different from the adult and may eat different food. Most beetle larvae are wormlike, but some look like tiny lizards. In most species, the larval stage lasts from a few weeks to a few months. The larvae of some June beetles may take five years to mature.

As a beetle larva grows, it completely fills its rigid exoskeleton. It then breaks out of the exoskeleton while forming a new, larger one. This process is known as *molting*. Beetle larvae molt from three to seven or more times.

The Pupa. When the larva molts for the last time, it transforms into a pupa. The pupa resembles an adult, but it is softer and different in color. In addition, the pupa has only small, padlike wings. Most beetles spend the pupal stage underground. This stage may last from a few days to an entire winter, depending on the species. During this time, adult organs form. When this process is complete, the pupa molts and the adult emerges.

The Adult has a short life and so must mate quickly. Then the female must find a place to lay eggs. Most beetles live as adults for several weeks or months. Adult females of some species may live only a few days.

Kinds of Beetles

Beetles make up the insect order Coleoptera. *Coleoptera* is a Greek word meaning *sheath wings*. It refers to a beetle's elytra, which form a *sheath* (cover) for much of the upper body. The order Coleoptera is the largest order of insects. Nearly 40 per cent of all insect species belong to it. The order is divided into about 150 families. This section describes some of the major beetle families. The scientific name of the family appears in parentheses after the common name.

Weevils (Curculionidae), also called *billbugs* and *snout beetles*, consist of more than 40,000 species. They are the largest family of beetles. The mouthparts of adult weevils are at the tip of a long snout used to bore into fruits, seeds, and other plant parts. The larvae are legless and feed inside fruits and nuts or are borers. Many weevils are crop pests. See Boll Weevil; Weevil.

Leaf Beetles (Chrysomelidae) total more than 25,000 species. Most leaf beetles can fly. When disturbed, however, many drop to the ground and play dead. Both the larvae and adults eat leaves and are serious crop pests. The potato bug, also called the *Colorado potato beetle*, is one of the most common pests. It causes much damage to potato crops. See Potato Bug.

Ground Beetles (Carabidae) number more than 20,000 species. The adults have long legs and long antennae. Most species hide during the day and search for food at night. Both the adults and larvae prey on other animals. Some species of ground beetles have been brought to the United States to prey on crop-eating insect pests. The bombardier beetle is an unusual ground beetle. It defends itself by squirting two chemicals from the end of its body. The chemicals mix to produce a hot puff of gas that can repel an enemy.

Rove Beetles (Staphylinidae) make up more than 20,000 species. Rove beetles have unusually short elytra, which make them look like other insects called *earwigs*. Earwigs, however, have sharp pincers at the tip of the abdomen. Some rove beetles turn up the tip of their abdomen as if they could sting. Most larvae and adults prey on other animals or eat dead or decaying materials.

Scarabs (Scarabaeidae) consists of about 20,000 species. Dung beetles and tumblebugs are scarabs. They feed on *dung* (solid body wastes of animals). They can shape a mass of dung into a ball and bury it in soil. Females lay one egg in the ball of dung. June bugs and Japanese beetles are two types of scarabs that eat crop plants. See Japanese Beetle; June Bug; Scarab.

Click Beetles (Elateridae) total about 8,000 species. These long, slender beetles jump or make a clicking sound if disturbed. They do this by means of a hooklike part that locks the first and second segments of the thorax. By building up pressure between these two body segments and then releasing the hook, a sudden body jerk and clicking sound is produced. Most larvae of click beetles are slender and have hard, ringlike body segments. These larvae are commonly called *wireworms*. The larvae of some species eat the roots and seeds of crop plants. See Click Beetle.

Predacious Diving Beetles (Dytiscidae) make up about 4,000 species. These beetles live in bodies of fresh water. They prey on snails, tadpoles, and small fish. The larvae, which also live in the water, have long, soft bodies. The adults swim by moving their hind legs together like oars. When under water, the adults breathe air trapped in their body hairs or beneath their elytra.

Ladybugs (Coccinellidae), also called *ladybirds* and *lady beetles*, number more than 4,000 species. Adult ladybugs have round bodies. Many are red, orange, or yellow and have black spots. The larvae look like miniature lizards and some are brightly colored. Both adults and larvae eat insects that attack trees, shrubs, and fruit and vegetable crops. See Ladybug.

Fireflies (Lampyridae), also called *lightning bugs*, total about 1,900 species. Most species produce a cool, chemical light in the abdomen through a process called *bioluminescence*. The fireflies produce this light to find one another during mating. Each species uses a special pattern of flashes to identify each other. Some adult fireflies do not feed. Others eat pollen or nectar. The larvae prey on snails and insects. See Firefly.

Scientific Classification. Beetles make up the order Coleoptera in the class Insecta and the phylum Arthropoda.

David J. Shetlar

Index

This index covers the contents of the 1986, 1987, and 1988 editions of SCIENCE YEAR, The World Book Science Annual.

Each index entry gives the edition year and a page number—for example, 88-123. The first number, 88, indicates the edition year, and the second number, 123, is the page number on which the desired information begins.

There are two types of entries in the index.

In the first type, the index entry (in **boldface**) is followed immediately by numbers:

Paleontology, 88-296, 87-294, 86-252

This means that SCIENCE YEAR has an article titled Paleontology, and that in the 1988 edition the article begins on page 296. In the 1987 edition, the article begins on page 294, and in the 1986 edition it is on page 252.

In the second type of entry, the boldface title is followed by a clue word instead of by numbers:

Doppler radar: tornado, Special Report, 86-87; meteorology, 88-285

This means that there is no SCIENCE YEAR article titled Doppler radar, but that information about this topic can be found in a Special Report in the 1986 edition on page 87. There is also information on this topic in the Meteorology article of the 1988 edition on page 285.

When the clue word is "il.," the reference is to an illustration only:

Black-footed ferret: il., 88-316

This means there is an illustration of this animal in the 1988 SCIENCE YEAR, on page 316.

The various "See" and "See also" cross-references in the index direct the reader to other entries within the index:

Horticulture. See **Agriculture; Botany;** and **Plant.**

This means that for the location of information on horticulture—look under the boldface index entries

Agriculture, Botany, and **Plant.**

Index

A

B

C

D

E

F

G

H

I

J

K

L

M

O

P

Q

R

S

T

U

V

W

X

Y

Z

Acknowledgments

The publishers of *Science Year* gratefully acknowledge the courtesy of the following artists, photographers, publishers, institutions, agencies, and corporations for the illustrations in this volume. Credits should read from top to bottom, left to right on their respective pages. All entries marked with an asterisk (*) denote illustrations created exclusively for *Science Year.* All maps, charts, and diagrams were prepared by the *Science Year* staff unless otherwise noted.

Cover ©Danny Lehman
4 Gary Gianni*; Sultan Al Otaibi, The Saudi Investment Bank
5 Joe VanSeveren*; Christopher Springmann, The Stock Market; © Runk/Schoenberger from Grant Heilman
8 Roberta Polfus*; H. Morgan, Science Source from Photo Researchers; Bath Archaeological Trust Limited
9 © Tom Bean; Photri; E. R. Degginger
10 David Molchos
13 Edward Slater, Southern Stock Photos; South Florida Water Management District
14 Sterling Dimmitt; Edward Slater, Southern Stock Photos; James H. Robinson
16 Caulion Singletary
17 Wendell Metzen, Bruce Coleman Inc., M. P. Kahl, Photo Researchers
18 Caulion Singletary
19 E. R. Degginger; Dan McCoy, Rainbow
23 Sara Woodward*; South Florida Water Management District; E. R. Degginger
24 David Molchos
26 – 37 Lawrence P. Clifford*
40 Lascaux Cave, Dordogne, France (© Jean Vertut)
42 Pêche-Merle Cave, Le Combel, France (© Jean Vertut)
44 Musée des Antiquités Nationales, Paris; © Randall White
45 Gary Gianni*
46 Lascaux Cave, Dordogne, France (© Jean Vertut); Pêche-Merle Cave, Le Combel, France (© Jean Vertut)
48 Lascaux Cave, Dordogne, France (© Jean Vertut)
51 Rouffignac Cave, Dordogne, France (© Jean Vertut)
52 Musée de l'Homme, Paris (SCALA/Art Resource); Musée des Antiquités Nationales, Paris; Le Tuc d'Audoubert Cave, Ariege, France (© Jean Vertut)
56 © Tom Hollyman, Photo Researchers
59 © Cynthia Johnson, Gamma/Liaison
60 Reuven Yagil, Ben-Gurion University
61 John Francis*
62 © Toni Angermayer, Photo Researchers
63 John Francis*
64 – 65 Reuven Yagil, Ben-Gurion University
66 Sultan Al Otaibi, The Saudi Investment Bank
69 © Danny Lehman
71 © Tom Bean
72 – 73 Yvonne Gensurowsky, Stansbury, Ronsaville, Wood Inc.*
74 © Christopher Johns
75 © Keith Gunnar, Bruce Coleman Inc.
76 – 78 Yvonne Gensurowsky, Stansbury, Ronsaville, Wood Inc.*
79 © Tom Bean, Aperture Photo Bank; Yvonne Gensurowsky, Stansbury, Ronsaville, Wood Inc.*
82 © Ron Sanford
83 – 85 Bath Archaeological Trust Limited
86 Brian Delf*
88 – 89 Bath Archaeological Trust Limited
90 Brian Delf*
92 – 93 Bath Archaeological Trust Limited
94 Brian Delf*
96 – 97 Bath Archaeological Trust Limited
100 – 103 Roberta Polfus*
104 Jet Propulsion Laboratory
105 – 106 Roberta Polfus*
108 – 112 Jet Propulsion Laboratory
113 Roberta Polfus*
116 – 126 John Zielinski*
128 Ocean Drilling Program, Texas A&M University
129 Copyright: Hachette—Guides Bleus
132 – 133 Joe Rogers*
134 – 138 Ocean Drilling Program, Texas A&M University
139 Scripps Institution of Oceanography
142 – 146 Eileen Mueller Neill*
149 Allan Tannenbaum, Sygma
150 Eileen Mueller Neill*
152 University of Pittsburgh; Illinois Department of Nuclear Safety
153 Eileen Mueller Neill*
156 – 167 Yvonne Gensurowsky, Stansbury, Ronsaville, Wood Inc.*
170 Grumman Corporation
173 Jeffrey Vock, Visions; Mark Greenberg, Visions
174 Konrad Hack*
176 – 177 Joseph Milioto*
178 Joseph Milioto* Grumman Corporation
179 Grumman Corporation; Photri
180 Joseph Milioto*; McDonnell Douglas Corporation
181 Boeing Commercial Airplane Company; Joseph Milioto*
182 Roy Ritola, © The Testor Corporation; Sikorsky Aircraft
183 U.S. Air Force
184 Joe VanSeveren*
187 © Peter B. Kaplan, Photo Researchers
188 © E. R. Degginger, Bruce Coleman Inc.
189 © Tardos Camesi, The Stock Market
190 Joe VanSeveren*
191 Joe VanSeveren*; Ralph Brunke*, © Marcello Bertinetti, Photo Researchers
192 © James Steinberg, Photo Researchers
193 CERN Laboratories
194 – 196 Joe VanSeveren*
198 George V. Kelvin*
201 IBM Corporation; Thomas J. Watson Research Center, IBM Corporation
202 George V. Kelvin*
203 IBM Corporation; Texas Instruments Inc.
204 H. Morgan, Science Source from Photo Researchers; Christopher Springmann, The Stock Market; Christopher Springmann, The Stock Market
205 IBM Corporation; David Madison, Bruce Coleman Inc.; Texas Instruments Inc.
206 George V. Kelvin*
208 Thomas J. Watson Research Center, IBM Corporation
210 Lawrence Berkeley Laboratory; Allan Walker, National Geographic Society; © John Chiasson, Gamma/Liaison
211 U.S. Department of Agriculture; National Radio Astronomy Observatory; U.S. Department of Agriculture
212 Terrence McCarthy, *The New York Times*
214 Tim McCabe, U.S. Department of Agriculture
215 Allan Walker, National Geographic Society; Trudy Rogers*
216 © John Nance, Magnum
219 Reprinted by permission from *Nature* Vol. 321 Copyright © 1986 Macmillan Journals Limited
222 Paola Villa, University of Colorado at Boulder
223 Anthony E. Marks, Department of Anthropology, Southern Methodist University
224 Richard Howitz © *Discover* Magazine, Time Inc.
225 National Optical Astronomy Observatories
227 David Sanders, Palomar Observatory, California Institute of Technology
228 AP/Wide World
230 Trudy Rogers*
231 National Radio Astronomy Observatory
233 Los Alamos National Laboratory
235 © Routledge and Kegan Paul; © Harvard University Press
236 Bettmann Archive
237 © Abrams; © Oxford University Press
238 – 242 U.S. Department of Agriculture
243 Raychem Corporation
245 Apple Computer, Inc.; © John Chiasson, Gamma/Liaison
246 From *The Wall Street Journal*—permission, Cartoon Features Syndicate
247 © Bob Nelson, Picture Group
248 *World Book* photo; AP/Wide World; AP/Wide World
249 AP/Wide World; AP/Wide World; UPI/Bettmann Newsphotos

250 Goodson, D.D.S., Forsyth Dental Center, Boston
252 Gail Shrader, Burroughs Wellcome Company
254 World Wildlife Fund
255 Robert Steneck
257 Marty Katz, NYT Pictures
258 Smart Card International, Inc.
259 Luma Telecom
261 Windstar Sail Cruises
263 S. Mettler, Sygma; Sygma; AP/Wide World
265 Trudy Rogers*; NASA
267 Keith V. Wood, University of California at San Diego
268 Trudy Rogers*; Albert Einstein College of Medicine, Yeshiva University
270 – 271 Trudy Rogers*
273 © Karen Jettmar, Gamma/Liaison
275 National Cancer Institute
277 Johns Hopkins University
278 Richard Malec, Technigraphic Studios; Trudy Rogers*; Reproduced with the permission of Timothy A. Sanborn, M.D., and the *Journal of Vascular Surgery,* 5:83, 1978; Trudy Rogers*; Candela Laser Corporation
280 Peter Gale, Sygma
283 Theodore Fujita, University of Chicago from National Center for Atmospheric Research
284 L. A. Frank and J. B. Sigwarth, University of Iowa
286 Robert Paz, California Institute of Technology
287 Mark Harmel, California Institute of Technology; H. David Shine and Richard L. Sidman, Harvard University
289 Mary Hightower, Linda Czisny, and Guenter W. Gross, Center for Network Neuroscience, North Texas State University
291 AP/Wide World
293 U.S. Department of Agriculture
295 Ken Able, Rutgers University; Trudy Rogers*
296 Barbara Laing, National Geographic Society; Sankar Chatterjee, Texas Tech University
297 Glen J. Kuban, © *Discover* Magazine, Time Inc.
299 Argonne National Laboratory
300 Visions
301 Frank W. Gayle, Reynolds Metal Company
303 K. S. Lutrell
304 Lawrence Berkeley Laboratory
307 From *The Wall Street Journal*—permission, Cartoon Features Syndicate
308 Trudy Rogers*
311 American Institute of Physics
312 Westinghouse Electric Corporation
315 From *The Wall Street Journal*—permission, Cartoon Features Syndicate
316 LuRay Parker, Wyoming Game and Fish Department
317 © Anthony Taber, reprinted from *Audubon* Magazine, the magazine of the National Audubon Society
318 U.S. Department of Agriculture
319 Joel Kingsolver
320 © Marty Stouffer Productions from Animals Animals
321 Steinkamp/Ballogg*; Sara Woodward*
323 Sara Woodward*
324 Olympus Optical Co., Ltd.
326 – 329 Sara Woodward*
331 Steinkamp/Ballogg*
332 – 335 Sara Woodward*
336 Gene Anthony*
337 Illinois Mathematics and Science Academy; Sandy Fritz
338 – 341 Gene Anthony*
342 – 343 Bruce Ames
344 – 345 Gene Anthony*
346 General Motors Corporation
347 – 348 Gene Anthony*
350 Bronx High School of Science; Sandy Fritz
351 The North Carolina School of Science and Mathematics
353 Louisiana School for Math, Science and the Arts; The North Carolina School of Science and Mathematics; Bronx High School of Science
354 Illinois Mathematics and Science Academy; Louisiana School for Math, Science and the Arts; Louisiana School for Math, Science and the Arts
355 Louisiana School for Math, Science and the Arts; The North Carolina School of Science and Mathematics; The North Carolina School of Science and Mathematics
356 – 357 © John McGrail
358 © Mark Weinkle
359 The North Carolina School of Science and Mathematics
360 – 361 Louisiana School for Math, Science and the Arts
362 Sandy Fritz
363 – 364 Illinois Mathematics and Science Academy
365 Hans Reinhard, Bruce Coleman Inc.

World Book Encyclopedia, Inc., offers a selection of fine porcelain collectible products including *For A Mother's Love Figurine* series. For subscription information write WORLD BOOK ENCYCLOPEDIA, INC., P.O. Box 3405, Chicago, IL 60654.